Ecological Studies, Vol. 110

Analysis and Synthesis

Edited by

O.L. Lange, Würzburg, FRG
H.A. Mooney, Stanford, USA
H. Remmert, Marburg, FRG†

Ecological Studies

Volumes published since 1989 are listed at the back of this book.

Lawrence S. Hamilton
James O. Juvik
F. N. Scatena
Editors

Tropical Montane Cloud Forests

With 102 illustrations

Springer-Verlag
New York Berlin Heidelberg London Paris
Tokyo Hong Kong Barcelona Budapest

Lawrence Hamilton
Islands and Highlands
Environmental Consultancy
RR #1, Box 1685A
Hinesburg, VT 05461
USA

F. N. Scatena
International Institute of
Tropical Forestry
USDA Forest Service
Río Piedras, PR 00928
USA

James O. Juvik
Department of Geography
University of Hawaii, Hilo
Hilo, HI 96720
USA

EAST-WEST CENTER
PROGRAM ON ENVIRONMENT

International Hydrological Programme

International Institute of Tropical Forestry

Library of Congress Cataloging-in-Publication Data
Tropical montane cloud forests / Lawrence S. Hamilton, James O. Juvik, and F. N. Scatena, editors.
p. cm. — (Ecological studies ; v. 110)
Includes bibliographical references and index.
ISBN-13:978-1-4612-7564-0
1. Cloud forest ecology. 2. Cloud forests. 3. Mountain ecology—Tropics. I. Hamilton, Lawrence S. II. Juvik, James O. III. Scatena, F. N. IV. Series.
QH541.5.C63T76 1994 94-12683
574.5'2642—dc20

Printed on acid-free paper.

An earlier version of this book was published by the East-West Center.

Production managed by Laura Carlson; manufacturing supervised by Jacqui Ashri.
Compostion by KP Company, Brooklyn, NY.

9 8 7 6 5 4 3 2 1

ISBN-13:978-1-4612-7564-0 e-ISBN-13:978-1-4612-2500-3
DOI: 10.1007/978-1-4612-2500-3

In Memory of Alwyn Gentry

The editors dedicate these proceedings to Dr. Alwyn Gentry, a participant in the symposium and a stalwart champion for tropical cloud forest conservation. His untimely death in an aircraft crash in Ecuador occurred as this publication was being edited.

Preface

Until relatively recently the valuable tropical montane cloud forests (hereafter usually referred to as TMCFs) of the world had scarcely come under the assaults experienced by the downslope montane and lowland forests. TMCFs are not hospitable environments for human occupation, and their remoteness (except in places near Andean high mountain settlements and in the Ethiopian Highlands) and difficult terrain have given them de facto protection. The adjacent upper montane rain forests have indeed been under assault for timber, fuelwood, and for conversion to grazing and agriculture for many decades, even centuries in the Andes, but true cloud forest has only come under exploitation as these lower elevational resources have disappeared. They have also been "nibbled" at from above where there have been alpine grasslands under grazing pressure. Increasingly now, however, these cloud forest ecosystems are being fragmented, reduced, and disturbed at an alarming rate. It is now becoming recognized that steps must be taken rapidly to increase our understanding of TMCF and to achieve their conservation, because:

- their water-capture function is extremely important to society;
- their species endemism is high;
- they serve as refugia for endangered species being marginalized in these environments by increasingly transformed lower elevation ecosystems;
- they are relatively little studied; yet, their value to science is extremely high;
- they have low resilience to disturbance;

and many other reasons, which will be discussed subsequently in this publication.

Concern over the need for pulling together much of the available knowledge about TMCF prompted us to organize and implement the symposium that produced the material presented in this proceedings.

We acknowledge a debt to the publication *Cloud Forests in the Humid Tropics* by Thomas Stadtmüller (Tokyo: United Nations University, and Turriabla, Costa Rica: Centro Agrónomico Tropical de Investigación y Ensenañza, 1987), which presented the subject's first bibliographic review at the global level. In this first worldwide symposium on TMCF, we aspired to bring together members of the scientific community carrying on investigations in these ecosystems and some of the managers of these areas. This was held in San Juan, Puerto Rico, from 31 May to 5 June 1993. While we indeed fell short of our goal (particularly in securing representation from African research) mainly due to funding limitations and busy schedules of potential participants, we have, we believe, assembled a substantial sample and have produced what can loosely be called a global state-of-the-knowledge review. The "core" group of symposium contributors and participants are listed following the table of contents. We express our gratitude to them for sharing their research, their experience, and their time. We are especially grateful that the World Wide Fund for Nature (International), through Adam Markham, brought a small group together for a mini-symposium that was incorporated into our larger activity. This group focused on the potential role of TMCFs as indicators of global climate change. Several members of this group participated substantially in our deliberations.

The publication that follows largely consists of the symposium/workshop output:

1. An introduction and synthesis chapter based on discussions in the workshop and material selected from the papers. A definition of TMCF is proposed. The annex in the introductory chapter lists known locations of TMCF.
2. A set of maps for the global regions: Oceania, Central America, and the Caribbean, South America, Africa, and South/Southeast Asia, showing the approximate locations of TMCF sites still known to exist by symposium participants. We hope that this will be the start of a more complete location map for these ecosystems. This information has been given to the World Conservation Monitoring Centre (WCMC). Readers of this publication who know of other TMCF sites are urged to send information to the Habitat/GIS Unit, World Conservation Monitoring Centre, 219 Huntington Road, Cambridge CB3 0DL, United Kingdom.
3. Three "overview" papers on TMCFs—the first, a general status report (by Doumenge et al.); the second, a review of the state-of-knowledge on TMCF hydrology and soils (by Bruijnzeel and Proctor); and the third, a global survey of cloud forest avifauna (by Long).

4. A series of papers (in alphabetical order by senior author's name) dealing both with research and management. The literature cited in these papers has not been standardized in order to expedite publication of these proceedings.
5. A bibliography for TMCF, which we hope will be of value to researchers and managers. Send any additional citations (in the same format) to Fred Scatena, International Institute of Tropical Forestry, U.S. Forest Service, Call Box 25000, Río Piedras 00928, Puerto Rico, USA.

We wish to acknowledge a few of the many individuals who helped to make the symposium and this publication possible. At the East-West Center, Senior Program Officer June Kuramoto receives the heartfelt thanks of all participants for her cheerful, effective, and often frustrating work in the logistics of financial support and travel that enabled most participants to attend the symposium. EWC Senior Editor Helen Takeuchi and Publications Assistant Dan Bauer provided their usual competent and laborious services in technical editing and formatting, and we owe them a huge debt of gratitude. We are also grateful to Kersten Johnson of the East-West Center for her assistance. Secretaries Tammy Jones and Gidget Tsui were indispensable in working on the various manuscripts. Any editorial errors in the publication are attributable to us, rather than to this fine support staff. At the International Institute of Tropical Forestry, Dr. Whendee Silver and Samuel Moya provided valuable work on both the symposium and the publication. At the meeting site, Helen Nuncie was a major force for sanity and helpfulness. Finally, we thank the UNESCO International Hydrological Programme for its financial and moral support for this venture.

Lawrence S. Hamilton
James O. Juvik
F. N. Scatena

Contents

Contributors

Andrés Acosta Baladón
P Canalejas
57/61, 7a, E-37001
Salamanca, Spain

Rene J. Beymer
International Institute of Tropical Forestry
USDA Forest Service
Call Box 25000
Río Piedras, PR 00928 USA

Jill Blockhus
Forest Conservation Programme
IUCN
28 Rue Mauverney
1196 Gland, Switzerland

Alejandro D. Brown
Laboratório de Investigaciones Ecológicas
Universidad Nacional de las Yungas
CC 34 (4107), Yerba Buena·
Tucumán, Argentina

L.A. Bruijnzeel
Faculty of Earth Sciences
Free University
De Boelelaan 1085
1081 HV Amsterdam
The Netherlands

Richard Campanella
Department of Geography and Anthropology
Louisiana State University
Baton Rouge, LA 70803 USA

Jaime Cavelier
Universidad de Los Andes
Departamento de Ciencias Biológicas
Apartado Aéreo 4976
Bogotá, Colombia

Daniel de Barcellos Falkenberg
Horto Botánico
Universidade Federal de Santa Catarina
Caixa Postal 476
Trindade, Florianopolis
CEP 88040-900 Brasil

Charles Doumenge
Forest Conservation Programme
IUCN
28 Rue Mauverney
1196 Gland, Switzerland

J.R. Flenley
Department of Geography
Massey University
Private Bag
Palmerston North, New Zealand

A.N. Gillison
CSIRO, Tropical Forest Research Centre
P.O. Box 780
Atherton, Queensland 4883
Australia

Don Gilmour
Forest Conservation Programme
IUCN
28 Rue Mauverney
1196 Gland, Switzerland

Alain Gioda
ORSTOM
B.P. 5045
34032 Montpellier, France

Roberto Espejo Guasp
Física, UCN
Casilla 1280
Antofagasta, Chile

Lawrence Hamilton
Program on Environment
East-West Center
1777 East-West Road
Honolulu, HI 96848 USA
(currently)
Islands and Highlands
Environmental Consultancy
RR#1, Box 1685A
Hinesburg, VT 05461 USA

Robert Hobdy
Hawai'i Department of Land and Natural Resources
54 South High Street
Wailuku, HI 96793 USA

James Juvik
Department of Geography
University of Hawai'i
Hilo, HI 96720 USA
(and East-West Center Consultant)

Kanehiro Kitayama
East-West Center
1777 East-West Road
Honolulu, HI 96848 USA

Mariella Leo
APECO
Parque José de Acosta 187
Magdalena, Lima 17
Perú

Blanca León
(Museo de História Natural, Lima, Peru)
c/o Department of Geography
University of Maryland
Baltimore County Campus
Baltimore, MD 21228 USA

Adrian Long
BirdLife International
Wellbrook Court
Girton Road
Cambridge CB3 0NA
United Kingdom

Lloyd L. Loope
Haleakalā National Park
P.O. Box 369
Makawao, HI 96768 USA

Cosmo Lush
97A Larkhall Rise
London SW4 6HR
United Kingdom

Jean Maley
ORSTOM
B.P. 5045
34032 Montpellier, France

Arthur Medeiros
Haleakalā National Park
P.O. Box 369
Makawao, HI 96768 USA

Mark D. Merlin
General Science Department
University of Hawai'i, Mānoa
Honolulu, HI 96822 USA

Dennis Nullet
Department of Geography
University of Hawai'i
Hilo, HI 96720 USA

Masahiko Ohsawa
Department of Ecology
Chiba University
Yoyoi-cho, Chiba
263, Japan

Samuel Penafiel
Department of Environment and Natural Resources
Region 3
San Fernando, Pampanga
Philippines 2000

Manuel Ruíz Pérez
Forest Conservation Programme
IUCN
28 Rue Mauverney
1196 Gland, Switzerland

J. Proctor
Department of Biological and Molecular Sciences
University of Stirling
Stirling FK9 4LA, United Kingdom

Bill Raynor
The Nature Conservancy
P.O. Box 216
Kolonia, Pohnpei 96941
Federated States of Micronesia

Fausto Sarmiento
Institute of Ecology
University of Georgia
Athens, GA 30602-2022 USA

F.N. Scatena
International Institute of Tropical Forestry
USDA Forest Service
Call Box 2500
Río Piedras PR 00928 USA

Whendee L. Silver
International Institute of Tropical Forestry
USDA Forest Service
Call Box 2500
Río Piedras PR 00928 USA

Clifford W. Smith
Department of Botany
University of Hawai'i, Mānoa
Honolulu, HI 96822 USA

José A. Vázquez-García
Instituto Manantlán de Ecología (MECBIO)
Universidad de Guadalajara
Apdo. Postal 1-3933
Guadalajara Jalisco 44100,
México
(currently)
Department of Botany
University of Wisconsin
420 Linden Drive
Madison, WI 53706 USA

Júlio César Voltolini
Departamento de Zoología
Universidade de São Paulo, Brasil

Dick Watling
Environmental Consultants (Fiji) Ltd.
P.O. Box 2041
Government Buildings
Suva, Fiji

Fiona Watt
International Institute of Tropical Forestry
USDA Forest Service
Call Box 25000
Río Piedras, PR 00928 USA

Peter L. Weaver
International Institute of Tropical Forestry
USDA Forest Service
Call Box 25000
Río Piedras, PR 00928 USA

Wolfgang Werner
Department of Geography
University of Heidelberg
Neuenheimer Feld 330
W-6900 Heidelberg, Germany

W. Arthur Whistler
Department of Botany
University of Hawai‘i, Mānoa
Honolulu, HI 96822 USA

Kenneth Young
Department of Geography
University of Maryland
Baltimore County Campus
Baltimore, MD 21228 USA

1. The Puerto Rico Tropical Cloud Forest Symposium: Introduction and Workshop Synthesis

Lawrence S. Hamilton, James O. Juvik, and F. N. Scatena

It is becoming increasingly recognized that the cloud forests of tropical mountains are high on the list of the world's most threatened ecosystems. In many countries or regions, their rate of loss exceeds that of the lowland tropical rain forests that have received much public concern. Perhaps 90 percent of these tropical montane cloud forests (TMCFs) in the northern Andes have been lost. During the symposium, for which this publication is the proceedings, visual evidence presented by Dr. Alwyn Gentry and others, and the verbal descriptions given by many participants, attested to the loss not only for the Andes, but for all parts of the world where TMCF occurs. The original estimate (and it was only an estimate) given by Persson (1974) of 50 million ha of cloud forests was probably on the high side. He suggested that TMCF constituted one-quarter of montane and submontane rain forest. We believe that TMCF is being lost at a rate considerably in excess of that of the much-abused, but much-publicized, lowland tropical rain forest. Indeed the first paper (Doumenge et al.) that follows indicates that annual forest loss in tropical hills and mountains is 1.1 percent compared with 0.8 percent for all forests of the tropics. Those on isolated volcanic mountains, especially where occurring on small oceanic islands, are particularly vulnerable, and their loss is a serious concern. Cutting of forest for firewood or for charcoal production, and conversion to agricultural uses (mainly grazing) are the principal villains. However, the threats and damaging activities are many. They will be discussed in this synthesis and in many of the papers that follow.

The hydrologic role of TMCF in water capture of "occult" or "horizontal" precipitation that is otherwise "lost" gives these ecosystems a value in terms of water resources quite distinct from other forests or types of land use (Stadtmüller 1987). However, this potentially important hydrologic role of TMCF is still poorly understood, and existing research needs to be reviewed, assessed, and synthesized. The paper by Bruijnzeel and Proctor, in this proceedings, tackles this formidable task. Moreover, those cloud forests on steep slopes, in high rainfall areas, at higher elevation, and with highly organic soils (often peats) have an important watershed protective role. The soils retain and "store" water, reducing runoff in these high rainfall locations, and the forests give erosion protection, because removal often results in catastrophic landslips (e.g., Daugherty 1973).

There is unusual biological wealth in these cloud forest ecosystems. Their biodiversity value has been underrated by those who consider the commonly accepted relationship that the number of tree species and lianas decreases with increasing altitude in the tropics. However, there is substantial evidence that the number of species of epiphytes, shrubs, herbs, and ferns increases with altitude in the humid tropics, so that total flora diversity does not compare all that unfavorably with the more publicity-highlighted tropical lowland rain forest (Wuethrich 1993). Moreover, the proportion of endemics is very high, whether they are considered locally or nationally (see, for instance, the paper by Leo where 32 percent of the Peruvian endemics are found in TMCF, and Long who reports that 260 of the world's nationally endemic species have cloud forest habitats).

Because of their unusual appearance, biological interest, and the fact that they may often be the only remaining natural "eyebrow" or "island remnant" of the once forest-clad mountains, they are even attaining a value for tourism, including bird-watching. In hot climates, they may offer desired respite of coolness. The Monteverde Cloud Forest Reserve in Costa Rica and the El Yunque in Luquillo National Forest are examples where substantial tourist use occurs (see Scatena, this volume).

Furthermore, the special characteristics of TMCFs make many of them excellent ecosystems and sites for monitoring global climate and air quality change impacts. A mini-symposium organized within our overall symposium by Adam Markham of World Wide Fund for Nature (WWF) International was convened to examine this question, and the participants concluded that this was indeed the case.

What Is Tropical Montane Cloud Forest?

Clear definition of tropical montane cloud forest is fraught with difficulty. This is well exemplified in several of the papers where authors deal with montane rain forests that grade into TMCF usually at their upper altitudinal limits (see, for instance, the Werner and the Young and Leon papers) in high rainfall environments. One paper (Kitayama) does examine the floristics of montane rain forest and its transition to TMCF to try to pull out floristic differences. On the other hand, some cloud forests exist and function as water-strippers in relatively dry environments where adjacent vegetation would have to be called dry forest or woodland (see Gioda et al., Juvik and Nullet, Smith, and Sarmiento). Moreover, some are scarcely on mountains (e.g., those on many of the small islands) and the Massenerhebung effect (mass elevation) is well known, so that combined with low inversion layers on some islands TMCF may occur as low as 300 m (e.g., see Watling and Gillison). Other papers point out the singularities of form, structure, or epiphytes in trying to understand this distinction and definition. The basic characteristic, however, that gives them such a high-value profile is that TMCF captures or strips water from clouds, and this, coupled with its low evapotranspiration "loss" of water, adds water to the hydrologic system. The overview paper by Bruijnzeel and Proctor deals with this hydrologic aspect of these ecosystems, which we have grouped under the title of "cloud forests." Despite all these difficulties, the need for definition, qualified though it may be, remains. The participants discussed various proposals, including one that the editors had brought to the symposium (which came in for substantial criticism). The synthesis working definition developed is set forth below.

> The tropical montane cloud forest is composed of forest ecosystems of distinctive floristic and structural form. It typically occurs as a relatively narrow altitudinal zone where the atmospheric environment is characterized by persistent, frequent, or seasonal cloud cover at the vegetation level. Enveloping clouds or wind-driven clouds influence the atmospheric interaction through reduced solar radiation and vapor deficit, canopy wetting, and general suppression of evapotranspiration. The net precipitation (throughfall) in such forests is significantly enhanced (beyond rainfall contribution) through direct canopy interception of cloud water (horizontal precipitation or cloud stripping) and low water use by the vegetation.
>
> In comparison with lower altitude tropical moist forest, the stand characteristics generally include reduced tree stature and increased stem density. Canopy trees usually exhibit gnarled trunks and branches; dense, compact crowns; and small, thick, and hard (sclerophyll) leaves. TMCF is also characterized by having a high proportion of biomass as epiphytes (bryophytes, lichens, and filmy ferns) and a corresponding reduction in woody climbers. Soils are wet and frequently waterlogged and highly organic in the form of mor humus and peat (histosol). Biodiversity in terms of tree species of herbs, shrubs, and epiphytes can be relatively high (considering the small areal extent) when compared with tree species–rich lowland rain forest. Endemism is often very high.
>
> TMCF occurs on a global scale within a wide range of annual and seasonal rainfall regimes (i.e., 500–10,000 mm/year). There is also significant variation in the altitudinal position of this mountain vegetation belt. For large, inland mountain systems, TMCF may typically be found between 2,000–3,500 m (Andes, Rwenzoris), whereas in coastal and insular mountains this zone may descend to 1,000 m (Hawai'i). Under exceptionally humid, marine, equatorial conditions, a TMCF zone may develop on steep, small island mountains at elevations as low as 500 m or even lower (Kosrae in Micronesia and Gau in Fiji).

The participants in the symposium offered a diverse list of local or professionally accepted names for TMCF in several languages: mossy forest, elfin forest, montane thicket, dwarf cloud forest, nuboselva, bosque montano nebuloso, selva de neblina, bosque nuboso, bosque de ceja, selva sublada, nebelwald, wolkenwald, forêt néphéliphile, forêt de nuage, unmu-rin, bosque anão, foresta nebular, mata nebular, matinha nebular, floresta fe neblina, floresta nuvígena, mata de neblina, matinha de altitude, floresta ombrófila densa montana e/ou alto-montana, floresta de altitude, floresta nublada, and floresta pluvial montana e/ou alto montana.

Stadtmüller (1987) suggested the helpful distinction between two general types of cloud forest as follows:

1. areas where the frequency of cloud cover occurs in combination with heavy and sometimes persistent orographic rainfall with high annual precipitation; these forests have a distinct canopy strata, a richness in epiphytes, and (often) a thick peat layer with a high potential for water retention and runoff control;
2. drier areas with more seasonal rainfall regimes where horizontal precipitation (cloud stripping) can represent a high component of the total precipitation (see review by Bruijnzeel and Proctor in this proceedings on the magnitude of this water capture).

A schematic diagram originated by Fred Scatena and refined during the working sessions attempts a conception of the relationships of various ecosystem properties and influences. It is offered as a stimulus to further elaboration (Figure 1).

Where Are the Tropical Montane Cloud Forests?

The location, identification (name), and extent of the various cloud forests are not well known. No detailed global or regional mappings have yet been produced. Stadtmüller (1987) in his landmark publication attempted "a general idea of the geographic distribution of cloud forests in the humid tropics," but his world map is so small and the areas so grossly indicated that it has minimal utility. LaBastille and Pool (1978) broke new ground in providing a list and a general location map for cloud forest reserves or potential reserves in the Caribbean and Middle America and supplied names for 18 such areas. Vásquez-García (this volume) has reproduced the LaBastille and Pool map, and then located other TMCF localities for Middle America and a portion of the Caribbean. These two sources were an excellent base for the information that was collected at the symposium for this particular region. Previous information on locations of cloud forest was either sporadically reported (i.e., Pacific Islands) or unassembled for the other regions, especially for Africa.

The first paper that follows, by Doumenge et al., presents a much-needed

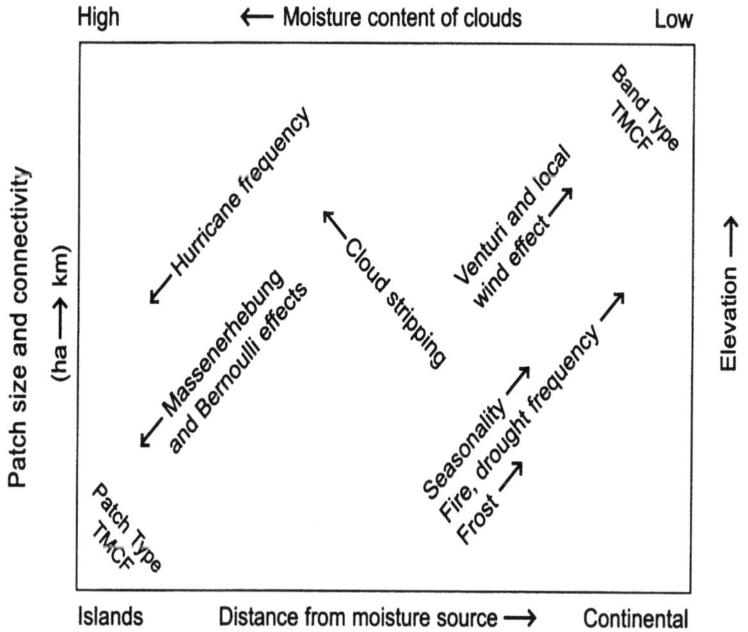

Figure 1. Conception of the relationships of various system properties that influence tropical montane cloud forests (TMCFs). Arrows indicate direction of increasing importance of system property.

general overview of TMCF, and with data from the World Conservation Monitoring Centre (WCMC) attempts a map of their general distribution. Since information was available in the GIS data bank on the location of only a few TMCF areas, the maps produced were for the larger categories of montane and submontane tropical rain forests (within which most, but not all, cloud forests would occur). As well as a global map, WCMC has produced more detailed maps for Africa, Southeast Asia, and South America, at scales more useful to the ultimate goal of demarcating the areal extent of existing TMCF (see maps in Doumenge et al.).

Our symposium workshop was able to provide much more specific information through the device of having participants name and locate on regional maps (1:12,000,000) the locations of TMCF known to them. A list of these cloud forests is given in the Annex to this paper, but it is still woefully incomplete and contains possible inaccuracies. Readers are encouraged to send any corrections or additions, with as much geographic detail as possible (especially coordinates, elevational limits, and area), to the Habitat/GIS Unit, World Conservation Monitoring Centre, 219 Huntingdon Road, Cambridge CB3 0DL, United Kingdom.

Using the information placed on maps by the participants, we produced a global map showing general concentrations of TMCF (Figure 2). More useful, however, are the series of regional maps showing the approximate location of TMCF concentrations (Figures 3–6). No map is included for the Pacific or Atlantic islands since these ocean areas are so vast and the islands so small that any symbol placed on a map would be meaningless. The list of islands with cloud forests (see Annex to this introduction) provides information on their location. Our larger scale symposium working maps that name and locate individual cloud forests, protected areas, and known remaining undisturbed areas have been sent to the World Conservation Monitoring Centre, who will maintain and improve the data set. Even with this crude level of detail, these areas are obviously mostly remnant fragments or small areas (as on islands), which require a major endeavor to achieve their protection or sustainable management. We propose an international "Conservation of Cloud Forests Campaign" that might well be extended to all cloud forests, and not just to those in the montane tropics.

Threats to Tropical Montane Cloud Forests

A major universal and pervasive problem in almost all geographic areas except Oceania (though it is a problem in Hawai'i) is *forest conversion to cattle or sheep grazing.*

The *extension of subsistence agricultural cropping* by local people threatens much TMCF. This is usually caused by a decline of soil productivity on lower marginal steep slopes already in a shifting or sedentary subsistence system, or by new immigrants from the lowlands pushing the traditional hill or mountain people successively higher (e.g., as in the Philippines, Thailand).

The opening up of TMCF lower edges for *commercial temperate vegetable production* (e.g., as in Sabah's Mount Kinabalu; Mount Data in the Philippines; Sri Lanka) has become a relatively recent threat.

In some areas there has been an *upward shift of the belt of horticulture*—tree crops such as coffee or cardamom (e.g., as in Sri Lanka, Cameroon, Philippine's Mount Kanlaon), flowers (e.g., as in Venezuela's El Avila), or berries (e.g., as in the Dominican Republic).

Most TMCF is under some pressure from *wood harvesting for fuelwood* (as in Cuba, Haiti), *charcoal* (e.g., as in the Venezuelan Andes), or for tea curing (as in Sri Lanka, Zaire).

Commercial logging is not an important threat due to inaccessibility and usually low-statured crooked trees. However, in those TMCFs where there are taller, straighter trees (areas of high normal precipitation with additional frequent cloud cover and in the boundary area between upper montane rain forest and TMCF), some commercial timber extraction has occurred and is a threat (e.g., Podocarps and high elevation oaks in Costa Rica). The roads associated with logging permit other interventions.

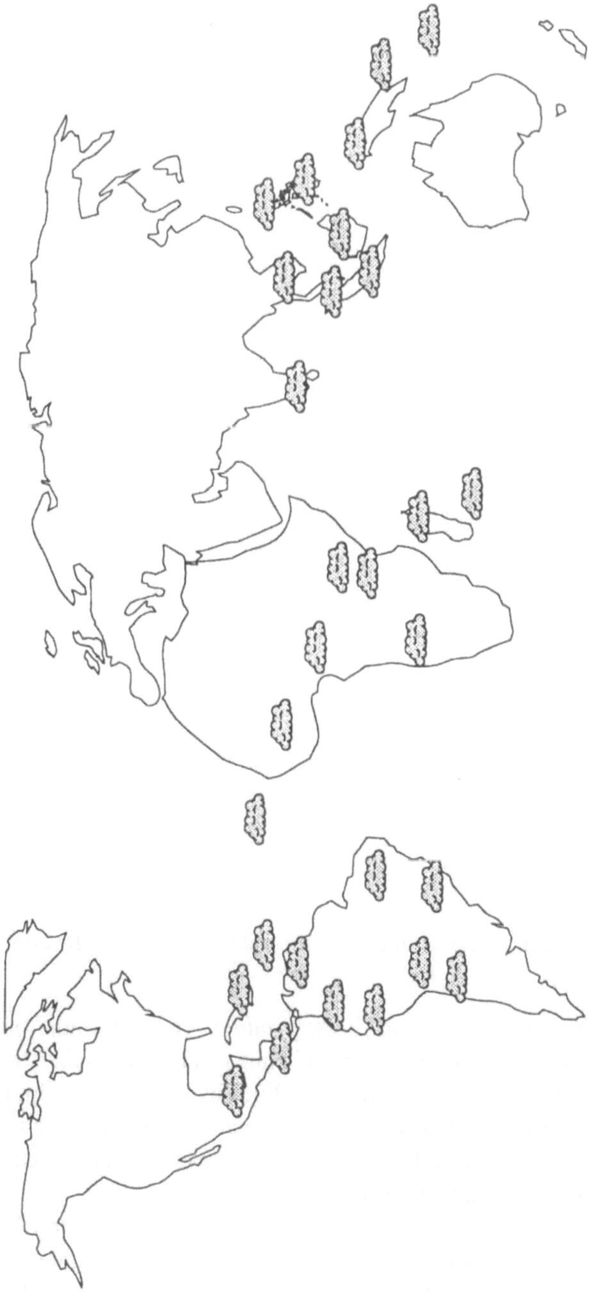

Figure 2. General concentrations of tropical montane cloud forest throughout the world.

Figure 3. Concentrations of tropical montane cloud forest in Africa.

Exploitation of nonwood forest products (e.g., orchids, bromeliads, medicinal plants, tree ferns, hearts of palm). This is a subsistence activity in most areas, but it has been commercialized by market development and middle persons in some areas (e.g., Hawai'i, Guadeloupe, Venezuela, Mexico, Philippines). Other nonwood forest products are the fauna, whose exploitation is designated as the hunting threat.

Hunting (or capture) for sport, subsistence, or commercial trade has, in some cases, been such a serious threat that it has now almost been eliminated due to protection efforts where species became officially threatened or endangered (e.g., the mountain gorilla in Zaire, Rwanda, Uganda; spectacled bear in central and northern Andes; parrots in the Caribbean). Tapir hunting is

Figure 4. Concentrations of tropical montane cloud forest in South and Southeast Asia.

still a problem in Central America, and so is sport hunting in the French Caribbean Islands. Subsistence bird-hunting is a problem in many Oceanic islands. Trade in reptiles and amphibians is on the increase, but the extent of this in TMCFs is unknown and its impact on the herpetofauna is not documented. Deliberate control hunting of alien ungulated (e.g., feral pigs in Hawai'i) can benefit TMCF conservation where animal damage is occurring.

Clearing for legal or illegal drug plant production (e.g., kava or sakau in many Pacific Islands, opium poppy in Colombia and Thailand), in addition to direct forest disturbance, also produces indirect impacts such as landslips (as in Pohnpei), herbicide application (as in Colombia), and development of degrading trails or vehicle tracks.

Associated with grazing and agricultural interventions (and perhaps hunting) in forest is *anthropogenic fire*. It is not reported as an important threat in the wet, generally nonflammable sites. However, in the seasonally drier occurrences of cloud forest, burning can be initiated (e.g., as in the Pacific

Figure 5. Concentrations of tropical montane cloud forest in Central America.

Islands). Fire in the lowlands during the season for burning grasslands or sugarcane can alter the chemistry of the cloud water, but this has not been well studied. Decreases in pH of rain have been reported in Monteverde Cloud Forest Reserve in Costa Rica and attributed to Pacific lowland farming activity.

Introduced (alien) plants and animals are especially serious threats in small, isolated island TMCFs (e.g., as in La Réunion, Galapagos, Canary Islands) or isolated mountains (volcanic mountains of Central America) but increasingly in the fragmented remnants of TMCF in mountain ranges (e.g., as in the Andean Cordillera or Ethiopian Highlands). Feral pigs in Hawai'i and Costa Rica, goats in Venezuela, or introduced grasses in Puerto Rico are examples of serious problems.

Mining and geothermal development are not major threats except in a few site-specific locations, which can be very damaging. This includes exploration damage, on-site extraction, pipelines or access roads, and off-site pollutants (such as mercury in the case of gold tailings at Mount Nimba, West Africa). Gem mining in Sri Lanka occurs in TMCF and gold mining in Irian Jaya (Indonesia). Geothermal well fields affect Mount Apo and Mount Makiling in the Philippines. Access roads that accompany mining in or close to TMCF

Figure 6. Concentrations of tropical montane cloud forest in South America and the Caribbean.

reduce the de facto protection from other disturbance activities once provided by site remoteness.

Tourism and recreation represent relatively recent but increasing threat to TMCF. Increasing tourism visitation to Mount Stanley in Uganda, Luquillo

National Forest in Puerto Rico, and Mount Makiling and Mount Pulog in the Philippines brings problems of trails and tracks and their sphere of influence (track deterioration and extension, litter, taking of plant and animal souvenirs, introduction of alien species). Where there is visitor accommodation, demand for firewood must be met or diverted (e.g., as in the Rwenzori Mountains). Golf course development in montane tropical areas has recently emerged as a new recreational land use (as on Sabah's Mount Kinabalu, and peninsular Malaysia's Cameron and Genting Highlands).

Telecommunication facilities and media transmitting stations are sited on summits of hills and mountains, and if there are cloud forests, these often coincide. Not only are these physically a visual blight to unusually scenic sites, but the infrastructure required and the frequent access needed for maintenance all provide impacts far beyond the actual site (e.g., landslides associated with the road to Pico del Este, Puerto Rico), including the biological baggage brought in and out by humans. Leakage of an oil storage tank in 1993 at Pico del Este resulted in seepage damage and the death of several trees in the downslope forest, indicating the potential for other harmful impacts of such installations. Fortunately, new technologies should make these facilities smaller and therefore less visually intrusive, and their installation and maintenance can be supported by helicopter transport, eliminating the need for destructive road construction.

Uncertain land tenure is an underlying and pervasive problem in most areas of the tropical world, particularly where there has been traditional indigenous ownership or uses that are officially superseded by a more recently developed state or nation with the designation of "government lands." Uncertainty of use rights or ownership generally fosters nonsustainable exploitation. This is a problem of special importance in the TMCF of Southeast Asia.

Global climate change and air quality deterioration have been thought to affect the tropics much less than in the temperate zones (e.g., ozone holes, increased temperature, acid precipitation). However, new global models show substantial effects on vegetation in tropical mountains from a significant temperature rise, and that such a rise may well occur. Acid precipitation from human nonindustrial activity can occur (i.e., fire). Tropical cities have rapidly developed their own horrendous air pollution situations, which are exported to nearby ecosystems (e.g., as in the Caracas and El Avila cloud forest). The impacts have not been studied, but future scenarios indicate that impacts could be severe on these ecosystems and that recovery from harmful events is slow (e.g., if global warming increases ocean surface temperatures, hurricanes become more frequent, more intense, and extend farther toward the North and South poles). Hurricanes are damaging to TMCF, especially on islands (e.g., the Caribbean islands, the Samoas). The sensitivity of TMCFs, and their current relative freedom from current climatic "noise" effects would seem to make them excellent sites for monitoring global climate and air quality changes.

Response to Threats

General

Any on-the-ground measures taken to ameliorate or eliminate the human actions that are degrading or destroying TMCF will depend for their viability on having the responsible people more fully informed of the values of the resource they are impacting. This includes not only local owners and users of land, but local politicians and government agencies who may be impacting TMCF with their own project activities. Efforts at local awareness-raising underlie all direct actions taken to reduce the threats from human impacts. Local or national non-governmental organizations (NGOs) may be very effective at this level if given reliable information by scientists and managers. The symposium participants also believe that an educational campaign is needed at the national and international levels, especially with decision-makers and politicians. We propose to the international conservation community the initiation of a "Cloud Forest Campaign" or highlighting TMCF (and other cloud forests) in "The Year of the Cloud Forest."

An urgent need underlying all responses is the development and implementation of a management plan for each TMCF unit, whatever the ownership. Such a plan, which needs to be proactive in anticipating problems and threats, is absolutely essential for an official protected area. The needed elements and procedures in developing forest land management plans have been fully developed elsewhere (for instance, see the management plan for the Luquillo Experimental Forest within the Caribbean National Forest referred to by Scatena in his paper). Only a few of the more important elements can be addressed in the following discussion.

Surveys and inventories should extend beyond the boundaries of the TMCF unit itself to include surveys of ownership, tenure, and use of adjacent lands. This will permit a better-informed initiation of proactive measures as well as responses to reduce or divert threats. Demographic information from surrounding communities may be very useful because much of the adverse human impact is driven by population growth, land hunger, and an increase in resource consumption. Local traditional knowledge about flora and fauna, their uses and legends are often a rich bank of important information and should be assiduously sought. Community involvement in the planning process is essential to achieving local support for the management policies and rules. This precept is much honored now in theory and verbiage, but infrequently well implemented.

New techniques and facilities for planning are available but are generally underutilized by land managers. They include time-sequence photos (air and ground), other remote sensing products of various kinds, and geographic information systems. These can be powerful tools. Finally, provisions for monitoring change are externally important in these "stressed" and slow-to-recover ecosystems. As well as measuring locally generated impacts, these may be good "miners' canaries" for indicating changes in global systems of weather, air quality, ozone levels, and related increases in ultraviolet radiation.

A number of more direct types of threat responses were suggested by the participants, some of which are being effected in various places with some success, and others that seem useful but not necessarily proven in practice. These responses address the major causes of degradation and fragmentation previously discussed.

1. *Grazing and agricultural/horticultural crops*. It has been recognized that in many situations fences, guards, and enforcement are not effective because local economic/demographic pressures are so great, usually from lower elevations (in a few instances from high elevation populations intervening from higher elevations). Nonetheless, boundaries of officially protected areas should be clearly identified and marked. Patrolling the boundaries works in some cases, or it at least slows the conversion process while other solutions are applied. The most frequently recommended and enduring solution is to attempt stabilization of these activities in already converted areas by making them more productive by agricultural intensification. This can take many forms, depending on the nature of the conversion pressure. In grazing systems, assistance in the development of improved pastures or integrating trees in sylvopastoral systems to provide leaf fodder in the drier season can be initiated. Carefully controlled gathering or collecting to enhance income from wood, nuts, fruit, etc., may be tried. In cropping systems, stabilization of shifting frontiers has been attempted by making fertilizer more available, increasing extension advisory service, reducing fertility loss by erosion control (technical advice and sometimes financial support), providing trees for establishing a more sedentary and sustainable agroforestry land use system, and so forth. TMCF managers should investigate the array of devices that have been used with at least partial success *outside* the area to be conserved in the so-called "buffer zone." Fencing of the TMCF should be considered in special cases.

2. *Wood cutting for fuel*. TMCF is often a source of wood for heating and cooking by local high-elevation inhabitants, whether for fuelwood or charcoal, or for sale to fuel-short settlements at lower elevations. This is admittedly a difficult pressure to ameliorate or eliminate. Shifting the pressure elsewhere by establishing wood-fuel plantations or incorporating good fuelwood species in agroforestry systems has been widely recommended in this kind of situation. Education about making available or even subsidizing more fuel-efficient cooking stoves or charcoal kilns is worth considering (not to mention the possible health benefits of more fuel-efficient stoves). In some cases, where other alternatives are infeasible, the response may have to involve providing controlled resource harvesting. Even dead-tree-only harvesting must be closely controlled for it is susceptible to abuse. Permitting the cutting of canopy trees for fuel removes their water-capture function, but within the management unit there may be some stands where water capture is low or nonexistent. In these cases some kind of priority zoning is needed, and any cutting confined to a multiple-use zone, with strict control. The high-interception stands and areas should be given priority as protected watershed forest only.

3. *Alien plant and animal invasion*. This threat is more serious on small

islands. Direct control measures are warranted where endemics are threatened by these invaders. In Hawai'i, feral pigs are fenced out of key areas, augmented with programs of trapping, hunting, and snaring (in remote areas). Biocontrol with insects and plants has been used effectively in some places, and might be considered in cloud forests being infested. Managers should press for stronger, and enforced, quarantine regulations. Where aliens occur nearby, any cloud forest areas disturbed by human actions (e.g., roads) or natural forces (e.g., hurricanes) should be carefully monitored and the invaders eradicated promptly while still at low levels. Managers should also consider restoration of any areas now exhibiting alien species, and planting native species to achieve results.

4. *Harvesting of nontimber forest products*. While this has been popularly touted as a "benign" use of tropical forests as an alternative to wood harvesting, there are many problems inherent in this utilization. There are many medicinals (leaves, bark, fruit, roots), orchids, bromeliads, ferns, amphibians, birds, and mammals that have market value and which are now collected or have the potential to be collected by locals on a small scale. Where population sizes and dynamics are known, this may be a regulated sustainable use of local economic significance. If properly regulated it may indeed help to maintain the existence of the whole system. However, it should be recognized that where substantial value becomes established, a commercial trade on a larger scale often follows (organized by middle-person entrepreneurs or even revenue-hungry governments). Here, danger of overexploitation exists and has often resulted in depletion of species or genera to the point of rarity or endangerment. With so many endemics in TMCF, there is a high risk of extirpation (e.g., Andean spectacled bear in many local regions and orchid species in many countries). When numbers become critically low, or preferably before that stage, there should be pressure for some kind of protection such as a listing of endangered species by the nation, and by IUCN if it is internationally at risk. This would, we hope, permit greater local enforcement measures and possible international support. A further step would be to seek restrictions on international trade (if that is the commercial market) through the Convention on International Trade in Endangered Species of Wild Fauna and Flora (CITES). In order to be effective, the country would have to be a signatory to the convention, and this might be a necessary first step to press for by managers.

5. *Increasing tourism*. While this was not reported to be a major problem by most of the symposium participants, there were clear warnings from a few where somewhat unexpectedly high levels of tourist visitation caught managers unprepared. Although the uses of the TMCF are almost entirely nonconsumptive (e.g., bird-watching, mountain-hiking, photography, and visitation to experience the cool misty climate), the side effects of having large numbers of people in these environments have to be recognized and planned for. It can represent an economic opportunity, if managed properly (e.g., as in Rwanda, Zaire, and Uganda for the mountain gorilla). However,

deteriorating trails have been experienced on Mount Stanley in the Rwenzoris, and hiker impact is being experienced on Mount Kinabalu in Sabah and on El Yunque in Puerto Rico. Demand for road access and accommodations for tourists are a threat, and managers need to do anticipatory management planning along these lines. Where commercial tourism is anticipated, it is prudent to develop regulations in advance, and in consultation with local people, in order to contain any local adverse cultural impacts and provide for revenue sharing (both to the forest and to the people). Managers should always consider training and employing local guides (e.g., the Peman Indians at Tepui Roraima in Venezuela).

Responses and anticipatory impact-reducing actions for the other threats previously presented are fairly self-evident. Mining seems to take land use precedence over all other uses because of its high national income-earning potential, and it is difficult or impossible to contain or divert damages. Ideally, environmental laws should be implemented to eliminate large spill-over, off-site effects if the activity is within or near a TMCF. Telecommunication siting has similar high-values and site-specific imperatives. Fire threat has obvious responses, as do illegal drug production and logging. In any of these situations, where roads are involved they are often a major villain and efforts should be directed at relocating them away from sensitive TMCF sites. If this is impossible, hydrologically/erosionally sound road design, construction, and maintenance must be demanded. Concurrently, managers will have to plan for an increase in many other uses and threats that invariably come with road access.

The threats to TMCF from global-type changes such as in air quality, hurricane frequency and duration, ultraviolet radiation, rainfall, and temperature are beyond the control of managers. But it is imperative that we come to better understand these ecosystems to enable us to learn the specific impacts of change in one or more of these "state factors." This understanding will be essential for any management response (or any restoration activity) to reduce the rate of ecosystem alteration that might prematurely cause loss of biodiversity or reduction in watershed function. Managers are obligated to foster scientific research and environmental monitoring in the areas under their care. Some of the research activities that should be encouraged are given in the next section.

Research Needs for Tropical Montane Cloud Forests

Outlined below is a list of general and specific research needs set forth by the symposium participants. The list has been sent to key international and regional forestry research centers, and to organizations that support forestry research.

General: More integrated (involving scientists from different disciplines) and more comprehensive research programs are required, including all principal ecosystem processes and elements (at least in hydrology, meteorology, vegetation, fauna, and soils) at a given site. The integrated research sites need to have a stable political environment since long-term research is needed (e.g., existing mountain protected areas such as national parks). Opportunity exists here for monitoring global climatic and air quality changes. Associated with the research should be co-workers from tropical countries, with training for young indigenous scientists.

Specific (in no particular order of priority):

- Conduct worldwide inventory of TMCFs and develop a data base for maintenance by the World Conservation Monitoring Centre (output from this symposium will be sent to WCMC, but scientists everywhere are encouraged to submit data).
- Map/inventory specific TMCFs including not only biophysical data but information on tenure of forest (and adjacent areas) and other socioeconomic matters that are likely to impact on the future welfare of the ecosystem under study.
- Promulgate standardized equipment, methodology, and information formats in order to make results more comparable and useful from site to site; develop a system to validate existing data sets; provide a means for updating the existing bibliography from this symposium (perhaps through the WCMC).
- Encourage and support more studies of the fauna of TMCFs, particularly of endemics.
- Conduct more studies of nutrient cycling in TMCF and encourage further research on role of epiphytes in this cycling.
- Establish a monitoring network of global climate and air quality change (warming, precipitation change, acid precipitation). Because these ecosystems are stressed, yet relatively less disturbed than other tropical forest ecosystems and represent the cloud-forest interface, they are good sites for establishment of such a network.
- Identify within TMCFs the individual processes highly sensitive to environmental change. Note also that trees of MCFs at extremes of tropics or of moisture supply may be good indicators.
- Conduct restoration ecology studies (for degraded sites) and tree-planting research (native species vs. exotics introduced as nurse crop). Include also native forest rates of recovery from natural or human disturbance.
- Determine best methods of buffering fragmented and disappearing TMCF; establish and manage buffer zones.
- Continue study of national or regional endemics in flora and fauna to prioritize TMCF sites for biodiversity conservation.

- Quantify TMCF cloud-water capture and establish economic value of cloud forests as sources of water.
- Establish climatic/topographic conditions and thresholds under which conversion of TMCFs results in water yield reduction.
- Conduct ecophysiology studies of how trees or other vegetation are adjusted to (coping with) their specialized and seemingly highly stressful environment.
- Determine sustainable levels (if any) of resources use for various products, especially nonwood products (orchids, tree ferns, etc.).
- Study impacts of ecotourism, road, trail, and boardwork construction; particularly in light of typical saturated organic soils.
- Synthesize existing paleo-ecological/paleo-climate data for reconstruction of past TMCF changes; expand paleo-ecological research.

References

Daugherty, H.E. 1973. The Montecristo cloud-forest of El Salvador—A chance for protection. *Biological Conservation* 59(1): 227–230.

LaBastille, A., and D. J. Pool. 1978. On the need for a system of cloud-forest parks in Middle America and the Caribbean. *Environmental Conservation* 5(3): 183–190.

Persson, R. 1974. *World Forest Resources.* Stockholm: Royal College of Forestry.

Stadtmüller, T. 1987. *Cloud Forests in the Humid Tropics. A Bibliographic Review.* Tokyo: United Nations University and Turrialba, Costa Rica: Centro Agronómico Tropical de Investigación y Enseñanza.

Wuethrich, B. 1993. Forests in the clouds face stormy future. *Science News* 144(2): 23.

Annex: Tropical Montane Cloud Forest Identified by Workshop Participants

Mexico and Central America

Costa Rica
Braulio Carrillo National Park
Hitoy Cerere Biosphere Reserve
La Amistad National Park
Monteverde Cloud Forest Reserve
Rincon del Viejo Volcan National Park
Tapanti Forest Reserve
Volcano Poas National Park
Volcano Chirripo National Park

El Salvador
Montecristo National Park

Guatemala
Quetzal Cloud Forest Reserve
Rio Dulce National Park
Trifinio National Park
Sierra de las Minas Biosphere Reserve

Honduras
Azul-Meambar National Park
Misoco Biological Reserve
Mixcure Wildlife Reserve
Montana Corralitos Wildlife Reserve
Montana El Armado Wildlife Reserve
Montana de Comayagua National Park
Montana de Puca Wildlife Reserve
Montana de Yoro National Park
Montana Verde Wildlife Reserve
Montecristo-Trifinio National Park
Montana El Chile Biological Reserve
Montana Yerba Buena Biological Reserve
Muralla-Los Higuerales Wildlife Reserve
El Pital Biological Reserve
El Guisayote Biological Reserve
Erapuca Wildlife Refuge
Pacayitas Biological Reserve
Celaque National Park
Cerro Azul National Park
Cusuco National Park
Santa Barbara National Park
Texigat Wildlife Reserve
Cordillera de Montecillos Biological Reserve
Guajiquiro Biological Reserve
Pico Pijol National Park
Comayagua National Park
Pico Bonito National Park
La Tigra National Park
Largo Yojoa Multiple Use Area
Agalta National Park
Yoro National Park
M. Uyuca, Azacualpa
Opalaca Biological Reserve
Yus-Carran Biological Reserve

Panama
Volcano Bara National Park, Panama
Chiriqui, Panama
Darien National Park and Highlands

Mexico
Saguaribo, Sonora
Surotato, Sinaloa
E. del Diablo, Durango
El Nayer, Nayarit
Cerro Viejo, Jalisco
Cuale-Talpa, Jalisco
Sierra de Manantlan Biosphere Reserve, Jalisco
Coalcoman, Michoacan
Atoyac, Guerrero
Qelchultenango-Guerrero
Loxicha, Oaxaca
Miahuatlan, Oaxaca
Huautla, Oaxaca
V. Nacional, Oaxaca
El Cielo, Tamaulipas
Xilitla, S.L. Potosi
Joya del Hielo, Queretaro
Chapulhuacan, Hidalgo
Zacatlamaya, Hidalgo
Nevado de Colima
Tancitaro
Toluca
Popocatepetl
Iztaczihuatl

La Malinche
Perote
San Martin
Santa Martha
Jitotol, Chiapas
Lagunas de Montebello, Chiapas
El Triunfo, Chiapas
Sierra Madre de Chiapas
Monte Bello, Chiapas
Teocelo, Veracruz
Valle de Mexico

Nicaragua
Saslay National Park
Macizo de Penas Blancas Wildland Area

Caribbean Islands

Greater Antilles

Cuba
Pico Turquino
Sierra Maestra

Dominican Republic
A.J. Bermudez National Park
Jose del Carmen Ramirez National Park
Sierra de Baoruco National Park
Ebano Verde National Reserve
Isabel de Torres National Reserve
Valle Nuevo National Reserve

Jamaica
Blue Mountain Ridges and Peaks

Puerto Rico
Caribbean National Forest
Toro Negro Commonwealth State Forest

Lesser Antilles
Nevis
Mt. Misery, St. Kitts
Soufriere and elsewhere, Guadeloupe
Morne Trois Pitons National Park, Dominica
Morne Diablotin National Park, Dominica
Mt. Pelee, Martinique
Mt. Gimie, St. Lucia
Soufriere, St. Vincent
Mt. St. Catherine, Grenada
El Aripo, Trinidad

South America

Argentina
Sierras Subandinas
Eastern Andean Cordillera
Sierra de Gonzalez (temperate cloud forest)

Bolivia
Amboro National Park
Carrasco National Park
Isiboro Secure National Park
Pilon-Lajas Biosphere Reserve
Ulla Ulla National Reserve
German Busch National Reserve

Brazil
Caparo National Park
Itiaia National Park
Pico da Neblina National Park
Serra da Bocaina National Park
Serra dos Orgaos National Park
Morro do Cuca, Serra dos Orgaos, Rio de Janeiro
Serra do Mar, in states of Santa Catarina and Parana
Aparados da Serra Geral, Santa Catarina and Rio Grande do Sur
Pico da Bandeira
Carangola
Serra do Mantiqueria

Colombia
Chingaza National Park
Cordillera de los Picachos National Park
Cueva de los Guacharos National Park
Las Hermosas National Park
Las Orquideas National Park
Parque Nacional Nevadi del Huila
Eastern slopes of Cordillera Oriental
Western Slope of Cordillera Occidental
El Cocuy National Park
Eastern slopes of El Cocuy
Cordillera de Merida
Cuenca del Rio Buritaca
Sierra de Macuira
Sierra Nevada de Santa Marta National Park
Sumapaz National Park
Nudo de Paramillo
Parque Nacional Tatama

Los Fallerones de Cali National Park
Los Nevados National Park
Macuira National Park
Munchique National Park
La Planada Reserve
Nudo de Pasto
Paramillo National Park
Pisba National Park
Purace National Park
Sierra de la Macarena National Park
Tama National Park
Tatama National Park
Iguaque Flora and Fauna Sanctuary

Ecuador
Cotacochin Cayapas Ecological Reserve
Cayambe Gear Ecological Reserve
Gran Sumaco Ecological Reserve
Maquipucuna Reserve
Santa Cruz, Galapagos
Isabella, Galapagos
Fernadina, Galapagos
Songay National Park
Podocarpus National Park
Pululahua Geobotanical Reserve

Peru
Eastern slopes of Peruvian Andes
Cordillera Sub-Andean
between Huallaga and Ucayali rivers
Cordillera del Condor at boundary
between Peru and Ecuador
Western slopes of Andes
along upper drainage of Rio Zana
Piura, Cajamarca

Venezuela
Sierra de San Luis National Park
Henri Pittier National Park
El Avila National Park
Pico de Neblina National Park
Cruz Carrillo National Park
Duida Marahuaca National Park
Guaramacal Mountain (Trujillo)

Africa and Associated Islands

Angola
Western escarpment
Cuanza Norte
Cuanza Sur

Canary Islands
Santa Cruz de la Palma
Gomera
Hierro
Tenerife
Las Palmas

Cameroon
Mt. Oku
Mt. Kupe
Bambuto mountains
Bakossi mountains
Cameroon mountains

Comoros
Grande Comore island
Anjouan island

Congo
Mayombe mountains

Equatorial Guinea
Bioko

Ethiopia
Mt. Guge

Gabon
Monts de Christal
Monts de Belinga
Sao Tome island
Principe island

Guinea
Mont Nimba

Kenya
Mt. Elgon
Mt. Kenya

Madagascar
Massif du Tsaratanana
Higher peaks along eastern escarpment

Mascarene Islands
Reunion
Mauritius

Nigeria
Dimland
Shebshi mountains
Alantika mountains

Rwanda
Nyungwe forest
Volcans National Park (Virungas)

Seychelles
Silhouette island
Mahe island

Sierra Leone
Loma Mansa

Tanzania
Mt. Kilimanjaro
Mt. Kungwe
Mt. Meru
Usambaras mountains
Eastern Highlands

Uganda
Rwenzori Mountains
Mt. Stanley National Park (Rwenzoris)
Mount Elgon Forest Park
Kidepo Valley National Park

Zaire
Mt. Kehuzi
Mt. Biega
Itambwe Mountains
Virungas
Virunga National Park

India and Southeast Asia

Burma
Chin hills

Borneo
Schwaner mountains
Muller mountains
Mt. Bukit Raya

Brunei
Mt. Bukit Pagon

China
Central Sichuan mountains
Wu Chin Shan, Hainan

India
Eastern Himalayas
Panyakunam mountains, Southern India
Ootacamund mountains, Southern India
Kampuchea
Mt. Phnum Aural
Malay
Khao Luang

Malaysia
throughout Malaysian peninsula, Sarawak, and Sabah
Sarawak-Borneo border
Crocker range, Sabah
Trus Madi range

Sri Lanka
Central mountains including:
Pidurutalagala mountain
Knuckles mountain

Thailand
Doi Angka
Southern border of Thailand and Kampuchea

Vietnam
Dalat plateau

Indonesia

Buru

Ceram

Java
throughout island including:
Gunung Pangrango
Gunung Gede
Gunung Kawi
Mahameru

Irian Jaya
Central New Guinean high-mountains
North New Guinean mountains
Van Rees mountains
Vogelkop Peninsula
Mamberamo mountains
Nassau range
Oranje mountains

Sulawesi (Celebes)
Gunung Klabat
Gunung Tangkoko
Molen-Graaff mountains
Quarles mountains
Bonthain

Sumatra
throughout island including:
Barisan Mountains
Mt. Kerinci

Philippines

Central Cordillera, Luzon
Mt. Pulog National Park, Luzon
Mt. Data National Park, Luzon
Mt. Maquiling, Luzon
Mt. Apo National Park, Mindanao
Mt. Canlaon National Park
Leyte Mountains, Visayas
Panay Mountains, Visayas
Ilocos region
Central Luzon region
Southern Tagalog
Bicol region
Iloiol-Panay region
Mindanao region
Mindoro island
Palawan island
Negros island
Panay island

Pacific/Oceania Group
(very small areas)

Cook Islands
Rarotonga

Federated States of Micronesia
Pohnpei
Kosrae

Fiji
Viti Lcvu, Mt. Tomaniivi Naturc Reserve
Taveuni
Vanua Levi
Gau
Kadaru - Cape Washington

French Polynesia
Marquesas Islands (high islands)
Society Islands (Tahiti, Mo'orea, Huahine, Ra'iatea)

Hawai'i
Kaua'i - Alakai
O'ahu - Mt. Ka'ala, part of Ko'olau ridge
Moloka'i - Kamakou, Pepeopae, Oloku'i
Maui - West Maui Mts., windward slopes of Haleakala
Hawai'i - Kohala, Mauna Kea, Mauna Loa, Hualalai
Lana'i - Lana'ihale

New Britain and New Ireland Islands

New Caledonia and the Loyalty Islands

Samoa, American
Tau
Olosega

Samoa, Western
Upolu
Savai'i

Solomons
Guadacanal
Vangung
Isabel
Kolowangara
Santa Cruz

Tonga
Late
Kau
Tafahi

Papua New Guinea (needs documentation)
Central Range, Eastern and Western Highlands
Mt. Wilhelm
Mt. Haagen

2. Tropical Montane Cloud Forests: Conservation Status and Management Issues

Charles Doumenge, Don Gilmour, Manuel Ruíz Pérez, and Jill Blockhus

With the collaboration of the World Conservation Monitoring Centre, Cambridge, UK, and Alain Gioda, ORSTOM, Montpellier, France

In the humid tropics, montane cloud forests are found between 500 and 3,500 m altitude (LaBastille and Pool 1978; Unesco 1981; Sosa 1987; Stadtmüller 1987; Fao 1992) with major occurrence between 1,200 and 2,500 m (Stadtmüller 1987). They may occur within the montane and submontane floristic zones.

Within the humid tropical zone, one can observe some vegetation gradients with increasing altitude: diminishing tree height, simplified stratification, smaller leaf size, more open understory, some floristic changes, and more epiphytes, mosses, and lichens (Whitmore 1975; Unesco 1981; White 1983; Jacobs 1988). Tropical montane forests are usually of low stature (White 1983; Letouzey 1985; Sayer, Harcourt, and Collins 1992) whereas submontane forests have greater height (20–30 m vs. 15–25 m), are richer in species, with more vascular epiphytes, and incorporate species from low and mid-altitude forests (Letouzey 1985; Thomas and Achoundong 1991).

In this paper, we will refer to tropical montane cloud forests (TMCF) in a topographic sense, as cloud forests lying on wetter mountainsides, ridges and saddles, whether in montane or submontane floristic zones. They can be defined as tropical montane or submontane forests characterized by persistent clouds or fogs, sometimes with small trees, often with abundant mosses, ferns, lichens, or flowering plants in trees and on soil surface.

Global Distribution

In Table 1 and Figures 1–3, we have summarized available information on the geographical distribution and overlap extent of montane and submontane forests. According to recent FAO estimates (Table 1), montane and submontane forests comprise more than 11 percent of the total tropical forests. According to Persson's (1974) estimates, cloud forests occur in a portion of this area and cover about 50 million ha, or one-fourth of the montane and submontane forests. These estimates should be considered as border figures because of a lack of precise data on the extent and distribution of TMCF within these broader montane and submontane forest categories. Cloud forests occur on humid mountain slopes where topo-climate conditions are favorable to the occurrence of ground-level clouds during a considerable period of the year. They occur more frequently in tropical America and Southeast Asia than in tropical Africa (Stadtmüller 1987). The reason for the difference is that there is a wider distribution of mountain ranges, subject to oceanic influences, in America and Asia than in Africa.

Apart from this general distribution, TMCF does appear as micro-occurrences, in locations where there is a cooler climate and a stronger influence of mists and clouds than might normally be expected. In Africa, some of these cloud forest micro-occurrences include isolated low elevation mountains exposed to wind and clouds such as the Belinga mountains of Gabon (800–900 m) and other topographical situations where low elevation passes or peaks are exposed to oceanic influences (e.g., in Congo, the Mayombe mountains, 500–700 m, [Cusset 1989; Dowsett-Lemaire 1991]); the Angola Highlands, where there is an extension of dense forests from southern Zaire, which also include important occurrences of cloud forests (White 1983; Stuart, Adams, and Jenkins 1990; Thomas and Achoundong 1991).

Table 1. Tropical montane and submontane forest cover and deforestation rate compared to the entire tropics

		Forest Cover 1990		Annual Deforestation 1981–90	
	Land Area (million ha)	Total (million ha)	Land Area (%)	Total (million ha)	% per Year
Total tropics	4,778.3	1,756.3	37	15.4	0.8
Hills and mountains	700.9	204.3	29	2.5	1.1

Source: FAO 1993.

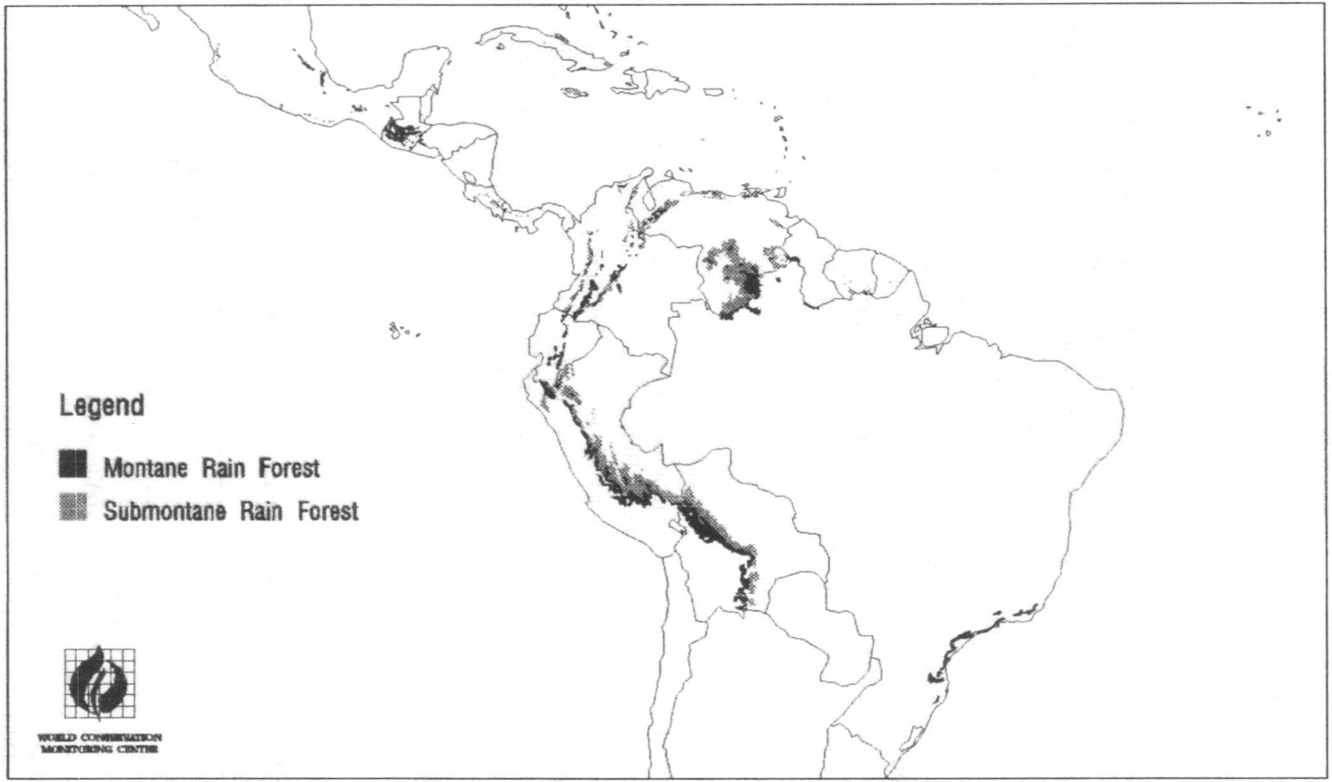

Figure 1. Distribution of tropical montane and submontane forests in Central and South America. Data are incomplete for the forest cover of Mexico, the Caribbean, and most of Central America and the southern countries of South America. *(Source: World Conservation Monitoring Centre.)*

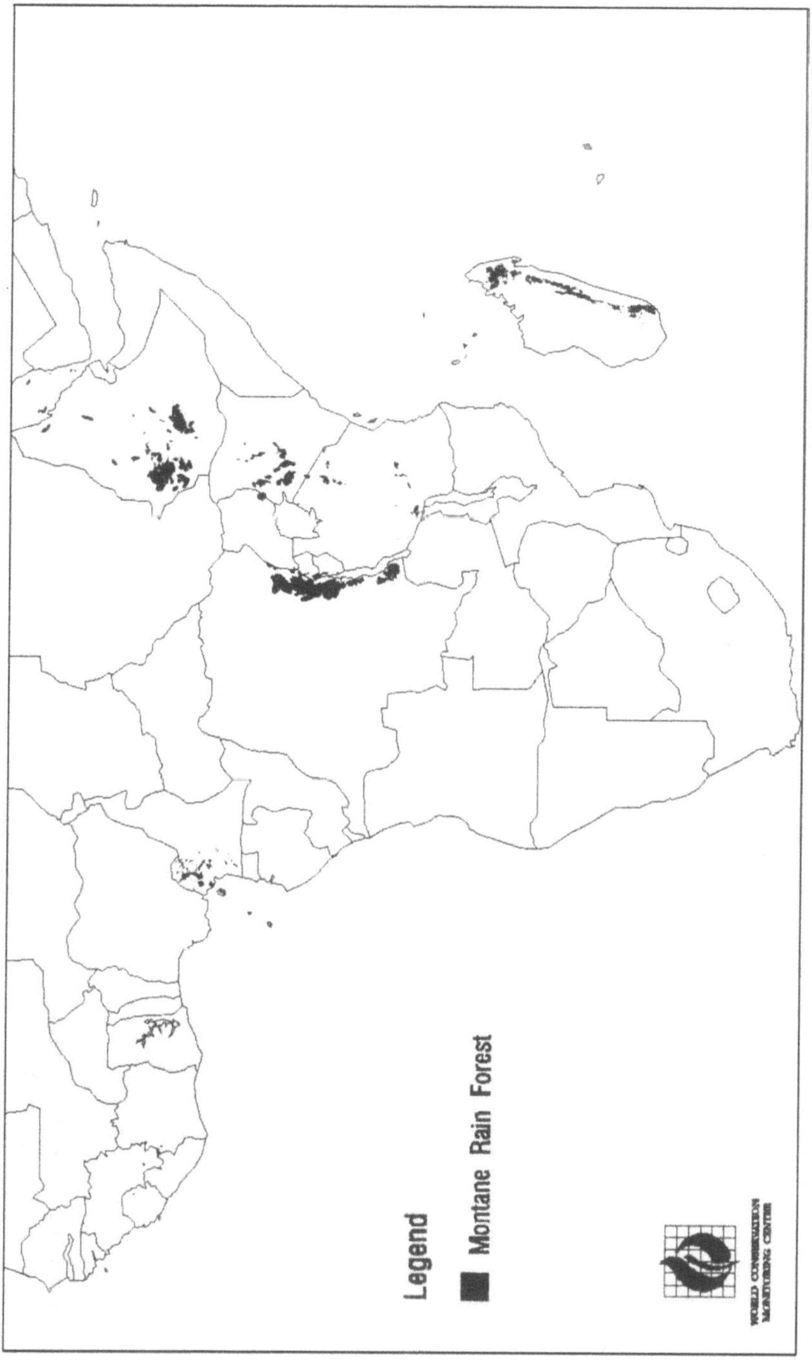

Figure 2. Distribution of tropical montane forests in Africa. *(Source: World Conservation Monitoring Centre.)*

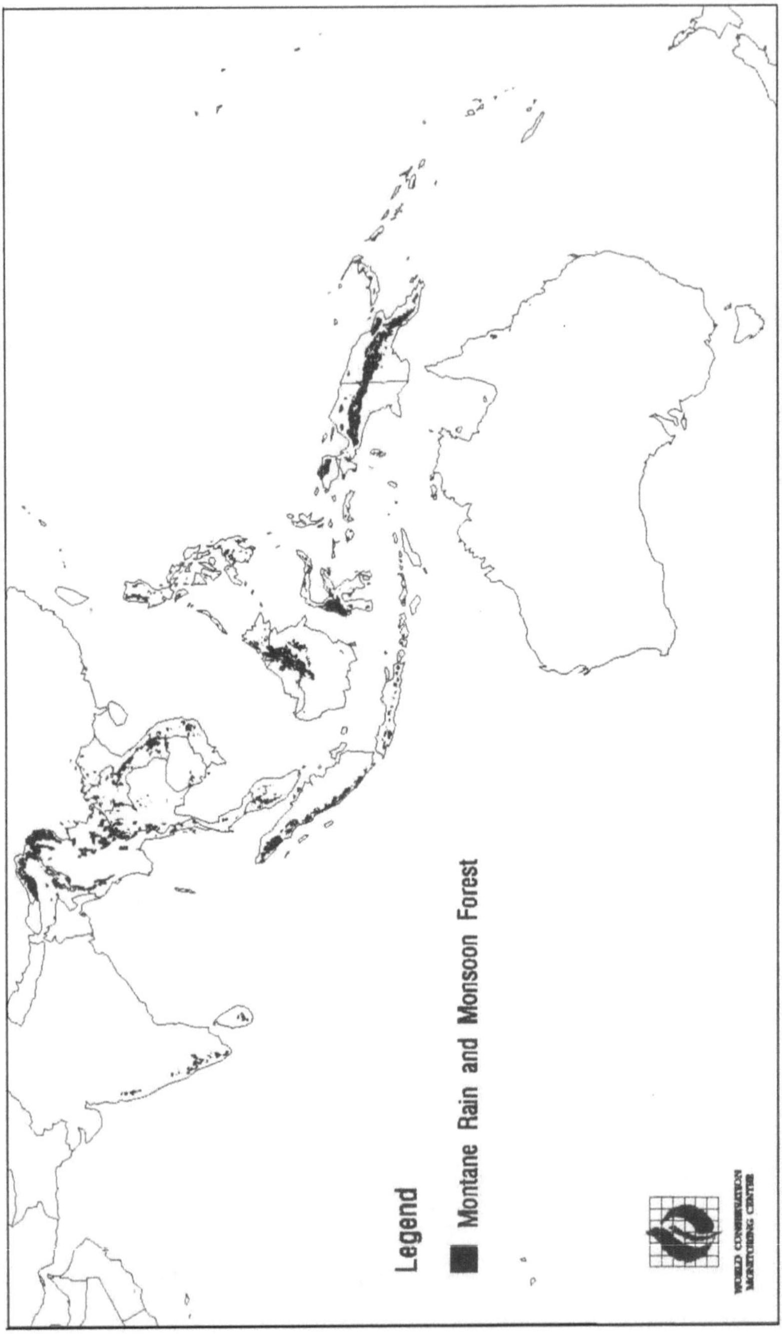

Figure 3. Distribution of tropical montane rain and monsoon forest in South and Southeast Asia and the Pacific. *(Source: World Conservation Monitoring Centre.)*

Values

Tropical montane cloud forests exhibit many values related to the uniqueness of these ecosystems, in terms of their diversity and endemism, and the functions they provide. These values encompass:

- *Watershed function*, including a role in the capture and transport of water and protection of soils against erosion (LaBastille and Pool 1978). Evidence suggests that cloud forests perform a watershed function that is somewhat different from that performed by noncloud forests. This difference relates to the presence of the cloud itself and the occurrence of occult precipitation. The tree crowns act to intercept wind-driven cloud moisture on leaves and branches that drips to the ground. The absolute increase in net precipitation is a result of the presence of trees. This can add to the groundwater and streamflow levels, but its precise effect on the hydrological cycle is difficult to determine. The impact will, in any case, vary from place to place depending on factors such as incidence of wind-driven clouds, density of the clouds and wind speed, size and orientation of mountains, altitude, type of vegetation, and other climatic variables (see Kerfoot 1968; Stadtmüller 1987).

This "cloud stripping" phenomenon may well double the effective rainfall received in dry seasons and increase the wet season rainfall total by about 10 percent (Vogelmann 1973; Bruijnzeel 1990). Experiments to measure the extent of occult precipitation in temperate cloud forest were carried out in the Snowy Mountains in southern Australia in woodlands close to the tree line in an area with a high incidence of wind-driven clouds (Costin and Wimbush 1961). The results indicated a significant increase in net precipitation under the canopy of scattered trees compared with locations between the canopies. This contrasts strongly with the results in noncloud forests where the opposite situation occurs. That is, in noncloud forests, tree canopies frequently intercept and evaporate 20 percent or more of the incident precipitation, and this is lost to the land component of the hydrological cycle.

It can be postulated that clearing cloud forests and replacing them with vegetation with a reduced "filtering" ability will decrease net precipitation. Indirect evidence for this comes from Sri Lanka, where large areas of the central highlands montane forests have been converted to tea estates. Long-term streamflow records have indicated a decline in water yield even though there has not been a corresponding trend of decline in rainfall (Madduma Bandara and Kurupuarachchi 1988). While correlation and causation are not always the same, it is a trend that is worth exploring further.

- *Vegetation elements*. In Sub-Saharan Africa, Stuart, Adams, and Jenkins (1990) postulated that 80 percent of about 4,000 plant species of the Afromontane region are endemic, some of them occurring in cloud forests. In cloud forests of Central America, endemism is high, especially in the Caribbean islands (LaBastille and Pool 1978). Throughout the tropics, the cloud forest zone provides habitat for notable species such as *Podocarpus* (Unesco 1981; White 1983; Sosa 1987; Jacobs 1988; FAO 1992) and tree ferns (Cyateaceae), which are common in this foggy environment (Stadtmüller 1987). In the neotropics, Luteyn (1989) determined that species of the family Ericaceae are most common in the cloud forest zone, although they are not restricted to it. The economic values of this vegetation type range from wood (often fuelwood or construction material) to medicinal plants and food plants (Torres et al. 1992), and to touristic values.

- *Fauna*. Possibly the best known cloud forest mammal is the Mountain Gorilla (*Gorilla g. beringei*), which in addition to its intrinsic value, is an important attraction for nature tourism in some countries of Central Africa (there is a similar case in South America with the Spectacled Bear). Cloud forests are also habitats for mammals and birds that play an important role in the functioning of the ecosystem, in terms of predation-dispersion of seeds and forest dynamics. Many species are valuable to local people as bushmeat, providing them with animal protein. In tropical Africa, the centers of endemism for birds largely coincide with mountainous areas, including cloud forests (Stuart, Adams, and Jenkins 1990); in South America montane areas and foothills were refuges during dry periods in the geological past, and are currently centers of diversity and endemism (Colinvaux 1989). In Costa Rica, cloud forests are covered with epiphytes, which "provide many birds with habitat and supplies of energy and nutrients," especially in the wet season (Nadkarni 1988).

Cloud forest environments in some areas have values for alternate uses, particularly where the soils are of high quality (LaBastille and Pool 1978). These include general agriculture, tea or coffee estates, pastures, and wood plantations (e.g., *Pinus* spp., *Eucalyptus* spp.).

Human Uses and Conservation Status

Many TMCFs suffer from human pressure, particularly in the submontane area (White 1983). However, it is difficult to assess precisely the global extent of such pressures in terms of their effects on structure and composition or disappearance of the cloud forests. A recent evaluation of tropical hill and mountain forest clearance by FAO (1993) estimated that the annual rate of deforestation was about 1.1 percent for 1981–1990 (Table 1). This is greater than the rate of deforestation for the tropical forest biome as a whole, estimated to be about 0.8 percent per year. LaBastille and Pool (1978) estimate that the cloud forests in Central America and the Caribbean are "probably disappearing faster than any other forest ecosystem."

A variety of nonwood forest products is harvested from cloud forests. For example, in the mountains of Central Africa, *Prunus africana* is heavily exploited for the medicinal properties of its bark and is in danger of extinction (Letouzey 1985). For other products, such as shoots of the edible mountain bamboo (*Arundinaria alpina*) in Uganda, it is difficult to assess whether or not harvesting is at a sustainable level. In heavily populated mountainous areas, the collection of fire wood (for heating and cooking) is a major contributor to forest degradation. Fires of human origin are of importance along forest edges (Thomas and Achoundong 1991). Logging, especially in submontane areas but also at higher altitudes for species such as *Podocarpus*, causes substantial changes to forest structure.

Hunting has a direct impact on fauna populations and an indirect impact on forest cover, as hunters regularly burn the vegetation at forest edges (and sometimes the forest itself) for ease of access and to concentrate grazing animals in areas of young vegetation.

Significant areas of cloud forests have been converted to plantations, especially in submontane areas. In Africa, agriculture is found as high as 2,400 m altitude and pastures as high as 2,000–3,000 m (Letouzey 1985) whereas in the Andes, agriculture has been developed as high as 3,700 m and pastures to well over 4,000 m (Morlon 1992). Tea and coffee estates are common in the montane forest zone of East Africa, and tea estates are widespread in the same zone in Sri Lanka. Cloud forests on steep slopes in South America have been converted to agriculture and pastures, resulting in severe erosion (Unesco 1981). In Costa Rica, the conversion of cloud forests into pastures, even when there is a colonization by shrub trees, has resulted in a vegetation complex with fewer epiphytes, smaller biomass, and lower diversity (Nadkarni 1988). Changes in land use, such as gold mining or iron ore extraction, can lead to locally severe land degradation.

Especially in islands or isolated forests, introduction of exotic plant and animal species has favored degradation of native ecosystems (La Reunion Island, Doumenge and Renard 1989; East Usambara Mountains, Tanzania, Hamilton and Bensted-Smith 1989).

Some cloud forests are nevertheless incorporated in the existing reserves in montane areas, but their extent and conservation status are difficult to assess because of fragmented information. Generally, areas with lesser known large species, but with numerous endemics, such as montane forests, have tended to be underrepresented in networks of protected areas (Stuart, Adams, and Jenkins 1990).

In some areas, local traditions favor conservation and sustainable use of certain forest resources. For example, in some parts of Cameroon Highlands, honey collection is an important activity for the local people and beehives are placed in trees such as *Schefflera* spp. Some forests or parts of forests are considered sacred by local people and are afforded protection (as occurs around Lake Oku in Cameroon).

Despite some notable examples of local protection, in most places high population density and the consequent demand for forest products and land for alternate uses are the most important factors that influence the conservation status of cloud forests. In some countries official government policies recognize the conservation value of such forests, and attempts are made to establish protected reserves. However, in heavily populated areas, current land pressures are likely to lead to the preservation of only small forest patches within a wider agricultural landscape. This is the case in montane Central Africa, where rich volcanic soils and high population densities favor conversion to agriculture and grazing lands.

Management Issues

Some important biophysical aspects should be considered when discussing management of TMCFs and converted land. The major issues are related to the role of vegetation in the hydrological cycle, long-term supply of forest products (both plants and animals), protection of soils, and maintenance of ecological processes. However, the details of the processes involved are poorly understood, and there is a lack of basic scientific data on the nature of the linkages.

The use of forest resources is increasingly driven by the immediate benefits that arise from these uses. There is often no real management of cloud forest resources based on rational assessment of the extent of goods and services that can be provided by well-managed forests. Some studies are being conducted to evaluate the carrying capacity of the environment with respect to resource use, but these studies have been recently established and are not yet fully operational.

It is frequently perceived that a lack of awareness and knowledge among local people is at the heart of unsustainable utilization of forest resources. However, it is rare that people living close to forests, and often dependent on them for their livelihood, are not aware of the importance of the forest. A more important constraint is often associated with the inability of local populations to take effective actions because of the presence of more powerful socioeconomic structures.

One essential element of effective management is the development of appropriate institutional structures to ensure that the wishes and desires of all those who have an interest in montane cloud forests can be considered. These interest groups may include a range of people, from those who depend on the forest for livelihood support to those who have an intellectual interest in forest conservation. However, the official decision-making process is normally in the hands of urban-based government officials and politicians who have the task of integrating sectoral interests with the broader interests of the state. There can also be, as in Eastern Zaire and the Cameroon Highlands, conflicts between traditional customary regulations governing the use of natural resources and state-controlled structures and laws that tend to focus on perceived national interests.

An additional influence, which often has a major impact on the way in which forests are managed, is that of in-migration. Examples can be seen in montainous parts of the Philippines where migrants, often from ethnic backgrounds different from the local indigenous populations, create situations that make effective management difficult.

Conservation and Sustainable Use: Some IUCN Experiences

The three projects described here are used as examples to illustrate a variety of approaches that may be suitable in advancing conservation objectives in areas subject to heavy pressure on forest resources. They are being implemented in montane and submontane areas where cloud forests occur, but where they cover only part of the project area. The primary goal of the projects is to ensure the conservation of forest resources by facilitating related development activities.

Mount Elgon Conservation and Development Project

The Mount Elgon Project in Eastern Uganda (financed by the Norwegian Agency for Development Cooperation) promotes a joint program of activities between the Ministry of Environment Protection of Uganda and IUCN. The long-term goal is to ensure conservation of natural forests on Mount Elgon while assisting community development activities where this development is directly linked to the sustainable use of forest products. The first phase of the project consisted of an extensive study and planning stage, combined with some immediate conservation measures. Phase two, under way, consists of implementation of a management plan for the forest and a long-term strategy aimed at promoting the sustainable use of natural resources in the area. Demarcation of the Mount Elgon Forest Park boundaries and improvement of the infrastructure and management of the park are also part of the activities. A study has recently started to determine bamboo (*Arundinaria alpina*) utilization patterns in the area. The plan will involve people who depend on harvesting bamboo shoots and stems in management of the bamboo resources. The goal is to continue bamboo harvesting in a sustainable manner while conserving the bamboo forests. Similar studies are planned to determine the nature of the interactions between people and other forest resources as a prelude to developing management procedures aimed at sustainable utilization.

Itombwe Mountains Project

In the Itombwe Mountains in Eastern Zaire, IUCN works closely with a governmental body, the Zairian Institute for Conservation of Nature, and with local people and NGOs. Pressures on the forests come from extension of pastures and agriculture, gathering of wood (for fuel and construction), hunting, and, in some places, gold mining. The approach selected is a participative one, bringing external expertise to work with local people and NGOs to facilitate planning and sustainable development. This participatory rural development process has permitted local people to gather information and analyze environmental and socioeconomic problems they are facing. The socioecomomic data are processed locally, and the results will be discussed with the rural people living in the area, through workshops to be held in selected villages. This action will form the basis for future project planning.

Although this process has just started, the cooperative mechanism has reinforced local awareness and local capacity to analyze and manage natural resources. This cooperative effort is vital in countries like Zaire where internal conflicts have resulted in disintegration of the administrative capacity of the official government. Some problems experienced include the necessity to send longer support missions than may be the case in normal circumstances, and the difficulty—in continuing with the analysis phase—of mobilizing funds to execute future on-ground activities in a country with political instability.

Native Andean Forests Conservation Project

A recent preliminary analysis by Intercooperation (an NGO) and IUCN of native Andean forests in Bolivia and Ecuador drew several conclusions: (1) only small patches of native montane forests remain; (2) these forest fragments are in danger of complete elimination due to pressure from uncontrolled colonization; (3) the official management institutions are ineffective with respect to protection of remaining forest areas. This analysis led to a program aimed at conservation and sustainable use of the native Andean forests by stabilizing the pressures on these forests and reclaiming degraded areas. The activities include establishing demonstration projects, strengthening institutional capacity, analyzing the legislative system, and enhancing regional coordination and knowledge sharing. This project is financed by and will be implemented by Intercooperation and the Swiss Technical Cooperation with local organizations and communities; IUCN will provide technical support.

Conclusions

The values and services provided by TMCFs should be fully recognized and taken into account in policy statements and management decisions. Management decisions should consider (1) the necessity of exploiting products within the carrying capacity of the ecosystems; (2) the value of forests in water capture and for maintaining soil stability (particularly important in areas where soils are prone to erosion); and (3) the necessity of conserving reservoirs for genetic material. The conversion of forests to other uses should be carried out only after careful impact analysis.

Such topics for consideration are important; however, they do not meet all the conditions necessary for sustainable use of natural resources. We have provided some details of possible approaches, with references to IUCN projects. A conclusion of these project experiences is that the major constraint to cloud forest conservation is awareness and commitment of all the participants and interest groups for the long-term conservation of these forests. It is necessary to build management processes that take into account the existing and potential conflicts to be resolved and to design a human-scale approach, which is more participative, favoring integration of concerns and cooperation (Poore 1992). Similar initiatives are being implemented around the world but should be further developed, extended, and adapted to different circumstances.

The diversity of situations implies a diversity of strategies in the field, but some basic principles that apply to managing mountain ecosystems can still be valid worldwide (see global guidelines in Dasmann and Poore 1979 and Poore 1992).

While a lack of biophysical knowledge is a constraint to understanding the nature of the hydrological and ecological processes operating in TMCFs, the major need is a conceptual framework within which to address the conservation needs of cloud forests globally, regionally, and nationally. We hope that the task of providing such a framework will be undertaken. Among points that need consideration are:

Globally

- listing of the unique features and values of TMCF, including a typology;
- evaluation of extent and distribution of cloud forests, and assessment of their conservation status;
- identification of the major threats to their conservation and of sustainable practices;
- definition of a global policy on cloud forests including prioritization of those areas where immediate conservation action is required.

Regionally

- identification of the transfrontier cloud forests that deserve coordinated actions between neighboring countries;
- establishment of cooperation processes to manage sustainably transfrontier cloud forests, and design of regional policies.

Nationally

- detailed mapping of cloud forest areas and assessment of their conservation value (locally and globally) and status (current land tenure, level of protection, threats to their integrity);

- development of a strategy to ensure that the values of cloud forests are recognized and fully considered in decision-making processes;
- implementation of management regimes that take account of the legitimate demands of local people for access to and usage of forest products in a sustainable way.

Acknowledgments. The authors thank their colleagues in Gland and around the world who provided valuable information and helpful comments on the drafts.

References

Bruijnzeel, L. A. 1990. *Hydrology of moist tropical forests and effects of conversion: A state of knowledge review.* Paris: Unesco.

Colinvaux, P. A. 1989. The past and future Amazon. *Scientific American* 68-74.

Costin, A. B., and D. J. Wimbush. 1961. Studies in catchment hydrology in the Australian Alps. 4) Interception by trees of rain, cloud and fog. CSIRO Australian Division of Plant Industry Technical Paper 16, Canberra, Australia.

Cusset, G. 1989. La flore et la végétation du Mayombe congolais. Etat des connaissances. In *Revue des connaissances sur la Mayombe*, ed. J. Sénéchal, Matuka Kabala, and F. Fournier, 103-136. Paris: Unesco.

Dasmann, R. F., and D. Poore. 1979. *Ecological guidelines for balanced land use, conservation and development in high mountains.* UNEP, Nairobi, IUCN, and WWF, Gland, Switzerland.

Doumenge, C., and Y. Renard. 1989. *La conservation des écosystèmes forestiers de l'île de la Réunion.* Gland, Suisse et Cambridge, Royaume-Uni: IUCN.

Dowsett-Lemaire, F. 1991. The vegetation of the Kouilou bassin in Congo. In *Flore et faune du bassin du Kouilou (Congo) et leur exploitation*, ed. R.J. Dowsett and F. Dowsett-Lemaire. *Tauraco Research Report* 4:17-51.

FAO. 1992. *Aménagement et conservation des forêts denses en Amérique tropicale.* FAO, Etude Forêts 101, Rome.

FAO. 1993. Summary of the final report of forest resources assessment 1990 for the tropical world. Paper prepared for the 11th COFO meeting, Rome.

Hamilton, A. C., and R. Bensted-Smith, eds. 1989. *Forest conservation in the East Usambara Mountains, Tanzania.* Gland, Switzerland, and Cambridge, UK: IUCN.

Jacobs, M. 1988. *The tropical rain forest: A first encounter.* Berlin and Heidelberg, Germany: Springer-Verlag.

Kerfoot, O. 1968. Mist precipitation on vegetation. *Forestry Abstracts* 29:8-20.

LaBastille, A., and D. J. Pool. 1978. On the need for a system of cloud-forest parks in Middle America and the Caribbean. *Environmental Conservation* 5(3): 183-190.

Letouzey, R. 1985. *Notice de la carte phytogéographique du Cameroun au 1:500 000 (1985). 2) M-SM: Région afro-montagnarde et étage submontagnard.* Toulouse, France: Institut de la Carte Internationale de la Végétation.

Luteyn, J. L. 1989. Speciation and diversity of Ericaceae in Neotropical montane vegetation. In *Tropical forests. Botanical dynamics, speciation and diversity*, ed. L.B. Holm-Nielsen, I.C. Nielsen, and H. Balslev, 297-310. London: Academic Press.

Madduma Bandara, C. M., and T. A. Kurupuarachchi. 1988. Land-use change and hydrological trends in the upper Mahaweli basin. For workshop on hydrology of natural and man-made forests in the hill country of Sri Lanka, Kandy. Unpublished.

Morlon, P., ed. 1992. *Comprendre l'agriculture paysanne dans les Andes Centrales.* Paris: INRA.

Nadkarni, N. M. 1988. Tropical rain forest ecology from a canopy perspective. In *Tropical rainforests: Diversity and conservation*, ed. F. Almeda and C.M. Pringle, 189-208. San Francisco: California Academy of Sciences and American Association for the Advancement of Science.

Persson, R. 1974. *World forest resources*. Stockholm: Royal College of Forestry.

Poore, D. 1992. *Guidelines for mountain protected areas*. Gland, Switzerland, and Cambridge, UK: IUCN Commission on National Parks and Protected Areas.

Sayer, J. A., C. S. Harcourt, and N. M. Collins, eds. 1992. *The conservation atlas of tropical forests. Africa*. Gland, Switzerland, and Cambridge, UK: IUCN.

Sosa, V. J. 1987. Generalidades de la región de Gómez Farías. In *El bosque mesofilo de montaña de Tamaulipas*, ed. H. Puig y R. Brach, 15-28. México: Instituto de Ecología.

Stadtmüller, T. 1987. *Cloud forests in the humid tropics. A bibliographic review*. United Nations University, Tokyo, and CATIE, Turrialba, Costa Rica.

Stuart, S. N., R. J. Adams, and M. D. Jenkins. 1990. *Biodiversity in Sub-Saharan Africa and its islands. Conservation, management and sustainable use*. IUCN, Occasional Papers of the IUCN Species Survival Commission 6, Gland, Switzerland.

Thomas, D. W., and G. Achoundong 1991. Montane forests of Western Africa. In AETFAT Congress.

Torres, H., R. Borel, N. Bustamante, and M. I. Centeno. 1992. *Usos tradicionales de arbustos nativos en el sur de Puno*. Puno, Perú: Intercooperation.

Unesco. 1981. *Vegetation map of South America. Explanatory notes*. Paris: Unesco.

Vogelmann, H. W. 1973. Fog precipitation in the cloud forests of Eastern Mexico. *BioScience* 23(2): 96-100.

White, F. 1983. *The vegetation of Africa*. Paris: Unesco.

Whitmore, T. C. 1975. *Tropical rain forests of the Far East*. London: Oxford University Press.

3. Hydrology and Biogeochemistry of Tropical Montane Cloud Forests: What Do We Really Know?

L. A. Bruijnzeel and J. Proctor

Arguably, montane "cloud forests" (MCFs) are among the least understood of humid tropical forest ecosystems as far as their water and nutrient dynamics are concerned (Whitmore 1990). This is in spite of the fact that TMCF is often found in important headwater areas that, although scattered, together occupied about 500,000 km^2 in the 1970s (Persson 1974). There is a growing recognition of the role of TMCF in supplying water to downstream areas during rainless periods (Zadroga 1981; Hamilton with King 1983; Stadtmüller and Agudelo 1990) and of their high degree of faunal and floristic endemism (La Bastille and Pool 1978; cf. Leo, this volume).

Stadtmüller (1987) produced an important introductory review, highlighting the confusing nomenclature of TMCF and focusing largely on hydrometeorological studies made in the Neotropics. The present contribution intends to expand Stadtmüller's review with respect to the water dynamics of TMCF.

With increasing altitude on wet tropical mountains, changes in forest structure and physiognomy occur, which have, in the words of Whitmore (1989), "puzzled and irritated" scientists for a long time. The principal changes are a decrease in forest stature and leaf size, and a tendency for the leaves to become thicker and harder (xeromorphic). A related phenomenon is the so-called Massenerhebung, or "telescoping" effect: the occurrence of stunted forests at lower elevations on small outlying mountains compared with those on large mountains (Richards 1952). Although the stunting of montane forests and the telescoping effect have received much study (see reviews by Leigh 1975; Grubb 1977, 1989; Bruijnzeel et al. 1993), both phenomena are still not fully explained. The suggestion of Grubb and Whitmore (1966) that the frequency of cloud or fog is the most important factor determining the stature of montane forest is supported by numerous descriptive studies (reviewed by Richards 1952; Lamprecht 1958; Van Steenis 1972; Stadtmüller 1987; Cavelier 1988). The second part of this paper discusses the available evidence of how a cloud cover may cause this effect, and includes recent results for Southeast Asian low-elevation MCF (Bruijnzeel et al. 1993). This then leads to a consideration of the cycling of nutrients in TMCF, and a brief review is made of atmospheric nutrient inputs, soils, and nutrients in litterfall, including the relatively understudied role of epiphytes in the cycling of nutrients in TMCF. Finally, a number of suggestions for further work on the water and nutrient dynamics of TMCF are made, stressing the need for multidisciplinary studies at carefully selected key locations.

Definition and Occurrence of TMCF

Grubb et al. (1963) distinguished the formations lower montane rain forest (LMF) and upper montane rain forest (UMF) that usually occur in sequence above evergreen lowland tropical rain forest. Grubb and Whitmore (1966) suggested LMF had "frequent" and UMF "long, persistent" cloud cover close to the ground, although this statement (like their definitions of LMF and UMF) was not fully quantified. Later, Grubb (1974) suggested that tall LMF might also occur on more-or-less fog-free locations. Montane cloud forests may belong to either LMF or UMF, and here "TMCF" is used for forests (including "elfin cloud forests"—ECF; Stadtmüller 1987) that are frequently covered in clouds or mist. It is recognized that quantitative criteria on what constitutes "frequent" clouds are lacking, but implicit in TMCF is a recognition of the important influence of clouds on a range of ecosystem processes.

Depending on latitude, the lower limit of TMCF on large mountains is generally between 1,500 and 2,500 m, whereas the upper limit is usually from 2,400 to 3,300 m. Local factors may cause cloud formation at much lower altitude (e.g., on small coastal or island mountains), while ECF may occur up to 3,900 m under favorable conditions (Stadtmüller 1987).

Hydrological Aspects

Horizontal Precipitation

While the importance of frequent low cloud to ecosystem functioning in TMCF is widely recognized, its quantification is notoriously difficult (Kerfoot 1968). A variety of terms has been coined to describe the movement of water onto vegetational surfaces via condensation or by direct contact of cloud droplets. In the following, this input of water will be called horizontal precipitation (HP) (Stadtmüller 1987).

The quantity of HP depends partly on vegetational factors and partly on climatic factors. The former include the following: height of the vegetation, canopy size and structure, biomass, arrangement and physical properties of leaves and epiphytes. Climatic factors include moisture content, drop sizes, velocity and direction of the passing air with respect to the orientation of the forested slope or ridge, and the duration of the whole process. Variations between locations will be large. Traditionally, two approaches have been followed in trying to quantify HP (Kerfoot 1968): (1) the use of "fog catchers" and (2) a comparison of amounts of canopy drip measured inside a stand with amounts of rainfall measured in the open. Both methods present problems of measurement and interpretation. Fog catchers, whether of the "gauze cylinder" or "wire harp" type (Nagel 1956; Ekern 1964), suffer from the inherent problem that each forest canopy represents a more-or-less unique situation that cannot be fully simulated. It is possible, however, to obtain an idea of the frequency and the relative importance of moisture contributions by low clouds with these instruments, although amounts of moisture trapped are rather sensitive to the height of the catcher with respect to the canopy or the ground (Ekern 1964; Cavelier and Goldstein 1989). Arguably, the second approach is superior as long as amounts of water intercepted and evaporated from the wetted canopy are taken into account. Without automated meteorological equipment, the procedure of separating daily throughfall totals for fog-free and fog-bound conditions (Kashiyama 1956; Harr 1982; Stadtmüller and Agudelo 1990) is the best method. However, since evaporation rates from a wet canopy may well differ for the two situations, the results will remain approximate (see also the next section).

More recently, a promising alternative approach has been developed by investigators studying rates of "occult" deposition of pollutants (Dollard and Unsworth 1983; Gallagher et al. 1988). This "gradient technique" combines continuous observations of the wind profile (and thus turbulence) above the intercepting surface with measurements of cloud water content (using either such advanced equipment as Knollenberg forward-scattering spectrometer probes or simple wire harps).

The amounts of HP reported from TMCF as determined with fog catchers are summarized in Table 1 whereas measurements of throughfall and rainfall are summarized in Table 2. In both cases, results are variable. For instance, annual totals of HP estimated with fog catchers range from 70 mm for an elfin cloud forest at 3,100 m in Venezuela (Cavelier and Goldstein 1989) up to 940 mm in LMF at 1,300 m in eastern Mexico (Vogelmann 1973). There is no positive correlation between annual totals of HP and ordinary rainfall, but relative values of HP tend to be lower at sites with high rainfall totals and during rainy seasons (Table 1). Cavelier and Goldstein (1989) compared amounts of HP between various low coastal mountains in northern Venezuela

Table 1. Amounts of horizontal precipitation (HP) in tropical montane cloud forest areas as determined by means of fog catchers (d, dry season only; w, wet season only)

Location	Altitude (m)	Vegetation type	HP (mm/d)	HP (% of rain)	Remarks
Colombia, Serrania de Maguira[1]	865	ECF	2.2	63/63d	Magel-type gauge, sampled every 14 d for 12 months
Costa Rica, Cerro Buenavista[2]	3500	Paramo	2.1	18	Two aluminum screens (3250 cm^2); 13 months; absolute value for dry days
Costa Rica, Balalaica[3]	1300	LMF	4.0 1.6w	33 15w	Three wire harps at 6 m; weekly samples for 6 months/9 weeks
Hawai'i, Mauna Loa[4]	1580 2530	Various Idem	2.1 1.9	30 68	Louvered screen; frequent readings over 1–2 years; windward slopes
Malaysia, Gunung Silam[5]	884	LMF	0.4	9	Nagel-type gauge, 6 weeks in dry season
Mexico, Sierra Madre[6]	1330	MCF	1.6w 0.6d	15w 85d	Nagel-type gauge, sampled weekly for 53 weeks, long dry season (40 weeks)
	1360	Idem	0.8w 0.4d	17w 60d	
	1900	Idem	0.6w 0.4d	14w 103d	
Puerto Rico, Pico del Oeste[7]	1050	ECF	0.9	7	Louvered screen plus tipping bucket, 258 days of observations
Venezuela[1]					See Colombian site
Cerro Sa Ana	815	ECF	1.4	32/66d	
Cerro Copey	987	ECF	1.3	11/9d	
El Zumbador	3100	ECF	0.2	4/19	

[1]Cavelier and Goldstein (1989); [2]Dohrenwend (1979); [3]Caceres (1981); [4]Juvik and Ekern (1978); [5]Bruijnzeel et al. (1993); [6]Vogelmann (1973); [7]Baynton (1969)

and a high-altitude inland location. While rainfall along the coast increased sharply from west to east, corresponding amounts of HP decreased, albeit less dramatically. Conversely, the inland site received very little HP (Table 1), which was interpreted by Cavelier and Goldstein (1989) in terms of differences in cloud water content (stratus in the Andes vs. cumuliform at the coast), and more persistent fog and higher wind speeds near the coast.

Relative amounts of throughfall recorded in TMCF are also variable (Table 2), ranging from about 70–80 percent of incident rainfall in tall LMF on inland mountains in Tanzania, Papua New Guinea, and Venezuela experiencing relatively light or virtually no cloud incidence (Lundgren and Lundgren 1979; Edwards 1982; Steinhardt 1979) to 130–180 percent in forests on exposed slopes and ridges near the coast (Weaver 1972; Stadtmüller and Agudelo 1990). The difference of 1,160 mm/year^{-1} between throughfall and rainfall determined (without correcting for evaporation of about 300 mm/year^{-1} intercepted rainfall) for an MCF on an exposed ridge in eastern Honduras by Stadtmüller and Agudelo (1990) is the highest HP reported to date (cf. Weaver 1972).

Ideally, amounts of water reaching the forest floor as stemflow should also be taken into account. Stemflow in LMF is generally low (e.g., Lundgren and Lundgren 1979; Steinhardt 1979; Caceres 1981), but larger proportions (5–10 percent of rainfall) have been observed in ECF, particularly those that are wind-exposed (Weaver 1972). The high tree density of these forests may be responsible for this (Baynton 1969). Interestingly, stemflow figures for lowland heath forest, which also has a high tree density (Whitmore 1990), were higher than those in adjacent ordinary lowland rain forests (Jordan 1978). Bruijnzeel (1989c) hypothesized that the many-stemmed nature of both elfin and heath forests might enhance nutrient availability through the funneling of larger volumes of water along the stems (see also the following).

Evaporation

Reference Evaporation Rates

Next to rainfall and HP, the rate of open-water evaporation (E_0) (Penman 1956) is a convenient measure of overall climatic conditions. Evaporation is expected to decrease with elevation as radiation, temperatures, and vapor pressure deficits are all reduced (Braak 1922; Troll 1956; Briscoe 1966). There have been suggestions that the most stunted types of MCF are found in areas with very low atmospheric evaporative demand, even to the extent that rates of water and nutrient uptake are claimed to become suppressed (Odum 1970; Weaver, Byer, and Bruck 1973; Leigh 1975; Ash 1987). However, average values of E_0 at sites carrying stunted MCF at low-to-intermediate elevations (e.g., Serrania de Macuira, Venezuela; El Yunque and Pico del Oeste, Puerto Rico; Blue Mountains, Jamaica; Gunung Silam, Malaysia) ranged between 2.0 and 3.75 mm/day^{-1} (Waterloo 1989). Although these values are lower than those recorded for tropical lowlands (typically around 5 mm/day^{-1}), one should remember that most of the daily evaporation totals in the TMCF belt will be attained during fog-free periods, usually in the mornings. Values that have been adjusted for this resemble more closely climatic conditions lower down the slopes (Grubb and Whitmore 1966; Langley 1985). As will be shown, water and nutrient uptake in some TMCF do seem suppressed, but not because of a lack of radiant energy or high aerial humidity.

Total Evapotranspiration

Total evaporation of moisture from a vegetated surface consists of three components: (1) evaporation of precipitation intercepted by the vegetation, E_i; (2) transpiration, E_t; and (3) evaporation from the ground layer, E_s. Since E_s has been shown to be very small in lowland rain forests (Jordan and Heuveldop 1981; Van der Plas 1992), it can be ignored for most montane forests. E_i and E_t then represent evaporation from wet and dry canopies, respectively.

The magnitude of E_i is determined by a number of climatic and "passive" vegetation factors, whereas E_t is regulated by plant stomatal control in response to changing conditions. Climatic factors include duration and intensity of precipitation (E_i only), wind speed, available energy, and vapor pressure deficit. "Passive" vegetational factors include height, roughness, size, arrangement, and total biomass of the canopy (Rutter 1975).

Before examining the values of E_i and E_t, the evidence on total evapotranspiration (ET) in TMCF is reviewed (Table 3). All but one study (San Eusebio, Venezuela) used the catchment water-balance technique to evaluate forest ET (Ward and Robinson 1989). A number of studies were made on volcanic terrain and must be treated with caution since they may have been influenced by subterranean water transfers (Gonggrijp 1941; Rijsdijk and Bruijnzeel 1990).

Those forests receiving much HP in Table 3 have ET values of 300–400 mm/year^{-1}, whereas those for LMF with only minor incidence of low cloud (e.g., the Colombian forest at 1150 m and the Indonesian forest) are about 1,200 mm/year^{-1}. These estimates for ET need to be adjusted for moisture inputs from HP. As indicated earlier, all estimates of HP are approximate, but reasonable estimates were available in a number of cases, either for the study site itself (e.g., the Philippine forest) or for a similar type of forest in the region (the Colombian forest at 2100 m, Vis 1986; the Costa Rican forest, Caceres 1981). The adjusted values ranged from 570 to 695 mm/year^{-1} (Table 3), thus remaining well below the average of 1,225 mm/year^{-1} derived for tropical montane forests that are little affected by low cloud (Bruijnzeel 1990).

The information given in Table 3 could usefully be expqnded by a systematic examination of streamflow and rainfall records collected routinely by government departments in countries where TMCF is found in the headwater areas of major river basins (cf. Aitken, Ribevy, and Brown 1972; Calvo 1986; Lugo 1986).

Evaporation of Intercepted Precipitation

Estimates of E_i are normally made by subtracting amounts of throughfall and stemflow from incident rainfall. As explained earlier, the unknown additional

Table 2. Measurements of throughfall (TF), stemflow (SF), and derived estimates of rainfall interception (E_i) in tropical montane forests (values expressed as percent of incident rainfall)

Location	Altitude (m)	Forest Type	MAP (mm)	TF (%)	SF (%)	E_i (%)	Remarks
Colombia[1]	1000	LMF	1600	78	—	22	6 roving gauges; 5 months
	1700	LMF	3150	76	—	24	idem; 15 months
	1950	LMF/MCF	2200	85	—	15	idem
	3000	UMF/MCF	1700	89	—	11	idem
Colombia[2]	2550	LMF/MCF	2115	87.5	—	12.5	20 fixed gauges; 12 months; weekly collections
	3370	UMF/MCF	1455	81	0.1	18.5	
Costa Rica[3]	1300	LMF/MCF	2510	101	0.4	?	20 fixed square metal collectors, sampled weekly for 6 months; cf. Table 1
Honduras[4]	1795	LMF/MCF	1500	94	—	6	4 fixed troughs; concave, convex, and ridge sites, resp.; daily observations for 12 months
				129	—	?	
				179	—	?	
Indonesia[5]	1750	LMF	3300	80	—	19	4 fixed square metal collectors; weekly samples; 12 months
Jamaica[6]	1020	LMF	3000	81	—	19	randomly located troughs; weekly (?) samples; 1 year

Jamaica[7]	1550	LMF/UMF	2600+	92+	—	<8	10–14 fixed gauges; 12 months; 2-week intervals
Mauritius[8]	550	LMF	3175	70/76	—	24/30	3 fixed gauges; 12 months; daily readings; values for 3 c.q. 2 gauges, resp.
New Guinea[9]	2500	LMF	3800	67	<1	32	8 fixed gauges in 4 plots each; 8 months; 2-week intervals
Philippines[10]	2200	LMF/MCF	3900	86	12?	2	4(?) troughs in 4 plots; 60 months; daily readings; value for SF suspect; TF varied per exposure and includes 9% fog drip
Puerto Rico[11] El Verde	450	LMF	2540	72	8	20	6 fixed troughs; 12 months; daily readings; low intensity
Puerto Rico[12] Pico del Este	1000	ECF	5400	115	5	−20	60 fixed collectors in each of 3 plots: windward, ridge and leeward; 8 months; daily readings
	1015		4800	125	10	−35	
	930		6000	96	3	1	
Tanzania[13]	1500	LMF	1230	78	<1	21	8 fixed gauges; 30 months; daily readings
Venezuela[14]	2300	LMF/MCF	1575	79.5	1	19.5	30 fixed gauges; 12 months; 1–2 readings per week

[1]Vis (1986); [2]Veneklaas and Van Ek (1990); [3]Caceres (1981); [4]Stadtmüller and Agudelo (1990); [5]Gonggrijp (1941); [6]Richardson (1982); [7]Tanner (1980a); [8]Vaughan and Wiehe (1947); [9]Edwards (1982); [10]Mamanteo and Veracion (1985); [11]Clements and Colon (1975); [12]Weaver (1972); [13]Lundgren and Lundgren (1979); [14]Steinhardt (1979)
MAP = mean annual precipitation

Table 3. Annual evapotranspiration (ET)* and transpiration (E_t)# vs. rainfall and elevation for selected montane forests

Location	Altitude (m)	Rainfall (mm/yr^{-1})	ET (mm)	E_t (mm)	Remarks
Colombia[1]	1150	1985	1265	830	LMF; E_i 22%[2]
	2100	2315	310	(285)	LMF/MCF; E_i 12%[2]
Costa Rica[3]	2400	2695	365	(270)	LMF/MCF; E_i 10%[4]
Indonesia[5]	1750	3305	1170	510	LMF; E_i 20%
Malaysia[6]	870	2500	(695)	(310)	LMF/MCF; E_i 15%; ET via site water balance; 6 weeks of observations
Philippines[7]	2350	3380	390	(250)	LMF/MCF; E_i 13%[8]
Venezuela[9]	2300	1575	980	675	LMRF/MCF?; E_i 20%; ET via energy balance
Puerto Rico[10,11]	1000	4700	—	75	ECF; cut shoot method, LAI[12] 2 m^2 m^{-2}

Adapted from Bruijnzeel (1990).
*Values for ET not corrected for contributions by horizontal precipitation
#E_t derived as net ET – E_i unless stated otherwise
[1]Herrmann (1971); [2]Vis (1986); [3]Calvo (1986); [4]Caceres (1981); [5]Gonggrijp (1941); [6]Bruijnzeel et al. (1993); [7]De los Santos (1981); [8]Mamanteo and Veracion (1985), taking stemflow at 1%; [9]Steinhardt (1979); [10]Weaver, Byer, and Bruck (1973); [11]Gates (1969); [12]Weaver et al. (1986)

input of moisture from the persistent low cloud precludes the use of this technique in TMCF. It seems there are no reliable estimates of amounts of intercepted and subsequently evaporated precipitation in "true" TMCF, only for LMF with relatively little cloud incidence (Table 2). Values under the latter conditions are about 25 percent of incident rainfall (range 15–35 percent, Table 2). It is difficult to extrapolate these values to foggier conditions because much will depend on the timing and duration of the fog (e.g., at night or during the day) as well as on possible contributions by advected energy (warmer air from lower elevations or the ocean).

An experimental design is needed that takes into account both temporal and spatial variations in throughfall and stemflow (Baynton 1969; Lloyd and Marques 1988; Veneklaas and Van Ek 1990) in combination with observations on occurrence and intensity of fog or rainfall; duration of canopy wetness; climatic factors needed to evaluate evaporation rates from a wetted canopy (Monteith 1965); and the time lag between the start of the fog and the start of canopy drip in relation to moisture status of epiphytic cover. Arguably, the data collected by Baynton (1969) and Weaver (1972) for ECF in Puerto Rico may be considered the most complete.

One of the outstanding features of MCF is an abundance of vascular and nonvascular epiphytes; the study of intercepted precipitation is complicated by this. The TMCF zone is a favorable environment for epiphytes because of the abundance of atmospheric moisture (Walter 1979). In tall-facies TMCF, the highest densities of epiphytes are found in the canopy, whereas in more stunted forests they also cover the bases of trunks and even rocks (Stadtmüller 1987). Wolf (1993), in a study of 20 epiphyte communities between 1,000 and 4,130 m in the Central Cordillera of Colombia, found a marked change in both species composition and biomass at about 2,400 m (Figure 1). These changes occurred at the transition from LMF to UMF and were accompanied by a dramatic rise in average daytime relative humidity levels, confirming the original contention of Grubb and Whitmore (1966).

Ecophysiological research confirms the importance of moisture availability to bryophytic growth. Hosokawa, Odani, and Tagawa (1964) demonstrated a positive relationship between photosynthetic rate and relative humidity at equal light conditions. Since bryophyte photosynthetic activity is relatively insensitive to low temperatures (Dilks and Proctor 1975), decreasing temperatures do not counteract the positive effects of high humidity at higher elevations where they are often luxuriant (Figure 1b). It is here that one would expect the strongest effects on rainfall and fog interception.

Values for epiphytic biomass in selected montane forests are given in Table 7 (see section on nutrients) and show substantial variation. Hofstede, Wolf, and Benzing (1993) interpreted these differences in terms of canopy structure, tree density and morphology, and disturbance of accumulated "aerial humus" by occasional cyclones (e.g., in Jamaica and Costa Rica) or torrential rains (lower elevations). Nonvascular epiphytes in particular have been shown to be capable of absorbing large volumes of rain and fog (up to 4–5 times

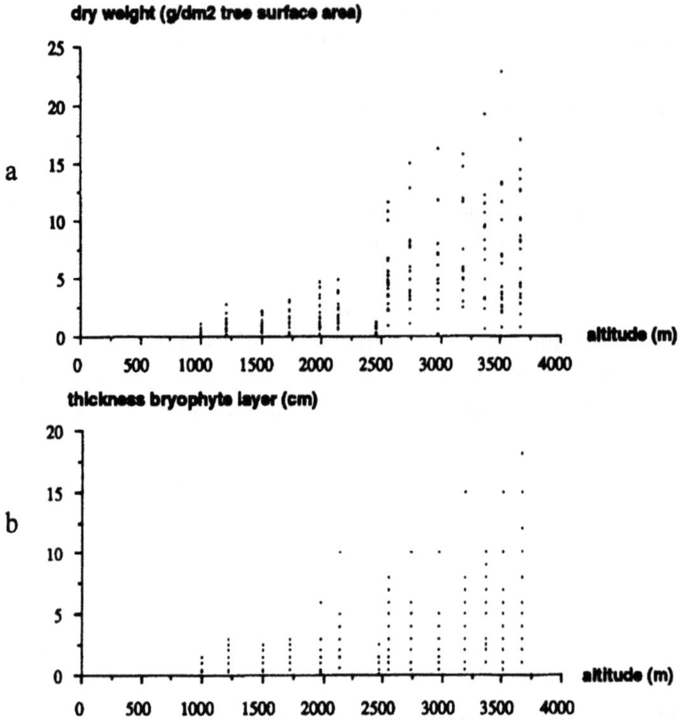

Figure 1. Biomass (a) and thickness (b) of the living epiphytic bryophyte layers as function of elevation in the Central Cordillera, Colombia *(after Wolf 1993).*

their dry weight; Pócs 1980; Nadkarni 1984), with subsequent losses via evaporation and drainage being rather slow (Veneklaas and Van Ek 1990; Veneklaas et al. 1990). Consequently, the magnitude of the so-called canopy saturation value of a forest (Rutter 1975) is expected to increase with increasing epiphyte loading. Preliminary estimates of the canopy saturation values for two montane forests in Tanzania with strongly contrasting epiphytic biomass (Table 7) were about 1.5 mm and 5.0 mm, respectively (Pócs 1980). Similarly, Veneklaas and Van Ek (1990) derived values of 2–5 mm for two MCFs in the Central Cordillera of Colombia, with the highest value in the mossier forest. By contrast, values for lowland rain forest are usually in the order of 0.8–1.0 mm (Calder, Wright, and Murdiyarso 1986; Lloyd et al. 1988; Hutjes, Wierda, and Veen 1990). Adjusting models of rainfall interception that have proven their value in tropical lowland forests (Bruijnzeel and Wiersum 1987; Lloyd et al. 1988; Hutjes, Wierda, and Veen 1990) for TMCF is an important challenge (cf. Veneklaas and Van Ek 1990).

Transpiration

Reliable quantitative information on rates of E_t in TMCF is scarce. By subtracting amounts of E_i from total ET for five LMFs with relatively little cloud incidence, Bruijnzeel (1990) obtained values of 510–830 mm/year^{-1} and found no clear relationship with elevation. Estimates derived in this way must be regarded as relatively crude, however, owing to the possibility of accumulated errors in the estimates of both throughfall (and hence E_i; Lloyd and Marques 1988) and ET (water-budget technique; Ward and Robinson 1989). While micro-meteorological techniques have yielded the best estimates for E_t of lowland rain forests (Shuttleworth 1988), these are less useful under the topographic and climatic gradients encountered in the TMCF, particularly in its stunted facies (cf. Baynton 1968, 1969; Monteith 1976).

The available evidence, representing a mixture of the above-mentioned indirect water-budgeting technique and plant physiological approaches, is presented in Table 3 along with the estimates of total ET discussed previously. Average values of E_t in MCF were about 300 mm/year^{-1} in most cases ($n = 4$) but seem to drop dramatically for ECF ($n = 1$). As indicated already, these estimates should be treated with caution, including the value quoted for ECF, which was based on rates of water absorption by cut shoots (Weaver, Byer, and Bruck 1973) and on leaf energy budget calculations involving a number of assumptions (Gates 1969).

The low rates of water uptake in stunted MCF have been the subject of much speculation (Gates 1969; Odum 1970; Weaver, Byer, and Bruck 1973; Grubb 1974, 1977; Ash 1987; Whitmore 1989; Bruijnzeel et al. 1993). Suggested explanations included toxic or anaerobic conditions in the soil, nutrient deficiency caused by reduced mineralization or extreme leaching, and adverse climatic conditions, notably reduced radiation and high atmospheric humidity (Bruijnzeel et al. 1993). It could be argued that in heavily fogged locations, transpiration is temporarily halted under conditions of a fully wetted canopy (Rutter 1975). Yet, there is evidence that rates of water uptake by these forests are also low in bright sunshine (Bruijnzeel et al. 1993). As already indicated for rates of open-water evaporation (E_o), the bulk of daily transpiration will be effected during such sunny spells, which typically total 3–6 h/day^{-1} at locations with tall LMF, dropping to about 3 h/day^{-1} at sites carrying more stunted forests (Grubb 1977). Bruijnzeel et al. (1993), working on Gunung Silam (a low coastal mountain in Sabah, Malaysia), expressed average values of E_t (determined by the site water-balance method) as a ratio of corresponding values of E_o for two LMFs of contrasting stature. They observed a much lower value (0.24) for the more stunted forest at 870 m (which was subjected to frequent low cloud) than for tall LMF situated just below the cloud cap at 680 m (0.51). The former value is markedly lower than those usually found for tall montane forests (0.5–0.7; Bruijnzeel 1988, 1989c). This suggests that the reduction in E_t / E_o is not caused by the difference in climatic conditions experienced by the two forests, at least not in a direct manner, because this effect is already taken into account in the estimation of E_o (Bruijnzeel et al. 1993). This raises the question as to why the trees in TMCF behave so differently from their counterparts just below the cloud cap.

Enhanced stomatal resistance in response to increases in atmospheric vapor pressure deficits has been reported for a number of tropical and warm temperate MCFs (Jane and Green 1985; Körner, Allison, and Hilscher 1983; Cavelier 1988). Interestingly, when both stomatal resistance and concurrent values of leaf water potential or rates of transpiration were determined, it was found that the increases in stomatal resistance were not accompanied by a reduction in leaf water potential (Green and Jane 1983). In addition, the increases in resistance were insufficient to prevent an increase in rates of transpiration during times of increased evaporative demand (Körner, Allison, and Hilscher 1983). In other cases, e.g., in the stunted UMF in Jamaica studied by Kapos and Tanner (1985), stomatal response to changing atmospheric conditions was less pronounced. All this suggests that the stomatal control observed for TMCF is not primarily aimed at preventing water stress. Neither does it reflect a different suite of evaporative conditions, as already demonstrated.

The physiological responses described above are typical of plants exposed to soil waterlogging, which leads to poorly developed root systems unable to respond fully to temporarily high atmospheric evaporative demands (Brouwer 1960; Bradford and Yang 1981) such as occur at higher elevations in the tropics during conditions of clear sky (Schmidt 1950; McVean 1974; Caldwell, Robberecht, and Billings 1980). In all studies of plant behavior in TMCF, the soils were wet for most of the time. Further evidence of persistently wet conditions in the soils of most TMCFs will be presented in the next section.

Some support for the view that transpiration in TMCF is suppressed may also be obtained from leaf anatomical observations. Leaves of UMF and ECF are distinctly more "xeromorphic" in appearance than leaves of lowland rain forest and LMF, the leaves becoming smaller, thicker, and harder with increasing elevation (Howard 1969; Grubb 1977; Buckley, Corlett, and Grubb 1980; Tanner and Kapos 1982; Körner, Allison, and Hilscher 1983; Cavelier 1988; Bruijnzeel et al. 1993). Yet, these "pachyphylls" (Grubb 1974) do not exhibit greater tolerance to desiccation (Buckley, Corlett, and Grubb 1980). In addition, they exhibit smaller numbers of stomata per unit leaf surface (Grubb 1977; Körner, Allison, and Hilscher 1983), which would be expected in view of the overall decrease in average atmospheric humidity deficits in UMF and ECF. Therefore, it would seem as though the xeromorphic appearance of the leaves in these forests reflects an inability of the trees to transpire at the potential rate, but not because of soil water deficiency (see also the next section). The resulting suboptimal cooling of the leaves may have led to the development of the xeromorphic characteristics, including the frequently observed trend toward vertical inclination of leaf surfaces that may reduce damage by overheating (cf. Medina, Sobrado, and Herrera 1978; Cavelier 1988; Bruijnzeel et al. 1993).

There is a need for more observations of transpiration in TMCF. Such studies could make use of a combination of measuring techniques like the heat-pulse velocity method (Durham and Hatton 1989; Olbrich 1991) and the deuterium tracing method (Calder 1992). In a study assessing the usefulness of the two techniques in estimating water uptake by eucalypts, Dye, Olbrich, and Calder

(1992) considered the two methods largely complementary. The heat-pulse velocity technique can provide detailed flow information over time scales of minutes, but estimates are sensitive to the size of wounds caused by the probes. It is particularly suitable for measurement of high flow rates. The deuterium method, on the other hand, gives estimates of the flow rate over periods of, typically, 2–4 days and is more suited and sensitive for measurement of low flow rates. Dye, Olbrich, and Calder (1992) suggested that the deuterium tracing method would be most useful for comparing a relatively longer-term mean transpiration rate against modeled cumulative transpiration based on instantaneous measurements obtained with other methods. In this way, any cumulative systematic error might be detected.

Heat-pulse velocity equipment was used in a recent short-term reconnaissance study comparing climatic and soil hydrological conditions in two forests of contrasting stature on Gunung Rakata (Krakatau), Indonesia. Two sets of probes were placed in a *Neonauclea* tree at 110 m (about 30-m-high lowland forest) and 700 m (about 15-m-high "mossy" forest) each. Both trees had a diameter at breast height of about 20 cm. Half-hourly observations of sapflow rates were made over a period of 3 weeks in conjunction with observations of soil water tensions and the climate above the canopy. As shown in Figure 2, sapflow rates on a sunny day during the dry season of 1992 were distinctly lower at 700 m than those at 110 m, despite very similar totals of incoming radiation. There was no soil water stress at either location. In addition, sapflow in the upper plot started later in the morning than at 110 m, presumably as a result of the canopy at 700 m having been wetted by nocturnal fog or low cloud. As the dry season progressed, the tree at 110 m gradually transpired less water, whereas that at 700 m exhibited a more or

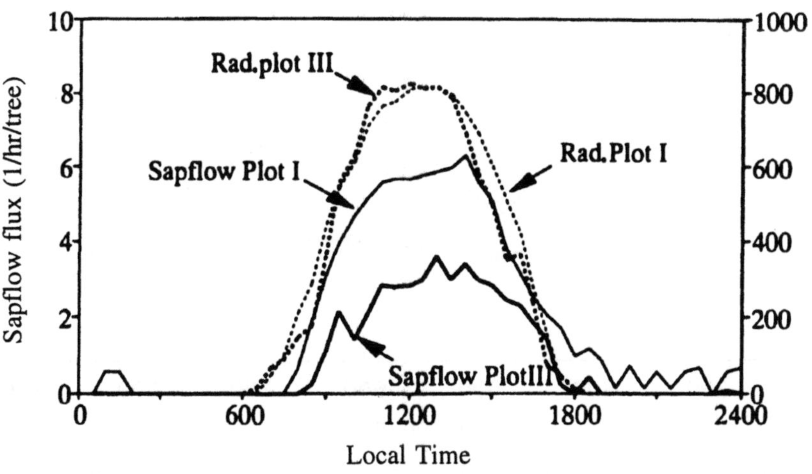

Figure 2. Daily course of short-wave radiation and sapflow in a *Neonauclea* tree at 110 m (plot I) and at 700 m (plot III) for a sunny day during the dry season on Rakata (Krakatau).

less constant pattern (L. A. Bruijnzeel and R. Hafkenscheid, unpublished). The reason for this different behavior is still unclear, but excessive soil moisture levels do not seem to play a major role on Rakata. While fluctuations in soil water tension were much larger at 110 m, the soil at 700 m could not be called very wet. Indeed, in view of its high permeability, it remains to be seen whether the soil at 700 m would ever become waterlogged. Observations of soil water tensions and sapflow during the rainy season are needed to resolve this question. Concentrations of soil nutrients were comparable in both forests (L. A. Bruijnzeel, unpublished).

Soil Water Regime

Although trees in UMF have been reported to die following severe droughts (Lowry, Lee, and Stone 1973; Werner 1988), all observations of soil water dynamics and leaf water potentials in MCF suggested that the trees would never experience severe soil-water deficits (Lyford 1969; Herrmann 1971; Hetsch and Hoheisel 1976; Dohrenwend 1979; Green and Jane 1983; Kapos and Tanner 1985; Jane and Green 1985; Cavelier 1989; Bruijnzeel et al. 1993). It could be argued that most studies never lasted more than 1 year and were too short to assess the effects of rare droughts. However, there is indirect evidence suggesting that the MCF on Gunung Silam, Malaysia, never experiences severe moisture deficit. As shown in Figure 3, amounts of leaf litter on the forest floor in July 1983 and July 1984 (Proctor et al. 1989) showed considerable differences between the 2 years for four forests situated below the usual level of cloud condensation (700 m asl), whereas only minor differ-

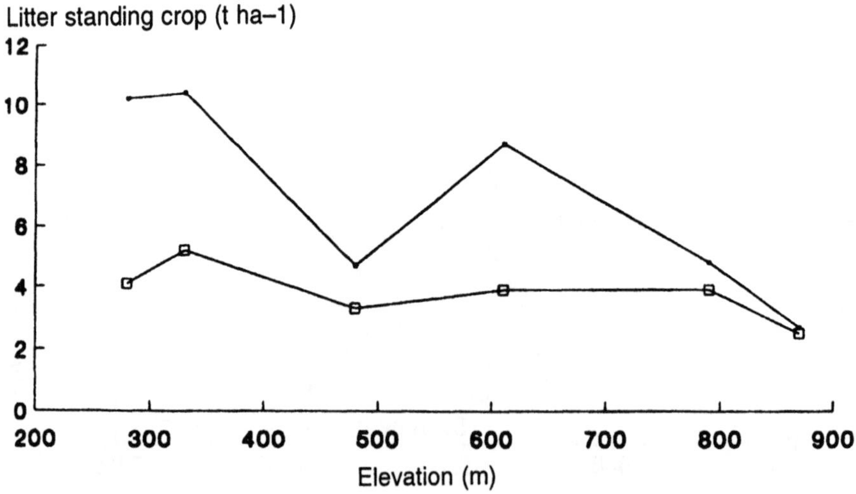

Figure 3. Quantity of leaf litter on the forest floor in six plots at a range of altitudes on Gunung Silam, Sabah, Malaysia, in July 1983 (•) and July 1984 (□) *(after Bruijnzeel et al. 1993).*

ences were found for forests at 790 m and 870 m. Prior to sampling in July 1983, the area had suffered one of the worst droughts in its history (Leighton and Wirawan 1986), which apparently caused extra shedding of litter in forests below but not within the cloud cap. New leaves that formed later in 1983 were probably still on the trees in July 1984, and hence the litter standing crop in the forests below the cloud cap were then lower. While visual evidence in 1984 (e.g., dried ball bryophytes at 790 m and higher) led Proctor et al. (1988) to believe that occasional drought could not be ruled out as a factor in explaining the stunted character of the forest within the cloud cap on Gunung Silam, the evidence presented in Figure 3 suggests otherwise.

The persistently wet conditions prevailing in TMCF compared with the more variable moisture regime of LMF with less cloud are illustrated in Figure 4. In the example from Colombia, where rainfall totals indicate the presence of 4 dry (<60 mm) months, soil water tensions in MCF rarely exceeded a pF value of 1.8, and then only in the uppermost part of the soil profile (Figure 4a). LMF at 1,150 m was largely fog-free and experienced pF values exceeding 2.5 throughout the profile for several months (Figure 4b). However, the tension classes used by Herrmann (1971) do not allow a distinction between various degrees of wetness below pF 1.8. Short-term observations of soil water tension made in LMF of contrasting exposure to fog incidence on Gunung Silam are therefore added for comparison (Figure 4c). The results fully confirm the previous tendencies of persistent wetness in the MCF and a sharp response to a lack of rainfall (e.g., between days 25 and 35) in the forest below the cloud cap.

The above contentions are further confirmed by morphological differences between the soils associated with the two forest types. Transect studies (e.g., Whitmore and Burnham 1969; Fölster and Fassbender 1978; Sevink 1984; Van Reuler 1987; Bruijnzeel et al. 1993) all noted major changes when passing into the zone of frequent low cloud, the most notable of these being increased organic matter content, darker colors, and increased abundance of hydromorphic characteristics. Relative abundance and size of mottles are well-known indicators of the frequency of waterlogged conditions, whereas their absence in an otherwise grey or greyish brown matrix is a sure sign of persistent waterlogging (Wadsworth and Bonnet 1951; Lyford 1969; Hetsch and Hoheisel 1976; Van Reuler 1987). Other indications of such conditions include a high frequency of aerial roots (Gill 1969; Cavelier 1988). On the other hand, no evidence of gleying was found in the stunted mossy forest on Rakata, although accumulation of organic matter was higher than below the cloud cap (Newsome 1986).

Streamflow

If annual streamflow totals are expressed as a ratio of incident rainfall, values for TMCF are among the highest reported for any tropical forest (Bruijnzeel 1990). This reflects the combined influence of extra moisture inputs via HP

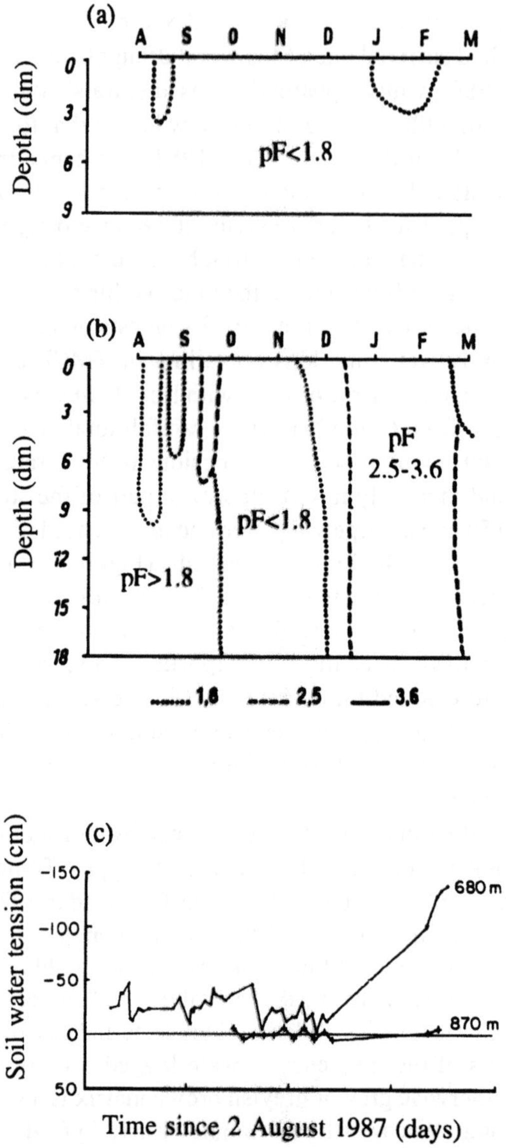

Figure 4. Variations in soil water tension with time in various montane forests: (a) cloud forest at 2,100 m, Colombia; (b) fog-free forest at 1,150 m, Colombia; (c) cloud forest and fog-free forest at 870 m and 680 m, respectively, on Gunung Silam, Sabah, Malaysia *(after Herrmann 1971a, b and Bruijnzeel et al. 1993c).*

and the low values for ET discussed earlier. As stressed by many (La Bastille and Pool 1978; Zadroga 1981; Hamilton with King 1983; Lugo 1986; Stadtmüller 1987), these high runoff ratios constitute one of the main reasons for the conservation of TMCF. While the importance of TMCF in headwater areas supplying water to densely populated downstream valleys is undisputed, it remains to be seen whether total water yield will decrease after clearing of TMCF. The hydrological consequences of forest clearance are likely to differ between locations, depending on the relative magnitudes of HP and ET associated with the original and the new vegetation cover. In addition, the degree to which topsoil infiltration capacity can be maintained after forest conversion constitutes a most important factor when assessing the likely impact on seasonal water yield.

It is possible that the change in total water yield following a conversion of TMCF to agricultural use will be minor. Total water use of crops or grass (ET) in the montane tropics roughly equals amounts of water transpired by forest total (Et) (Bruijnzeel 1990), whereas the data on HP and E_i in Tables 1 and 2 suggest that both vary between 5 and 20 percent of incident rainfall for most locations. If contributions of HP under forested conditions were very high (and E_i relatively low—e.g., Stadtmüller and Agudelo 1990), there would be a good possibility of a decrease in total streamflow after conversion. But in cases where contributions of HP are modest, one can be almost certain that forest clearing will cause soils to become wetter and water yield to increase (Steinhardt 1979; Blackie 1979). A different picture may be obtained for the change in seasonal water yield, particularly if infiltration capacities degrade after forest clearing (Bruijnzeel 1989a; cf. Vis 1986). Experimental evidence is lacking, however, and filling this gap in our knowledge is perhaps the most important challenge facing hydrologists working in TMCF areas. In the meantime, there may be scope in a comparison of long-term records of streamflow (and rainfall) collected routinely by government departments with documented changes in land use (cf. Fleming 1986; Lugo 1986).

Biogeochemical Aspects

A variety of reasons has been offered to explain the stunting of montane forests. These include (1) periodic water shortage (despite frequent cloud cover), especially in shallow and stony soils (Van Steenis 1972; Proctor et al. 1988); (2) saturated soils and impeded root respiration (Wadsworth and Bonnet 1951; Hetsch and Hoheisel 1976); (3) reduced leaf temperatures associated with low radiation inputs (Grubb 1977); (4) limited nutrient uptake due to (a) climatically controlled reduction in transpiration rates (Odum 1970; Weaver, Byer, and Bruck 1973), (b) extreme soil acidity or low fertility or both (Van Steenis 1972; Grubb and Tanner 1976), often coupled with (c) reduced decomposition and mineralization rates (Edwards and Grubb 1977; Marrs et al. 1988; Heaney and Proctor 1989); and (5) exposure to strong winds (Jaffe 1980; Lawton 1982; Sugden 1986). Whitmore (1989) believes that there is no single uniform set of causes of forest zonation and its compression on small, wet tropical mountains. Bruijnzeel et al. (1993) examined the various hypotheses in the light of their own data for Gunung Silam (884 m) in Sabah, Malaysia, and those for other locations. As shown earlier in this paper, reasons (1) and

(4a) can be dismissed, whereas wind does rarely play a significant role at equatorial latitudes (unlike the belt affected by the trade winds). Also, reason (3) was shown not to apply on Gunung Silam (Bruijnzeel et al. 1993). On the other hand, we have seen that the soils of most (but not all) TMCFs are continuously wet to very wet, and in the following we shall examine reasons (4b) and (4c).

Soils

Chemical analytical data for several TMCF soils are shown in Table 4. It can be seen that there is a wide range of soil types and that no generalizations about the soil chemistry are free of exceptions. pH_{H2O} ranges from 3.0 to 7.0. Total organic carbon and total nitrogen are often high and increase with altitude, but there are a number of instances in TMCF where the soil has lower than 5 percent organic carbon and less than 0.3 percent total nitrogen. Total phosphorus has rarely been measured, but values for extractable phosphorus range from 1.1 to 78 $\mu g/g^{-1}$ dry soil and are usually relatively high with no clear trend to decrease with altitude. Exchangeable bases and percentage base saturation are usually low, but not invariably so. Few measurements have been made of rates of nitrogen mineralization and nitrification. Marrs et al. (1988) showed that both had a clear decline with altitude on Volcán Barva, Costa Rica. However, rates in various Jamaican UMFs measured by Tanner (1977a) at 1,550 m were relatively high, except for a negative nitrogen mineralization rate in his "Wet Slope" forest. In Hawai'i, Vitousek, Matson, and Turner (1988) showed a range of negative rates of nitrogen mineralization, but these were less negative with increasing altitude. They observed no nitrate ions in any of the soils they analyzed. On Krakatau, rates of nitrification in lowland rain forest and mossy LMF at 700 m were almost equal, although numbers of mineralizing bacteria were lower in the mossy forest (L.A. Bruijnzeel and R. Zoomer, unpublished).

Our conclusion is that there is no overriding soil chemical factor that uniquely characterizes TMCF, although micronutrient supply remains little investigated. The possibility of aluminum toxicity is raised by the high concentrations of this element in the leaves of Venezuelan MCF (mean 1640 $\mu g/g^{-1}$ dry matter aluminum; Grimm and Fassbender 1981) and in those of a Sri Lankan MCF (mean 1040 $\mu g/g^{-1}$; Werner and Balasubramaniam 1988). However, Gautam-Basak and Proctor (1983) found much lower foliar aluminum in leaves of a stunted MCF (mean 230 $\mu g/g^{-1}$) compared with those of a larger stature MCF (mean 860 $\mu g/g^{1}$) on Gunung Mulu, Sarawak. Similarly, leaves of stunted *Vaccinium* trees on Gunung Gedeh, Java, had similar concentrations of aluminum (but much higher concentrations of polyphenols) as tall *Vaccinium* trees at a lower elevation (L. A. Bruijnzeel and W. H. O. Ernst, unpublished). Often the soils of TMCF are of too high or low pH for aluminum to have an effect on plant growth. Iron and manganese toxicities associated with waterlogging may occur but waterlogging is by no means a universal feature of TMCF (Bruijnzeel et al. 1993). Further observations are desirable.

Litterfall and Limiting Nutrients

A summary of litterfall measurements that have been made in TMCF is given in Table 5. These measurements show that the quantities of litterfall are relatively low compared with those for lowland forests (cf. values in Proctor 1984). Nutrient concentrations in the litterfall are variable, but a distinct trend to lower nitrogen (but not phosphorus) concentrations can be observed in those cases where measurements have been made at two or more altitudes on the same mountain (Figure 5). However, when comparisons are made between element concentrations in fresh leaves and litterfall (Table 6), they indicate proportionally more retranslocation of phosphorus than nitrogen back into the branches before leaf abscission.

These observations on litterfall nutrients seem generally in agreement with the idea supported by fertilization experiments (Tanner et al. 1990; Tanner, Kapos, and Franco 1992) that nitrogen or phosphorus, or both, may limit the growth of some tree species in cloud forests. On the other hand, Bruijnzeel et al. (1993) showed that the net uptake (i.e., that associated with biomass increment) of nutrients in a stunted MCF on Gunung Silam was only a small fraction of the corresponding amounts of nutrients (including nitrogen and phosphorus) arriving at the forest floor in rainfall and throughfall, whereas corresponding losses in drainage water were small. An excess supply of nutrients over net uptake was also demonstrated for stunted UMF in Jamaica by Bruijnzeel (1989b), using data on rainfall chemistry and rates of net nutrient uptake from Tanner (1977b) and Tanner (1985), respectively. Similarly, high atmospheric nutrient inputs (Trinidad Pizarro 1985) and very low tree growth rates (Weaver et al. 1986) suggest an even higher excess of nutrients over net uptake for ECF in the Luquillo Mountains, Puerto Rico. Conversely, rates of net uptake in a tall-statured montane forest in Venezuela were similar to or higher than corresponding amounts of nutrients in rainfall (Steinhardt 1979).

Bruijnzeel et al. (1993) have stressed the possibility that foliar phenolic compounds may play a role in MCF stunting on Gunung Silam and elsewhere. Generally speaking, nitrogen stress and water deficit are among the most important factors affecting the concentration of phenolic metabolites in plant tissue (Horner, Gosz, and Cates 1988), but neither of these were thought to be decisive in the case of Gunung Silam (Bruijnzeel et al. 1993). Nevertheless, concentrations of polyphenols in fresh leaf litter were much higher in the stunted forest on Gunung Silam than in tall-statured forest just below the cloud cap. This contrast in concentrations was not observed for mature litter, and the polyphenols thus can be expected to enter the soil during rain. Here they may have a toxic effect on roots and interfere with various metabolic processes. Polyphenols have been shown to affect photosynthetic rates (Einhellig et al. 1970; Patterson 1981; Kuiters and Sarink 1987), cell division

Table 4. The soil types and their chemical features under a range of tropical montane cloud forests

Place	Altitude	Soil Type	Horizon & Depth (cm)	pH(log units)	C (%)	N (%)	P_{total} (ppm)	P_{extra} (μg/g^{-1})	K (m equivs 100/g^{-1})	Na	Ca	Mg	CEC	BS (%)	Authors
Colombia,	2500	—	—	4.0	—	—	—	—	0.3	0.1	1.2	0.8	—	—	Veneklaas (1990)
Cordillera Central	3550	—	—	4.7	—	0.9	—	—	0.3	0.2	1.3	1.3	—	—	
Colombia, Macuira	600	—	A, 0–20[a]	—	1.76	0.26	2.0	—	0.3	0.9	2.0	2.4	13.3	—	Cavelier (1988)
				—	1.40	0.22	4.0	—	1.1	0.7	4.8	3.2	13.9	—	
Costa Rica,	1500	Humitropept	A_h, 0–22	3.2	12.0	—	—	12.9	0.38	0.38	1.72	0.94	100.7	3.4	Grieve, Proctor, and Cousins (1990)
Volcán Barva	2000	Tropofibrist	O_f, 0–15	3.9	16.2	—	—	4.7	0.34	0.31	0.56	0.51	88.3	1.9	
	2600	Vitrandept	O_h, 0–10	3.3	17.2	—	—	55.7	1.30	0.31	17.0	9.8	113	25.1	
Indonesia, Seram,	1000	Typic Humitropept	A_h, 0–11	4.5	14.8	0.4	600	4.9	0.02	0	0.05	0.05	1.6	6.0	Payton (1993)
Gunung Kobipoto	1470	Histic Humaquept	O, 0–20	3.6	39.0	2.1	—	65.5	0.01	0	0.02	0.01	1.4	5.0	
Indonesia, Seram,	2100	Humic Kanhapludult	E_b, 0–12	3.2	37.0	2.5	—	77.9	0.11	0.04	0.22	0.00	—	—	Payton (1993)
Gunung Binaia	2400	Lithic Tropofolist	H/A_h, 0–12	3.5	11.0	0.7	—	29.3	0.02	0.05	0.27	0.005	3.8	8.9	
	2820	Lithic Rendoll	A_h, 0–15	7.0	12.0	1.3	1800	—	0.02	0.02	6.92	0.20			

Indonesia,	1750	Aquept	0–18	—	25.7	—	—	—	—	—	—	—	13.5	2	Van Reuler (1987)
Kalimantan,	2150	Aquept	0–33	4.3	13.3	—	—	—	—	—	—	—	12.4	2	
Bukit Raya															
Jamaica,															
'Mor Ridge'	1550	—	0–10	3.0	47.4	1.6	—	—	0.90	0.69	0.39	11.3	185	7	Tanner (1977a)
'Mull Ridge'	1550	—	0–8	3.7	29.2	1.7	—	—	1.38	0.51	6.05	6.1	120	12	
'West Slope'	1550	—	0–10	4.0	3.9	0.35	—	—	0.80	0.18	6.18	3.9	29	38	
'Gap'	1550	—	0–10	4.4	9.0	0.52	—	—	0.32	0.08	1.57	0.44	37	7	
Malaysia,	1310	Organic Mulu Series	O_i, 0–15	3.9	13.1	0.77	280	37	0.48	0.10	0.11	0.30	45.4	2.2	Tie et al. (1979)
Sarawak,															
Gunung Mulu	1930	Organic Mulu Series	O_e, 0–15	3.4	42.5	1.81	360	78	0.56	2.82	0.05	0.72	73.8	5.6	
	2090	Organic Mulu Series	O_e, 0–45	3.0	43.2	1.40	—	68	0.05	0.29	0.12	0.68	52.8	1.1	
	2370	Organic Mulu Series	O_i, 0–30	3.2	39.9	1.14	—	38	0.26	0.20	0.68	0.90	81.4	1.7	
Malaysia,	2350	—	2–10	3.9	34.6	1.96	—	37	1.17	0.23	0.51	8.17	78.7	13	Kitayama (1992)
Sarawak,	2600	—	2–10	3.9	38.0	0.98	—	4	1.17	0.42	4.08	7.21	79.6	16	
Gunung	2800	—	3–20	4.6	7.7	0.45	—	13	0.26	0.10	0.40	0.27	9.8	10	
Kinabalu	3000	—	1–5	5.4	15.5	0.83	—	9	0.63	0.20	18.3	2.42	33.4	64	
	3200	—	5–10	3.9	11.6	0.91	—	28	0.68	0.18	1.07	1.09	34.3	9	
	3400	—	2–30	4.7	8.8	0.56	—	11	0.40	0.21	3.41	0.47	22.1	20	
Malaysia, Sabah,	790		0–15	5.6	8.5	—	—	1.1	0.17	0.08	0.9	5.4	105	6.1	Proctor etal.(1988)
Gunung Silam	870		0–15	4.0	22.9	0.40	—	16.8	0.53	0.41	1.2	5.6	105	7.4	Waterloo (1989)

continued

Table 4. *continued*

Place	Altitude	Soil Type	Horizon & Depth (cm)	pH(log units)	C (%)	N (%)	P_{total} (ppm)	P_{extra} (µg/g^{-1})	K (m equivs 100/g^{-1})	Na	Ca	Mg	CEC	BS (%)	Authors
USA, Hawai'i	1675	—	—	—	—	—	—	6	0.74	—	5.75	1.1	—	—	Vitousek, Matson,
	1585	—	—	—	—	—	—	8	1.67	—	21.0	2.6	—	—	and Turner (1988)
Venezuela, Parque	1170	Typic Dystropept	0–11	4.9	3.8	0.34	—	6	0.18	0.08	0.5	0.7	16.6	8.8	Zinck (1986)
Nacional "Henri Pittier"	1160	Typic Tropohumult	0–15	4.7	6.4	0.44	—	6	0.21	0.09	0.7	0.7	21.9	7.8	
	1680	Typic Tropohumult	0–11	4.4	4.3	0.34	—	9	0.26	0.07	0.7	0.6	14.3	11.4	
	1670	Typic Tropophumult	0–17	4.8	3.6	0.25	—	8	0.27	0.06	0.8	0.9	15.2	13.3	
	1580	Typic Tropophumult	0–16	4.7	3.8	0.29	—	3	0.31	0.17	1.33	0.40	15.7	14.1	
Venezuela,															
Santa Ana	600	—	A, 0–12[a]	—	2.51	0.29	7.0	—	0.6	0.9	9.3	4.9	17.9	88	
				—	4.29	0.48	4.0	—	0.6	1.7	4.1	2.4	19.6	45	
Copey	600	—	A, 0–12[a]	—	2.3	0.28	6.0	—	0.2	0.3	1.6	2.5	17.1	27	
				—	2.5	0.21	6.0	—	0.1	0.2	1.4	1.6	14.3	23	
Zumbador	3100	—	A, 0–25[a]	—	20.6	1.10	31.0	—	0.6	0.2	0.3	0.3	80.0	1.8	
				—	20.5	0.70	28.0	—	0.4	0.2	0.1	0.2	84.0	1.1	Cavelier (1988)
Venezuela, Andes Occidentales	2300	—	0–20	3.5	5.5	0.45	500	—	0.43	0.1	1.3	0.59	17.5	14	Grimm and Fassbender (1981)

[a] Two samples analyzed in each case and reported separately in the present table.
— = no data available

in fine roots (Vaughan and Ord 1990), ion uptake (Glass 1973, 1974; Kobza and Einhellig 1987; Kuiters and Sarink 1987), and stomatal aperture (Einhellig and Kuan 1971). Some or all of these interferences may be at work in stunted TMCFs (low nutrient uptake, low E_t). Further work is needed to test the validity of the "toxin resistance" hypothesis.

It is uncertain what causes the high concentrations of polyphenols in the leaves of stunted TMCFs. Bruijnzeel et al. (1993) found a weak inverse relationship between concentrations of total nitrogen and total phenolics, but the relationship differed between sites, suggesting that other factors may also play a role. Interestingly, leaf phenolic compounds such as flavonoids were also produced in larger quantities by plants subjected to increasing doses of UV-B radiation. The flavonoids are thought to play a role as internal "filters" that absorb and reduce the UV flux reaching certain plant tissues. Plant response to enhanced UV-B radiation included stunting and changes in leaf anatomy (Teramura 1983; Caldwell, Teramura, and Tevini 1989) that resemble those found in many TMCFs, i.e., small, thick, and hard leaves (cf. the section on transpiration). At first sight, increased intensities of UV-B may not be able to explain the occurrence of stunted MCF with xeromorphic leaves at low elevations, because intensities of UV-B only become sufficiently higher at elevations of 2,500–3,000 m (Caldwell, Robberecht, and Billings 1980). However, the finding that plants usually exposed to low levels of visible light and growing under well-watered conditions (both conditions that are common to TMCF) tended to be more sensitive to enhanced UV-B radiation than plants from environments with (more constant) high levels of visible light (Teramura 1983; Sullivan, Teramura, and Ziska 1992; Ziska, Teramura, and Sullivan 1992) may be pertinent. After all, the one factor that TMCFs have in common regardless of elevation is a soil that is persistently wet or very wet. The connections between soil water regime, root development, water and nutrient uptake, and radiation regime, particularly that of UV-B, in TMCF merit further attention (cf. Flenley 1992, and this volume).

Epiphytes and Their Effect on Nutrient Cycling

Epiphytes are usually conspicuous in MCF and contribute a substantial mass of living and dead organic matter (including humus) to the tree canopy. Epiphyte mass is difficult to quantify, but some estimates are included in Table 7. The epiphyte mass is only a small proportion of that of the total biomass for MCF, but when compared with the foliar biomass, it can assume substantial proportions. There is no evidence that epiphytes obtain mineral nutrients directly from the host trees, but they can be regarded as nutritional "pirates" (Benzing and Seeman 1978) in that they can intercept nutrients in aerosols, litterfall, rainfall, throughfall, and animal droppings. In the absence of epiphytes, these nutrients would become more quickly available to the trees. Epiphytes can, in the words of Nadkarni (1984), become "major participants in the impoundment and movements of mineral elements in a forest ecosystem." Some

Table 5. Concentrations (as % oven-dry weight) of nitrogen, phosphorus, potassium, sodium, calcium, and magnesium in (a) leaf litterfall, (b) total small litterfall, and (c) epiphytic litterfall from cloud (or high montane >2500 m) forests

Location	Altitude (m)		Litterfall Mass	N	P	K	Na	Ca	Mg	Authors
Colombia, Cordillera Central	2500	a	4.6	1.11	0.074	0.89				Veneklaas (1990)
		b	7.0	1.16	0.086	0.84				
	3350	a	2.8	0.78	0.040	0.35				
		b	4.6	0.73	0.044	0.31				
Costa Rica, Monteverde	1500	a	4.8	1.47	0.08	0.12		1.91	0.27	Nadkarni and Matelson (1992a, b)
		b	7.0	1.32	0.09	0.17		1.64	0.21	
		c	0.5	1.50	0.10	0.02		0.84	0.16	
Costa Rica, Volcán Barva	2000	a	4.8	0.90	0.054	0.22	0.03	1.1	0.19	Heaney and Proctor (1989)
		b	5.8	0.79	0.046	0.21	0.03	1.05	0.17	
	2600	a	4.6	0.68	0.053	0.30	0.02	1.04	0.23	
		b	5.3	0.64	0.049	0.28	0.02	0.98	0.21	
India, Western Ghats		b	2.1	1.1	0.062	0.52	—	1.60	0.24	Venkataraman and Chinnamani (1978)

Jamaica,	1550									Tanner
'Mor Ridge'		a	4.9	0.60	0.018	0.23	0.13	0.58	0.33	(1977b, 1980a)
		b	6.6	0.59	0.020	0.22	0.11	0.51	0.29	
'Mull Ridge'		a	5.3	0.84	0.025	0.59	0.087	0.91	0.32	
		b	5.5	0.90	0.027	0.71	0.085	0.90	0.31	
'Wet Slope'		a	4.4	0.66	0.039	0.41	0.099	1.01	0.42	
		b	5.6	0.62	0.037	0.37	0.089	0.97	0.37	
'Gap'		a	5.5	0.93	0.038	0.59	0.11	0.95	0.28	
		b	6.5	0.90	0.037	0.53	0.10	0.85	0.26	
Malaysia, Sabah, Gunung Silam	790	a	3.66	1.05	0.017	0.10	0.1	0.68	0.33	Proctor et al. (1988)
		b	5.53	1.08	0.019	0.09	0.08	0.68	0.29	
	870	a	3.32	0.83	0.020	0.09	0.09	0.58	0.30	
		b	4.80	0.86	0.023	0.10	0.08	0.57	0.26	
Malaysia, Sarawak, Gunung Mulu National Park	1310	a	5.7	0.81	0.021	0.37	0.0051	0.19	0.18	Proctor, Anderson, and Vallack (1983)
		b	11	0.78	0.023	0.28	0.0051	0.19	0.15	
	1860	a	2.3	0.74	0.026	0.18	0.010	0.19	0.20	
		b	3.6	0.79	0.030	0.16	0.011	0.16	0.17	
Puerto Rico, Luquillo Mountains	1000	a	2.45	0.77	0.024	0.14	—	0.53	0.24	Weaver et al. (1986)
Venezuela, Andes Occidentales	2300	a	3.4	1.2	0.062	0.57	0.0041	0.73	0.26	Fassbender and Grimm (1981)
		b	7.0	0.99	0.057	0.47	0.0039	0.62	0.21	

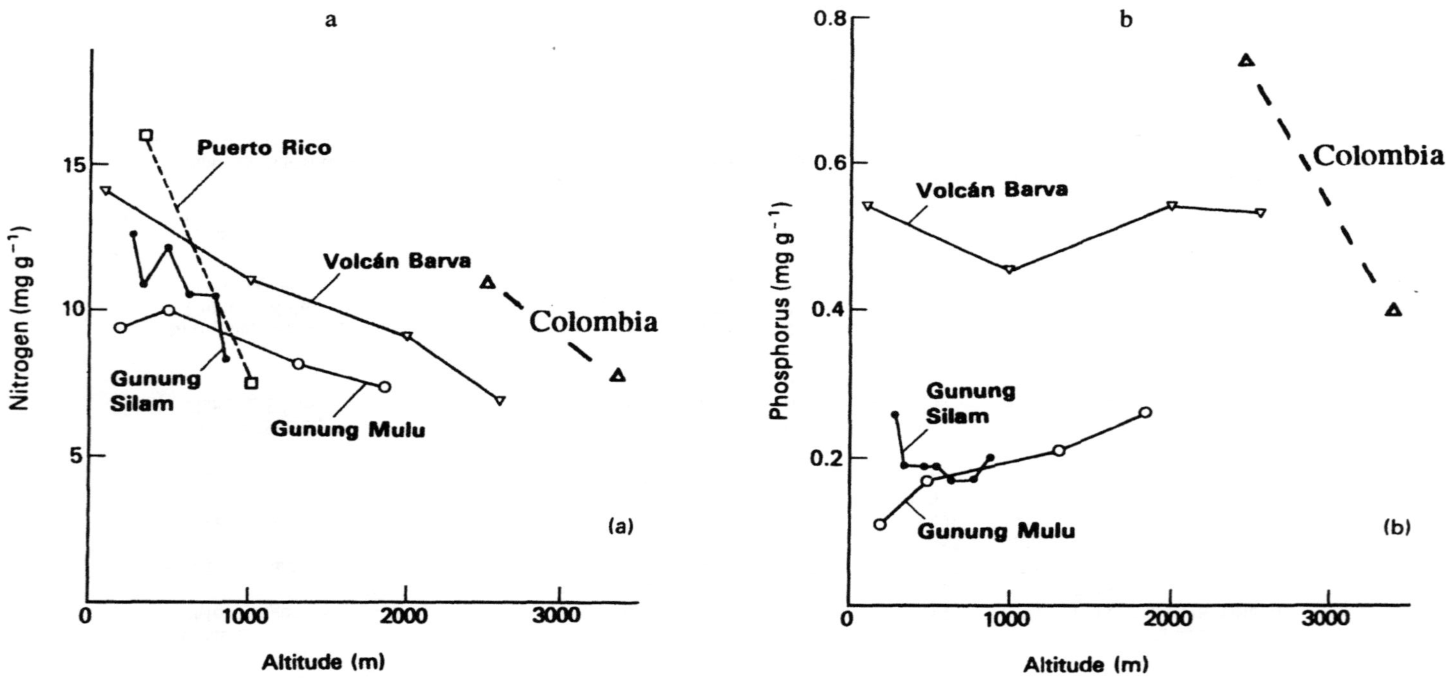

Figure 5. Changes in (a) nitrogen concentrations and (b) phosphorus concentrations in leaf or nonwoody litterfall in studies made at a range of altitudes on one mountain *(after Heaney and Proctor 1989; data for Colombia from Veneklaas 1990).*

Table 6. Element concentration, nitrogen/phosphorus quotients, and nitrogen and phosphorus retranslocation of (a) live foliage and (b) leaf litter in tropical montane forests. Retranslocation is calculated as (element concentration in mature leaves – element concentration in litterfall) ÷ (element concentration in mature leaves)

Location	Altitude (m)		Element Conentrations			Retranslocation		Authors
			N(%)	P(%)	N/P	N(%)	P(%)	
Colombia, Cordillera Central	2550	a	1.8	0.13	14			Veneklaas (1990)
		b	1.1	0.07	15	39	45	
	3370	a	1.5	0.09	16			
		b	0.9	0.03	28	39	65	
Costa Rica, Monteverde	1500	a	1.97	0.14	14			Nadkarni and Matelson (1992a)
		b	1.47	0.08	18	25	42	
Jamaica,	1550							Tanner (1977a, b)
'Mor Ridge'		a	1.05	0.05	21			
		b	0.60	0.018	33	43	64	
'Mull Ridge'		a	1.61	0.07	23			
		b	0.84	0.025	34	48	64	
'Wet Slope'		a	1.27	0.08	16			
		b	0.66	0.039	17	48	51	
'Gap'		a	1.76	0.10	18			
		b	0.93	0.038	25	47	62	
Malaysia, Sabah, Gunung Silam	790	a	1.7	0.05	32			Proctor et al. (1989)
		b	1.1	0.02	62	39	69	
	870	a	1.3	0.05	29			
		b	0.8	0.02	42	38	57	
Venezuela, Andes Occidentales	2300	a	1.74	0.08	22			Fassbender and Grimm (1981)
		b	1.20	0.06	20	31	25	

Table 7. The mass (kg/ha^{-1}) of live and dead (including organic matter and humus) epiphytes and their contents (kg/ha^{-1}) of mineral nutrients in tropical montane cloud forest

Location	Altitude (m)	Mass	N	P	K	Na	Ca	Mg	Authors
Colombia	3700	44000	443	35	103	—	108	—	Hofstede, Wolf, and Benzing (1993)
Colombia	3370	12000	—	—	—	—	—	—	Veneklaas (1990)
Costa Rica, Monteverde	1700	4730	105	3	23	7	12	4	Nadkarni (1984)
Jamaica,									Tanner (1980b, 1985)
'Mor Ridge'	1550	2800	25	2	22	—	18	9	
'Mull Ridge'	1550	500	—	—	—	—	—	—	
Puerto Rico, Pico del Este	1000	7360	—	—	—	—	—	—	Weaver (1972)
	1015	4350	—	—	—	—	—	—	
	930	4750	—	—	—	—	—	—	
Tanzania, Uluguru Mountain									Pócs (1980)
Mossy elfin forest	2120	14000	—	—	—	—	—	—	
Submontane rainforest	1415	2130	—	—	—	—	—	—	

epiphytes can also fix nitrogen or harbor nitrogen-fixing symbionts, although the fluxes involved are still speculative. For epiphylls (those epiphytes that are restricted to the surfaces of leaves), Bentley and Carpenter (1984) calculated from N^{15} studies that they could contribute up to 25 percent of host leaf nitrogen in a rainforest understory. Bentley (1987) described the conditions that might influence the epiphyll nitrogen fixation, but concluded that "the contribution of new nitrogen by epiphylls to an ecosystem is probably fairly small."

Estimates of the nutrient pool contained in epiphytes are given in Table 7. Nadkarni (1984) estimated that for a range of montane forests, the "mineral capital contained in epiphytes is up to 45% of that which is held in the foliage." More recently, Hofstede, Wolf, and Benzing (1993) have shown that the proportions were much higher in the Colombian cloud forest. Here the epiphytes held 7.4 times more nitrogen, 6.5 times more phosphorus, 3.5 times more potassium, and 2.8 times more calcium than the leaves. The epiphyte nutrients can be transferred to other ecosystem pools by crown wash, herbivory litterfall of the epiphytes themselves, and in some cases, by host-tree canopy-root systems (Nadkarni 1981). The canopy-root systems are not a universal feature of TMCF, however. A high microbial activity in epiphyte-derived canopy organic matter of a Costa Rican cloud forest has been found by Vance and Nadkarni (1990). They made a preliminary estimate that 0.7 kg/ ha^{-1} of nitrogen were immobilized in microbial tissue and considered that "because of their high biomass and activity, microorganisms in the canopy could be important in regulating nitrogen availability in this forest." This conclusion seems justified, at least as far as nitrogen availability to epiphytes is concerned.

The question of epiphyte litterfall has been dealt with by Nadkarni and Matelson (1992b). The authors found that a specially designed sampling system was necessary to give a reasonably good estimate of the highly clumped epiphyte litterfall. They found the quantities of nutrients contained in this litterfall to be about 1–8 percent of corresponding amounts in foliar litterfall and pointed out that epiphyte litterfall shows substantial differences in decomposition rates from the litterfall of terrestrially rooted plants.

Research Perspective

One must conclude that epiphytes can be an important component in the nutrient cycling of TCF but that much remains to be discovered about them.

Although TMCF structure is very characteristic, there does not seem to be any environmental factor (except for the frequency of low cloud) shared by all TMCFs. Therefore, research focusing on a single factor is unlikely to provide generally valid answers. Attention to TMCF ecology has greatly increased in the last two decades, but surprisingly little information is available on the key processes determining productivity, viz. carbon economy and the influence of climatic and soil conditions thereupon. Also, while forest structure and productivity are functionally related, different factors may underlie low stature and low productivity. An approach that explicitly separates effects on forest structure and functioning is required if causal relationships are to be identified. The study of the forest carbon balance may offer a convenient framework to assess effects on specific abiotic and biotic factors on productivity, and for reinterpreting results of research undertaken in the past. High priority should be given to the estimation of canopy photosynthesis and the way it is influenced by radiation climate, and to the quantification of respiratory losses, especially those associated with nutrient uptake.

The linkages between soil water regime, root development and morphology, and (rapid) fluctuations in radiation (including UV-B) and fog incidence, and water and nutrient uptake in TMCF merit further attention. The frequency of occurrence of excess soil water has been little studied in TMCF, either alone or in relation to root morphological adaptations (e.g., aerial roots, occurrence of root lenticels, etc.) and oxygen deficiency. The postulated connection between levels of phenolics in leaves and UV-B irradiation requires field observations and experimentation. Finally, the role of cloud water chemistry (notably its acidity and sulphate concentrations) requires further study, particularly in view of the importance that is usually attached to the role played by acidic cloud water (in combination with various photochemical oxidants) in the widespread decline in forest productivity in the temperate zone (Schier and Jensen 1992). Information on cloud water chemistry in TMCF is now beginning to be collected (e.g., Weathers et al. 1988; Bruijnzeel et al. 1993), but its impact on tree physiological behavior has remained unstudied so far.

These aspects could probably be dealt with best in a network of multidisciplinary studies conducted at a limited number of sites for which there is sufficient floristic and climatic background information. Such key sites need to be carefully chosen because the study of the role of atmospheric and edaphic factors in TMCF is often hindered by the change in community floristics that usually accompanies the transition from tall forest below the clouds to smaller statured forest within the clouds (Brown 1919; Van Steenis 1972; Proctor et al. 1988). Fortunately, there are a few locations carrying cloud forest where this floristic transformation is incomplete, such as on Rakata Island, Indonesia (Whittaker, Bush, and Richards 1989); Margarita Island, Venezuela (Sugden 1986); and Maui Island, Hawai'i (Kitayama and Mueller-Dombois 1993). Similarly, certain species are known to occur in both cloud forest and lower elevation forest type, with corresponding changes in tree stature (e.g., *Micropholis garciniaefolia* in Puerto Rico: W.L. Silver, personal communication). Arguably, the study of the physiological behavior of such species in combination with observations of climatic and edaphic conditions along the elevational gradient might be particularly instructive (Whitmore 1989). In addition, similar studies in adjacent forests of contrasting stature but subjected to the same overall climatic conditions, such as the UMRFs in Jamaica described by Tanner (1977), could be equally enlightening. Finally, the fact that there are structural and ecological similarities between UMRF and Heath forests (Richards 1952; Whitmore 1989) could also be used as a starting point for comparative studies.

Summary

Tropical montane cloud forest (TMCF) covered about 500,000 km^2 in the early seventies but is disappearing rapidly in parts of Latin America and Southeast Asia to make way for other land uses. Despite their importance for the continued supply of water to densely populated lowlands, the hydrological functioning of TMCF has been studied far less than that of lowland rain forest. The first part of this paper reviews TMCF hydrology.

The most notable changes in forest physiognomy with increasing elevation are a decrease in stature and a tendency for the leaves to become more xeromorphic. Both the stunting of montane forests and the related Massenerhebung effect have remained unexplained. The second part of the paper reviews the evidence.

An important feature of TMCF is the stripping by the vegetation of moisture from passing low cloud (horizontal precipitation—HP). (Its quantification at the ecosystem level presents major difficulties, exacerbated by the presence of epiphytes.) Additions of moisture via HP may reach hundreds of millimeters per year. Typical values range between 5 and 20 percent of ordinary rainfall but may exceed the latter under more seasonal conditions.

The few studies of evapotranspiration in TMCF indicate annual totals of 300–400 mm, i.e., well below corresponding values for fog-free montane forests (1,100–1,300 mm/year^{-1}). Correcting for "occult" contributions of precipitation, evapotranspiration varies between 570 and 770 mm/year^{1}. More data are needed. Evidence is accumulating that transpiration rates in TMCF are also low (0.7–1.4 mm/day^{-1}), even during periods of bright sunshine.

Converting TCMF to pasture may influence total water yield and seasonal distribution of flow negatively, because of the loss of moisture inputs via HP, although experimental evidence is mostly lacking on the associated threshold value for HP. Deterioration of infiltration characteristics after clearing could cause further degradation of the flow regime. Research to this end should be given high priority.

A variety of reasons has been proposed to explain the stunting of TMCF. While it is acknowledged that wind may be a factor in nonequatorial latitudes, none of the current hypotheses is capable of explaining the occurrence of stunted forest at low altitudes in all cases published to date. Nutrient budgeting computations (and fertilization experiments) have shown that stunted TMCF takes up only a small fraction of the nutrients arriving at the forest floor via water-bound pathways. The excess supply over net uptake contradicts to the widely established idea that nitrogen is a key limiting element in TMCF.

Living foliage and fresh leaf litter in a number of stunted TMCFs contained high concentrations of polyphenols. Upon washing into the soil, the polyphenols may interfere with photosynthesis, cell division in fine roots, transpiration, and ion uptake. High levels of phenolic compounds in leaves may play a role as internal "filters," absorbing and reducing the flux of ultraviolet-B (UV-B), radiation reaching certain plant tissues. The interrelationships between soil water regime (notably excess water), root development, uptake of water and nutrients, radiation regime (including UV-B), and photosynthesis require more study.

Apart from their often high degree of endemism and other ethical reasons, the demonstrated low productivity and water use of TMCF are strong arguments for preserving these forests.

Acknowledgments. The hydrological part of this paper was written during a visit by the senior author to the University of Stirling. The financial support of the Netherlands Organization for Scientific Research (NWO) and the British Council, as well as John Proctor's immaculate hospitality, are gratefully acknowledged. Sampurno Bruijnzeel thanks Dr. Lawrence Hamilton for the invitation to compile this paper and the East-West Center for enabling him to present it at the San Juan meeting. Special thanks are due to Prof. Dr. W. H. O. Ernst, Raimond Hafkenscheid, H. J. M. Nelissen, and Rik Zoomer for their help in expediting the material collected on Rakata.

References

Aitken, P., Ribeny, F.M.J., and Brown, J.A.H. 1972. The estimation of mean annual runoff over the territory of Papua New Guinea. *Civil Engineering Transactions (Australia)* 14:49-56.

Ash, J. 1987. Stunted cloud forest in Taveuni, Fiji. *Pacific Science* 41:191-199.

Baynton, H.W. 1968. The ecology of an elfin forest in Puerto Rico. 2. The microclimate of Pico del Oeste. *Journal of the Arnold Arboretum* 49:419-430.

Baynton, H.W. 1969. The ecology of an elfin forest in Puerto Rico. 3. Hilltop and forest influences on the microclimate of Pico del Oeste. *Journal of the Arnold Arboretum* 50:80-92.

Bentley, B.L. 1987. Nitrogen fixation by epiphylls in a tropical rainforest. *Annals of the Missouri Botanical Garden* 74:234-241.

Bentley, B.L., and Carpenter, E.J. 1984. Direct transfer of newly-fixed nitrogen from free-living epiphyllous microorganisms to their host plant. *Oecologia* 63:52-56.

Benzing, D., and Seeman, J. 1978. Nutritional piracy and host tree decline. *Selbyana* 2:133-148.

Blackie, J.R. 1979. The water balance of the Kericho catchments. *East African Agricultural and Forestry Journal* 43:55-84.

Braak, C. 1922. The climate of the Dutch East Indies. *Verhandelingen van het Koninklijk Magnetisch en Meteorologisch Observatorium te Batavia* no. 8 (1):1-497.

Bradford, K.J., and Yang, S.A. 1981. Physiological responses to waterlogging. *Horticultural Science* 16:3-8.

Briscoe, C.B. 1966. Weather in the Luquillo Mountains of Puerto Rico. *Forest Service Research Paper ITF*-3. Institute of Tropical Forestry, Rio Piedras, Puerto Rico.

Brouwer, R. 1960. The influence of aeration of the root environment on the growth of

bean plants. *Communications of the Soil Fertility Institute, Haren* 107:11-21.
Brown, W.H. 1919. *Vegetation of Philippine mountains.* Bureau of Printing, Manila.
Bruijnzeel, L.A. 1988. Estimates of evaporation in plantations of *Agathis dammara* Warb. in south-central Java, Indonesia. *Journal of Tropical Forest Science* 1:145–161.
Bruijnzeel, L.A. 1989a. (De)forestation and dry season flow in the humid tropics: a closer look. *Journal of Tropical Forest Science* 1:229-243.
Bruijnzeel, L.A. 1989b. Nutrient content of bulk precipitation in south central Java, Indonesia. *Journal of Tropical Ecology* 5:187-202.
Bruijnzeel, L.A. 1989c. Nutrient cycling in moist tropical forests: the hydrological framework. In *Mineral nutrients in tropical forest and savanna ecosystems*, ed. J. Proctor, 383-415. Blackwell Scientific Press, Oxford.
Bruijnzeel, L.A. 1990. *Hydrology of moist tropical forests and effects of conversion: a state of knowledge review.* IHP-UNESCO Humid Tropical Programme, Paris.
Bruijnzeel, L.A., and Wiersum, K.F. 1987. Rainfall interception by a young *Acacia auriculiformis* plantation forest in West Java: application of Gash's analytical model. *Hydrological Processes* 1:309-319.
Bruijnzeel, L.A., Waterloo, M.J., Proctor, J., Kuiters, A.T., and Kotterink, B. 1993. Hydrological observations in montane rain forests on Gunung Silam, Sabah, Malaysia, with special reference to the "*Massenerhebung*" effect. *Journal of Ecology* 81:145-167.
Buckley, R.C., Corlett, R.T., and Grubb, P.J. 1980. Are the xeromorphic trees of tropical montane rain forests drought-resistant? *Biotropica* 12:124-136.
Caceres, G. 1981. *Importancia hidrologica de la intercepcion horizontal en un bosque muy humedo premontano en Balalaica, Turrialba, Costa Rica.* MSc thesis, University of Costa Rica, Turrialba.
Calder, I.R. 1992. Deuterium tracing for the estimation of transpiration from trees. Part 2. Estimation of transpiration rates and transpiration parameters using a timeaveraged deuterium tracing method. *Journal of Hydrology* 130: 27-35.
Calder, I.R., Wright, I.R., and Murdiyarso, D. 1986. A study of evaporation from a tropical rainforest - West Java. *Journal of Hydrology* 89:13-31.
Caldwell, M.M., Robberecht, R., and Billings, W.D. 1980. A steep latitudinal gradient of ultraviolet-B radiation in the arctic-alpine life zone. *Ecology* 61:600-611.
Caldwell, M.M., Teramura, A.H., and Tevini, M. 1989. The changing solar ultraviolet climate and the ecological consequences for higher plants. *Trends in Ecology and Evolution* 4:363-367.
Calvo, J. 1986. An evaluation of Thornthwaite's water balance technique in predicting stream runoff in Costa Rica. *Hydrological Sciences Journal* 31:51-60.
Cavelier, J. 1988. *The Ecology of elfin cloud forests in Northern South America.* Dissertation submitted for the annual research fellowship competition, Trinity College, University of Cambridge, Cambridge.
Cavelier, J. 1989. *Root biomass, production and the effect of fertilization in two tropical rain forests.* PhD thesis, University of Cambridge.
Cavelier, J., and Goldstein, G. 1989. Mist and fog interception in elfin cloud forests in Colombia and Venezuela. *Journal of Tropical Ecology* 5:309-322.
Clements, R.G., and Colon, J.A. 1975. The rainfall interception process and mineral cycling in a montane rain forest in eastern Puerto Rico. In *Mineral cycling in southeastern ecosystems*, ed. F.G. Howell, J.B. Gentry, and M.H. Smith, 813-823. Technical Information Center, Energy Research Development and Administration, Washington, D.C.
De los Santos, A.E. 1981. *Water budget and nutrient fluxes in mossy forest on Mt. Data, Mt. Province, Philippines.* MSc thesis, University of the Philippines at Los Baños, Laguna, the Philippines.

Dilks, T.J.K., and Proctor, M.C.F. 1975. Comparative experiments on temperature responses of bryophytes: assimilation, respiration and freezing damage. *Journal of Bryology* 8:317-336.

Dohrenwend, R.E. 1979. Hydrologic behavior at the top of a tropical mountain. *Research Note* 29. Michigan Technological University, l'Anse, Michigan.

Dollard, G.J., and Unsworth, M.H. 1983. Field measurements of turbulent fluxes of wind-driven fogs to a grass surface. *Atmospheric Environment* 17:775-780.

Durham, I.H., and Hatton, T.J. 1989. A powerful heat pulse logger system. In *Hydrology and Water Resources Symposium*, 477-478. Proceedings of the Christchurch Symposium 1989. University of Canterbury, Christchurch, New Zealand.

Dye, P.J., Olbrich, B.W., and Calder, I.R. 1992. A comparison of the heat pulse method and deuterium tracing methods for measuring transpiration from *Eucalyptus grandis* trees. *Journal of Experimental Botany* 43:337-343.

Edwards, P.J. 1982. Studies of mineral cycling in a montane rain forest in New Guinea. V. Rates of cycling in throughfall and litter fall. *Journal of Ecology* 70:807-827.

Edwards, P.J., and Grubb, P.J. 1977. Studies of mineral cycling in a montane rain forest in New Guinea. I. The distribution of organic matter in the vegetation and soil. *Journal of Ecology* 65:943-969.

Einhellig, F.A., and Kuan, L.Y. 1971. Effects of scopoletin and chlorogenic acid on stomatal aperture in tobacco and sunflower. *Bulletin of the Torrey Botanical Club* 98:155-162.

Einhellig, F.A., Rice, E.L., Risser, P.G., and Wender, S.H. 1970. Effects of scopoletin on growth, CO_2 exchange rates, and concentration of scopoletin, scopolin, and chlorogenic acids. *Bulletin of the Torrey Botanical Club* 97:22-33.

Ekern, P.C. 1964. Direct interception of cloud water on Lana'ihale, Hawai'i. *Soil Science Society of America Proceedings* 28:419-421.

Fassbender, H.W., and Grimm, U. 1981. Ciclos bioquímicos en un ecosistema forestal de los Andes Occidentales de Venezuela II. Producción y descomposición de los residuous vegetales. *Turrialba* 31:39-48.

Fleming, T.H. 1986. Secular changes in Costa Rican rainfall: correlation with elevation. *Journal of Tropical Ecology* 2:87-91.

Flenley, J.R. 1992. Ultraviolet-B insolation and the altitudinal forest limit. In *The nature and dynamics of forest-savanna boundaries*, ed. P. Furley, J. Proctor, and J.A. Ratter, 273-282. Chapman & Hall, London.

Fölster, H., and Fassbender, H.W. 1978. Untersuchungen über Bodenstandorte der humiden Bergwälder in der nördlichen Andenkordillere. In *Wissenschaftliche Tagung "Landnutzung in den humiden Gebieten der Tropen,"* 101-110. Institute of Soil Science and Forest Nutrition, Göttingen.

Gallagher, M.W., Choularton, T.W., Morse, A.P., and Fowler, D. 1988. Measurements of the size dependence of cloud droplet deposition at a hill site. *Quarterly Journal of the Royal Meteorological Society* 114:1291-1303.

Gates, D.M. 1969. The ecology of an elfin forest in Puerto Rico. 4. Transpiration rates and temperatures of leaves in cool humid environment. *Journal of the Arnold Arboretum* 50:93-98.

Gautam-Basak, M., and Proctor, J. 1983. Micronutrients, aluminium, silicon and ash in leaf litterfall from forests in Gunung Mulu National Park, Sarawak. *Malaysian Forester* 46:224-232.

Gill, A.M. 1969. The ecology of an elfin forest in Puerto Rico. 5. Aerial roots. *Journal of the Arnold Arboretum* 50:197-209.

Glass, A.D.M. 1973. Influence of phenolic acids on ion uptake. I. Inhibition of phosphate uptake. *Plant Physiology* 51:1037-1041.

Glass, A.D.M. 1974. Influence of phenolic acids on ion uptake. III. Inhibition of potassium absorption. *Journal of Experimental Botany* 25:1104-1113.

Gonggrijp, L. 1941. Evaporation from montane forest in West Java at an altitude of 1,750-2,000 m above sealevel. *Tectona* 34:437-447.

Green, T.G.A., and Jane, G.T. 1983. Diurnal patterns of water potential in the evergreen cloud forests of the Kaimai Ranges, North Island, New Zealand. *New Zealand Journal of Botany* 21:379-389.

Grieve, I.C., Proctor, J., and Cousins, S.A. 1990. Soil variation with altitude on Volcán Barva, Costa Rica. *Catena* 17:525-534.

Grimm, U., and Fassbender, H.W. 1981. Ciclos biquímicos en un ecosistema forestal de los Andes Occidentales de Venezuela. I. Inventario de las reservas orgánicas y minerales (N, P, K, Ca, Mg, Mn, Fe, Al, Na). *Turrialba* 31:27-37.

Grubb, P.J. 1974. Factors controlling the distribution of forest types on tropical mountains; new facts and a new perspective. In *Altitudinal Zonation in Malesia*, ed. J.R. Flenley, 1-25. University of Hull, Department of Geography Miscellaneous Series 16.

Grubb, P.J. 1977. Control of forest growth and distribution on wet tropical mountains: with special reference to mineral nutrition. *Annual Review of Ecology and Systematics* 8:83-107.

Grubb, P.J. 1989. The role of mineral nutrients in the tropics: a plant ecologist's view. In *Mineral nutrients in tropical forest and savanna ecosystems*, ed. J. Proctor, 417-439. Blackwell Scientific Press, Oxford.

Grubb, P.J., Lloyd, J.R., Pennington, T.D., and Whitmore, T.C. 1963. A comparison of montane and lowland rain forest in Ecuador. I. The forest structure, physiognomy and floristics. *Journal of Ecology* 51:567-601.

Grubb, P.J., and Tanner, E.V.J. 1976. The montane forests and soils of Jamaica: a reassessment. *Journal of the Arnold Arboretum* 57:33-68.

Grubb, P.J., and Whitmore, T.C. 1966. A comparison of montane and lowland rain forest in Ecuador. II. The climate and its effects on the distribution and physiognomy of the forests. *Journal of Ecology* 54:303-333.

Hamilton, L.S., with P.N. King. 1983. *Tropical forested watersheds. Hydrologic and soils response to major uses or conversions*. Westview Press, Boulder, Colorado.

Harr, R.D. 1982. Fog drip in the Bull Run municipal watershed, Oregon. *Water Resources Bulletin* 18:785-789.

Heaney, A., and Proctor, J. 1989. Chemical elements in litter in forests on Volcán Barva, Costa Rica. In *Mineral nutrients in tropical forest and savanna ecosystems*, ed. J. Proctor, 255-271. Blackwell Scientific Press, Oxford.

Herrmann, R. 1971. Die zeitlichen Anderung der Wasserbindung im Boden unter verschiedenen Vegetationsformationen der Höhenstufen eines tropischen Hochgebirges (Sierra Nevada de Sta. Marta, Kolumbien). *Erdkunde* 25:90-102.

Hetsch, W., and Hoheisel, K. 1976. Standorts- und Vegetationsgliederung in einem Tropischen Nebelwald. *Allgemeine Forst- und Jagd Zeitung* 147:200-209.

Hofstede, R.G.M., Wolf, J., and Benzing, D.H. 1993. Epiphytic biomass and nutrient status of a Colombian Upper Montane Rain Forest. *Selbyana* (in press).

Horner, J.D., Gosz, J.R., and Cates, R.G. 1988. The role of carbon-based plant secondary metabolites in decomposition in terrestrial ecosystems. *American Naturalist* 132:869-883.

Hosokawa, T., Odani, N., and Tagawa, T. 1964. Causality of the distribution of corticulous species in forests with special reference to the eco-physiological approach. *Bryologist* 67:396-411.

Howard, R.A. 1969. The ecology of an elfin forest in Puerto Rico. 8. Studies of stemgrowth and form and of leaf structure. *Journal of the Arnold Arboretum* 50:225-267.

Hutjes, R.W.A., Wierda, A., and Veen, A.W.L. 1990. Rainfall interception in the Tai forest, Ivory Coast: application of two simulation models. *Journal of Hydrology* 114:259-275.

Jaffe, M.J. 1980. Morphogenetic responses of plants to mechanical stimuli or stress. *BioScience* 30:239-243.

Jane, G.T., and Green, T.G.A. 1985. Patterns of stomatal conductance in six evergreen tree species from a New Zealand cloud forest. *Botanical Gazette* 46:413-420.

Jordan, C.F. 1978. Stemflow and nutrient transfer in a tropical rain forest. *Oikos* 31:257-263.

Jordan, C.F., and Heuveldop, J. 1981. The water budget of an Amazonian rainforest. *Acta Amazonica* 11:87-92.

Juvik, J.O., and Ekern, P.C. 1978. *A climatology of mountain fog on Mauna Loa, Hawai'i island*. Technical Report no. 118, Water Resources Research Center, University of Hawai'i, Honolulu.

Kapos, V., and Tanner, E.V.J. 1985. Water relations of Jamaican upper montane rain forest trees. *Ecology* 66:241-250.

Kashiyama, T. 1956. Decrease of sea-fog density by a model shelterbelt. In *Proceedings of the 12th Congress of the International Union of Forestry Research Organizations, Oxford*, November.

Kerfoot, O. 1968. Mist precipitation on vegetation. *Forestry Abstracts* 29:8-20.

Kitayama, K. 1992. An altitudinal transect study of the vegetation on Mount Kinabalu, Borneo. *Vegetatio* 102:149-171.

Kitayama, K. and Mueller-Dombois, D. 1993. An altitudinal transect analysis of the windward vegetation on Haleakalā, a Hawaiian island mountain. *Phytocoenologia* 23:20 pp.

Kobza, J., and Einhellig, F.A. 1987. The effects of ferulic acid on the mineral nutrition of grain sorghum. *Plant & Soil* 98:99-109.

Körner, C., Allison, A., and Hilscher, H. 1983. Altitudinal variation of leaf conductance and leaf anatomy in heliophytes of montane New Guinea and their interrelation with microclimate. *Flora* 174:91-135.

Kuiters, A.T., and Sarink, H.M. 1987. Effects of phenolic acids on growth, mineral composition and chlorophyll content of several deciduous and coniferous trees. *Zeitschrift fur Pflanzenernahrung und Bodenkunde* 150:94-98.

La Bastille, A., and Pool, D.J. 1978. On the need for a system of cloud-forest parks in Middle America and the Caribbean. *Environmental Conservation* 5:183-190.

Lamprecht, H. 1958. Der Gebirgs-Nebelwald der venezolanischen Anden. *Schweizerische Zeitschrift für Forstwesen* 2:1-26.

Langley, A.M. 1985. *Altitudinal changes in climate and litter on Gunung Silam, a small ultrabasic mountain in Sabah, East Malaysia*. BSc thesis, University of Stirling.

Lawton, R.O. 1982. Wind stress and elfin stature in a montane rain forest tree: an adaptive explanation. *American Journal of Botany* 69:1224-1230.

Leigh, E.G. 1975. Structure and climate in tropical rain forest. *Annual Review of Ecology and Systematics* 6:67-86.

Leighton, M., and Wirawan, N. 1986. Catastrophic drought and fire in Borneo tropical rain forest associated with the 1982-83 El Niño southern oscillation. In *Tropical rain forest and the world atmosphere*, ed. G.T. Prance, 75-102. Westview Press, Boulder, Colorado.

Lloyd, C.R., Gash, J.H.C., Shuttleworth, W.J., and Marques, A. de O.F. 1988. The measurement and modelling of rainfall interception by Amazonian rain forest. *Forest and Agricultural Meteorology* 43:277-294.

Lloyd, C.R., and Marques, A. de O.F. 1988. Spatial variability of throughfall and stemflow measurements in Amazonian rain forest. *Forest and Agricultural Meteorology* 42:63-73.

Lowry, J.B., Lee, D.W., and Stone, B.C. 1973. Effects of drought on Mount Kinabalu. *Malayan Nature Journal* 26:178-179.

Lugo, A.E. 1986. Water and the ecosystems of the Luquillo experimental forest. *General Technical Report* SO 63. Institute of Tropical Forestry, Rio Piedras.

Lundgren, L., and Lundgren, B. 1979. Rainfall interception and evaporation in the Mazumbai forest reserve, West Usambara Mountains, Tanzania. *Geografiska Annaler* 61A:157-177.

Lyford, W.H. 1969. The ecology of an elfin forest in Puerto Rico. 7. Soil, root, and earthworm relationships. *Journal of the Arnold Arboretum* 50:210-224.

Mamanteo, B.P., and Veracion, V.P. 1985. Measurements of fogdrip, throughfall and stemflow in the mossy and Benguet pine (*Pinus kesiya* Royle ex Gordon) forests in the upper Agno river basin. *Sylvatrop, the Philippines Forest Research Journal* 10:271-282.

Marrs, R.H., Proctor, J., Heaney, A., and Mountford, M.D. 1988. Changes in soil nitrogen-mineralization and nitrification along an altitudinal transect in tropical rain forest in Costa Rica. *Journal of Ecology* 76:466-482.

McDowell, W.H., Sanchez, C.G., Asbury, C.E., and Ramos Pérez, C.R. 1990. Influence of sea salt aerosols and long range transport on precipitation chemistry at El Verde, Puerto Rico. *Atmospheric Environment* 24A:2813-2821.

McVean, D.N. 1974. Mountain climates of the southwest Pacific. In *Transactions of the third Aberdeen-Hull symposium on Malesian ecology*, ed. J.R. Flenley, 47-57. Department of Geography, University of Hull Miscellaneous Series No. 16.

Medina, E., Sobrado, M., and Herrera, R. 1978. Significance of leaf orientation for leaf temperature in an Amazonian sclerophyll vegetation. *Radiation and Environmental Biophysics* 15:131-140.

Monteith, J.L. 1965. Evaporation and environment. In *The state and movement of water in living organisms*. Symposium of the Society of Experimental Biology 19:205-234.

Monteith, J.L. (ed.) 1976. *Vegetation and the atmosphere. Volume 2. Case studies*. Academic Press, New York.

Nadkarni, N.M. 1981. Canopy roots: convergent evolution in rainforest nutrient cycles. *Science* 214:1023-1024.

Nadkarni, N.M. 1984. Epiphyte biomass and nutrient capital of a neotropical elfin forest. *Biotropica* 16:249-256.

Nadkarni, N.M., and Matelson, T.J. 1992a. Biomass and nutrient dynamics of epiphytic litterfall in a neotropical montane forest, Costa Rica. *Biotropica* 24:24-30.

Nadkarni, N.M., and Matelson, T.J. 1992b. Biomass and nutrient dynamics of fine litter of terrestrially rooted material in a neotropical montane forest, Costa Rica. *Biotropica* 24:113-120.

Nagel, J.F. 1956. Fog precipitation on Table Mountain. *Quarterly Journal of the Royal Meteorological Society* 82:452-460.

Newsome, D. 1986. Aspects of soil surface characteristics on Rakata and Anak Krakatau. In *Krakatoa centenary expedition 1983 final report*, ed. M. Bush, P. Jones, and K. Richards, 134-156. Department of Geography, University of Hull Miscellaneous Series No. 33.

Odum, H.T. 1970. Rain forest structure and mineral cycling homeostasis. In *A tropical rain forest*, ed. H.T. Odum and R.F. Pidgeon, H3-52. United States Atomic Energy Commission, Oak Ridge, Tennessee.

Olbrich, B.W. 1991. The verification of the heat pulse velocity technique for estimating sap flow in *Eucalyptus grandis*. *Canadian Journal of Forest Research* 21:836-841.

Patterson, D.T. 1981. Effects of allelopathic chemicals on growth and physiological responses of soybean (*Glycine max*). *Weed Science* 29:53-59.

Rutter, A.J. 1975. The hydrological cycle in vegetation. In *Vegetation and the atmosphere*, ed. J.L. Monteith, 111-154. Academic Press, New York.

Schier, G.A. and Jensen, K.F. 1992. Atmospheric deposition effects on foliar injury and foliar leaching in red spruce. In *Ecology and decline of red spruce in the Eastern United States,* eds. C. Eagar and M.B. Adams, 271-294. Springer-Verlag, New York.

Schmidt, F.H. 1950. On the distribution of sunshine in Java. *Verhandelingen van het Meteorologisch en Geofysisch Observatorium te Djakarta* 42:1-77.

Sevink, J. 1984. An altitudinal sequence of soils in the Sierra Nevada de Santa Marta. In *Studies on tropical Andean ecosystems. Volume 2. La Sierra Nevada de Santa Marta (Colombia)*, ed. T. van der Hammen and P.M. Ruiz, 131-138. J. Cramer, Berlin.

Shuttleworth, W.J. 1988. Evaporation from Amazonian rain forest. *Philosophical Transactions of the Royal Society Series B* 233:321-346.

Stadtmüller, T. 1987. *Cloud forests in the humid tropics: a bibliographic review*. The United Nations University, Tokyo, and Centro Agronomico Tropical de Investigacion y Ensenanza, Turrialba, Costa Rica.

Stadtmüller, T., and Agudelo, N. 1990. Amount and variability of cloud moisture input in a tropical cloud forest. *International Association of Hydrological Sciences Publication* 193:25-32.

Steinhardt, U. 1979. Untersuchungen über den Wasser- und Nährstoffhaushalt eines andinen Wolkenwaldes in Venezuela. *Göttinger Bodenkundliche Berichte* 56:1-185.

Sugden, A.M. 1986. The montane vegetation and flora of Margarita Island, Venezuela. *Journal of the Arnold Arboretum* 67:187-232.

Sullivan, J.H., Teramura, A.H., and Ziska, L.H. 1992. Variation in UV-B sensitivity in plants from a 3,000 m elevational gradient in Hawai'i. *American Journal of Botany* 79:737-743.

Tanner, E.V.J. 1977a. *Mineral cycling in montane rain forests in Jamaica*. PhD thesis, University of Cambridge.

Tanner, E.V.J. 1977b. Four montane rain forests of Jamaica: a quantitative characterization of the floristics, the soils and the foliar mineral levels, and a discussion of the interrelations. *Journal of Ecology* 65:883-918.

Tanner, E.V.J. 1980a. Litterfall in montane rain forests of Jamaica and its relation to climate. *Journal of Ecology* 68:833-848.

Tanner, E.V.J. 1980b. Studies on the biomass and productivity in a series of montane rain forests in Jamaica. *Journal of Ecology* 68:573-588.

Tanner, E.V.J. 1985. Jamaican montane forests: nutrient capital and cost of growth. *Journal of Ecology* 73:553-568.

Tanner, E.V.J., and Kapos, V. 1982. Leaf structure of Jamaican upper montane rainforest trees. *Biotropica* 14:16-24.

Tanner, E.V.J., Kapos, V., and Franco, W. 1992. Nitrogen and phosphorus fertilization effects of Venezuelan montane forest trunk growth and litterfall. *Ecology* 73:78-86.

Tanner, E.V.J., Kapos, V., Freskos, S., Healey, J.R., and Theobald, A.M. 1990. Nitrogen and phosphorus fertilization of Jamaican montane forest trees. *Journal of Tropical Ecology* 6:231-238.

Teramura, A.H. 1983. Effects of ultraviolet-B radiation on the growth and yield of crop plants. *Physiologia Plantarum* 58:415-427.

Tie, Y.L., Baillie, I.C., Phang, C.M.S., and Lim, C.P. 1979. *Soils of Gunung Mulu National Park.* Department of Agriculture, Sarawak.

Trinidad Pizarro, R. 1985. La composicion quimica del agua de alluvia y su influencia en el pH. In *Abstracts of the 12th Symposium on Natural Resources, 11 December 1985,* p. 10. Puerto Rico Department of Natural Resources, San Juan.

Troll, C. 1956. Der Klima- und Vegetationsaufbau der Erde im Lichte neuer Forschungen. *Journal of the Academy of Science, Mainz*:216-229.

Vance, E.D., and Nadkarni, N.M. 1990. Microbial biomass and activity in canopy organic matter and the forest floor of a tropical cloud forest. *Soil Biology and Biochemistry* 22:677-684.

Van der Plas, M.C. 1992. *Hydrological characteristics of the forest floor in undisturbed and logged-over rain forest, Danum Valley, Sabah, Malaysia*. MSc thesis, Free University, Amsterdam.

Van Reuler, H. 1987. Soil studies in the Bukit Raya nature reserve. In Report of the 1982-1983 Bukit Raya expedition, ed. H.P. Nooteboom, 7-23. Rijksherbarium, Leiden, the Netherlands.

Van Steenis, C.G.G.J. 1972. *The mountain flora of Java*. E.J. Brill, Leiden, the Netherlands.

Vaughan, D., and Ord, B. 1990. Influence of phenolic acids on morphological changes in roots of *Pisum sativum*. *Journal of the Science of Food and Agriculture* 52: 289-299.

Vaughan, R.E., and Wiehe, P.O. 1947. Studies on the vegetation of Mauritius. IV. Some notes on the internal climate of the upland climax forest. *Journal of Ecology* 34:126-136.

Veneklaas, E.J. 1990. *Rainfall interception and above-ground nutrient fluxes in Colombian montane tropical rain forest*. PhD thesis, University of Utrecht, Utrecht, the Netherlands.

Veneklaas, E.J., and Van Ek, R. 1990. Rainfall interception in two tropical montane rain forests, Colombia. *Hydrological Processes* 4:311-326.

Veneklaas, E.J., Zagt, R.J., Van Leerdam, A., Van Ek, R., Broekhoven, A.J., and Van Genderen, M. 1990. Hydrological properties of the epiphyte mass of a montane tropical rain forest. *Vegetatio* 89:183-192.

Venkataramanan, C., and Chinnamani, S. 1978. A preliminary note on the return of nutrient by the leaf litter of wet (montane) evergreen Shola forests of Nilgiris. *Indian Forester* 104:450-456.

Vis, M. 1986. Interception, drop size distributions and rainfall kinetic energy in four Colombian forest ecosystems. *Earth Surface Processes & Landforms* 11:591-603.

Vitousek, P.M., Matson, P.A., and Turner, R.A. 1988. Elevation and age gradients in Hawaiian montane rainforest: foliar and soil nutrients. *Oecologia* 77:565-570.

Vogelmann, H.W. 1973. Fog precipitation in the cloud forests of Eastern Mexico. *BioScience* 23:96-100.

Wadsworth, F.H., and Bonnet, J.A. 1951. Soil as a factor in the occurrence of two types of montane forest in Puerto Rico. *Caribbean Forester* 12:67-70.

Walter, H. 1979. *Vegetation und Klimazonen*. Ulmer Verlag, Stuttgart.

Ward, R.C., and Robinson, M. 1989. *Principles of hydrology*. McGraw Hill, London.

Waterloo, M.J. 1989. *A hydrological study of the mass elevation effect on Gunung Silam, a small coastal ultrabasic mountain in Sabah, East Malaysia*. MSc thesis, Free University, Amsterdam.

Weathers, K.C. et al. 1988. Cloud water chemistry from ten sites in North America. *Environmental Science and Technology* 22:1018-1026.

Weaver, P.L. 1972. Cloud moisture interception in the Luquillo Mountains of Puerto Rico. *Caribbean Journal of Science* 12:129-144.

Weaver, P.L., Byer, M.D., and Bruck, D.L. 1973. Transpiration rates in the Luquillo Mountains of Puerto Rico. *Biotropica* 5:123-133.

Weaver, P.L., Medina, E., Pool, D., Dugger, K., Gonzales-Liboy, J., and Cuevas, E. 1986. Ecological observations in the dwarf cloud forest of the Luquillo Mountains of Puerto Rico. *Biotropica* 18:79-85.

Werner, W.L. 1988. Canopy dieback in the Upper Montane rain forests of Sri Lanka. *GeoJournal* 17:245-248.

Werner, W.L., and Balasubramaniam, S. 1988. Structure and dynamics of the upper montane rain forests of Sri Lanka. In *Tropical forests in transition*, ed. J.G. Goldammer, 165-171. Birkhäuserverlag, Basel, Switzerland.

Whitmore, T.C.W. 1989. Tropical forest nutrients, where do we stand? A *tour de horizon*. In *Mineral nutrients in tropical forest and savanna ecosystems*, ed. J. Proctor, 1-13. Blackwell Scientific Press, Oxford.
Whitmore, T.C. 1990. *An introduction to tropical rain forests.* Clarendon Press, Oxford.
Whitmore, T.C., and Burnham, C.P. 1969. The altitudinal sequence of forest and soils on granite near Kuala Lumpur. *Malayan Nature Journal* 22:99-118.
Whittaker, R.J., Bush, M.B., and Richards, K. 1989. Plant recolonization and vegetation succession on the Krakatau islands, Indonesia. *Ecological Monographs* 59:59-123.
Wolf, J. 1993. *Ecology of epiphytes and epiphyte communities in montane rain forests, Colombia.* PhD thesis, University of Amsterdam.
Zadroga, F. 1981. The hydrological importance of a montane cloud forest area of Costa Rica. In *Tropical agricultural hydrology,* ed. R. Lal and E.W. Russell, 59-73. J. Wiley, New York.
Zinck, A. 1986. Los Suelos. Caracteristicas y fragilidad de los suelos en ambiente de Selva Nublada: el ejemplo de Rancho Grande. In *La Selva Nublada de Rancho Grande Parque Nacional "Henri Pittier,"* ed. O. Huber, 31-66. Fondo Editorial Acta Cientifica Venezolana, Caracas.
Ziska, L.H., Teramura, A.H., and Sullivan, J.H. 1992. Physiological sensitivity of plants along an elevational gradient to UV-B radiation. *American Journal of Botany* 79:863-871.6.75/8.

4. The Importance of Tropical Montane Cloud Forests for Endemic and Threatened Birds

Adrian J. Long

Rampant growth in human population and economic development is resulting in the exhaustion of the planet's natural resources at a speed that threatens the earth's biodiversity. Biota are being lost (including species with significant economic or medicinal potential); in some cases even before they have been identified by science. An urgent conservation priority must be to halt the destruction of natural ecosystems and promote their protection. However, biodiversity is not distributed evenly over the surface of the earth, and as a matter of urgency, conservationists should be identifying areas where habitat destruction is most likely to result in global extinctions.

Areas that hold assemblages of species found nowhere else on earth are critical to biodiversity conservation planning, and the richer they are in such species, the more significant their claim on our attention. BirdLife's Biodiversity Project (ICBP 1992; BirdLife, in prep.) has advanced the analysis of centers of avian endemism throughout the world by collecting data on all land birds that have had, in historical times, a total global breeding range estimated at less than 50,000 km^2 ("restricted-range species"), which is about the size of Costa Rica. The 50,000 km^2 range size criterion is arbitrary, but produces a manageable sample of species that are most vulnerable to habitat destruction and need some form of protection. The use of this size as the threshold was influenced by the work of Terborgh and Winter (1983), who used it to map the distributions of 155 Colombian and Ecuadorian bird species in order to locate areas of endemism and demonstrate the need for protected areas for species with small ranges.

The BirdLife Biodiversity Project considers of primary importance those areas that support two or more species entirely confined to them, and refers to them as Endemic Bird Areas (EBAs). EBAs were identified by a combination of a multivariate statistical analysis and an investigation of the habitat and altitudinal requirements of the bird species under review. Simultaneously, patterns of endemism in other life forms were investigated and shown to be closely congruent with those that emerged from the avian analysis (ICBP 1992; Thirgood and Heath, 1994).

The study showed that more than 25 percent of all bird species have a breeding range of 50,000 km^2 or less, and are distributed through 147 countries, which is over 62 percent of all countries. A total of 221 EBAs were identified; over 95 percent of restricted-range birds (2,564 species) occur solely in these areas. The total area occupied by EBAs is relatively small (less than 6,500,000 km^2), with all EBAs being confined to just 4.5 percent of the world's land area. A more memorable figure is that 20 percent of the world's bird species are confined to just 2 percent of the world's land area.

An important feature of the project is that geo-referenced point-locality data have been gathered for every restricted-range species, except for those occurring on small islands when presence or absence data only were collected. The database file holds over 50,000 locality records (of which 87 percent have been assigned coordinates). These data provide great versatility because they offer the chance to examine the distribution of restricted-range species at a variety of scales. It is possible therefore to not only use the data to make comparisons between areas or species at the global or continental level, but also to investigate further a particular habitat type, a smaller area, or a single species (Crosby, 1994).

Here I aim to show the flexibility of the information gathered for the BirdLife Biodiversity Project. A survey of the world's EBAs and restricted range-species in relation to tropical montane cloud forests (TMCFs) is presented by giving global and regional overviews of those EBAs that hold cloud forests and listing those species that are mainly cloud forest specialists. Information is presented mainly in a tabular form, but distribution maps are provided for most of the regions as well. The species tables are given so that the dataset is open for further refinement. A specific example of data application in establishing conservation priorities is shown for the TMCFs of northern South America.

Methods: Coding Species for Tropical Montane Cloud Forests

Information on restricted-range species is stored in two databases, one holding individual records for each locality, and a second storing general information on a species in a single record. In the latter there is information stored on the habitat and altitudinal requirements, data which were gathered from regional handbooks, field guides, and other references.

The database does not specifically code the restricted-range bird species for their occurrence in TMCFs, but for the purposes of this analysis it was possible to use the habitat and altitudinal information to place the species into the following habitat categories: (1) the species is confined to TMCF; (2) the species' main habitat is TMCF; (3) the species is found in TMCF, but other habitats are equally or more important; (4) the species may occur occasionally in TMCF; (5) the species does not occur in TMCF; (6) the habitat of the species is unknown.

The definition for a TMCF followed that of the East-West Center (Hamilton, Juvik, and Scatena, this volume), which states:

> The tropical montane cloud forest is composed of forest ecosystems of distinctive floristic and structural form. It typically occurs as a relatively narrow altitudinal zone where the atmospheric environment is characterized by persistent, frequent, or seasonal cloud cover at the vegetation level. . . . The net precipitation (throughfall) in such forests is significantly enhanced (beyond rainfall contribution) through direct canopy interception of cloud water (horizontal precipitation or cloud stripping) and low water use by the vegetation.
>
> . . . Canopy trees usually exhibit gnarled trunks and branches; dense, compact crowns; and small, thick, and hard (sclerophyll) leaves. TMCF is also characterized by having a high proportion of biomass as epiphytes (bryophytes, lichens, and filmy ferns) and a corresponding reduction in woody climbers. . . . Endemism is often very high.

The information stored in the habitat field of the database was not always explicit enough to distinguish TMCF, as defined above, from other montane forests, but it was possible to code species that do not occur in this habitat. The problem of assigning species to the habitat categories relevant to TMCF (categories 1 to 3) was exacerbated by authors using different definitions for montane forest and cloud forests. Consequently, for the preparation of this review, the opinions of experts were sought for areas and species where information was incomplete or confusing, thus avoiding the problems of variation in terminology in the literature. The reviewers and the areas that they advised on are listed in the acknowledgments.

Results

EBAs and Tropical Montane Cloud Forests: Global Overview

Restricted-range species using TMCF are found in EBAs throughout tropical and subtropical areas of the world (Figure 1). Seventy-nine (34 percent) of all EBAs have at least one species using TMCF, but for many EBAs the proportion that are TMCF species is very small, and this is the case especially on many of the island EBAs (see Tables 2, 4, 6, 8, 10, and 12 for details on each of the 79 EBAs).

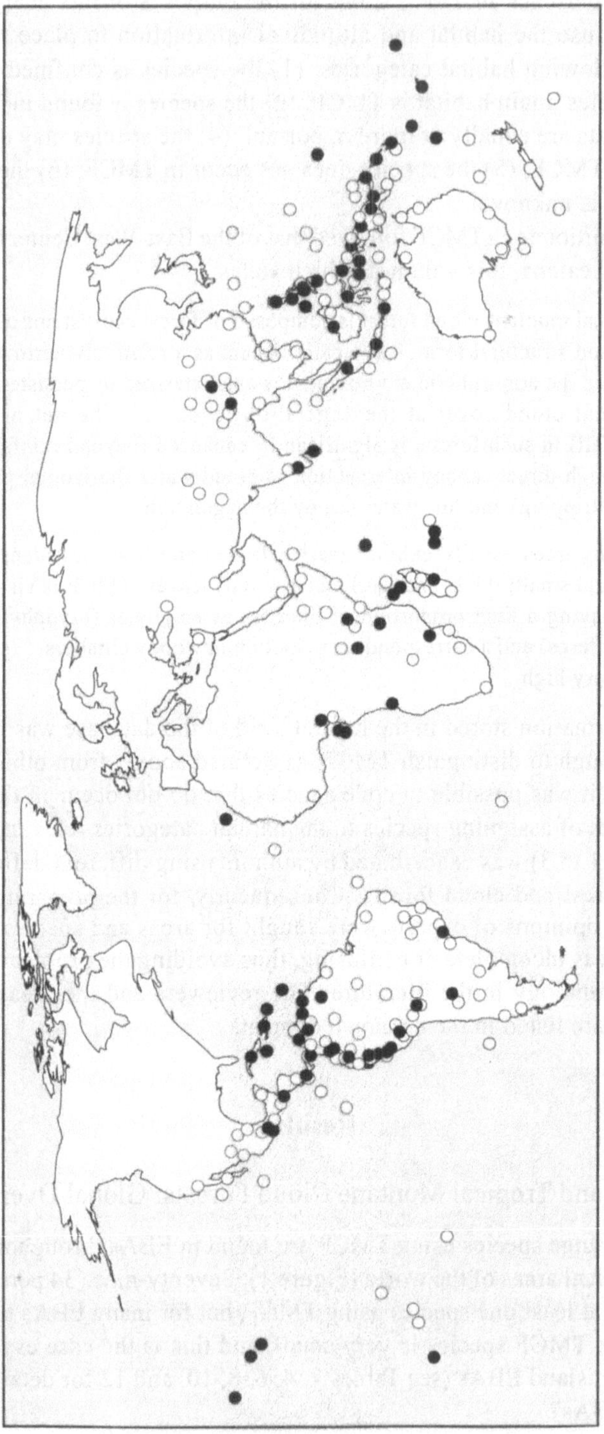

Figure 1. Global distribution of Endemic Bird Areas showing those that have at least one restricted-range species occurring in cloud forest (filled circle) and those that do not (open circles).

Table 1 shows that there are 255 restricted-range species (10 percent of all restricted-range species: 2,609 as identified in ICBP 1992) coded in categories 1 and 2, and a further 315 (12 percent) are in category 3, representing 2.73 percent or 5.97 percent of the total number of bird species in the world. The number of bird species confined to the world's TMCFs would undoubtedly increase if species with ranges greater than 50,000 km^2 were added. TMCFs cover a smaller percentage of the world's surface than would be expected from the proportion of bird species found in them: it has been estimated that TMCFs cover 500,000 km^2 (see Stadtmüller 1987), which is about 0.3 percent of the world's surface.

The regional breakdown of the restricted-range species in the different habitat categories shows that there is little difference between regions in the proportion of non–cloud forest species (habitat category 5) or perhaps found occasionally in cloud forest (habitat category 4), especially when the categories are pooled together for each region. However, South America clearly has more species and a higher proportion of its restricted-range species confined to TMCF habitats compared to the other regions.

Regional Overviews

The parenthetical alphabet and figures in the text refer to codes applied in the BirdLife Biodiversity Project to identify EBAs individually (see ICBP 1992), and these codes can be matched with the regional tables (2, 4, 6, 8, 10, 12, and 14) and figures (2–5 and 7).

EBAs and Cloud Forests in Africa and Associated Islands

Most of the EBAs in Africa are discrete and share few restricted-range species. There are 22 EBAs found on continental Africa (BirdLife, in prep.), including 3 more than originally identified in ICBP (1992), with a further 19 EBAs on the Indian and Atlantic ocean islands, including 5 in Madagascar. Information on the EBAs holding cloud forest species is summarized in Table 2, and species for which cloud forest is the major habitat are listed in Table 3 and shown in Figure 2.

Several mountain ranges in continental Africa have significant areas of montane forest, and the most important for cloud forest species are the Albertine Rift mountains (C20), the Eastern Arc mountains (C24), and the Cameroon mountains (C04). These EBAs also have the highest totals of restricted-range species for the continent. Other cloud forests are found on the Angola escarpment and in southern Ethiopia, but their importance for cloud-forest-dwelling species is unclear. For instance, the EBA in Angola (C08) includes lowland, submontane, and montane species, but could not be subdivided further due to the lack of bird records and information on their habitat requirements.

Along the upper slopes of the eastern escarpment and northern massif of Madagascar lies an important band of cloud forest for restricted-range spe-

Table 1. Detailed breakdown for the restricted-range species found in EBAs in a particular region, the total number of species in each of the six habitat categories (expressed as number of species and as a percentage of the total number of species in the region).

Region	Total Species	Habitat Categories* 1	%	2	%	3	%	4	%	5	%
African EBAs	326	1	0.3	13	5.9	69	21.3	60	15.5	183	52.9
Asian EBAs	186	0	0	17	7.5	27	14.5	9	4.8	133	71.5
Southeast Asian and New Guinean island EBAs	619	12	1.9	56	8.9	73	11.5	82	13.0	396	62.7
Pacific island EBAs	391	15	3.7	20	4.9	29	7.2	49	12.2	278	69.1
Caribbean EBAs	149	0	0	7	4.7	31	20.0	23	15.3	88	58.6
Central American EBAs	184	5	2.7	11	5.9	35	19.0	32	17.3	101	56.4
South American EBAs	679	21	3.1	77	11.2	51	7.5	98	14.3	432	63.3
Total for restricted-range species in EBAs of the regions listed above	2,534**	54	2.1	201	8.0	315	12.3	353	13.3	1,611	62.4

*The habitat categories are (1) the species is confined to tropical montane cloud forest; (2) the species' main habitat is cloud forest; (3) the species is found in cloud forest, but other habitats are equally or more important; (4) the species probably occurs occasionally in cloud forest; (5) the species does not occur in cloud forest; (6) the habitat of the species in relation to cloud forest is unknown. It should be noted that the regional tables showing the number of species in each habitat category for each EBA with at least one cloud forest species (Tables 2, 4, 6, 8, 10, 12, and 14) will not correspond with the regional breakdowns given in this table, since restricted-range species can occur in more than one EBA.

**This figure does not include the 27 species (<1.5 percent of all restricted-range species) for which habitat was unknown (category 6).

A *chi*-square test performed on this contingency table showed a significant difference from the expected values ($c^2 = 132.4$, $df = 24$, $p < 0.001$).

A *chi*-square test performed on three habitat categories (1 and 2 pooled = cloud forest species; 3 left the same = cloud forest species, in part; 4 and 5 pooled = non–cloud forest species) also showed a significant difference from the expected values ($c^2 = 89.4$, $df = 12$, $p < 0.001$).

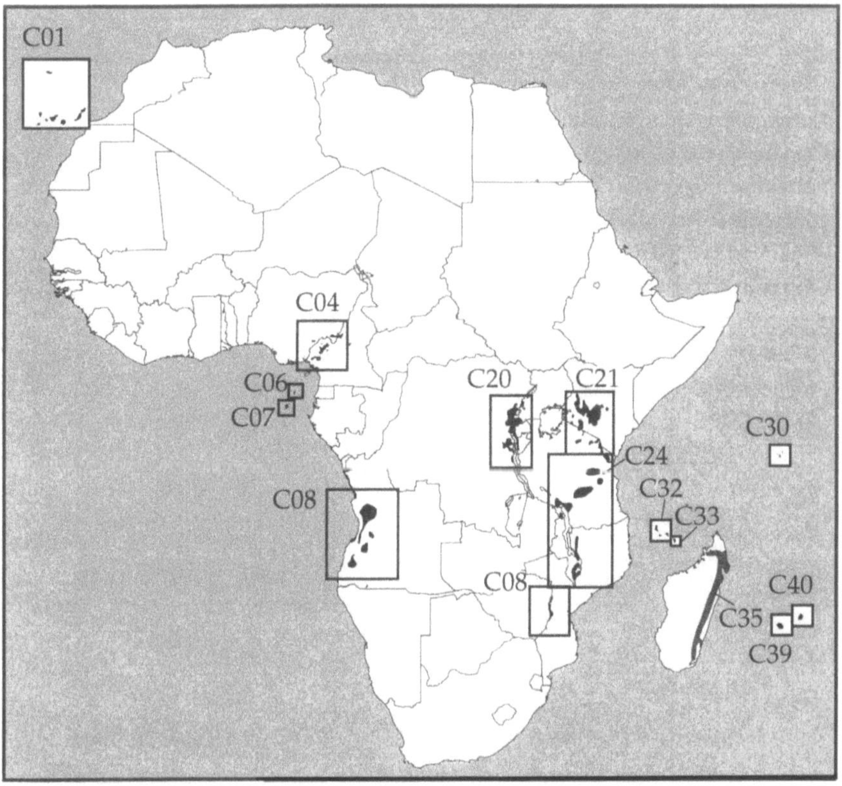

Figure 2. The Endemic Bird Areas in Africa and associated islands that have at least one species found in cloud forest.

cies. Only a few species are confined to this habitat (Table 3), but several of the other endemics included in this EBA occur in cloud forest (M. I. Evans, pers. com.). Cloud forests occur on the Canary Islands, São Tomé, Príncipe, Mahé in the Seychelles, the Comoros, and Mauritius and Réunion, but there are few species restricted to them (see Tables 2 and 3).

Many species of the Indian Ocean islands already have become extinct due to forest destruction, introduced predators, and hunting, with islands such as the Mascarenes (EBAs C39, C40, C41) having suffered a high number of bird extinctions, most notably the loss of the dodo (*Raphus cucullatus*) from Mauritius. Lowland forests tend to have been cleared first, but now on many islands the montane forests are being encroached upon. Since most of the island species show wider habitat and altitudinal preferences, the cloud forest areas are now very important refuges for any forest-dwelling species.

Table 2. Breakdown of the number of species for each of the six habitat categories for those EBAs in Africa and associated islands that have at least one species found in cloud forest. The habitat categories are listed in Table 1.

	Habitat Category					
Endemic Bird Area	1	2	3	4	5	6
Canary Islands and Madeira (C01)	0	2	0	0	8	0
Cameroon mountains (C04)	0	2	9	11	6	0
Príncipe (C06)	0	1	2	1	5	0
São Tomé (C07)	0	2	5	4	10	0
West Angola (C08)	0	1	2	1	11	0
Albertine Rift mountains (C20)	0	0	12	26	1	0
Kenyan mountains (C21)	0	0	2	1	5	0
Eastern Arc mountains (C24)	1	2	13	5	8	0
Southeast African mountains (C26)	0	0	3	2	2	0
Granite Seychelles (C30)	0	0	1	0	11	0
Comoro Islands (C32)	0	7	5	0	1	0
Mayotte (C33)	0	1	2	2	1	0
East Madagascan humid forests (C35)	0	3	11	0	4	0
Réunion (C39)	0	0	6	0	1	3
Mauritius (C40)	0	0	7	1	1	2

Table 3. Restricted-range species confined mainly to TMCF (categories 1 and 2) in Africa and associated islands

Species Name	English Name	Cloud	EBA
Columba bollii	Bolle's Laurel Pigeon	2	C01
Columba trocaz	Trocaz Pigeon	2	C01
Columba sjostedti	Cameroon Olive-pigeon	2	C04
Tauraco bannermani	Bannermann's Turaco	2	C04
Zosterops ficedulina	Príncipe White-eye	2	C06, C07
Columba thomensis	São Tomé Olive-pigeon	2	C07
Nectarinia ludovicensis	Montane Double-collared Sunbird	2	C08
Malaconotus alius	Uluguru Bush-shrike	2	C24
Sheppardia montana	Usambara Akalat	1	C24
Sheppardia sharpei	Sharpe's Akalat	2	C24
Otus pauliani	Grand Comoro Scops-owl	2	C32
Hypsipetes parvirostris	Comoro Bulbul	2	C32
Turdus bewsheri	Comoro Thrush	2	C32
Nesillas brevicaudata	Grand Comoro Brush-warbler	2	C32
Nesillas mariae	Moheli Tsikirity	2	C32
Humblotia flavirostris	Humblot's Flycatcher	2	C32
Columba polleni	Comoro Olive Pigeon	2	C32, C33
Atelornis crossleyi	Rufous-headed Ground-roller	2	C35
Phyllastrephus cinereiceps	Grey-crowned Tetraka	2	C35
Crossleyia xanthophrys	Madagascar Yellowbrow	2	C35

This is the case for Madagascar and for the Comoros. There are 105 species endemic to Madagascar but about half of these were excluded from the BirdLife Biodiversity Project because their historical ranges are too large (ICBP 1992). These included a number of species that formerly occurred throughout the humid evergreen forests of eastern Madagascar, which originally covered approximately 110,000 km^2, but which now covers less than 38,000 km^2 (Green and Sussman 1990). Thus, the cloud forests within the East Madagascan humid forests' EBA (C35) are very important for restricted-range species included within the Biodiversity Project criteria, but also for other forest species with contracting ranges. The extent of habitat destruction in this EBA is reflected by Collar and Stuart's (1985) listing of 12 of its 17 species as threatened. A similar pattern exists in the Comoros with almost complete lowland forest clearance leaving forest confined to only the steepest slopes and interior of the volcanic craters of the islands (Louette and Stevens 1992), and 6 of its 14 endemic species being threatened (Collar and Stuart 1985).

EBAs and TMCFs in Continental Asia and Associated Islands

There are 20 EBAs found on continental Asia, with a further 7 EBAs on associated islands. Information on the EBAs holding cloud forest species is summarized in Table 4, and species for which cloud forest is the major habitat are listed in Table 5. Cloud forest is restricted to only a few small areas of the continent.

The topography of continental Asia is especially complex in the high-altitude Qinghai-Tibet plateau of central Asia, and along the Himalayas and isolated mountains ranging across Indochina. Around the plateau exists a mosaic of tropical and subtropical forests at low and middle altitudes, and temperate forests and alpine habitats at higher altitudes. Several of the isolated areas of forest, particularly in the subtropical and temperate zones, have their own unique groupings of species and constitute EBAs, but the wettest part within the area and consequently the only zone having cloud forest is the central Sichuan mountains' EBA (D12).

TMCFs are restricted to the eastern part of the Himalayas where clouds bringing heavy monsoon rains from the south are trapped by the mountains.

Table 4. Breakdown of the number of species for each of the six habitat categories for those EBAs in continental Asia and associated islands that have at least one species found in TMCF. The habitat categories are listed in Table 1.

	Habitat Category					
Endemic Bird Area	1	2	3	4	5	6
Sri Lanka (D05)	0	0	6	4	13	0
Eastern Himalayas (D08)	0	17	5	1	2	0
Central Sichuan mountains (D12)	0	0	9	1	0	0
Western Ghats mountains (D29)	0	0	7	0	2	0

The rain forests of this region are extremely rich, both in restricted-range species and overall diversity. Other isolated areas of montane rain forest with some areas of TMCF are located on the coastal flanks of the southern tip of the Western Ghats (D04), running parallel to the western edge of the Indian subcontinent, and in southwestern Sri Lanka (D05), but these EBAs do not have species confined solely to TMCFs.

Large areas of Asia historically have been inaccessible to Western scientists, for political and logistical reasons, and the status of their avifaunas is therefore relatively poorly known. This includes some of the most important areas of cloud forest in the eastern Himalayas of northeastern India and regions such as eastern Myanmar (= Burma), and southern Cambodia, where TMCFs probably occur but whose importance has not been assessed for birds. However, habitat destruction is occurring in all the known cloud forests that are important for restricted-range birds such as in central Sichuan, the Western Ghats, and Sri Lanka.

EBAs and Cloud Forest in Southeast Asian Islands and New Guinea

Virtually the whole of the Philippines, Indonesia, and New Guinea are divided into EBAs. This reflects the high biodiversity of the region. For instance, Indonesia, comprising some 17,500 islands of varying size and origin, has a rich avifauna of over 1,500 species including 411 restricted-range species, the highest total for any country in the world. Information on the EBAs holding cloud forest species is summarized in Table 6, and species for which TMCF is the major habitat are listed in Table 7 and their distribution is shown in Figure 3.

Areas holding restricted-range species are found on the larger Philippine islands of Luzon and Mindanao, with smaller areas on Mindoro, Palawan,

Table 5. Restricted-range species confined mainly to TMCF (categories 1 and 2) in continental Asia

Species Name	English Name	Cloud	EBA
Arborophila mandellii	Chestnut-breasted Partridge	2	D08
Tragopan blythii	Blyth's Tragopan	2	D08
Harpactes wardi	Ward's Trogon	2	D08
Brachypteryx hyperythra	Rusty-bellied Shortwing	2	D08
Actinodura waldeni	Streak-throated Barwing	2	D08
Alcippe cinerea	Yellow-throated Fulvetta	2	D08
Alcippe ludlowi	Ludlow's Fulvetta	2	D08
Garrulax austeni	Brown-capped Laughingthrush	2	D08
Garrulax virgatus	Striped Laughingthrush	2	D08
Heterophasia gracilis	Grey Sibia	2	D08
Heterophasia pulchella	Beautiful Sibia	2	D08
Spelaeornis badeigularis	Rusty-throated Wren-babbler	2	D08
Spelaeornis caudatus	Rufous-throated Wren-babbler	2	D08
Stachyris oglei	Snowy-throated Babbler	2	D08
Tickellia hodgsoni	Broad-billed Warbler	2	D08
Sitta victoriae	White-browed Nuthatch	2	D08
Actinodura nipalensis	Hoary-throated Barwing	2	D08, D28

Table 6. Breakdown of the number of species for each of the six habitat categories for those EBAs in the Philippines, Indonesia, and New Guinea that have at least one species found in TMCF. The habitat categories are listed in Table 1.

	Habitat Category					
Endemic Bird Area	1	2	3	4	5	6
Luzon mountains (E01)	0	4	5	5	26	0
Mindoro (E03)	0	1	1	2	5	0
Negros and Panay (E04)	1	1	0	2	10	0
Palawan (E06)	0	1	1	1	16	0
Mindanao mountains (E08)	0	11	7	7	1	0
Bornean mountains (E10)	0	3	7	16	2	0
Sumatra and Peninsular Malaysia (E11)	0	2	16	16	4	0
Javan and Balinese mountains (E13)	0	14	13	7	1	0
Sulawesi (E21)	0	7	16	3	22	6
Banggai and Sula islands (E22)	0	0	2	1	14	0
Buru (E23)	1	2	2	4	20	0
Seram (E24)	4	2	5	8	12	0
Vogelkop mountains (E27)	0	3	4	7	6	0
North New Guinean mountains (E29)	0	1	3	2	2	0
Adelbert and Huon mountains (E31)	1	2	4	2	2	0
Central New Guinean high mountains (E32)	7	13	6	14	15	0

Negros, and Panay. Some areas of these islands are poorly known. For instance, a new species, the Panay striped-babbler (*Stachyris latistriata*) was recently discovered and is confined to elfin forests of Panay (Gonzales and Kennedy 1990).

There are important areas of cloud forest in the mountains of the Greater Sundas. Peninsular Malaysia and Sumatra (E11), Borneo (E10), and Java (E13) each have unique species in their montane forests. Cloud forests in Wallacea are more restricted. The Lesser Sundas are drier with a dry season from February to October, and cloud forest is found only on Lombok, with few bird species present (White and Bruce 1986)—the only restricted-range cloud forest species being *Zoothera horsfieldi* found principally in the Greater Sundas. In northern Wallacea, Sulawesi and Seram have particularly important cloud forests for bird species, and smaller areas are found on Buru and Obi.

New Guinea is rich in restricted-range species and there are four montane forest EBAs, each with a number of cloud forest species. The largest EBA is the Central New Guinean mountains, with 20 species restricted mainly to its cloud forests (Table 7).

Many highly threatened EBAs exist in the Philippines and Indonesia. Both these countries have suffered very substantial losses of forest, including the almost complete deforestation of whole islands in the Philippines, such as Cebu (E05), Negros and Panay (E04), and Mindoro (E03). The montane forests have been somewhat less impacted than the lowland forests, and, as in the case of Madagascar and the Comoros, they offer the only refuges for some

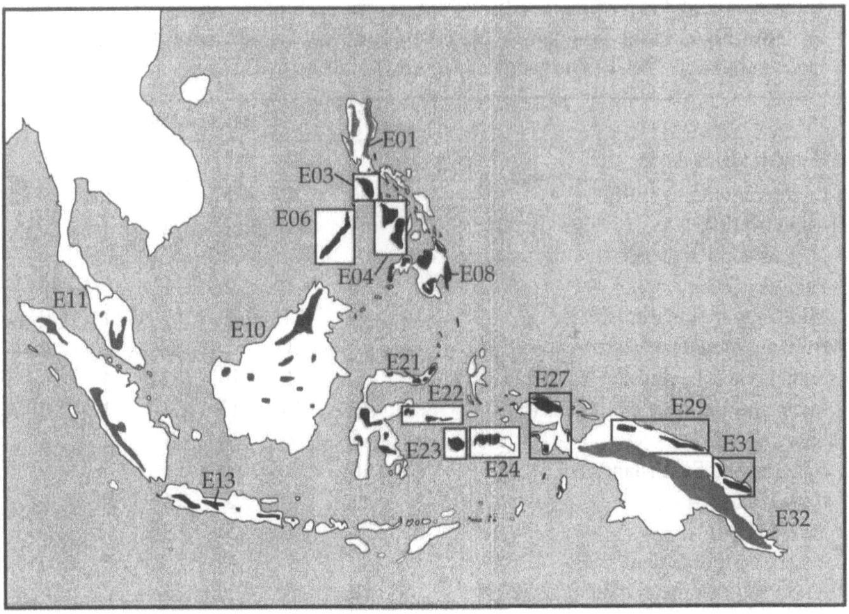

Figure 3. The Endemic Bird Areas in the Philippines, Indonesia, and New Guinea that have at least one species found in cloud forest.

forest-dwelling species. By contrast, the forests of New Guinea remain relatively intact compared to many other parts of the world, and few cloud forest species are listed as threatened there.

Pacific Islands

The Pacific Ocean occupies over one-third of the earth's surface, and within this vast expanse of sea are scattered more than 10,000 relatively small islands. The island EBAs of the Pacific are quite variable in form, ranging from the low-lying coral atolls of the Tuamotu Archipelago (F28), which rise to only 110 m, to volcanic islands of considerable altitude such as Hawai'i (F25) at over 4,000 m. However, most of the islands are oceanic in origin, never having had past continental connections. As a result of this isolation and the difficulty of animal dispersal from continents, they have a relatively depauperate land avifauna (Diamond 1984). There are 30 EBAs within the region, and information on those holding cloud forest bird species is summarized in Table 8 and illustrated in Figure 4. Species for which TMCF is the major habitat are listed in Table 9. Although 12 EBAs have cloud forest species, the number restricted to this habitat is very small in relation to the total number of species. This reflects the trend in which many of the isolated island bird species become generalists, being able to occupy a variety of habitats including scrub, plantations, and gardens, as well as forest.

A notable exception is the Hawaiian Islands (represented by EBAs F24

Table 7. Restricted-range species confined mainly to TMCF (categories 1 and 2) in the Philippines, Indonesia, and New Guinea

Species Name	English Name	Cloud	EBA
Prioniturus montanus	Luzon Racket-tail	2	E01
Pitta kochi	Whiskered Pitta	2	E01
Dicaeum anthonyi	Flame-crowned Flowerpecker	2	E01, E08
Pyrrhula leucogenis	White-cheeked Bullfinch	2	E01, E08
Otus mindorensis	Mindoro Scops-owl	2	E03
Stachyris latistriata	Panay Striped-babbler	1	E04
Stachyris nigrorum	Negros Striped-babbler	2	E04
Stachyris hypogrammica	Palawan Striped-babbler	2	E06
Prioniturus waterstradti	Mindanao Racket-tail	2	E08
Otus mirus	Mindanao Scops-owl	2	E08
Rhinomyias goodfellowi	Slaty-backed Jungle-flycatcher	2	E08
Rhipidura nigrocinnamomea	Black-and-cinnamon Fantail	2	E08
Aethopyga boltoni	Apo Sunbird	2	E08
Aethopyga primigenius	Grey-hooded Sunbird	2	E08
Hypocryptadius cinnamomeus	Cinnamon Ibon	2	E08
Lophozosterops goodfellowi	Black-masked White-eye	2	E08
Basilornis miranda	Apo Myna	2	E08
Bradypterus accentor	Friendly Warbler	2	E10
Urosphena whiteheadi	Bornean Stubtail	2	E10
Arachnothera juliae	Whitehead's Spiderhunter	2	E10
Pericrocotus miniatus	Sunda Minivet	2	E11, E13
Seicercus grammiceps	Sunda Warbler	2	E11, E13
Otus angelinae	Javan Scops-owl	2	E13
Megalaima armillaris	Orange-fronted Barbet	2	E13
Megalaima corvina	Brown-throated Barbet	2	E13
Cochoa azurea	Javan Cochoa	2	E13
Alcippe pyrrhoptera	Javan Fulvetta	2	E13
Crocias albonotatus	Spotted Crocias	2	E13
Garrulax rufifrons	Rufous-fronted Laughingthrush	2	E13
Tesia superciliaris	Javan Tesia	2	E13
Rhipidura euryura	White-bellied Fantail	2	E13
Rhipidura phoenicura	Red-tailed Fantail	2	E13
Aethopyga eximia	White-flanked Sunbird	2	E13
Lophozosterops javanicus	Javan Grey-throated White-eye	2	E13
Scolopax celebensis	Sulawesi Woodcock	2	E21
Ptilinopus fischeri	Red-eared Fruit-dove	2	E21
Cataponera turdoides	Sulawesi Thrush	2	E21
Malia grata	Malia	2	E21
Cyornis hoevelli	Blue-fronted Flycatcher	2	E21
Hylocitrea bonensis	Olive-flanked Whistler	2	E21
Myza sarasinorum	White-eared Myza	2	E21
Bradypterus castaneus	Chestnut-backed Bush-warbler	2	E21, E23, E24
Gymnophaps mada	Long-tailed Mountain-pigeon	2	E23, E24
Zoothera dumasi	Moluccan Thrush	1	E23, E24
Eos semilarvata	Blue-eared Lory	1	E24
Lophozosterops pinaiae	Grey-hooded White-eye	1	E24
Lichmera monticola	Seram Honeyeater	1	E24
Rallina rubra	Chestnut Forest-rail	2	E27, E29, E32
Psittacella modesta	Modest Tiger-parrot	2	E27, E32
Ptiloprora perstriata	Black-backed Honeyeater	2	E27, E32

(continued)

Table 7. *(continued)*

Species Name	English Name	Cloud	EBA
Melidectes foersteri	Huon Melidectes	2	E32
Eurostopodus archboldi	Ribbon-tailed Astrapia	1	E32
Astrapia mayeri	Stephanie's Astrapia	1	E32
Astrapia stephaniae	Loria's Bird-of-paradise	2	E32
Cnemophilus loriae	Crested Bird-of-paradise	2	E32
Cnemophilus macgregorii	Brown Sicklebill	1	E32
Epimachus meyeri	King-of-Saxony Bird-of-paradise	1	E32
Pteridophora alberti	Mountain Nightjar	2	E32
Ptiloprora guisei	Rufous-backed Honeyeater	1	E31
Aegotheles archboldi	Archbold's Owlet-nightjar	2	E31, E32
Androphobus viridis	Papuan Whipbird	1	E31, E32
Acanthiza murina	Papuan Thornbill	2	E32
Colluricincla umbrina	Sooty Shrike-thrush	2	E32
Eulacestoma nigropectus	Wattled Ploughbill	1	E32
Daphoenositta miranda	Black Sitella	2	E32
Melidectes belfordi	Belford's Melidectes	2	E32
Melidectes fuscus	Sooty Melidectes	2	E32
Archboldia sanfordi	Sanford's Bowerbird	2	E32

and F25) where a large proportion of the endemic avifauna is confined to the wet montane *'ohi'a* forests (including cloud forest). The importance of these islands for cloud forest bird species is increased further because eight species belong to monotypic genera with another six polytypic genera being confined to these islands.

A high proportion of threatened species exists in many of the Pacific EBAs, with 15 extant Hawaiian cloud forest species listed as threatened by Collar and Andrew (1988). Indeed, global and local extinctions have been common throughout the Pacific region, with 9 species extinct from the island of Hawai'i (F25) alone, and 4 species extinct and 2 species locally extinct from the Soci-

Table 8. Breakdown of the number of species for each of the six habitat categories for those EBAs in the Pacific islands that have at least one species found in TMCF. The habitat categories are listed in Table 1.

	Habitat Category					
Endemic Bird Area	1	2	3	4	5	6
Micronesian islands (F04)	0	2	0	2	16	0
New Britain and New Ireland (F07)	0	0	7	6	43	0
Solomon Islands (F10)	0	1	3	16	45	1
San Cristobal (F11)	0	1	1	9	23	0
Vanuatu and the Santa Cruz islands (F13)	1	0	3	2	26	1
New Caledonia and the Loyalty islands (F14)	0	0	1	7	24	0
Samoan Islands (F15)	0	1	1	8	11	0
Fijian Islands (F16)	0	2	2	6	26	1
Northwestern Hawaiian Islands (F23)	0	0	1	0	5	0
Hawaiian Islands (F24)	12	4	8	0	3	3
Hawai'i (F25)	3	11	8	0	4	0
Lower Cook Islands (F29)	0	0	1	1	6	1

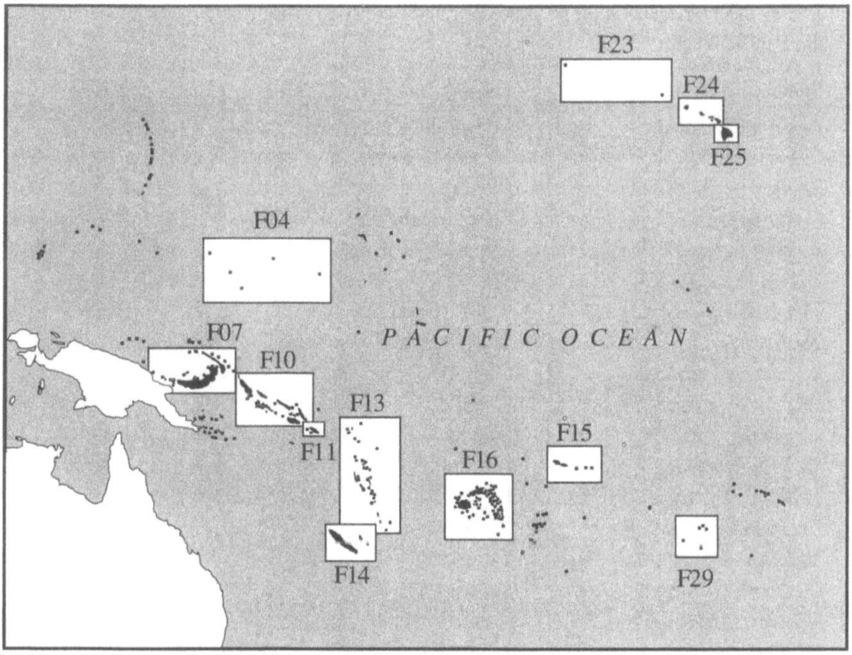

Figure 4. The Endemic Bird Areas in the Pacific islands that have at least one species found in cloud forest.

ety Islands (F27) since the 1700s (Johnson and Stattersfield 1990). This high rate of loss reflects the vulnerability of species on small islands to the agency of introduced predators, browsing animals, competitors, and avian diseases.

EBAs and TMCFs in the Caribbean

Nearly every island in the Caribbean is included within, or forms, an EBA, with a total of five in the region. Each of the EBAs is characterized in part by cloud forest (Table 10), but for only a few of the restricted-range species is this the principal habitat (Table 11). A notable exception, *Dendroica angelae*, is endemic to Puerto Rico. This species was discovered to science only in 1971, 45 years after any other bird discovery in the West Indies. This was doubtless in part due to its restricted habitat requirements which confine it to the elfin forest on mountain ridges and summits.

Important single-island EBAs are Jamaica and Puerto Rico, both supporting high numbers of restricted-range species (34 and 26, respectively) in relation to their size (10,990 km^2 and 8,900 km^2, respectively), but few of them are cloud forest specialists (Table 11). The Eastern Caribbean EBA (A26) incorporates 14 political units extending from the Virgin Islands south to Grenada. Several of the islands include their own single-island endemics, and the EBA (A26) has five species with TMCF as their main habitat (Table 11).

The cloud forests have been less affected by clearing than other habitats in the Caribbean, and consequently there are only two threatened species from this habitat, namely *Amazona imperialis* and *Leucopeza semperi*. *Amazona*

Table 9. Restricted-range species confined mainly to TMCF (categories 1 and 2) in the Pacific Ocean islands

Species Name	English Name	Cloud	EBA
Rukia longirostra	Long-billed White-eye	2	F04
Rukia ruki	Truk White-eye	2	F04
Phylloscopus amoenus	Sombre Leaf-warbler	2	F10
Cettia parens	Shade Warbler	2	F11
Aplonis santovestris	Mountain Starling	1	F13
Zosterops samoensis	Samoan White-eye	2	F15
Charmosyna amabilis	Red-throated Lorikeet	2	F16
Trichocichla rufa	Long-legged Thicketbird	2	F16
Myadestes lanaiensis	Oloma'o	1	F24
Myadestes myadestinus	Kama'o	1	F24
Myadestes palmeri	Puaiohi	1	F24
Moho bishopi	Bishop's 'O'o	1	F24
Moho braccatus	Kauai 'O'o	1	F24
Drepanis funerea	Black Mamo	1	F24
Loxops caeruleirostris	'Akeke'e	1	F24
Melamprosops phaeosoma	Po'ouli	1	F24
Oreomystis bairdi	Kaua'i Creeper	1	F24
Palmeria dolei	Akohekohe	1	F24
Paroreomyza maculata	O'ahu Creeper	1	F24
Viridonia parva	'Anianiau	2	F24
Viridonia stejnegeri	Kaua'i 'Amakihi	2	F24
Hemignathus lucidus	Nukupu'u	2	F24, F25
Loxops coccineus	Akepa	2	F24, F25
Psittirostra psittacea	'O'u	1	F24, F25
Chaetoptila angustipluma	Kioea	2	F25
Moho nobilis	Hawai'i 'O'o	2	F25
Chloridops kona	Kona Grosbeak	2	F25
Ciridops anna	Ula-ai-hawane	2	F25
Drepanis pacifica	Hawai'i Mamo	1	F25
Hemignathus wilsoni	Akia pola'au	2	F25
Loxioides bailleui	Palila	2	F25
Oreomystis mana	Hawai'i Creeper	2	F25
Rhodacanthis palmeri	Greater Koa-finch	2	F25
Viridonia sagittirostris	Greater 'Amakihi	1	F25
Corvus hawaiiensis	Hawaiian Crow	2	F25

imperialis is endemic to the montane rain forests of Dominica and numbered less than 100 in 1992 (Collar et al. 1992). Almost nothing is known about *Leucopeza semperi*, which is or was confined to montane undergrowth on St. Lucia and has eluded almost all recent efforts to find it.

EBAs and TMCFs in Central America

There are 22 EBAs in this region, three on islands and the rest on the mainland. The topography of the region is complex: a series of mountain chains effectively separating the Pacific lowlands from the Atlantic or Caribbean

Table 10. Breakdown of the number of species for each of the six habitat categories for those EBAs in the Caribbean that have at least one species found in TMCF. The habitat categories are listed in Table 1.

	Habitat Category					
Endemic Bird Area	1	2	3	4	5	6
Cuba and the Bahamas (A22)	0	0	2	2	23	1
Jamaica (A23)	0	2	10	5	18	0
Hispaniola (A24)	0	1	5	7	21	0
Puerto Rico (A25)	0	1	11	7	7	0
Eastern Caribbean (A26)	0	5	9	10	14	0

Table 11. Restricted-range species confined mainly to TMCF (categories 1 and 2) in the Caribbean

Species Name	English Name	Cloud	EBA
Nesopsar nigerrimus	Jamaican Blackbird	2	A23
Myadestes genibarbis	Rufous-throated Solitaire	2	A24, A23, A26
Dendroica angelae	Elfin-woods Warbler	2	A25
Amazona imperialis	Imperial Parrot	2	A26
Cyanophaia bicolor	Blue-headed Hummingbird	2	A26
Catharopeza bishop	Whistling Warbler	2	A26
Leucopeza semperi	Semper's Warbler	2	A26

lowlands. EBAs are situated in the lowlands on both of these slopes and in the higher mountainous areas.

Cloud forests are found throughout the mountains of Central America, especially on slopes facing the northerly trade winds (Figure 5). However, TMCFs in the mountains of Mexico to the north of the Isthmus de Tehuantepec are very limited in distribution with pine, pine-oak, and pine-oak-*Liquidamber* forests being the most common vegetation formations. South of the isthmus, however, more extensive cloud forests exist, especially in the Sierra Madre in Chiapas, Mexico, and Guatemala, in the Cordillera de Talamanca of Costa Rica and Panama, and in the Darién highlands of Panama and Colombia.

The mountain ranges of Central America have been split into seven EBAs. Five of these have restricted-range species requiring cloud forest habitat (Table 12). The Costa Rican and Panamanian highlands EBA (A18), with 52 species, has one of the highest species numbers in the world, although it is only 30,000 km^2 in size. Cloud forest is the principal habitat for 10 of the species (Table 13), and most of the other endemics can be found in this habitat.

The montane forests of Central America are relatively well protected with several large biosphere reserves such as El Triunfo in Chiapas, Mexico, and La Amistad in the Cordillera de Talamanca of Costa Rica and Panama. However, cloud forests are being cleared, such as in the Sierra Madre del Sur of Guerrero and Oaxaca EBA of Mexico (A12), which has 10 restricted-range species, including 4 that are threatened through habitat loss, namely *Eupherusa cyanophrys*, *E. poliocerca*, *Lophornis brachylopha*, and *Cyanolyca mirabilis*.

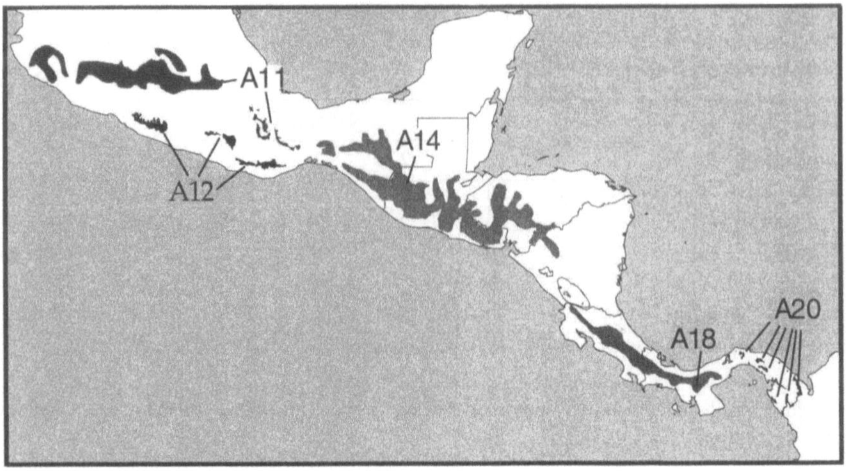

Figure 5. The Endemic Bird Areas in Central America that have at least one species found in cloud forest.

EBAs and TMCFs in South America

South America has more than 3,200 land bird species; its 52 EBAs is higher than any other continental area. Moreover, this continent has more EBAs holding species requiring TMCF habitat than any other continental area. There are 18 EBAs partly composed of TMCF, and 15 of these have species for which cloud forest is the principal habitat (Tables 14 and 15). The Andes hold the majority of the continent's cloud forest EBAs, but there are others in isolated mountains such as the Santa Marta mountains (B08), the Paria peninsula (B03), the Tepuís (B02), and in the Serra do Mar of southeastern Brazil (Figure 6a).

A Focus on the Threatened Cloud Forest Species in Northern South America

BirdLife maintains the Red List for threatened birds. EBAs with concentrations of red-listed, restricted-range species are of critical conservation priority as they represent areas where multiple extinctions are imminent. Collar et al.

Table 12. Breakdown of the number of species for each of the six habitat categories for those EBAs in Central America that have at least one species found in TMCF. The habitat categories are listed in Table 1.

	Habitat Category					
Endemic Bird Area	1	2	3	4	5	6
Central Mexican highlands (A11)	0	0	0	3	5	0
Sierra Madre del Sur (A12)	0	0	3	1	5	0
North Mesoamerican highlands (A14)	1	1	7	4	6	0
Costa Rican and Panamanian highlands (A18)	1	9	18	18	3	0
Darién highlands (A20)	4	1	3	4	1	0

Table 13. Restricted-range species confined mainly to TMCF (categories 1 and 2) in Central America

Species Name	English Name	Cloud	EBA
Oreophasis derbianus	Horned Guan	1	A14
Lampornis sybillae	Green-breasted Mountain-gem	2	A14
Chamaepetes unicolor	Black Guan	2	A18
Odontophorus leucolaemus	Black-breasted Wood-Quail	2	A18
Geotrygon costaricensis	Buff-fronted Quail-Dove	2	A18
Semnornis frantzii	Prong-billed Barbet	2	A18
Thripadectes rufobrunneus	Streak-breasted Treehunter	2	A18
Scytalopus argentifrons	Silvery-fronted Tapaculo	2	A18
Myiodynastes hemichrysus	Golden-bellied Flycatcher	2	A18
Troglodytes ochraceus	Ochraceous Wren	2	A18
Tangara dowii	Spangle-cheeked Tanager	2	A18
Otus clarkii	Bare-shanked Screech Owl	1	A18, A20
Odontophorus dialeucos	Tacarcuna Wood-quail	2	A20
Margarornis bellulus	Beautiful Treehunter	1	A20
Chlorospingus inornatus	Pirre Bush-Tanager	1	A20
Chlorospingus tacarcunae	Tacarcuna Bush-Tanager	1	A20

Table 14. Breakdown of the number of species for each of the six habitat categories for those EBAs in South America that have at least one species found in TMCF. The habitat categories are listed in Table 1.

	Habitat Category					
Endemic Bird Area	1	2	3	4	5	6
Tepuís (B02)	0	15	4	16	4	2
Cordillera de Caripe and Paria peninsula (B03)	5	6	0	1	1	0
North Venezuelan mountains (B04)	1	11	1	2	3	0
Mérida mountains (B06)	3	11	3	3	8	0
Santa Marta mountains (B08)	0	2	3	11	5	0
Eastern Andes of Colombia (B10)	2	6	3	7	15	0
Cauca valley (B12)	1	3	7	7	5	0
Chocó (B14)	2	9	11	18	23	1
Central Andes of Colombia and Ecuador (B17)	0	9	2	6	2	0
Eastern Andes of Ecuador (B18)	0	0	2	5	7	0
Ecuadorian dry forests (B20)	1	0	1	1	52	0
North Peruvian cloud forests (B21)	0	1	2	2	0	0
East Cordilleran ridgetop forests (B24)	1	3	2	1	0	0
East Peruvian cordilleras (B25)	7	9	2	6	0	0
Western Andes of Peru (B27)	0	0	1	2	30	0
Eastern Andes of Peru (B29)	0	0	1	4	5	0
Upper Bolivian yungas (B33)	2	8	4	3	1	0
Southeast Brazilian mountains (B53)	0	2	8	7	6	0
Boliviano-Tucumán *yungas* (B57)	0	0	5	3	1	0

Table 15. Restricted-range species confined mainly to TMCF (categories 1 and 2) in South America

Species Name	English Name	Cloud	EBA
Crypturellus ptaritepui	Tepuí Tinamou	2	B02
Campylopterus hyperythrus	Rufous-breasted Sabrewing	2	B02
Automolus roraimae	White-throated Foliage-gleaner	2	B02
Roraimia adusta	Roraiman Barbtail	2	B02
Myrmothera simplex	Brown-breasted Antpitta	2	B02
Myrmotherula behni	Plain-winged Antwren	2	B02
Lipaugus streptophorus	Rose-collared Piha	2	B02
Pipreola whitelyi	Red-banded Fruiteater	2	B02
Phylloscartes chapmani	Chapman's Tyrannulet	2	B02
Phylloscartes nigrifrons	Black-fronted Tyrannulet	2	B02
Todirostrum russatum	Ruddy Tody-flycatcher	2	B02
Myioborus albifacies	White-faced Redstart	2	B02
Myioborus cardonai	Saffron-breasted Redstart	2	B02
Myioborus castaneocapillus	Tepuí Redstart	2	B02
Nannopsittaca panychlora	Tepuí Parrotlet	2	B02, B03
Campylopterus ensipennis	White-tailed Sabrewing	1	B03
Hylonympha macrocerca	Scissor-tailed Hummingbird	1	B03
Premnoplex tatei	White-throated Barbtail	1	B03
Diglossa venezuelensis	Venezuelan Flower-piercer	2	B03
Basileuterus griseiceps	Grey-headed Warbler	1	B03
Myioborus pariae	Yellow-faced Redstart	1	B03
Syndactyla guttulata	Guttulated Foliage-gleaner	2	B03, B04
Pipreola formosa	Handsome Fruiteater	2	B03, B04
Chlorostilbon alice	Green-tailed Emerald	2	B03, B04, B06
Pharomachrus fulgidus	White-tipped Quetzal	2	B03, B04, B08
Synallaxis castanea	Black-throated Spinetail	2	B04
Grallaricula loricata	Scallop-breasted Antpitta	2	B04
Phylloscartes venezuelanus	Venezuelan Bristle-tyrant	2	B04
Odontophorus columbianus	Venezuelan Wood-quail	2	B04, B06
Pyrrhura hoematotis	Red-eared Parakeet	2	B04, B06
Phylloscartes flaviventris	Rufous-lored Tyrannulet	2	B04, B06
Pauxi pauxi	Helmeted Curassow	1	B04, B06, B10
Grallaria excelsa	Great Antpitta	2	B04, B06, B10
Pyrrhura rhodocephala	Rose-headed Parakeet	2	B06
Heliangelus spencei	Mérida Sunangel	2	B06
Grallaria chthonia	Tachíra Antpitta	1	B06
Grallaria griseonucha	Grey-naped Antpitta	2	B06
Hemispingus reyi	Grey-capped Hemispingus	2	B06
Hemispingus goeringi	Slaty-backed Hemispingus	2	B06
Myioborus albifrons	White-fronted Redstart	2	B06
Hapalopsittaca amazonina	Rusty-faced Parrot	2	B06, B10
Grallaricula cucullata	Hooded Antpitta	1	B06, B12
Odontophorus atrifrons	Black-fronted Wood-quail	2	B08, B10
Odontophorus strophium	Gorgeted Wood-quail	2	B10
Pyrrhura calliptera	Brown-breasted Parakeet	2	B10
Grallaria kaestneri	Cundinamarca Antpitta	1	B10
Coeligena prunellei	Black Inca	2	B10, B12
Grallaria alleni	Moustached Antpitta	2	B12
Aglaiocercus coelestis	Violet-tailed Sylph	1	B14

(continued)

Table 15. *(continued)*

Species Name	English Name	Cloud	EBA
Eriocnemis mirabilis	Colourful Puffleg	2	B14
Haplophaedia lugens	Hoary Puffleg	2	B14
Heliodoxa imperatrix	Empress Brilliant	2	B14
Margarornis stellatus	Fulvous-dotted Treerunner	2	B14
Grallaria flavotincta	Yellow-breasted Antpitta	2	B14
Entomodestes coracinus	Black Solitaire	2	B14
Bangsia aureocincta	Gold-ringed Tanager	1	B14
Bangsia melanochlamys	Black-and-gold Tanager	2	B14
Chlorospingus flavovirens	Yellow-green Bush-tanager	2	B14
Semnornis ramphastinus	Toucan Barbet	2	B14, B12
Hapalopsittaca fuertesi	Azure-winged Parrot	2	B17
Hapalopsittaca pyrrhops	Red-faced Parrot	2	B17
Eriocnemis nigrivestis	Black-breasted Puffleg	2	B17
Metallura baroni	Violet-throated Metaltail	2	B17
Metallura odomae	Neblina Metaltail	2	B17
Grallaria gigantea	Giant Antpitta	2	B17
Grallaria milleri	Brown-banded Antpitta	2	B17
Grallaria rufocinerea	Bicoloured Antpitta	2	B17
Buthraupis wetmorei	Masked Mountain-tanager	2	B17
Pyrrhura orcesi	El Oro Parakeet	1	B20
Penelope barbata	Bearded Guan	2	B21
Xenoglaux loweryi	Long-whiskered Owlet	1	B24
Heliangelus regalis	Royal Sunangel	2	B24
Hemitriccus cinnamomeipectus	Cinnamon-breasted Tody-tyrant	2	B24
Henicorhina leucoptera	Bar-winged Wood-wren	2	B24, B25
Metallura eupogon	Fire-throated Metaltail	2	B25
Metallura theresiae	Coppery Metaltail	2	B25
Aulacorhynchus huallagae	Yellow-browed Toucanet	1	B25
Thripophaga berlepschi	Russet-mantled Softtail	2	B25
Grallaria blakei	Chestnut Antpitta	1	B25
Grallaria carrikeri	Pale-billed Antpitta	1	B25
Grallaria przewalskii	Rusty-tinged Antpitta	1	B25
Scytalopus macropus	Large-footed Tapaculo	1	B25
Anairetes agraphia	Unstreaked Tit-tyrant	2	B25
Buthraupis aureodorsalis	Golden-backed Mountain-tanager	2	B25
Hemispingus rufosuperciliaris	Rufous-browed Hemispingus	1	B25
Nephelornis oneillei	Pardusco	2	B25
Otus marshalli	Cloud-forest Screech-owl	1	B25, B33
Iridosornis jelskii	Golden-collared Tanager	2	B25, B33
Iridosornis reinhardti	Yellow-scarfed Tanager	2	B25, B33
Andigena cucullata	Hooded Mountain-toucan	1	B33
Cranioleuca albiceps	Light-crowned Spinetail	2	B33
Cranioleuca marcapatae	Marcapata Spinetail	2	B33
Grallaria erythrotis	Rufous-faced Antpitta	2	B33
Thryothorus eisenmanni	Inca Wren	2	B33
Hemispingus calophrys	Orange-browed Hemispingus	2	B33
Hemispingus parodii	Parodi's Hemispingus	2	B33
Drymophila genei	Rufous-tailed Antbird	2	B53
Tijuca condita	Grey-winged Cotinga	2	B53

a.

b.

Figure 6. Distribution of restricted-range species using cloud forest in South America. This has been made by mapping all the distributional data for restricted-range species placed in habitat categories 1, 2, and 3. Figure 6a shows those species that are mainly cloud forest dwellers (1 and 2), whereas 6b maps those species in categories 1, 2, and 3. The greater spread of datapoints for 6b reflects the wider variety of habitats in which category 3 species are found.

(1992) recently published an exhaustive compilation of the threatened birds of the Americas, including information on distribution, population, ecology, threats, conservation measures taken, and measures proposed for 327 species.

By looking at those species that are both threatened and principally restricted to TMCF, it is possible to identify cloud forest areas deserving more attention. Figure 7 details the distribution of threatened TMCF species in Venezuela, Colombia, Ecuador, and northern Peru in relation to both the overall cloud forest distribution in these countries and the EBAs. Information on the threatened TMCF species of this region is listed in Table 16, together with the

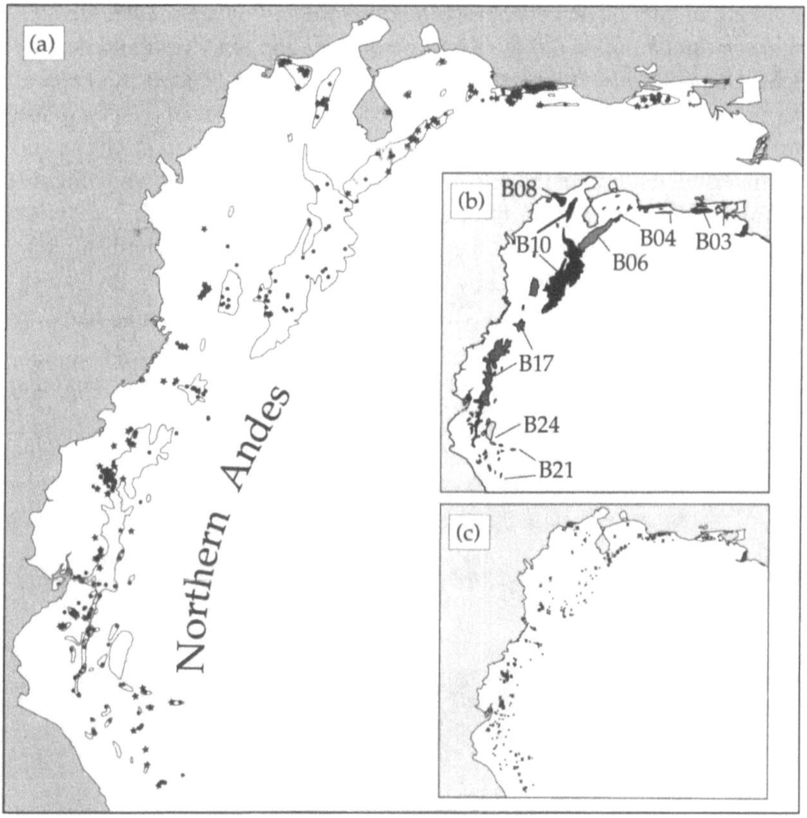

Figure 7. The distribution of cloud forest species (categories 1 and 2) in northern South America, showing those areas where threatened cloud forest species are found (black datapoints) and those where threatened species are not (stars) in relation to the main EBAs (solid line polygons) holding cloud forest species. The figure should be studied in conjunction with Table 16. The inset figures show the same data but separated out from each other, 7b giving the EBAs and their codes (full names of the EBA are given in Table 16), and 7c showing those areas where threatened cloud forest species are found (black datapoints) and those where threatened species are not (x's).

EBAs in which they occur, the types of threat impinging on them, and their conservation status.

Habitat loss was found to be the most important of 17 types of threat in an analysis of the threatened species listed for the Americas (Collar et al. 1992), affecting over 75 percent of the total (Collar 1992). This trend is supported in the northern Andean EBAs with habitat loss being the cause of threat for most of the 22 red-listed species in cloud forest habitat categories 1 and 2 (Table 16). Indeed, 38 (11.6 percent) of all 327 threatened species of the Americas are cloud forest species confined to South America.

The Cordillera de Caripe and Paria peninsula EBA (B03) in northern Venezuela has a significant number of restricted-range species that are also threatened, largely through habitat destruction. It covers only ca. 4,000 km^2, consisting of two disjunct mountain ranges: the low-lying coastal mountains (highest peaks 1,300 m) of the Paria peninsula, and the Cordillera de Caripe (highest peaks 2,600 m) located farther inland to the southwest (Figure 8). This EBA is rich in restricted-range species with a total of 13 (including 5 wholly endemic) that inhabit TMCF. All 5 species that are confined to the area are listed as threatened (Table 16) and are currently known from only a handful of localities within the two mountain ranges. Indeed, *Hylonympha macrocerca* and *Myioborus pariae* are restricted to the Paria peninsula mountains and *Basileuterus griseiceps* is confined to the Cordillera de Caripe. A sixth restricted-range species, *Campylopterus ensipennis*, is also considered threatened, but its range extends to the island of Tobago.

The whole area is a priority for conservation action because of extensive

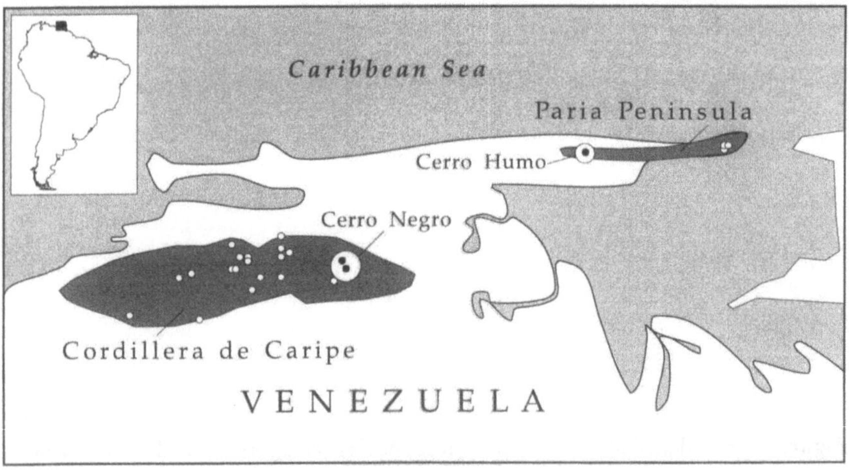

Figure 8. The Paria Peninsula and Cordillera de Caripe EBA (shaded polygons), and the restricted-range species locality records (open circles). At least two sites are needed to conserve all restricted-range species in the EBA, the two most important being Cerro Humo and Cerro Negro.

forest clearing. Conservation is needed at more than one site in the EBA as there is no single locality that holds all the threatened species. Particularly important sites are Cerro Humo in the central Paria peninsula and Cerro Negro in the Cordillera de Caripe. Both are located in protected areas— Cerro Humo in the Paria Peninsula National Park and Cerro Negro in the El Guácharo National Park, but currently these reserves offer little protection and habitat destruction is still proceeding at an alarming rate.

Conclusions

Identifying key sites is essential for the conservation of biological diversity, and it is clear that many of the world's TMCFs are important for restricted-range and threatened bird species. BirdLife is making further additions and adjustments to its databases on biodiversity and threatened species and eventually hopes to store information on all bird species. The application of the Biodiversity Project methodology to the more widespread species would identify new areas and further refine those already highlighted by the project. Nevertheless, the Biodiversity Project and associated threatened species analysis already have laid the foundation for work to begin on improving the conservation status of many key areas.

Acknowledgments. I thank Nigel Collar for extensive comments and ideas on earlier drafts of this paper and Colin Bibby, Melanie Heath, Alison Stattersfield, and David Wege for reviewing earlier drafts. The following experts helped: (1) in the Americas: J. Fjeldså, Peru and Bolivia and especially the Boliviano-Tucuman area; M. Robbins, the Chocó and Ecuador; R. S. Ridgely, Panama and Ecuador; S. N. G. Howell, Sierra Madre del Sur of Mexico; F. R. Lambert, southeast Brazil; D. C. Wege, Tepuís; (2) for Africa: P. Atkinson, São Tomé and Príncipe; D. McNiven, Cameroon mountains; G. A. Allport, Kenyan and Albertine Rift mountains; J. Lovett, Eastern Arc mountains and Western Angola; M. I. Evans, Comoros and Madagascar; R. Safford, Comoros, Madagascar, Réunion, and Mauritius; and (3) for Asia and the Pacific: J. C. Eames, Vietnam and western Himalayas; M. J. Crosby, China and eastern Himalayas; P. A. J. Morris, Western Ghats and Sri Lanka; J. Bowler, Seram; A. J. Stattersfield, New Guinea.

Table 16. Threatened species occurrence in seven EBAs in northern South America for which TMCF is a major habitat

EBAs and Species	Other EBAs	BDP?	Hab. Cat.	Reason for Status	Status
Cordillera de Caripe and Paria Peninsula (B03)					
Campylopterus ensipennis	—	Y	1	habitat loss	Vulnerable/Rare (10)
Hylonympha macrocerca	—	Y	1	habitat loss; tiny range size	Vulnerable/Rare (10)
Premnoplex tatei	—	Y	1	habitat loss	Vulnerable/Rare (10)
Basileuterus griseiceps	—	Y	1	habitat loss	Endangered (2)*
Myioborus pariae	—	Y	1	habitat loss; trade	Endangered (2)
Diglossa venezuelensis	—	Y	2	habitat loss	Indeterminate (7)
North Venezuelan mountains (B04)					
Pauxi pauxi	B06, B10	Y	1	habitat loss; hunting	Endangered (2)
Laterallus levraudi	—	Y	5	pollution and pesticides	Vulnerable (9)
Merida mountains (B06)					
Pauxi pauxi	B04, B10	Y	1	habitat loss; hunting	Endangered (2)
Grallaria chthonia	—	Y	1	tiny range size	Endangered (3)
Grallaricula cucullata	—	Y	1	habitat loss	Vulnerable/Rare (10)
Hemispingus goeringi	B12, B13 B10	Y	2	habitat loss	Vulnerable/Rare (10)
Hapalopsittaca amazonina	—	Y	2	habitat loss	Vulnerable/Rare (10)
Amazilia distans		Y	5	habitat loss?	Vulnerable (9)
Eastern Andes of Colombia (B10)					
Pauxi pauxi	B04, B06	Y	1	habitat loss; hunting	Endangered (2)
Grallaria kaestneri	—	Y	1	?	**
Hapalopsittaca amazonina	B06	Y	2	habitat loss	Vulnerable/Rare (10)
Odontophorus strophium	—	Y	2	habitat loss; hunting	Vulnerable/Rare (10)
Pyrrhura calliptera	—	Y	2	habitat loss; trade; hunting	Vulnerable/Rare (10)
Coeligena prunellei	—	Y	2	habitat loss	Vulnerable/Rare (10)
Pseudodacnis hartlaubi	B12	Y	3	habitat loss	Vulnerable/Rare (10)
Ognorhynchus icterotis	B17	Y	4	habitat loss	Endangered (3)
Clytoctantes alixii	B09	Y	5	habitat loss	Endangered (3)
Podiceps andinus	—	Y	5	habitat loss; pollution; intro. predators	Endangered/Extinct (4)
Rallus semiplumbeus	—	Y	5	habitat loss, pollution; disturbance; hunting	Vulnerable/Rare (10)
Amazilia castaneiventris	—	Y	5	habitat loss	Vulnerable (9)
Cistothorus apolinari	—	Y	5	habitat loss; pollution	Rare (11)
Thryothorus nicefori	—	Y	5	tiny range size	Indeterminate (6)
Touit stictoptera	B13, B18, B25	N	3	habitat loss?	Insufficiently known (12)
Central Andes of Colombia and Ecuador (B17)					
Hapalopsittaca fuertesi	—	Y	2	habitat loss	Endangered (1)
Eriocnemis nigrivestis	—	Y	2	habitat loss	Endangered (1)

Metallura baroni	—	Y	2	habitat loss	Rare (11)
Grallaria gigantea	—	Y	2	habitat loss	Insufficiently known (12)
Grallaria milleri	—	Y	2	habitat loss	Endangered/Extinct (4)
Buthraupis wetmorei	—	Y	2	habitat loss	Vulnerable/Rare (10)
Bolborhynchus ferrugineifrons	—	Y	3	habitat loss	Vulnerable/Rare (10)
Metallura odomae	—	Y	3	habitat loss	Rare (11)
Ognorhynchus icterotis	B10	Y	4	habitat loss	Endangered (3)
Leptosittaca branickii	?	N	3	habitat loss; hunting	Vulnerable/Rare (10)
Agriornis andicola	?	N	5	unknown	Vulnerable/Rare (10)
North Peruvian cloud forests (B21)					
Penelope barbata	—	Y	2	habitat loss; hunting	Vulnerable/Rare (10)
Hapalopsittaca pyrrhops	—	Y	2	habitat loss	Endangered (2)
Leptosittaca branickii	?	N	3	habitat loss; hunting	Vulnerable/Rare (10)
East cordilleran ridgetop forests (B24)					
Xenoglaux loweryi	—	Y	1	tiny range size	Insufficiently known (12)
Heliangelus regalis	—	Y	2	tiny range size	Insufficiently known (12)
Hemitriccus cinnamomeipectus	—	Y	2	tiny range size	Vulnerable (9)
Wetmorethraupis sterrhopteron	—	Y	3	habitat loss	Vulnerable (9)
Herpsilochmus parkeri	—	Y	4	habitat loss	Endangered (2)

For each EBA the threatened species are listed, and for each species, information is given on its occurrence in other EBAs; whether it is a restricted-range species (column BDP? with Y = yes and N = no); its habitat category (codes are explained in Table 1); the reasons for its threatened status; and its IUCN status (International Union for Conservation of Nature and Natural Resources).

The reasons for a species' threatened status were taken from Collar (in prep.). The numbers in parentheses refer to categories of threat published in Collar et al. (1992), which are: (1) situation critical—action urgent; (2) situation serious—action urgent; (3) situation critical—action urgent when population found; (4) situation terminal—action urgent if population extant; (5) situation serious but action in progress; (6) situation unclear—action urgent if taxonomic status confirmed; (7) conflicting evidence—possibly urgent; (8) bird perhaps in need if and when found; (9) bird largely unprotected and needing attention; (10) bird with populations only partly protected; (11) bird largely protected, but for which vigilance is needed; (12) bird for which further protection is desirable.

* *Basileuterus griseiceps* was listed as Indeterminate (7) in Collar et al. (1992), but recent surveys indicate that the species should be upgraded to endangered.

** *Grallaria kaestneri* was described to science after the publication of Collar et al. (1992), but subsequently Collar (1992) has stated that the species is clearly threatened.

References

BirdLife. In prep. A global directory of endemic bird areas (EBAs).

Collar, N. J. 1992. A red data book for the Americas. *World Birdwatch* 14(4): 8–9.

Collar, N. J., and P. Andrew. 1988. *Birds to watch: The ICBP world checklist of threatened birds*. Techn. Publ. 8. Cambridge, UK: International Council for Bird Preservation.

Collar, N. J., L. P. Gonzaga, N. Krabbe, A. Madroño Nieto, L. G. Naranjo, T. A. Parker, and D. C. Wege. 1992. *Threatened birds of the Americas: The ICBP red data book*. 3d ed. (part 2). Cambridge, UK: International Council for Bird Preservation.

Collar, N. J., and S. N. Stuart. 1985. *Threatened birds of Africa and related islands: The ICBP/IUCN red data book*. 3d ed. (part 1). Cambridge, UK: International Council for Bird Preservation and International Union for the Conservation of Nature.

Collar, N. J., and S. N. Stuart. 1988. *Key forests for threatened birds in Africa*. Monogr. 3. Cambridge, UK: International Council for Bird Preservation.

Collar, N. J., D.C. Wege, and A.J. Long, In prep. Patterns and causes in the endangerment of the New World avifauna.

Crosby, M. J. 1994. Mapping the distributions of restricted-range birds to identify global conservation priorities. In R. I. Miller, ed., *Mapping the diversity of nature*. London, UK: Chapman & Hall.

Diamond, J. 1984. Biogeographic mosaics in the Pacific. In F. J. Radovsky, P. H. Raven, and S. H. Sohmer, eds., *Biogeography of the tropical Pacific*, pp. 1–14. Spec. Publ. 72. Honolulu: Bishop Museum.

Dowsett, R. J. 1986. Origins of the high altitude avifauna of tropical Africa. In F. Vuilleumier and M. Monasterio, eds., *High altitude tropical biogeography*, pp. 557–585. Oxford: Oxford University Press.

Gonzales, P. C., and R. S. Kennedy. 1990. A new species of *Stachyris* babbler (Aves: Timaliidae) from the island of Panay, Philippines. *Wilson Bull.* 102(3): 367–379.

Green, G. M., and R. W. Sussman. 1990. Deforestation history of the eastern rain forests of Madagascar from satellite images. *Science* 248:212–215.

ICBP. 1992. *Putting biodiversity on the map: Priority areas for global conservation*. Cambridge, UK: International Council for Bird Preservation.

Johnson, T. H., and A. J. Stattersfield. 1990. A global review of island endemic birds. *Ibis* 132:167–180.

Louette, M., and J. Stevens. 1992. Conserving the endemic birds on the Comoro Islands; I: General considerations on survival prospects. *Bird Conserv. Internatn.* 2:61–80.

Stadtmüller, T. 1987. Los bosques nublados en el trópico húmedo: Una revision bibliográfica. San José, Costa Rica: Universidad de las Naciones Unidas.

Terborgh, J., and B. Winter. 1983. A method for siting parks and reserves with special reference to Colombia and Ecuador. *Biol. Conserv.* 27:45–58.

Thirgood, S. J., and M. F. Heath. 1994. Global patterns of endemism and the conservation of biodiversity. In P.L. Forey, C. J. Humphries, and R. I. Vane-Wright, eds., *Systematics and conservation evaluation*. Oxford: Clarendon Press.

White, C. M. N., and M. D. Bruce. 1986. *The checklist of the birds of Wallacea (Sulawesi, the Moluccas and Lesser Sundas Islands, Indonesia): An annotated checklist*. Checklist no. 7. London: British Ornithologists' Union.

5. Ecology and Conservation of the Argentine Montane Forest

Alejandro Diego Brown

In Argentina, the montane subtropical cloud forest is located in the extreme north of the country and is within the Atlantic forest in Misiones Province (in the northeast) and the *yungas* or Tucumano-Bolivian forest in the northwest of the country and in southern Bolivia (Departments of Chuquisaca and Tarija).

The Argentine northwest represents the southernmost extension of the Central Andes (Brush 1982) with heights ranging from 200 m asl (Chaco Plain) to 6,723 m asl (Cerro Llullaillaco), generating a vertical zonation of climatic conditions incorporating a great variation in temperature and rainfall (Figure 1).

The montane moist forests or *yungas* occur between 300 and 2,500 m asl, and include three main vegetation units: subtropical forest (premontane and montane), cloud forest, and foggy grassland (Cabrera 1976). These units develop on the mountain range called the Sierras Subandinas (Feruglio 1946), as well as on the eastern Andean Cordillera slopes.

Throughout northwest Argentina, 80 percent of the rainfall occurs in summer, in contrast with the dry winter-spring season. A relationship exists between the height of the mountain to be overcome by humid winds coming from the west and the distribution and amount of rainfall to windward and leeward. If the mountain is below 1,000 m, maximum rainfall occurs leeward of the mountain top, given the tendency of the wind to keep rising (Sierra de Lumbrera). On mountains around 2,000 m, rainfall occurs mainly windward, although a rainfall increase leeward is also recorded (Sierra de Gonzalez). In

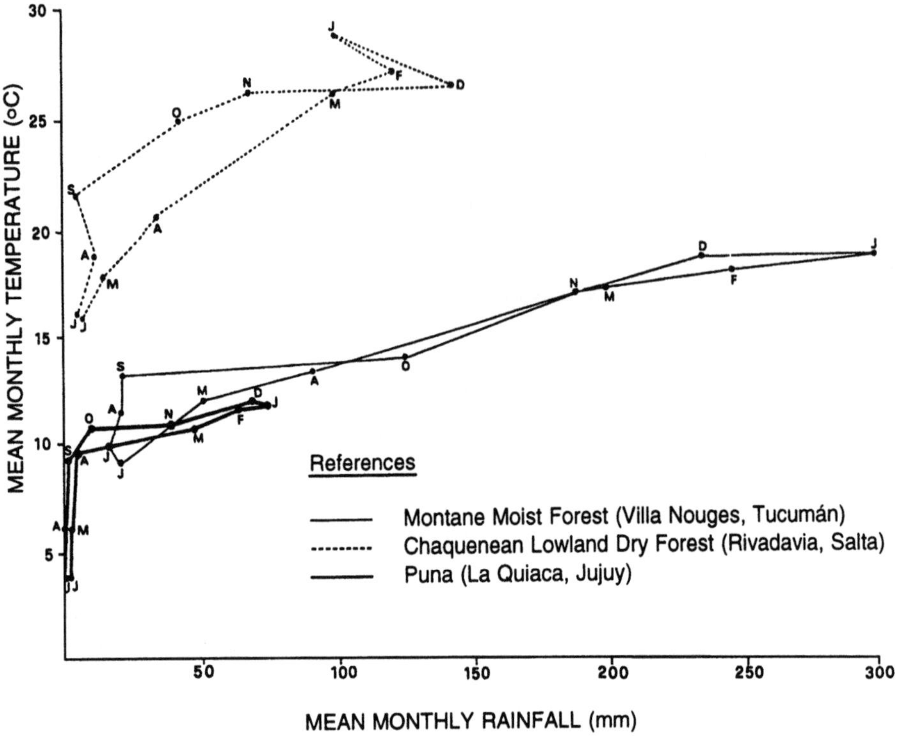

Figure 1. Climatic diagram of towns located at different altitudinal levels in the Argentine northwest: Villa Nougués, Tucumán (1,338 m asl, 1,500 mm annual rainfall); Rivadavia, Salta (205 m asl, 592 mm); and La Quiaca, Jujuy (3,458 m asl, 313 mm).

mountain areas above 3,000 m, rainfall increase occurs only windward with the mountain ranges acting as barriers (Sierra de Calilegua); nevertheless, precipitation to leeward is still significant. For mountains above 5,000 m asl (Sierras de Santa Victoria, Zenta, and Aconquija), heavy rainfall occurs on the slopes facing the winds. On leeward slopes, there is a significant "rain shadow" effect and annual precipitation does not exceed 200 mm/yr (Bianchi 1981).

In a succession of intermontane valleys and hills varying in height, the landscape becomes more complex, with humid slopes forming windward, and dry slopes leeward. The *yungas* develop in the altitudinal level (300 to 2,500 m) where the heaviest orographic rainfall occurs (1,000 to 3,000 mm/yr).

The objective of this paper is to synthesize existing information about the geographic distribution of the Argentine *yungas*, its principal floristic characteristics, the human impacts on this forest type, and to propose areas of conservation priority.

Geographic Distribution and Floristic Composition of the Argentine *Yungas*

Subtropical Forest

Both the premontane and montane subtropical forest in the Argentine northwest can be subdivided into three main geographical areas: a northern region with greater floristic richness, and a central and southern region. The division between the northern and central regions runs through the Chaco forest of the San Francisco and Bermejo river valleys, thus presenting a natural discontinuity (Figure 2). The central region includes two sectors at around 24° S latitude encompassing the Sierras de Centinela, Santa Barbara, and Lumbrera; and the southern region includes a hilly area in southern Salta and Tucumán (Brown and Ramadori 1989).

Premontane Subtropical Forest

This forest type includes two main communities. One is the "Selva de Tipa y Pacará" (*Tipuana tipu* and *Enterolobium contortisiliquum* are codominant Leguminosae species) of the lower and drier areas in the *yungas* of Tucumán and southern Salta (Cabrera 1976). These forests have been largely replaced by sugarcane plantations. The other forest community is called "Selva de Palo Blanco y Palo Amarillo" (*Callycophyllum multiflorum* [Rubiaceae] and *Phyllostyllon rhamnoides* [Ulmaceae] are the dominant tree species), and it occurs in the lower altitudinal levels of the central and northern sectors. Epiphytism is less frequent than in montane forest, and all the species are drought tolerant.

Montane Subtropical Forest

The northern region of the montane subtropical forest encompasses the northernmost areas in Argentina, and it extends from the upper course of the Bermejo River in Bolivia (22° S) down to the Zapla hills in the neighborhood of Jujuy City (24° S). Characteristic tree species in this sector are *Amburana cearensis, Lonchocarpus lilloi* (Leguminosae), *Tabebuia lapacho* (Bignoniaceae), *Nectandra pichurim* (Lauraceae), *Athyanna weinmanifolia* (Sapindaceae), *Ficus maroma* (Moraceae), *Cedrella angustifolia* (Meliacea), *Croton densiflorus* (Euphorbiaceae), and *Miconia molybdea* (Melastomataceae). Epiphytes are abundant, principally bromeliads (*Vriesia icterica*, an endemic species), orchids (*Epidendron* and *Isochilus*), and ferns. The understory is dominated by ferns (*Asplenium tucumanensis, Bolbitis serraticifolia, Adiantum pectinatum, Hypolepis repens*), heliconias, and bamboos. Patches of tree ferns (*Nephelea*) develop at heights between 1,000 and 1,200 m.

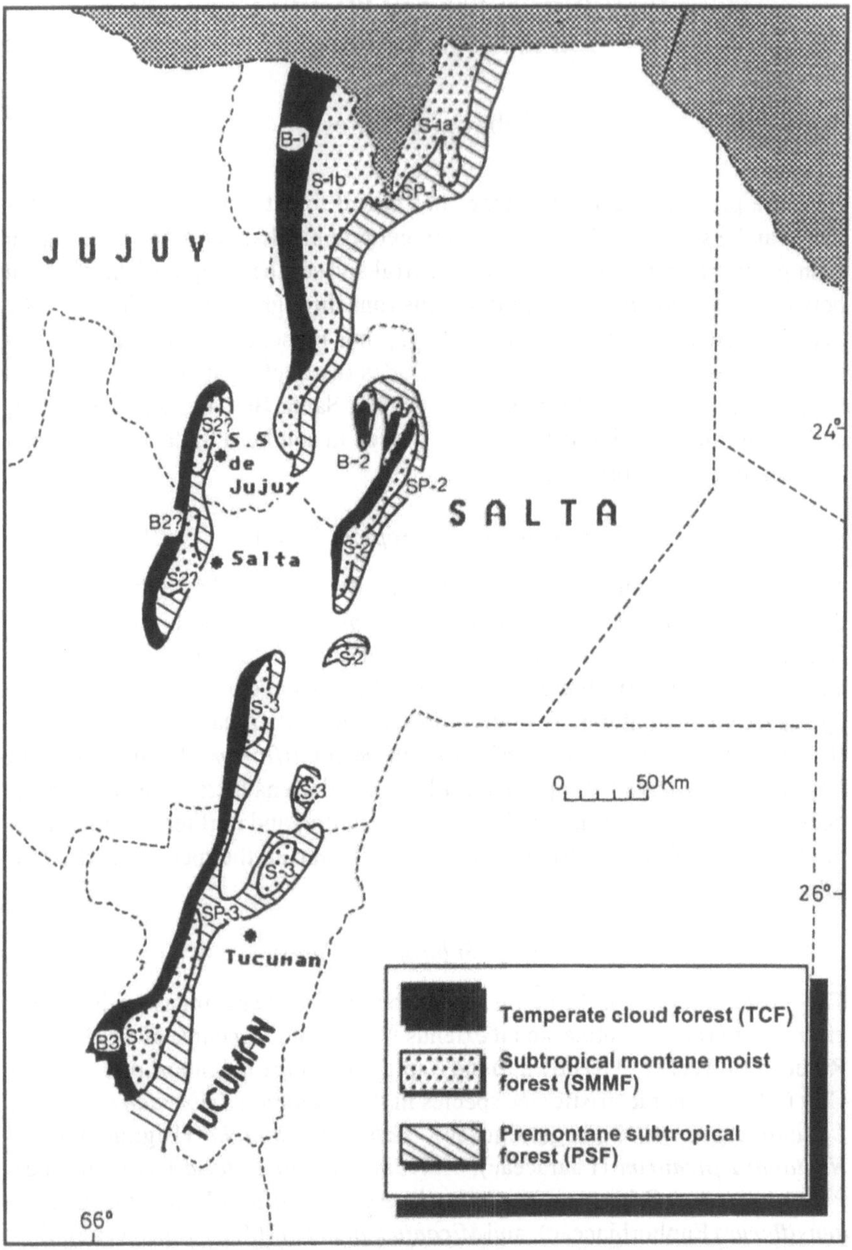

Figure 2. *Yungas* distribution in the Argentine northwest. B–1: TCF northern sector; S–1: SMMF northern sector; SP–1: PSF northern sector. In the remaining symbols, "2" represents the central sector and "3" the southern sector.

The central region of the montane subtropical forest is closely related to the forest in the southern region, its main feature being the presence of species that have the southern border distribution (e.g., *Trichilia hyeronimii, Ocotea puberula, Myroxylon peruiferum*, and *Patagonula americana*). The percentage of trees with vascular epiphytes ranges between 50 and 80 percent (Brown 1990) with a mean of 25 tree species per hectare (Brown, Chalukian, and Malmierca 1985).

A feature of the montane forest both in the central and the southern regions is its subdivision into two altitudinal levels: the Basal Forest between 600 and 900 m, and the Mirtaceas Forest between 1,000 and 1,500 m (Meyer 1963; Hueck 1978; Brown, Chalukian, and Malmierca 1985). The last level constitutes the lower parts of the true cloud forest, and all the trees are covered by moss and epiphytic ferns (*Polypodium, Asplenium, Elaphoglossum, Campyloneurum*, and *Hymenophyllum*). This central region constitutes an independent unit from other northwest montane forest because it is separated by xerophytic forest all along its perimeter (Figure 2). Floral and physiognomic composition of the southern region montane forest occurring in the hills of Tucumán and southern Salta is very similar to the central region of the montane forest.

Cloud Forest

Cloud forest is located altitudinally above the subtropical forest, from 1,500 m asl up to 1,900 m at El Rey National Park (Sierra de Gonzalez); and to 2,500 m in Tucumán Province and the eastern Cordillera (Sierras de Santa Victoria y Zenta in Salta and Jujuy provinces). Two characteristic forest communities can be found here: Pino del Cerro mixed forest (*Podocarpus parlatorei* dominates) and Aliso (monospecific forest of *Alnus acuminata*).

The latitudinal distribution of these forests is discontinuous, and the same division as in the subtropical forest applies, although floristic composition is relatively uniform latitudinally (Brown and Ramadori 1989). In the *Podocarpus* mixed forest, other important genera include *Juglans, Cedrella, Oreopanax, Prunus, Blepharocalyx, Ilex, Viburnum, Berberis, Escallonia, Clusia*, and *Sambucus*. Epiphytes are very important, principally mosses and ferns. The understory is open and dominated by ferns (*Dryopteris paralelogramma, Blechnum penna-marina, Asplenium lorentzii, Polystichum montevidense*, and *Asplenium tucumanensis*). Bamboo (*Chusquea*) is also common.

These forests represent the highest elevation tree communities and grade upward into foggy grassland. The forests' upper limits appear to be dynamic through time and strongly affected by human activity, mainly forest clearing and burning to widen and maintain pasturelands. At present, along the ecotone boundary with the grassland, there is a mosaic of features probably representing different successional stages of forest development. *Alnus* forest represents a pioneer stage among specific communities, which later may be invaded by *Podocarpus*. As it develops further, *Alnus* may be replaced by subtropical montane forest species that give rise to a mixed forest. *Podocarpus parlatorei* then becomes the dominant or codominant species with differentially aged individual trees that may vary by more than 1 m of dbh (diameter at breast height).

Foggy Grassland

At the highest elevations is an ecozone known as the foggy grassland. Cabrera (1976) has referred to this as the *praderas montañas* (montane prairies) within the *yungas* phytogeographic province of the montane forest. These grasslands are located between 2,500 m and the mountain summits at 3,000 m. They are observed to have large *Puya* sp. colonies that resemble, but at a smaller scale, the colonies of *Espelletia* and *Puya* of the Perú, Ecuador, and Venezuela *paramo* (plateau). They have very little in common with the "puya" vegetation.

Human Activities in the Present

Based on human activity, the *yungas* can be subdivided into three main zones: intensive agriculture, forestry exploitation, and grazing lands.

Intensive Agriculture

Intensive agriculture (sugarcane, banana, citrus trees, and tobacco) has replaced much of the premontane subtropical forest. Other agricultural activities operating at a smaller scale include production of crops such as mango, avocado pear, vegetables, as well as small timber plantations (*Pinus, Eucalyptus*). Until their recent conversion to agriculture, these forests were exploited for valuable timber species including *Cedrella, Juglans, Amburana,* and *Cordia*, as well as others having a lesser value (*Callycophyllum, Tabebuia, Phyllostyllon, Chlorophora, Piptadenia*). As agricultural activities have expanded, extensive irrigation systems have been developed to support crops during the dry season (May–October). These irrigation works depend exclusively on perennial streams flowing from the *yungas*. Large human settlements have also been established in this area and population exceeds two million permanent inhabitants, plus an additional migrant labor force employed in seasonal agricultural activities (e.g., sugar harvesting). Premontane subtropical forests represent about 25 percent of the *yungas* and constitute an environmental unit clearly different from the rest of the northwest subtropical forest. They are not yet represented in formal protected areas, even though they are important for irrigation water.

Forestry Exploitation

Forestry exploitation in the montane subtropical forest is based almost exclusively on *Cedrella* (two species). In more accessible areas other species such as *Tipuana, Piptadenia, Myroxylon,* and *Fagara,* and even some Lauraceae are included. No significant human settlements occupy this altitudinal strip, although the montane forest is seasonally used (in winter) for grazing by people who live at higher elevations. Small citrus, maize, and coffee plantations (usually of less than 1 ha) occur sporadically. Some petroleum exploitation also occurs within this zone.

The montane forests constitute the most widely protected type (ca. 200,000 ha). They are represented in three national parks—El Rey (44,000 ha), Calilegua (76,000 ha), and Baritú (72,000 ha)—and in one reserve—the Parque Biológico Sierra de San Javier (17,000 ha) of the University of Tucumán.

Grazing Lands

Extensive grazing lands have been developed within the cloud forest (sometimes converted into anthropogenic grasslands) and in the foggy grasslands, in a wide altitudinal range from the upper limit of the continuous forest line (1,500–2,000 m) up to the vegetation limit (4,000 m). The main human activities are cattle ranching and small-scale shifting agriculture on steep slopes (between 40 and 80 percent). Crops are diverse and include maize, potato, peanut, *achoscha* (*Cyclantera* sp., Cucurbitaceae), and *yacón* (*Polymnia* sp., Compositae). There is significant human settlement in this ecological zone. Overgrazing is the main environmental problem. Even though existing protected areas include samples of this environment, they are too small and isolated to provide any substantive, long-term conservation value.

Conservation Action Plan

An important requisite for the long-term conservation of these montane forests is to reduce further forest fragmentation and to keep the connections between the existing reserves. Also needed is the development of new reserves in key biological areas. In order to achieve this latter objective, two correlated criteria should be considered: biodiversity and climatic stability. In this way, and because their state of conservation depends on the human groups whose activities are potentially destructive, we have selected and defined as a priority conservation area that sector found between the Baritú and Calilegua national parks (Figure 3). This area covers about 700,000 ha (about 200,000 ha of cloud forest). The land is at present mostly under private tenancy and is subject to exploitation for wood and to agricultural pressure by local farmers who lease small plots. Often squatter use (rather than formal land rental) takes place, causing conflicts between landowners, who are afraid that their prop-

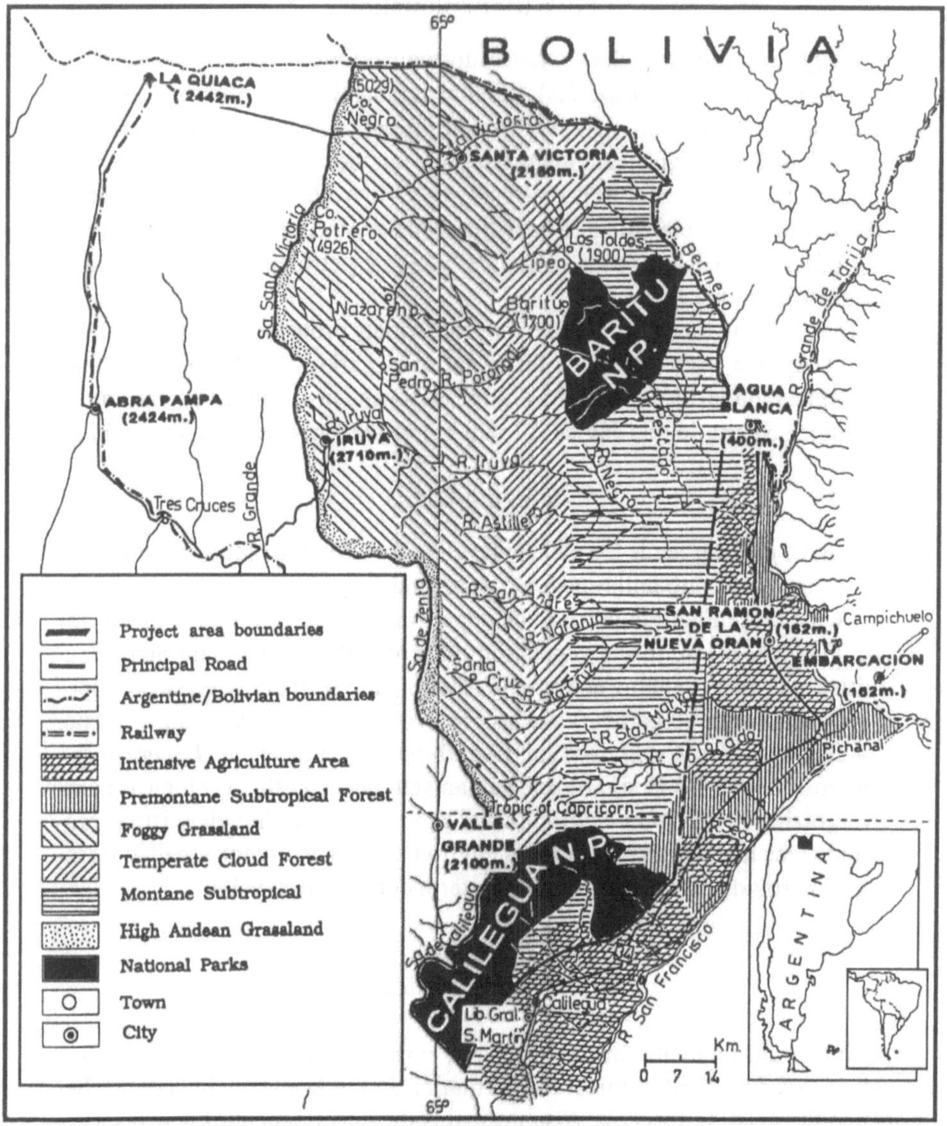

Figure 3. Priority conservation area of the montane forest in northwest Argentina.

erty may deteriorate, and the expanding landless population. On a positive note, this northern Argentine montane forest continues into Bolivia, occupying many hundred thousands of hectares (about 250,000 ha is included in the Tariquia Forestry Reserve). This points to the importance of transnational conservation and development efforts, incorporating forest management concepts into the activities of small timber producers and integrating the forest with cattle and agricultural production via agroforestry and silvipastoral systems.

Acknowledgments. This research was supported with funds from the Consejo Nacional de Investigaciones Científicas y Tecnológicas (CONICET), the World Wildlife Fund (WWF-US), and the Universidad Nacional de Tucumán. My symposium participation was supported by the East-West Center (Hawai'i).

References

Bianchi, A. 1981. *Las precipitaciones en el noroeste argentino.* INTA. 388 pp.

Brown, A. D. 1990. El epifitismo en las Selvas Montanas del Parque Nacional El Rey, Argentina. *Rev. Biologia Tropical* 38(2A): 155–166.

Brown, A. D., S. C. Chalukian, and L. Malmierca. 1985. Estudio florístico-estructural de un sector de Selva Semidecidua del noroeste argentino.I.Composición florística, densidad y diversidad. *Darwiniana* 26:27–41.

Brown, A.D., and E. D. Ramadori. 1989. Patrón de distribución, diversidad y características ecológicas de especies arbóreas de las selvas y bosques montanos del noroeste de la Argentina. In *Anales VI Congreso Forestal Argentino*, pp. 177–181.

Brush, S. 1982. The natural and human environment of the Central Andes. *Mountain Research and Development* 2:19–38.

Cabrera, A. 1976. Regiones fitogeográficas argentinas. *Enciclopedia Agricultura y Jardineria.* 85 pp.

Feruglio, E. 1946. *Los sistemas orográficos de la Argentina.* Soc. Estudios Geográficos de la Rep. Arg. Ed Coni, Bs. As. 542 pp.

Hueck, K. 1978. *Los Bosques de Sudamerica.* Sociedad Alemana de Cooperación Tecnica. 476 pp.

Meyer, T. 1963. Estudios sobre la selva tucumana. *Opera Lilloana* 19:1–140.

6. The Role of GIS in Evaluating Contour-Based Limits of Cloud Forest Reserves in Honduras

Richard Campanella

In August 1987, the National Congress of Honduras designated most of the country's tropical montane cloud forest (TMCF) as protected (Congreso Nacional 1987). Decree 87-87, which created the 37 reserves (Figure 1), set forth strict conservationist measures for each unit and stipulated that the core zone limits follow the 1,800 m, 2,000 m, or 2,100 m contour, depending on further studies. Five years later, the issue of delimitation is still unsettled for most of the reserves, due largely to the dearth of recent forest coverage data and the disagreements among planners concerning the "how" and "where" of delimitation. This study examines the role of a geographic information system (GIS) in evaluating the methodology and placement of limits; specifically, it presents a GIS-based technique that evaluates the viability of contour-based limits for Honduras's cloud forest reserves, using the Guisayote Biological Reserve as a study area.

Contours as Limits

The TMCF of Honduras is a moist, verdant broadleaf jungle usually occurring from 1,800 m asl up to the summit of the mountain (Agudelo C. 1987; Congreso Nacional 1987). In its virgin state, cloud forest is probably best

---

Reprinted, with permission, from the *1993 ACSM/ASPRS Annual Convention and Exposition Technical Papers,* vol. 1, copyright 1993, by the American Society for Photogrammetry and Remote Sensing: "The Role of GIS in Evaluating Contour-Based Limits of Cloud Forest Reserves in Honduras," Richard Campanella, pp. 61–69.

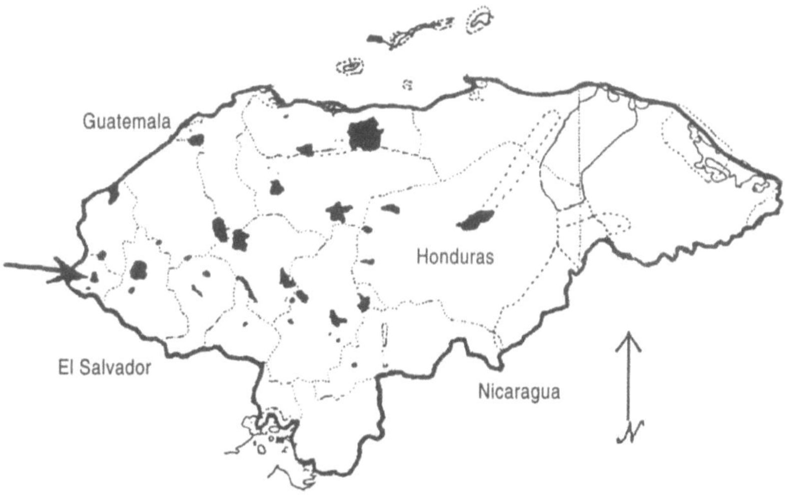

Figure 1. Honduras, Central America. The shaded areas indicate locations of most of the reserves, refuges, and parks created by the Decree 87-87. The Guisayote Biological Reserve is indicated by the arrow. Honduras is about 400 miles wide, east to west. *Asociacion Hondurena de Ecologia, 1992.*

delimited by contour, as cloud forest is altitudinally determined and contours are a measure of altitude. However, most reserves have experienced significant incursions of land conversion (deforestation) far into the forest, some of which penetrate the forest in a parallel (to the contours) fashion and some in a perpendicular fashion. Areas where land conversion has occurred parallel to the contours of the cloud forest are still conducive to the contour-based delimitation technique stipulated by law, but those areas where incursion has penetrated the forest perpendicularly may or may not be suitable to delimitation by contour.

Many factors should and will influence the final decision as to the placement of the reserve limits, but it is argued that the decision-making process must be predicated on a clear understanding of the geography of each reserve's cloud forest and incursions, to determine *whether*, and if so, *which* contours would best serve as limits. The GIS-based technique presented here evaluates the viability of contour-based limits for the Guisayote Biological Reserve, a 40 km^2 mountain ridge straddling the continental divide in extreme western Honduras (see arrow Figure 1). The technique introduced here is based on the author's experiences as the first planner and promoter assigned to this reserve from 1988 to 1990. Although the data pertain only to Guisayote, the technique is applicable to any cloud forest reserve, or to any situation that involves the delimitation of an altitudinally determined resource.

Collection of Data

This study utilizes four panchromatic 9" by 9" aerial photos taken by the Bolivian Air Force under contract with the Honduran military on the morning of 23 January 1992. The photos were taken with a 6-inch lens at 20,000 ft, recording at a 1:40,000 scale. The 1:50,000 scale Ocotepeque and Conception quadrangles of the Inter-America Geological Survey (IAGS) map series were used as base maps and were photographically enlarged to match the scale of the aerial photos and printed on transparency. This process eliminated the need for a stereo zoom transfer scope and facilitated the interpretation of the photos and is highly recommended for microlevel photo interpretation of rugged terrain.

Features on the photos were classified as *cloud forest* or *not cloud forest*, with cloud forest defined as forest occurring above 1,800 m in a geographically contiguous area. This definition, which groups together second-growth and virgin forest as well as broadleaf and needleleaf trees, is satisfactory because this study distinguishes those lands which belong in the nature reserve (forested areas) from those which do not belong (deforested areas); it does not purport to distinguish forest composition. The definition is limited to the area above 1,800 m to conform to the stipulations of Decree 87-87, although it is apparent from the aerial photos that the cloud forest ecosystem occurs as low as 1,600 m in some places.

Role of GIS

Because the technique introduced below is intended for continued application to the other cloud forest reserves in Honduras, the selection of a GIS in this project is as important as its role in the analysis. PC Arc/Info software was chosen for three reasons: (1) it runs on DOS-based microcomputers, which are scarce but available in agency offices in Honduras; (2) it is a powerful and well-documented software; and (3) it was recently acquired by CAFOR, a German development organization operating out of the National Forest Science School in Siguatepeque, Honduras. GIS projects such as this, designed for use in developing countries, must make great efforts to conform to the level of technology available in the host country.

In the arc digitizing system (ADS) environment of PC Arc/Info, the 100 m contours above 1,800 m on the base maps were digitized and divided into primary watersheds and secondary drainage basins. Each polygon was labeled according to its location and cleaned under Arcedit environment and the Clean command. Topology was established under the Build command, and area and perimeter were determined for each polygon in the Tables environment. Through the Transform command, the data were converted to ground measurements based on the UTM coordinate system (Figure 2). This process

Figure 2. **A:** Digitized map of the Guisayote Biological Reserve. Outermost lines are 1800 m contours; innermost lines form the 2,300 m contours. Lines were digitized through the ADS environment of PC Arc/Info. **B:** Outline of cloud forest in the Guisayote area, as derived from the aerial photographs. Outline was digitized through ADS using the same ticks and scale of map A. **C:** Intersected contours and forest outline. Areal extent in meters was determined for each contour interval within each drainage basin. Note that only those cloud forest polygons that landed in contours above 1,800 m were analyzed, to conform to the stipulations of the Decree 87-87. The intersection was performed with the Intersect command in PC Arc/Info.

was repeated for the cloud forest coverage map derived through the aerial photographs. Through the Intersect command, the cloud forest map was overlaid onto the contour polygons, a simple process because the base maps and the photo-derived forest maps are the same scale (1:40,000). The Tables environment and the Frequency command were used to determine the areal extent of cloud forest within each contour polygon (ESRI 1990).

These measurements would probably be impossible without the use of a GIS. Many of the contour polygons and forest coverage polygons are too small and irregular for a planimeter and results would be extremely tedious to produce, very inaccurate, and impossible to update and revise. Use of a GIS permits accurate measurements, the incorporation of additional data layers (such as land tenure, agriculture, settlement patterns, and other issues relevant to delimitation), and the important capability of revision.

Methodology

The technique to determine the viability of contour-based limits involves the measurement of *forest quantity*, which generally decreases with altitude, and *forest quality*, which generally increases with altitude. Forest quantity is defined as the total areal extent of cloud forest in three levels of analysis: (1) each of the nine drainage basins, which pertain to (2) two major watersheds, which belong to (3) the Guisayote Biological Reserve as a whole.[1] Percent of forest quantity is calculated as the areal extent of cloud forest above n^{th} contour divided by the total cloud forest quantity in each area at the three levels of analysis, multiplied by 100.

Forest quality refers to the *intactness* of the TMCF and is calculated as the areal extent of cloud forest above n^{th} contour divided by the total land area above n^{th} contour, multiplied by 100. A high percent indicates a high-quality, intact forest above n^{th} contour; a low percent signifies a low-quality, dispersed[2] forest above n^{th} contour.

Values for forest quantity and forest quality calculated for each level of analysis are then organized and graphed for each 100 m contour interval between 1,800 m and the summit, at about 2,325 m (Figure 3).

Analysis

The data derived from the aerial photos, processed through the GIS, and organized in the graphs now can be analyzed to evaluate the viability of contour-based limits for each basin, watershed, and the Guisayote Biological Reserve as a whole. To do this, a *threshold contour* will be determined for each area.

Threshold contour is the name given to the contour (x axis on the graphs) that corresponds to the point of intersection at which the forest quantity line is equal to the forest quality line. Depending on the magnitude of the percent (y axis), this intersection will indicate whether contours are an appropriate delimitation technique, and if so, which contour is optimal for use as a nature reserve limit, given the scope of the study. The stronger the magnitude of the percent (y axis), the more conducive is the area to contour-based limits. For example, the Rio Idolo (Basin 2 in the graphs) has a rating of 84 percent at the 1,900 m contour interval, signifying an area that, if delimited as a reserve at 1,900 m, would incorporate 84 percent of the basin's total cloud forest and would produce a reserve that is 84 percent covered with forest (and 16 percent covered with pasture, croplands, and other competing land uses). Relative to the rest of the mountain, this is considered a strong rating, and confirms the viability of a contour-based limit (at 1,900 m) for this particular basin. Rio Blanco (Basin 3), with a 43 percent rating at 2,000 m, is probably not conducive to contour-based limits, as both forest quantity and quality are relatively low at the contour where they are both maximized. Planners should choose an alterative delimitation method here.

The cutoff percent for contour-conduciveness is an arbitrary figure that should be set by the Honduran planners and agency personnel involved in the management of these reserves. The stratification given here is preferred by

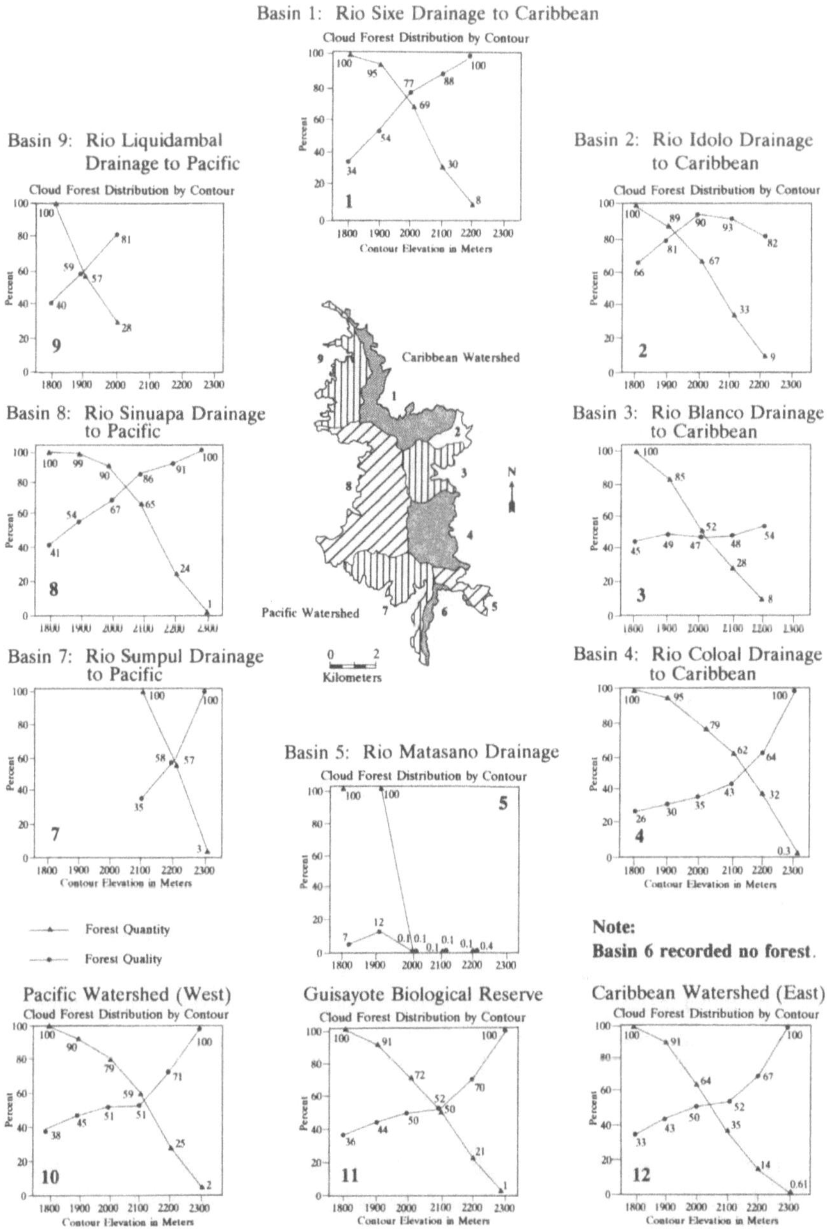

Figure 3. Drainage basins and watersheds in the Guisayote Biological Reserve. Basins 1 to 5 form the Caribbean watershed; Basins 6 to 9 form the Pacific watershed. Graphs 1–12 track the decreasing forest quantity and increasing forest quality as a hypothetical reserve limit is moved upslope in each basin (Graphs 1–9), watershed (Graphs 10 and 12), and the reserve as a whole (Graph 11). Interpretation is as follows (using Graph 4/Basin 4 as an example): If a reserve is delimited at the 1,800 m contour (*x*-axis), it will include 100% of the basin's cloud forest, but the resulting reserve will be only 26% covered by forest (*y*-axis), and therefore 74% occupied by pastures, crops, and other uses. If a reserve is delimited at 2,000 m, it will include less forest—79% of the total forest—but it will be more intact, at 35%. See Analysis for further explanation.

the author from personal experiences and impressions but is not necessarily the best classification.

Rating (%)	Analysis
0–50	Not conducive to contour-based limits
50–60	Marginally conducive to contour-based limits
60–100	Conducive to contour-based limits

The following conclusions are based on these criteria.

1. The Rio Sixe Drainage (Basin 1) has a rating of 73 percent at a threshold contour of 1,900 m. It is **conducive** to a contour-based limit, at the 1,900 m level.
2. The Rio Idolo Drainage (Basin 2) has a rating of 84 percent at a threshold contour of 1,900 m. It is **conducive** to a contour-based limit, at 1,900 m.
3. The Rio Blanco Drainage (Basin 3) has a rating of 48 percent at a threshold contour of 2,000 m. It is **not conducive** to a contour-based limit.
4. The Rio Coloal Drainage (Basin 4) has a rating of 50 percent at a threshold contour of 2,100 m. It is **marginally conducive** to a contour-based limit, at 2,100 m.
5. The Rio Matasano Drainage (Basin 5) has a rating of 0 percent at a threshold contour of 2,000 m. It is **not conducive** to a contour-based limit. This unusual distribution indicates the presence of an island of forest that resides within one contour interval, surrounded by other land uses—a situation clearly incompatible with contour-based delimitation, as confirmed by the rating.
6. The Rio Jaralon Drainage (Basin 6) has no cloud forest within it ridges. It is **not conducive** to a contour-based limit.
7. The Rio Sumpul Drainage (Basin 7) has a rating of 58 percent at a threshold contour of 2,200 m. It is **marginally conducive** to a contour-based limit, at 2,200 m.
8. The Rio Sinuapa Drainage (Basin 8) has a rating of 77 percent at a threshold contour of 2,000 m. It is **conducive** to a contour-based limit, at 2,000 m.
9. The Rio Liquidambal Drainage (Basin 9) has a rating of 58 percent at a threshold contour of 1,900 m. It is **marginally conducive** to a contour-based limit, at 1,900 m.
10. The Caribbean Watershed (Basins 1 to 5 combined) has a rating of 51 percent at a threshold contour of 2,000 m. It is **marginally conducive** to a contour-based limit, at 2,000 m.
11. The Pacific Watershed (Basins 6 to 9 combined) has a rating of 54 percent at a threshold contour of 2,100 m. It is **marginally conducive** to a contour-based limit, at 2,100 m.

12. The Guisayote Biological Reserve (Basins 1 to 9 combined) has a rating of 53 percent at a threshold contour of 2,000 m. It is marginally conducive to a contour-based limit, at 2,000 m.

Note that although the evaluation of contour-based limits is an important aspect of this study, the geographic information provided by the other points on each graph (aside from the point of the intersection) and the map itself are perhaps even more significant, in that they inform planners of the location and concentration of cloud forest at a variety of levels. Such information is useful in project planning, promotional and educational activities, and for the prioritization of areas for the allocation of conservationist resources. Regarding this last item, note that areas with low threshold contours (1,800 m or 1,900 m) and very high ratings may be considered highest-priority areas, as their forests are both large and intact. Areas with high threshold contours (2,100 m or 2,200 m) and very low ratings are probably too small and too dispersed to warrant the investment of time, money, and effort necessary for management.

Conclusion

The GIS-based technique presented here provides a quantifiable and repeatable method of understanding the geography of cloud forest on the recently designated reserves of Honduras. It is on this basic understanding that a foundation for reasonable delimitation and good management may be built. The role of GIS in the project is concise but open to expansion: it serves as the data collection center and measurement tool, with the possibilities of incorporating other layers of data—land tenure, wildlife habitat, hydrological systems, recoverable lands—into the formula for delimitation.

Acknowledgment. Acquisition of the aerial photos was made possible by a field trip to Honduras partially funded by the Robert C. West Field Research Award, Louisiana State University.

Notes

1. The Guisayote Biological Reserve was subdivided into these three levels of analysis to increase accuracy (since cloud forest may be distributed very differently from basin to basin, albeit in the same reserve), to extend usefulness to municipal governments seeking to protect their water sources, and to detect any possible orographic effects east and west of the continental divide.
2. A dispersed forest necessarily implies a preponderance of pasture, crops, and other land uses incompatible with a nature reserve, and is therefore ranked as "low-quality" cloud forest in the eyes of a wildlands planner.

References

Agudelo C., Nelson. 1987. *Ecosistemas terrestres de Honduras.* Tegucigalpa, Honduras: Asociacion Hondurena de Ecologia.

Congreso Nacional de Honduras. 1987. Decreto 87-87. Tegucigalpa, Honduras: La Gaceta.

ESRI (Environmental Science Research Institute, Inc.). 1990. PC Arc/Info Manuals and Users Guides. Redlands, CA: ESRI.

7. Reforestation with the Native Tree *Alnus acuminata*: Effects on Phytodiversity and Species Richness in an Upper Montane Rain Forest Area of Colombia

Jaime Cavelier

Worldwide tropical deforestation currently is proceeding at the rate of around 7.5 x 10^6 ha/year (Whitmore 1990), including about 600,000 ha/year in Colombia. Estimations of annual deforestation rates in Colombia range from 1.7 percent (WRI 1990) to 2.8 percent (Myers 1989). In contrast, reforestation programs, with mainly introduced species such as *Pinus patula* and *Cupressus lucitanica*, account for only 2,500 ha/year (Carton de Colombia, pers. com.). Reforestation programs are rarely carried out with native species.

Currently only about 27 percent of the once widespread Colombian upper and lower montane rain forests remain in the Andes, Sierra Nevada de Santa Marta, and isolated mountains such as Serrania de Macuira (Guajira) and Cerro Tacarcuna (Choco) (Cavelier and Etter, unpubl.). Most of these remaining montane forests are on the eastern slopes of the Cordillera Oriental and the western slopes of the Cordillera Occidental. The clearing of montane rain forest usually results in the replacement of native vegetation by pastures or cultivated fields. Pasture areas rarely revert to secondary forest because of continued grazing pressure.

This paper presents results of a 15-year regeneration/reforestation scheme in which high altitude pastures on the western slope of the Central Cordillera were either abandoned or planted with the native *Alnus acuminata* (or introduced species such as *Pinus patula* and *Cupressus lucitanica*). The regeneration of the native upper montane rain forest under the nonmanaged *Alnus* plantation is compared with the secondary growth of the abandoned pastures. An adjacent, undisturbed area of upper montane rain forest was also studied as a reference or control for regeneration rates in the disturbed study sites (Figures 1, 2).

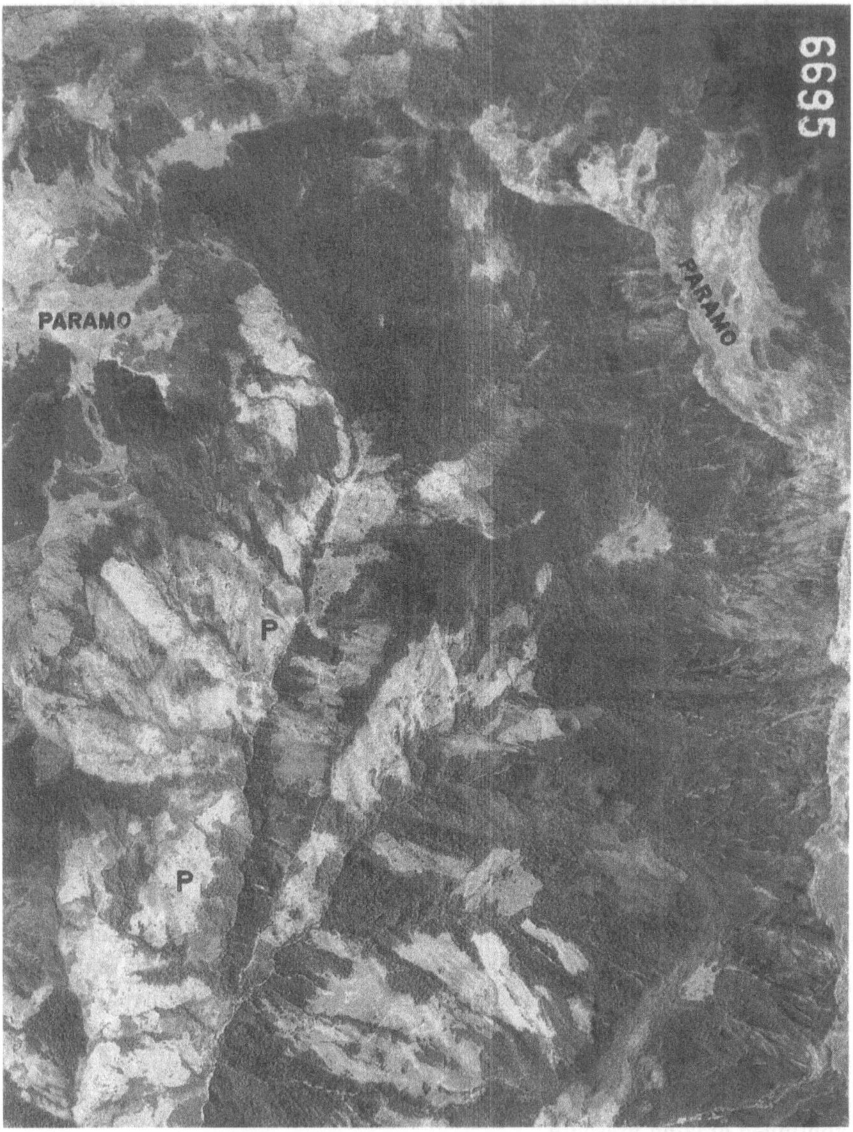

Figure 1. Aerial photograph of the upper watershed of Rio Quindio, western slope of the Central Cordillera. Note pasture areas (P) in white. There are also native open grasslands above the tree line (paramo). Photo taken February 1959 (picture no. 6695, flight M547, envelope 393, IGAC). Scale: 1:30,000 (approx.). *Source: Instituto Geografico Agustin Codazzi, Bogotá.*

Materials and Methods

Study Site

This study was carried out at the western slope of the Central Cordillera of the Colombian Andes— specifically, at Finca Guayaquil (3,100 m) in the upper watershed of Rio Quindio. Most of the area is the property of the Corporacion Autonoma Regional del Quindio (CRQ), a regional corporation charged with preservation and management of the natural resources of the Departamento del Quindio.

The areas around Finca Guayaquil support three main vegetation types: primary upper montane rain forest (cloud forest), plantations of native (*Alnus acuminata*) and introduced species (*Cupressus lucitanica* and *Pinus radiata*), and secondary forest. The primary montane forest is dominated by large (>100 cm diameter at breast height [dbh]) *Weinmannia* spp. (25–30 m tall). Although they are usually found on the steepest and more remote slopes, some relatively large forest patches occur at about 3,400 m on relatively flat terrain near the field station of the Centro de Investigaciones y Formacion Ambiental para la Zona Andina (CIFAC). The *Alnus acuminata* plantations were established during 1977–78 on abandoned agricultural and grazing lands (Carlos A. Gonzales, pers. com.). The secondary forests are the result of abandoned pastures that were left to natural regeneration. They are the same age (15 years) as the *Alnus* plantations, providing an opportunity for comparison of regeneration processes. The purpose of establishing the plantations was to protect mountain water sources for both domestic (Armenia, the region's major city) and agricultural use (coffee processing).

Short-term climatological data are available from a weather station near the study site (Finca Navarco, elev. 2,600 m). The total annual rainfall in 1990 was 1,953 mm (156 days with rain), distributed bimodally with drier periods in January/February and from July to September. Mean annual temperature is 13.7° C (Agudelo 1991). Both rainfall and temperature at the Finca Guayaquil study site (3,100 m) could be expected to be somewhat lower than at Finca Navarco, given the difference in altitude (ca. 500 m).

Sampling Methods

Vegetation sampling was carried out in each of the three forest types: *Alnus* plantation, secondary forest, and primary forest. A total area of 1/10 ha was sampled using 10 plots, 10 m by 10 m. In each plot, diameter at breast height and total height were measured for all stems >2.5 cm dbh. Botanical samples were collected from each individual for identification. Because the survey was carried out only once (April 1992), most of the species were not in flower or fruit and botanical identification (to the species level) was sometimes impossible.

In each forest type, three soil surface samples (0–10 cm) were also col-

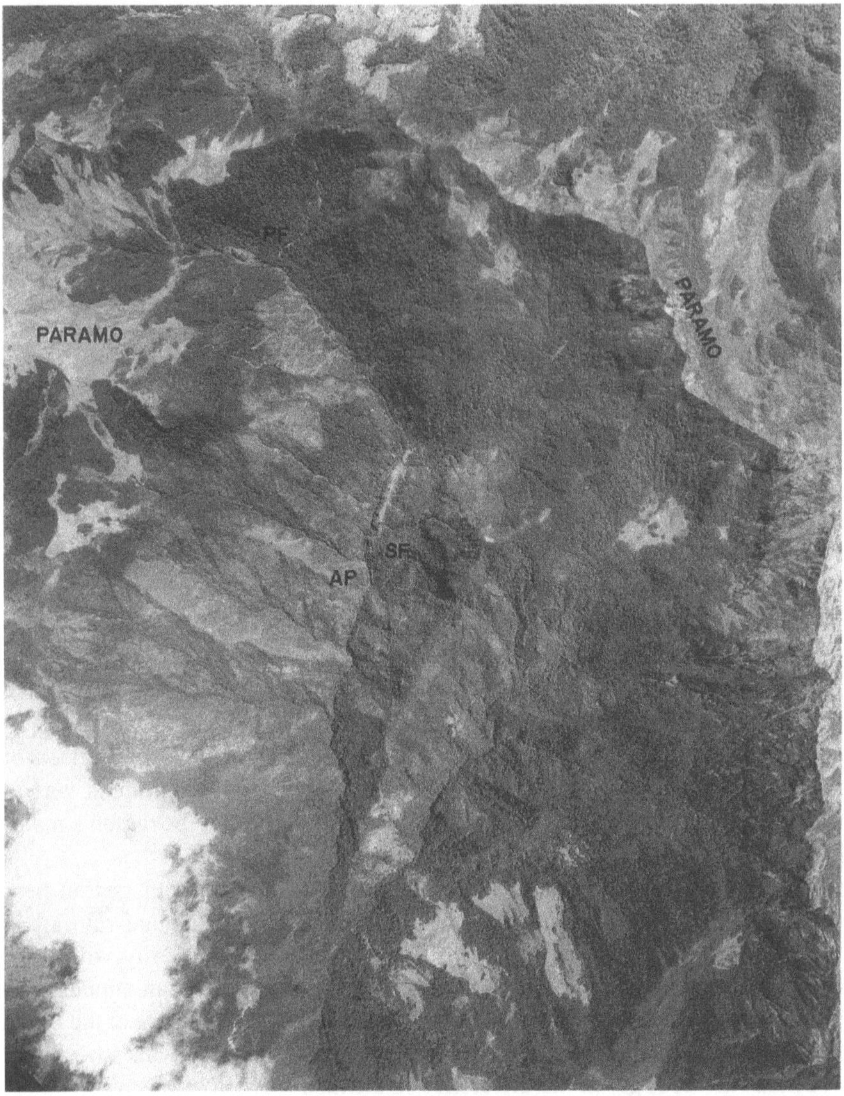

Figure 2. Aerial photograph of the upper watershed of Rio Quindio, western slope of the Central Cordillera. Compare with photo in Figure 1 for pasture areas. The study sites are marked with AP (*Alnus* plantation), SF (secondary forest), and PF (primary forest). Picture taken during December 1989 (picture 109, flight C2268, envelope 33726, IGAC). Scale: 1:30,000 (approx.). *Source: Instituto Geografico Agustin Codazzi, Bogotá.*

lected. Soils were analyzed for cation exchange capacity, total bases, percentage base saturation, exchangeable calcium, magnesium, potassium and sodium, pH, Al, texture, and soil water content (percent dry weight). The pools of ammonium and nitrate were also measured in each soil sample using solutions of 2M KCl. Available phosphate was extracted with NH_4F (Cavelier 1992).

Results

Forest Structure

The tallest forest was the primary upper montane forest with *Weinmannia glabra* trees 25–30 m tall and diameters up to 1 m (Table 1). The *Alnus acuminata* forest was also tall with trees up to 20 m. No other species reaches the canopy in this forest. The smallest trees occurred in the secondary forest, with canopies of 8 10 m in height, but exhibiting greater tree species diversity than in the other two forests.

The basal area was similar between the *Alnus* plantation and the secondary forest, and significantly lower than the primary forest (Table 1). Because the *Alnus* plantation has taller trees, this means that this forest has a higher biomass than the secondary forest. Total density of trees >2.5 cm dbh was similar between the secondary and primary upper montane rain forests.

Forest Composition

Alnus acuminata *Plantation*

In addition to *Alnus acuminata*, at this site a total of 26 species was found distributed in 16 families (Table 2). The canopy of this forest (ca. 20 m high) was entirely composed of the *Alnus* trees planted 15 years ago. Under the canopy there was an understory composed of elements of the primary and

Table 1. Characteristics of the forests studied in the upper watershed of Rio Quindio, Finca Guayaquil (3,100 m). Values for all trees >2.5 cm dbh measured in 10 plots, 10 m by 10 m (1/10 ha).

	Tree height (m)	Basal area (m^2/ha)	Stem density (no./ha)
Alnus plantation	20	24.1	31.5
Secondary forest	8–10	26.5	59.8
Primary forest	25–30	72.2	51.1

secondary forests, including *Bocconia frutescens* (the most common species), *Viburnum* spp. (Caprifoliaceae), *Solanum* spp., and *Cestrum* spp. (Solanaceae). Melastomataceae, the most common family in the secondary forest, was represented by only one species (*Miconia jahnnii*) and very few individuals. *Weinmannia glabra* (Cunoniaceae), common in the primary forest, was very poorly represented in the *Alnus* plantation. There were species of the genus *Verbesina*, which occurred in the *Alnus* plantation but not in the primary forest. Where the understory was dominated by *Chusquea* sp. (Poaceae) and *Uncina hamata* (Cyperaceae), the density of woody plants was lower. *Thelypteris* sp. (Polypodiaceae) was also common in the understory.

Table 2. List of families and species collected in the *Alnus acuminata* plantation at Finca Guayaquil (3,100 m) in the upper part of the Rio Quindio watershed, property of the Corporacion Autonoma Regional del Quindio (CRQ). The list includes all the individuals (>2.5 cm dbh) collected in 1/10 ha (312 individuals).

Family	Species	No. Individuals
Polypodiaceae	*Thelypteris* sp. 1	1
Betulaceae	*Alnus acuminata*	77
Boraginaceae	*Tournefortia fuliginosa* H.B.K.	9
Brunelliaceae	*Brunellia cf. goudotii* Tul.	1
Caprifoliaceae	*Viburnum* sp. 1	6
	Viburnum sp. 2	1
	Viburnum cf. pichinchense Benth.	1
	Viburnum aff. jamesonni (Oerst.) Killip & Smith	3
	Viburnum triphyllum Benth.	1
Chlorantaceae	*Hedyosmum aff. racemosum* (R. et. P.) G. Don	1
Compositae	*Verbesina* sp. 1	80
	Verbesina cf. nudipes Blake	6
	Verbesina arborea H.B.K.	9
Cunoniaceae	*Weinmannia glabra* L.F.	1
Gentianaceae	*Lehmanniella* sp. 1	1
Melastomataceae	*Miconia jahnnii* Pittier	1
Monimiaceae	*Siparuna cf. buddleiaefolia* (Benth.) A.D.C.	1
Myrsinaceae	*Geissanthus serulatus* Mez.	3
Papaveraceae	*Bocconnia frutescens* L	32
Polygalaceae	*Monnia cf. angustata* Tr. et Pl.	2
Solanaceae	*Aenistus* sp. ?	3
	Solanum sp. 1	3
	Cestrum sp. 2	2
	Cestrum sp. 3	1
	Cestrum acharaceum Francey	1
	Dunalia solanace H.B.K.	1
Theaceae	*Freziera canescens* Bonp.	1
Indeterminate		63
Families = 16		
Species = 26		

Secondary Forest

In this forest type, there were 33 species distributed within 19 families (Table 3). The canopy of the secondary forest was dominated by Melastomataceae of the genus *Miconia* (*M. jahnnii* and *M. glesoniana*) and by Theaceae of the

Table 3. List of families and species collected in the 15-year-old Secondary Forest at Finca Guayaquil (3,100 m) in the upper part of the Rio Quindio watershed, property of the Corporacion Autonoma Regional del Quindio (CRQ). The list includes all the individuals (>2.5 cm dbh) collected in 1/10 ha (598 individuals).

Family	Species	No. Individuals
Aquifoliaceae	*Ilex* sp. 1	1
Boraginaceae	*Tournefortia fuliginosa*	1
Brunelliaceae	*Brunellia cf. goudotii* Tul.	16
Caprifoliaceae	*Viburnum cf. pichinchense* Benth.	12
	Viburnum sp. 1	1
Chlorantaceae	*Hedyosmum aff. racemosum* (R. et P.) G. Don	3
Clethraceae	*Clethra* sp. 1	1
Compositae	*Verbesina* sp. 1	1
	Baccharis nitida (R. et P.) Pers.	1
	Eupatorium sp. 1	2
	Eupatorium popayanense Hierom.	3
	Gynoxis aff. verrucosa Wedd.	1
Cunoniaceae	*Weinmannia glabra* L.F.	15
	Weinmannia balbisiana H.B.K.	26
Ericaceae	*Gaultheria bracteata* (Cav.) G. Don	3
	Psammisia sp. 2 (?)	1
Melastomataceae	*Miconia jahnnii* Pittier	178
	Miconia sp. 1	4
	Miconia sp. 3	8
	Miconia glesoniana Wurdack	14
Myrsinaceae	*Geissanthus quindiuensis* Mez.	1
	Geissanthus serulatus Mez.	1
Papaveraceae	*Bocconnia frutescens*	1
Polygalaceae	*Monnia cf. angustata* Tr. et. Pl.	1
Rosaceae	*Hesperomeles aff. heteophylla* (H.de P.) Hook	2
Rubiaceae	*Palicourea* sp. 1	1
Saxifragaceae	*Escallonia myrtilloides* L.F.	3
Solanaceae	*Solanum ovalifolium* Dund.	1
Theaceae	*Freziera reticulata* H. & B.	169
	Freziera canescens Bonp.	24
	Freziera aff. candicans Tul.	2
	Laplacea sp. 1	1
Winteraceae	*Drymis winteri*	19
Indeterminate		80
Families = 19		
Species = 33		

genus *Freziera* (*F. reticulata, F. canescens,* and *F. aff. candicans*). There were also individuals of *Brunellia cf. goudotii* (Brunelliaceae) and species of the genus *Viburnum* (Caprifoliaceae). Under this canopy, there was an understory with a few individuals of *Wein-mannia* (*W. glabra* and *W. balbisiana*), which are common in the primary forest. In the 1/10 ha sample there was only one individual of *Bocconia frutescens*, the most abundant subcanopy species in the *Alnus* plantation.

Primary Forest

In this forest type there were 40 species, distributed in 18 families (Table 4). The canopy was dominated by trees of *Weinmannia glabra*. In the understory, there were trees of *Hedyosmum aff. granizo* (Chlorantaceae), *Escallonia myrtilloides* (Saxifragaceae), and *Miconia* spp. (Melastomataceae).

Soils

The soils of the primary forest were more acidic (pH = 3.6) than at the *Alnus* plantation (pH = 4.6) and the secondary forest (pH = 5.5). High soil acidity at the primary forest was associated with high exchangeable aluminum, which is usually soluble under pH = 4.5 (Table 5). The greatest difference between the physical properties of the soils was the gravimetric water content. The highest values were found at the primary forest (71.8 percent) and the lowest at the *Alnus* plantation (8.8 percent).

The soils of the primary forest showed the highest cation exchange capacity (CEC). This seems to be the result of the higher carbon content in this forest (29.7 percent) in relation to the *Alnus* plantation soils (14 percent) and the secondary forest (15.3 percent). In contrast to the high CEC of the soils at the primary forest, the base saturation was the lowest. The *Alnus* plantation showed the highest concentration of Ca and Mg, as well as the highest base saturation (Table 6). The soils of the primary forest showed the highest concentration of ammonium, nitrate, and phosphates (Table 7).

A general perspective of the soils suggests that the most fertile soils are those in the primary forest (in spite of the high acidity and exchangeable aluminum), followed by the soils in the secondary forest and the *Alnus* plantation.

Discussion

Effect of *Alnus* Plantation on Floristic Diversity

The planting of *Alnus* trees, the abandonment of pastures, and the preservation of primary montane forest in the upper part of the Rio Quindio watershed 15 years ago were the beginning of an interesting long-term experiment to compare the regeneration of montane forest under natural conditions with that in areas planted with a native species and subsequently abandoned.

Table 4. List of families and species collected in the primary upper montane rain forest at Finca Guayaquil (3,100 m) in the upper part of the Rio Quindio watershed, property of the Corporacion Autonoma Regional del Quindio (CRQ). The list includes all the individuals (>2.5 cm dbh) collected in 1/10 ha (295 individuals).

Family	Species	No. Individuals
Cyatheaceae	*Cyathea aff. caracasana*	1
	Cyathea sp. 1	3
Actinidaceae	*Saurauia brachybothrys* Turcz.	13
Aquifoliaceae	*Ilex* sp. 1	3
	Ilex sp. 2	1
Araliaceae	*Oreopanax aff. discolor* (H.B.K.) Decne et Pl.	1
	Oreopanax aff. floribundum (H.B.K.) Decne Pl.	4
Caprifoliaceae	*Viburnum* sp. 1	1
	Viburnum sp. 2	2
	Viburnum cf. pichinchense Benth.	6
	Viburnum aff. jamesonni (Oerst.) Killip & Smith	1
Chlorantaceae	*Hedyosmum aff. granizo* Cuatr.	1
Compositae	*Vernonia* sp. 1	1
	Liabum sp. 1	2
	Mikania cf. stuebelii Hierom.	1
Cunoniaceae	*Weinmannia glabra* L.F.	22
Ericaceae	*Psammisia* sp. 1	1
	Disterigma acuminata (H.B.K.) Ndzu	1
Lauraceae	*Persea ferruginea* H.B.K	2
	Nectandra sp. ?	3
Melastomataceae	*Miconia jahnnii* Pittier	5
	Miconia orcheotoma Naud.	1
	Miconia sp. 1	1
	Miconia sp. 2	2
	Miconia aff. asperrima Tr.	4
	Miconia sp. 3	82
	Miconia glesoniana Wurdack	57
Myrsinaceae	*Myrsine dependens* (R. et. P.) Spaengof.	4
	Geissanthus quindiuensis Mez.	2
	Geissanthus serulatus Mez.	6
Rosaceae	*Hesperomeles glabrata* H.B.K.	1
	Hesperomeles aff. lanuginosa R. de P.	1
Rubiaceae	*Psychotria cf. boqueronensis* Wernh.	5
Saxifragaceae	*Escallonia myrtilloides* L.F.	2
Solanaceae	*Cestrum* sp. 1	2
	Cestrum sp. 2	42
	Solanum sp. 1	4
	Sessea corymbosa Goudot	1
Theaceae	*Laplacea* sp. 1	1
Verbenaceae	*Duranta* sp. 1	2
Families = 18		
Species = 40		

Table 5. Mean values (n = 3; ± 1SEM) of soil texture, pH, aluminum, and gravimetric soil water content (%) of the soils of the *Alnus* plantation (AP), secondary forest (SF), and primary forest (PF) around Finca Guayaquil (3,100 m), upper watershed of Rio Quindio (CRQ)

	Sand	Silt	Clay	pH	Al	H_2O
		(%)		(1:1)	(meq/100g)	(%)
AP	54.6 (5.0)	38.6 (4.1)	6.6 (3.0)	4.6 (0.28)	1.3 (0.6)	8.8 (4.0)
SF	66.6 (3.0)	26.6 (2.3)	6.6 (1.1)	5.5 (0.30)	0.2 (0.3)	20.8 (4.7)
PF	72.6 (4.1)	21.3 (3.0)	6.0 (3.4)	3.6 (0.20)	5.6 (1.7)	71.8 (9.6)

Table 6. Mean values (n = 3; ± 1SEM) of chemical properties of the soils in the *Alnus* plantation (AP), secondary forest (SF), and primary forest (PF) around Finca Guayaquil (3,100 m), upper watershed of Rio Quindio (CRQ). Cation exchange capacity (CEC), total bases (TB), and total saturation (TS)

	CEC	TB	Ca	Mg	K	Na	TS (%)
			(meq/100g)				
AP	44.0 (2)	11.3 (3.5)	9.8 (3.1)	0.9 (0.5)	0.5 (0.1)	0.1 (0.0)	25.7 (7)
SF	58.0 (8)	30.6 (2.8)	24.5 (1.7)	5.0 (3.3)	0.9 (0.1)	0.1 (0.0)	51.9 (10)
PF	107.5 (34)	18.6 (2.5)	13.9 (2.0)	3.3 (0.5)	1.2 (0.3)	0.2 (0.0)	18.1 (4)

Table 7. Mean values (n = 3; ± 1SEM) of the concentration of NH_4, NO_3, PO_4 and carbon in soils at the *Alnus* plantation (AP), secondary forest (SF), and primary forest (PF) around Finca Guayaquil (3,100 m), upper watershed of Rio Quindio (CRQ)

	NH_4	NO_3	PO_4	C (%)
		(μg/g dry weight)		
AP	9.3 (2.3)	11.8 (0.4)	4.1 (0.5)	14.0 (0.9)
SF	11.9 (1.0)	16.5 (1.1)	5.2 (0.1)	15.3 (3.2)
PF	28.0 (4.4)	50.7 (1.8)	14.7 (0.6)	29.7 (10.0)

The results of this study show that species diversity (evenness of distribution of the individual trees among the species) is more affected than species richness (number of species) in areas planted with the native *Alnus acuminata*. First, the canopy of the plantation consists solely of the planted trees, whereas in the secondary forest, several species make up the canopy. These include *Miconia* (Melastomataceae), *Freziera* (Theaceae), *Brunellia* (Brunelliaceae), and *Viburnum* (Caprifoliaceae). Second, in the understory of the *Alnus* plantation most of the individuals are *Bocconnia frutescens*, whereas in the understory of the secondary forest there are several species, including small trees of *Weinmannia glabra* and *W. balbisiana*, the two most abundant trees of the canopy in the primary forest.

There are several hypotheses to explain the low level of species richness and diversity in the *Alnus* plantation understory. First, it is known that enrichment of the soils by means of fertilization, particularly with nitrogen, results in a reduction in phytodiversity. Because *Alnus* is known to have association with the nitrogen-fixing actinomycete *Frankia*, one would expect to find higher nitrogen concentrations in the *Alnus* soils and thus lower plant diversity. Nevertheless, soil ammonium is not significantly different from the secondary forest soils, and soil nitrate is significantly lower than in the secondary and primary forests (Table 7). It is possible that because the *Alnus* plantation soils are not waterlogged, N fixation is depressed.

Second, the relatively poor regeneration of the forest in the understory of the *Alnus* plantation seems not to be related to low light levels. Indeed, light levels in the *Alnus* plantation are higher than in the secondary and primary forests (the trees were planted in rows 5 m from each other). The crowns of the *Alnus* trees are very small, allowing abundant light to reach the forest floor.

Third, the reduction in plant diversity in the *Alnus* plantation may be partly caused by a reduction in the number of bird species that visit these forests. In a 15-year-old *Alnus* plantation at Finca El Bosque (3,100 m) very similar to the one at Finca Guayaquil, 30 bird species were captured and/or observed, in contrast to 57 in a nearby secondary forest. Twenty-five (25) of these 57 bird species were observed in both forests (M. Alvarez, pers. com.). From the species observed in the *Alnus* plantation, at least five (*Turdus fuscater, Tangara vassori, Anisognatus igniventris, A. lacrimosus,* and *Diglosa cynea*) are known to feed and disperse Melastomataceae (M. Alvarez, pers. com.). In spite of this, in the *Alnus* plantation only one species and very few individuals of this family were found. This result suggests that dispersal of seeds is not a major limiting factor for forest regeneration. It is likely that the seeds of Melastomataceae and other families are being dispersed to the *Alnus* plantations, but they do not germinate or the seedlings die soon after germination.

Finally, the most likely explanation of the reduction of diversity in the *Alnus* plantation is a change in the soil chemistry, particularly regarding secondary compounds. It is known that trees can have allelopathic effects on understory plants (Goel and Sareen 1986) as well as on canopy trees (Fisher 1980). In the case of *Alnus*, the product of fermentation could be resulting in the allelopathic effects on other plants. Some of these products include ferulic, siringic, and cumaric acids (Rice 1984). Nevertheless, since the soils are not saturated or waterlogged, it is very unlikely that fermentation occurs in these soils and therefore unlikely that these products are responsible for the allelopathic effects. An alternative explanation, also related to allelopathy, may involve compounds found in leaves such as esterols, flavonoids, tannins, and other phenolic compounds (Barrera 1986). These secondary products may reach the forest floor via throughfall, inhibiting germination or the growth of seedlings. Experimental work is needed to elucidate the mechanisms relating secondary compounds and allelopathic effects with *Alnus*.

Effect of *Alnus* Plantation on the Water Balance of the Watershed

The gravimetric water content of the soils suggests that the *Alnus* plantation retains less water in the soil than the secondary and primary forests. This result must be considered carefully, since a small difference in soil bulk density may compensate for differences in relative water content measured gravimetrically. Because neither the carbon content nor the texture at the *Alnus* plantation and the secondary forest soils are significantly different, differences in soil relative water content (%) may represent true differences in volumetric water content and thus in the capacity to retain water.

If the water content at the *Alnus* site turns out to be significantly lower than in the secondary forest, this would have important implications for the management of the forest that was planted for the conservation of the water resources of the upper watershed of the Rio Quindio. Perhaps abandoning the pastures to natural forest regeneration is the best management strategy for the conservation of both water and biodiversity in the area.

Conclusions

1. Reforestation with the native tree *Alnus acuminata* has strongly influenced both species richness and diversity in the subcanopy. There may also be effects on soil water content and thus on the water balance of the entire watershed.
2. Forest development through natural, secondary succession on abandoned agricultural and grazing pastures may be better for both water and species conservation in upper montane rain forest, as compared with planting *Alnus* trees. Native secondary succession will be most rapid where pasture patches are small (on the order of 0.5 to 2.0 ha) and adjacent to forest seed banks.

Acknowledgments. I would like to thank Julian Serna, head of the Corporacion Autonoma Regional del Quindio (CRQ) for his support during the fieldwork. I also thank Aureliano Sabogal, Fernan Castaño, Ruben Arcila, Rene Lozada, John E. Vera, Carlos A. Gonzales, and the students that helped during the fieldwork, particularly those who helped process the plant material and entered and processed the data. Without them this work never could have been done. Thanks also to Gilberto Maecha for helping with the identification of the plant material.

References

Agudelo, C.A. 1991. *Estudio floristico y climatico del Cañon Quindio*. Informe inedito, Instituto de Ciencias Naturales, Universidad Nacional de Colombia, Bogota.

Barrera, C. 1986. *Analysis fitoquimico preliminar y determinacion abortiva de las hojas de* Alnus acuminata. Tesis de la Facultad de Farmacia (inedita), Universidad Nacional de Colombia, Bogota.

Cavelier, J. 1992. Fine root biomass and soil properties in a semideciduous and a lower montane rain forest in Panamá. *Plant and Soil* 142:187–201.
Fisher, R. F. 1980. Allelopathy phytotoxicity of chemicals from plants: A potential cause of regeneration failure of forest trees. *Journal of Forestry* 78:346–348.
Goel, U., and T. S. Sareen. 1986. Allelopathic effect of trees on the understory vegetation. *Acta Botanica Indica* 14:162–166.
Myers, N. 1989. *Deforestation rates in tropical forests and their climatic implications*. Friends of the Earth, London.
Rice, E. L., ed. 1984. *Allelopathy*. 2d ed. Academic Press, San Francisco.
Whitmore, T. C. 1990. *An introduction to tropical rain forests*. Clarendon Press, Oxford.
WRI (World Resources Institute). 1990. *World resources: 1990–1991*. Oxford University Press, New York.

8. The Montane Cloud Forest in Southern Brazil

Daniel de Barcellos Falkenberg and Júlio César Voltolini

> The time has come to undertake more consistent biogeographical research in the Brazilian tropical mountains, especially studies better able to integrate physical, ecological, bioclimatic and biological facts. . . . In fact, the information available on the biogeography of the tropical mountains of Brazil is *very restricted and fragmented*. In the first instance, the Brazilian mountains are only local massifs, the residual relief of summits or high elongated mountain ridges (*Serra do Mar, Serra Geral*, or high coastal mountain ridges). . . . Each sector . . . is a field in itself for the study of the contacts between masses of vegetation of diverse natural domains, including high altitude zones of vegetation belonging to the nuclei and transitional facies of the different flora that had competed for area and altitude (Ab'Saber 1989; the emphases are ours).

The tropical montane cloud forest (TMCF, or simply cloud forest, because all Brazilian cloud forests are montane) occurs at altitudes of around 700 m above sea level and upwards, and is largely associated with the highlands of the Serra do Mar mountains, in the states of Santa Catarina (SC) and Paraná, and with the Aparados da Serra Geral (or simply Aparados) in the states of SC and Rio Grande do Sul (RS).

This paper presents a compilation of data from a series of phytogeographic studies developed by the Brazilian botanist Balduíno Rambo, jointly with several field observations and collections made by ourselves, along with other field and bibliographic information. Our objective is to present a brief regional description of TMCF for the states of Rio Grande do Sul (RS) and Santa

Catarina (SC). Although restricted in its geographic focus, we believe this is the first study to synthesize information on Brazilian cloud forests.

Literature Review

The distinctive montane cloud forest formation was largely ignored in books dealing with Brazilian vegetation until the 1970s. Lindman (1900) failed to distinguish cloud forest in his pioneer study on the vegetation of Rio Grande do Sul. Wettstein (1904), Romariz (1964), and Joly (1970) also did not clearly differentiate TMCF in their vegetation classification schemes. Maack (1968:221) suggests that in the Serra do Mar from Paraná, the rain-tropical-- subtropical forest changes at 1,150 m above sea level to the cloud forest that occurs up to an altitude of 1,350 m where grassland with isolated shrubs replaces the cloud forest. Martinelli and Bandeira (1989) consider the cloud forest a localized formation between the hillside Atlantic forest and the high altitude fields ("campos de altitude") in the Morro do Cuca, Serra dos Órgãos (Rio de Janeiro state).

Other authors, including Veloso and Góes-Filho (1982) and Veloso et al. (1991), have briefly characterized the vegetation of the Brazilian cloud forest and named it upper montane dense ombrophilous forest (popularly, "mata nuvígena"). However, in these two publications they give quite conflicting statements on the canopy statue of those forests (i.e., in the earlier publication an average canopy height of 5 m is indicated, compared with 20 m in the more recent article).

There has also been some confusion over the biogeographic placement of Brazilian cloud forest. Rizzini (1979) considers TMCF as part of the "Atlantic forest" (in either the montane rain forest or *Araucaria* formations), whereas Cabrera and Willink (1980) are vague as to TMCF placement in either the Paranaense or Atlantic provinces.

The literature indicates no clear consensus regarding the best way to treat Brazilian cloud forest, and there is not even agreement on its underlying subtropical (Wettstein 1904) or tropical affinities (Takhtajan 1986, in the latitudinal limit of his Neotropical Kingdom). Hueck (1972) and Schnell (1987) considered the south Brazilian cloud forest as a part of the tropical rain forest occurring on mountain slopes (Atlantic forest). However, important Brazilian botanists (Rambo, Klein, Reitz) have considered this cloud forest as a particular kind of vegetation; the expression "mata nebular" or "matinha nebular" seems to have been used for the first time by Rambo (1949). Klein (1960:29–31) treats the cloud forest as a group of pioneer associations (*Siphoneugenietum, Myrceugenietum*, etc.) within the *Araucaria* formation (as Rambo [1951] did earlier), but he later ignores the cloud forest in his recent phytogeographic synthesis of south Brazil (Klein 1984).

Some Physiographic Aspects

In the states of Santa Catarina (SC) and Rio Grande do Sul (RS), the cloud forests occur generally at altitudes between 700 and 1,600 m in the highlands of Serra do Mar (NE corner of SC) and along the Aparados da Serra Geral (SE part of SC and NE corner of RS). The origins of the Serra do Mar date back to the Precambrian age; all rocks exhibit metamorphic character, were formed and reach their southernmost limit in the extreme north of SC. The Serra Geral is formed from volcanic rocks (basaltic) of the Cretaceous age, and this lithology underlies the planalto das Araucarias. The easternmost limit of this vast plateau is characterized by escarpments, vertical cliffs, and some deep, steep-walled valleys (canyons), several of which exhibit relative relief of 1,000–1,200 m, giving the appearance of deep cuts and giving rise to the Portuguese name Aparados ("cuts").

These regions are covered almost daily by orographic cloud or fog (popularly named "cerração" or "viração"), caused by the ascent of warm and moist oceanic winds that cool and condense at higher altitudes. The annual rainfall is generally above 1,500 mm. The annual mean temperature is around 14°–16° C or less (10° C in the Morro da Igreja, in the south of SC). The mean temperature of January (warmest month) can reach 20°–22° C, and the mean temperature of July (coolest month) frequently is less than 10° C. Generally there are more than 15 nights with temperature less than 0° C in a year; the annual mean occurrence of frost is only about 10 nights in the Serra do Mar but more than 30 nights in the Aparados.

The soils generally are poorly developed, shallow, acid, and organic, with much litter on the surface (Rambo 1953, 1956a).

Delimitation of the Cloud Forest

The concept of TMCF adopted in the present work is essentially the same as developed by Rambo (1949, 1951, 1953, 1956a, b) for south Brazil. Later, this same definition was adopted in Santa Catarina by Reitz (1965—"matinha nebular"), Veloso and Klein (1968), and Klein (1978—"floresta nebular"); in Paraná by Maack (1968—"matinha de neblina"); and in Rio Grande do Sul more recently by Baptista et al. (1978).

In southeastern Brazil, this forest type occupies the highlands in the complex of the Serra do Mar mountains and also the plateau border (Aparados) of the Serra Geral, above the upper altitudinal limits of the luxuriant Atlantic tropical rain forest (about 700–1100 m in south Brazil and 1,200–1,500 m in southeast Brazil).

The size of each region covered by cloud forest is variable. In several places, the cloud forest forms a narrow belt along the Aparados; in others (e.g., Urubici), there are several great (ca. 1 km^2) spots of cloud forest; in some places where the escarpments are less vertical (only 50°–70°) or have some plateau or stair, there are generally small spots of cloud forest.

In the present study, the TMCF is considered as a transition between Atlantic forest ("mata Atlântica"—lower and middle slopes of coastal mountains) and *Araucaria* forest ("pinhal"—from higher altitudes), or as an intermediary belt between Atlantic forest and grassland. Grassland is also refered to as "campo de altitude" (sensu Martinelli and Bandeira 1989), when it occurs at high elevations on coastal ridges, or "campo planaltino" (Rambo 1953:187), when it occurs on interior plateaus (planalto meridional, planalto das Araucárias, or planalto Sul-Brasileiro). "Pinhal" occurs essentially in the inland planalto das Araucárias, in proximity with "campo planaltino." Our expression "campo planaltino" includes dry and wet grasslands, rocky grasslands, marshes, and peat bogs. The physiognomy and floristics of the peat bogs in the Aparados da Serra Geral show clear relationships with Andean and subantarctic regions (Pfadenhauer and Boechat 1981).

The cloud forests do not form a single homogeneous vegetational type but, rather, a group of related floristic communities with relatively similar physiognomies, under the influence of altitude, degree of exposure to the oceanic winds, local topography, variation in substrate, and local microclimatic influences.

Vegetation

There are no published phytosociological studies or floristic inventories dealing specifically with TMCFs of southern Brazil, but there are several sample collections (Reitz 1965; Herbaria FLOR, PACA, HBR, ICN, MBM), and the principal vascular species are known (Reitz 1965-1989). The senior author is currently preparing a checklist for the Morro da Igreja (Aparados da Serra Geral, SC).

Rambo (1949, 1953, 1956b), Hueck (1972), Baptista et al. (1978), and Klein (1978, 1980) provide good characterizations of the vegetation. Veloso and Klein (1968), Veloso and Góes-Filho (1982), Leite and Klein (1990), and Joly, Leitão Filho, and Silva (1991) also contain important information relative to these forests.

In general, TMCF is a dense and shadowy formation, dominated by small trees (6–8 m) and shrubs, frequently without a clear distinction of strata, and with an incredible abundance and diversity of epiphytes. Canopy species are dominated by *Myrtaceae* (*Siphoneugena reitzii, Myrceugenia euosma, M. bracteosa, Myrceugenia* spp., *Gomidesia sellowiana, G. palustris*) and/or *Drimys brasiliensis,* generally with tortuous trunks, rigid branches, roundish crowns, and generally small, discolored, glabrous, and coriaceous leaves. *Mimosa scabrella, Clethra scabra, Tibouchina sellowiana, Lamanonia* sp., *Ilex kleinii* (apparently endemics in SC), and *I. microndonta* are additional, less common canopy species.

There are some floristic differences between cloud forests of the Aparados and the Serra do Mar (Klein 1978), which will be discussed below. Several tree species occur in only one of the two regions, although they may also occur in other geographic areas or vegetation types. There are also some endemic species that apparently are restricted to one or the other of these two TMCF regions. Even where species are shared between the two regions, there are great variations in the relative dominance of canopy species (e.g., *Drimys brasiliensis* occurs with much lower frequency in São Francisco de Paula).

In the Serra do Mar, small trees such as *Myrcia rupicola* (also in "pinhal"), *Myrciaria cordifolia* (also in "mata Atlântica"), *Psidium spathulatum*, and *Myrceugenia leptorhyncha* are thought to be south Brazilian endemics, while other trees, for example, *Schoepfia brasiliensis, Weinmannia discolor, Eugenia myrtifolia, E. handroana, E. pluriflora, Myrcia obtecta, Ilex theezans,* and *I. pseudobuxus* have a wide geographic range. Localized endemics to the Aparados TMCF include *Mimosa balduinii, Crinodendron brasiliense,* and *Euplassa nebularis*. Other important cloud forest trees of more widespread geographic distribution include *Inga lentiscifolia, Weinmannia humilis*, and *Azara uruguayensis*.

There are many shrub species in the cloud forest, including *Berberis laurina, Clethra scabra, C. uleana, Escallonia* spp., *Weinmannia* spp., *Miconia ramboi, Rhamnus sphaerosperma, Tibouchina dubia, Leandra quinquedentata, Myrcia lajeana, Myrceugenia regnelliana, M. nothorufa,* and *M. myrcioides*. *Mimosa coniflora* is believed endemic to the cloud forest of Serra do Mar, while *Tibouchina reitzii* and *Leandra reitzii* (both also in "mata Atlântica") are shrubs apparently restricted to SC (but not found in the Aparados) and to Paraná. Shrub species endemic to the Aparados include *Tibouchina ramboi, Escallonia petrophila, Mimosa taimbensis, M. aparadensis* (synonym of *M. pilulifera* for Lins 1984), *M. ramentacea, Buddleja kleinii, B. cestriflora, Berberis kleinii* (treated as *B. montana* by Smith 1962), *Myrceugenia ramboi*, and *M. venosa*. *Myrceugenia macrosepala* and *Dicksonia sellowiana* are more widely distributed species, but have not been reported from the cloud forest of the Serra do Mar.

Lianas are also frequent in the TMCF and include *Fuchsia regia* (also epiphyte), *Senecio desiderabilis* (idem), *Griselinia ruscifolia* (idem), *Mikania paranensis*, and *Valeriana scandens*. Bamboo (*Chusquea mimosa*) also occurs with great abundance in both regions.

The herbaceous layer is composed of *Pamphalea araucariophila, Hydrocotyle quinqueloba, Croton migrans,* several *Melastomataceae, Plagiogyria semicordata, Hymenophyllum magellanicum, Selaginella* spp., and bryophytes (*Ptychomnion fruticetorum, Breutelia* spp.). In the Serra do Mar, severe terrestrial Bromeliaceae (*Vriesea hoehneana, Canistrum cyathiforme*, and the very rare endemic *Aechmea apocalyptica*) are present. Other species provide dominant ground cover in the Aparados—*Gunnera manicata* (not found in the Serra do Mar), together with *Hydrocotyle itatiaiensis, Eryngium smithii,* and *Leandra luctatoris* (endemic in the Aparados).

The TMCF woody plants are densely covered by vascular epiphytes, including *Hatiora rosea* and *Schlumbergera obtusangula* (the latter occurring only in the Serra do Mar), and several Bromeliaceae such as *Vriesea philippocoburgii* and the rare night-flowering *Aechmea kleinii,* endemic to the Aparados. Other epiphytes include *Pleurothallis* spp., *Sophronitis coccinea, Oncidium concolor, Rodrigueziella gomezoides* (Orchidaceae), *Nematanthus australis*, *Grammitis organense*, as well as many other ferns. Mosses (*Leiomela piligera, Schlotheimia* spp., *Rigodium* spp., and *Prionodon densus*), liverworts (*Lepidozia, Herbertus, Lepicolea, Anoplolejeunea, Lejeunea*), and lichens (*Heterodermia, Cladonia, Sticta,* and *Usnea*) are also abundant in the epiphyte biomass. These plants occur not only on trunks and branches but also on leaves, and are important to water retention in the canopy.

Pteridophytes are conspicuous in the Brazilian TMCF. The majority of the species are either exclusively terrestrial (*Asplenium langsdorffii, A. martianum, A. araucarietii*) or exclusively epiphytic (*Grammitis schenckii, Hymenophyllum fucoides, H. rufum*), although there are some species (*Hymenophyllum polyanthos, H. caudiculatum*) with varied behavior (Bueno and Senna 1992). Sehnem (1979), summarizing his several works with south Brazilian ferns, cited 15 species considered exclusive to the cloud forest (one is endemic: *Polypodium subinaequale*); 67 species (18 endemics) are also found in "pinhal," or grassland. Bueno and Senna (1992) presented many species also found in the cloud forest. Schnell (1987:286) calculated a "pteridophytic index," which showed that ferns represent 15 percent of the species in the Serra dos Órgãs. In the south Brazilian cloud forest, it is possible that this index will also show the great importance of ferns.

In a taxonomic survey of an order of xylophilous fungi from a reserve in the Aparados, Silveira and Guerrero (1991) identified 38 species. *Datronia mollis* and *Junghuhnia nitida* were cited for the first time for Brazil; *Trichaptum sector, Polyporus tricholoma, Phellinus wahlbergii, P. gilvus*, and *Ganoderma applanatum* were abundant in the region studied (which was not exclusively cloud forest).

In regions that were disturbed several or many years ago, dense populations of *Tibouchina sellowiana* or *Mimosa scabrella* are generally the dominant species. *Senecio* spp., *Baccharis* spp., *Eupatorium* spp. (as the endemic *E. laxicephalum*), other *Asteraceae, Solanum* spp., and *Pteridium aquilinum* frequently appear in places where the cloud forest was recently removed. If such an area were burned, *P. aquilinum* is generally the dominant species. In the recovery of rocky places with great humidity in the Aparados, Klein (1978) cites *Gunnera manicata* as very important.

Several plants cited in this work (*Fuchsia, Berberis, Escallonia, Drimys, Griselinia, Plagiogyria*) were used as evidence to confirm the generic similarities evidenced by Rambo (1951, 1956b) between this flora and the Andean flora. Our study was mainly floristic, but we found many resemblances between the south Brazilian and the Central American cloud forests, through some situations described and many genera cited by Lawton and Dryer (1980) for Costa Rica.

In many places in the Aparados, there are interesting ecotones between cloud forest and "pinhal." Sometimes, mainly in Rio Grande do Sul, seedlings of *Araucaria angustifolia* occur in association with cloud forest species. Such cases represent a transition between "pinhal" and cloud forest.

Also in the Aparados, between the cloud forest and the grassland, a shrubby or subshrubby belt generally occurs, formed by several species of Ericaceae (*Leucothoe, Gaylussacia, Gaultheria*) and Melastomataceae (mainly *Tibouchina, Leandra, Miconia*), as well as *Baccharis* spp., *Mimosa* spp., and *Croton* spp.

At the crest (600–900 m above sea level) of several mountains (Morro do Baú, Morro Spitzkopf, etc.) inside the region covered by Atlantic forest in Santa Catarina (Klein 1988), there are vegetations with many physiognomic resemblances with the cloud forest. As the floristic composition is relatively different from the cloud forest and the fog probably is less frequent, we considered them as part of the Atlantic forest.

In São Francisco de Paula, Canela, and Gramado (Rio Grande do Sul), there are regions frequently covered by fog, but the vegetation is different from the cloud forest characterized in this work. Future studies are necessary to characterize this vegetation, which appears to show many contributions from "pinhal" and Atlantic forest.

Faunal Composition

There are no published faunal surveys for the cloud forest in southern Brazil. Information on ecology, behavior, and natural history is limited and based on chance field observations (tracks, fecal pellets, carcasses, and sightings) by scientists working in the cloud forest (J. Albuquerque, A. Cimardi, M. Mazzolli, and J. C. Voltolini).

There are cloud forest records for puma (*Felis concolor*), crab-eating fox (*Cerdocyon* spp.), coati (*Nasua nasua*), skunk (*Conepatus* sp.), tapir (*Tapirus terrestris*), deer (*Mazama* sp.), armadillo (*Dasipus* sp.), agouti (*Dasyprocta* sp.), harpy eagle (*Harpia harpyja*), tinamous (*Tinamus* sp.), and others. Some species that use this forest as a refuge (e.g., puma, harpy eagle, and tapir) are highly endangered because of hunting and extensive habitat degradation in other parts of their range (Bernardes, Machado, and Rylands 1990; AAZPA and IUCN 1992).

The puma exemplifies the overall faunal situation. The farmers and ranchers living around reserves practice selective "defaunazation." Hunting pressure focused on medium-size animals (e.g., armadillo, agouti, and paca) has been more intensive. This selective hunting and related habitat degradation decrease available prey for the puma. Consequently, there are several records of puma attacks on cattle, with the result that puma are frequently pursued and killed by farmers. Despite this example, there are few data on local fauna, and this area of research demands immediate attention in view of current hunting pressures.

Conservation Status and Priorities

Brazilian environmental legislation gives good protection to the cloud forests, because they are included in several conservation units and generally are in places considered "permanent preservation region" by the Forest Code enacted in 1965. Unfortunately, the Brazilian legislation is generally not obeyed.

Localization of the cloud forest on mountain tops or in the plateau border serves partially to isolate it from anthropogenic degradation. Despite this, some "reserves" and surrounding areas have been cleared frequently by farmers to create pasture and obtain firewood. The cloud forest also has been cut for conversion to homogeneous plantations of *Pinus*, mainly along the Aparados (e.g., in Serra da Rocinha, Fortaleza Canyon). Another problem is the heavy extraction of tree fern ("xaxim"—*Dicksonia sellowiana*) for use in floriculture and gardening; this exploitation is on the increase.

At the end of the winter (August, September), a major part of the "campos planaltinos" is burned by farmers to accelerate grass regrowth. Many fires reach the cloud forests above the grasslands, causing destruction of the border areas and effecting unknown (but probably tragic) damage on the living organisms and abiotic characteristics.

In 1951, a huge fire burned the grasslands and cloud forests from southern Santa Catarina up to northeastern Rio Grande do Sul. The area was extensively destroyed, and Klein (1988) affirms that dozens of species probably became extinct from this fire. Because we have few data about the consequences of that fire, it is not clear whether permanent deforestation occurred or whether the cloud forest can regenerate itself (which may take centuries); but the species richness, the ecosystem structural complexity, and species' interdependencies almost certainly were disrupted.

With all these problems, it is impossible to understand why Teixeira and Coura Neto (1986:567) consider the anthropogenic action in this forest as "inexpressive." The changes and influences of human use described by Saunders, Hobbs, and Margueles (1991) certainly must be considered in any ecological analysis or conservation project for this region.

There is currently no true conservation program targeting TMCF in Brazil. The major part of this forest type in Santa Catarina and Rio Grande do Sul is located outside the protected areas (e.g., national or state parks, biological reserves, ecological stations). Even "reserves" containing some cloud forests are not totally established and protected.

As part of a conservation strategy developed by the Brazilian government, some cloud forest reserves have been demarcated. However, they exist only as "paper reserves" in governmental decrees and in publications or maps from government agencies. These reserves have no funding or site protection.

Perhaps the best example of a "paper reserve" is the São Joaquim National Park, with 49,300 ha in Santa Catarina (Câmara 1991), including the Morro da Igreja and protecting cloud forest, grassland, and "pinhal." It was enacted/ "created" in 1961 by Federal Decree no. 50992 (Klein 1988), but remains unprotected even now. There is virtually no government control: the human population living inside the park has grown; the government has not regularized ownership of any park land; there is still no fixed boundary or demarcation; and deforestation is increasing everyday (the "pinhal" was completely destroyed, as well as several regions with cloud forest).

Other "reserves" in Santa Catarina that include cloud forest are not much different. The best is probably the Serra Furada State Park, which now has 1,329 ha of relatively demarcated land in Grão Pará and Orleans (only 280 ha are really governmental proprietorship) and one fixed government controller (E. Amaral, pers. com.). Another is the "recreated" and amplified Aguaí State Biological Reserve (formerly joined with the Serra Furada), which includes 7,672 ha in Meleiro, Nova Veneza, and Siderópolis, but which has no regularization of landownership, no demarcation, and no fixed controller. Hunting is intense inside the reserve (E. Amaral, pers. com.). Both are in the Aparados and also protect the Atlantic forest.

Although Klein (1981) describes the occurrence of cloud forests in the Serra do Tabuleiro State Park (SC), we are not familiar with this area because accessibility is difficult. This park, created in 1975 with 90,000 ha, is the greatest in Santa Catarina and Rio Grande do Sul, but only a small part has actually been established.

There is no conservation unit including the cloud forests from the Serra do Mar in Santa Catarina (SC), but the Forest Code protects the majority of them.

In Rio Grande do Sul (RS), Câmara (1992) cites the Serra Geral State Biological Reserve, created in 1982 with 1,700 ha demarcated (Fundação SOS Mata Atlântica 1992). We do not know this reserve. It is localized in Terra de Areia and includes cloud forest, but an illegal road was made inside the reserve, facilitating increased deforestation (L. Baptista, pers. com.). Also in RS, The Serra Geral National Park, which apparently exists only in decree and perhaps includes cloud forest, was "created" in 1992.

The only relatively well-established reserve that includes cloud forest is the Parque Nacional de Aparados da Serra, with an area of about 10,250 ha and confined to Cambará do Sul in RS and Praia Grande in SC. The park also includes the "campo planaltino," "pinhal," and "mata Atlântica" ecozones (Baptista et al. 1978). It was created, and is reasonably well-maintained, mainly because of the scenic and much-visited Itaimbezinho Canyon. The park has not been completely established (Fundação SOS Mata Atlântica 1992) because several landowners were not indemnified until recently, and cattle and fire occurrences are still common inside the park. Klein (1988) affirms that more than 100 rare or endemic species are threatened with extinction in this park.

Conclusions

Little is yet known about species diversity in the TMCF of southern Brazil. Therefore, we need basic biological research and inventories as a prerequisite for ranking conservation action.

Protection of remnant TMCF is essential to preserve what is surely a unique component of Brazilian biodiversity.

Because of the situation described in this paper, we suggest some basic priorities for southern Brazilian cloud forests:

1. accurately delimit the total area of remaining cloud forests (primary and disturbed) and catalog the remaining flora and fauna of this unique forest type;
2. improve current forest reserve administration and supporting governmental institutions;
3. expand funding for basic ecological and management research in existing reserves;
4. fully establish the existing "paper reserves" and create new reserves;
5. develop environmental education programs applicable for schools and communities living near cloud forests; and
6. encourage ecotourism in existing reserves (e.g., catching a glimpse of a puma) to generate income for management actions and TMCF protection.

Acknowledgments. We are grateful to Bruno Irgang for invaluable discussions and for accompanying us in the field. Jorge Waechter and Terezinha Paulilo critically read the manuscript and made many corrections and suggestions. Marcos Sobral, T. Paulilo, J. Waechter, Paul Berry, Renzo Bassanetti, Leonor Souza, Eduardo Amaral, Jorge Albuquerque, Luís Baptista, Marcelo Mazzolli, Roberto Palacios, and Ana Cimardi were very important in different stages of this work. We thank Dr. Lawrence Hamilton for the invitation to attend the Montane Cloud Forests Symposium.

References

AAZPA and IUCN. 1992. *Field action plan, 1991 and 1992.* Virginia: Smithsonian Institution.

Ab'Saber, A.N. 1989. Introduction. In *Campos de altitude*, G. Martinelli and J. Bandeira, pp. 22–35. Rio de Janeiro: Index.

Baptista, L.R.M., B.E. Irgang, J.F.M. Valls, and J.L. Waechter. 1978. Parque Nacional dos Aparados da Serra. Levantamento da vegetação. Unpublished.

Bernardes, A.T., A.B.M. Machado, and A.B. Rylands. 1990. *Fauna brasileira ameaçada de extinção. Brazilian fauna threatened with extinction.* Belo Horizonte: Fundação Biodiversitas.

Bueno, R.M., and R.M. Senna. 1992. Pteridófitas do Parque Nacional dos Aparados da Serra. v. 1, Região do Paradouro. *Caderno de Pesquisa Sér. Bot.* 4(1): 5–12.

Cabrera, A.L., and A. Willink. 1980. *Biogeografía de América Latina.* 2d ed. Washington, D.C.: OEA.

Câmara, I.G. 1991. Conservação da mata atlântica. In *Mata atlântica/Atlantic rain forest*, pp. 158–183. s.l.: Index and Fundação S.O.S. Mata Atlântica.

Câmara, I.G. 1992 (1991). *Plano de ação para a mata atlântica*. São Paulo: Fundação SOS Mata Atlântica.

Fundação SOS Mata Atlântica. 1992. *Dossiê mata atlântica 1992*. São Paulo: Fundação SOS Mata Atlântica.

Hueck, K. 1972. *As florestas da América do Sul*. São Paulo: Polígono and Editora da Universidade de Brasília.

Joly, A.B. 1970. *Conheça a vegetação brasileira*. São Paulo: Editora da Universidade de São Paulo and Polígono.

Joly, C.A., H.F. Leitão Filho, and S.M. Silva. 1991. O patrimônio florístico. In *Mata atlântica/Atlantic rain forest*, pp. 94–125. s.l.: Index and Fundação S.O.S. Mata Atlântica.

Klein, R.M. 1960. O aspeto dinâmico do pinheiro brasileiro. *Sellowia* 12(12): 17–44, 4 pl.

Klein, R.M. 1978. *Mapa fitogeográfico do Estado de Santa Catarina*. Itajaí: SUDESUL, FATMA and Herbário Barbosa Rodrigues.

Klein, R.M. 1980. Ecologia da flora e vegetação do Vale do Itajaí (continuação). *Sellowia* 32(32): 165–389.

Klein, R.M. 1981. Fitofisionomia, importância e recursos da vegetação do Parque Estadual da Serra do Tabuleiro. *Sellowia* 33(33): 5–54.

Klein, R.M. 1984. Aspectos dinâmicos da vegetação do Sul do Brasil. *Sellowia* 36(36): 5–54.

Klein, R.M. 1988. Parques Nacionais e Estaduais; Reservas Biológicas Estaduais e particulares, bem como a Estação Ecológica do Estado de Santa Catarina. *Estudos de Biologia* 18:11–29.

Klein, R.M. 1992/1990. *Espécies raras ou ameaçadas de extinção. Estado de Santa Catarina*. v.1, Mirtáceas e Bromeliáceas. Rio de Janeiro: IBGE.

Lawton, R., and V. Dryer. 1980. The vegetation of the Monteverde Cloud Forest Reserve. *Brenesia* 18:101–116.

Leite, P.F., and R.M. Klein. 1990. Vegetação. In *Geografia do Brasil*, v.2, Região Sul, pp. 113-150. Rio de Janeiro: IBGE.

Lindman, C.A.M. 1900. *Vegetationen i Rio Grande do Sul*. Stockolm: Nordin & Josephson.

Lins, D.M.T. 1984. *Mimosa* (Leguminosae—Mimosoideae) série *Lepidotae* no Rio Grande do Sul. *Boletim do Museu Botânico Kuhlmann* 7(3): 3–59.

Maack, R. 1968. *Geografia física do Estado do Paraná*. Curitiba: Banco de Desenvolvimento do Paraná, UFPR and IBPT.

Martinelli, G., and J. Bandeira. 1989. *Campos de altitude*. Rio de Janeiro: Index.

Pfadenhauer, J., and S.C. Boechat. 1981. Vegetation und Ökologie eines Sphagnum-Moores in Südbrasilien. *Vegetatio* 44:177–187.

Rambo, B. 1949. A flora de Cambará. *Anais Botânicos do Herbário Barbosa Rodrigues* 1(1): 111–135.

Rambo, B. 1951. O elemento andino no pinhal riograndense. *Anais Botânicos do Herbário Barbosa Rodrigues* 3(3): 7–39.

Rambo, B. 1953. História da flora do planalto riograndense. *Anais Botânicos do Herbário Barbosa Rodrigues* 5(5): 185–232.

Rambo, B. 1956a. *A fisionomia do Rio Grande do Sul*. 2d ed. Porto Alegre: Livraria Selbach.

Rambo, B. 1956b. A flora fanerogâmica dos Aparados riograndenses. *Sellowia* 7/8(7): 235–298.

Rambo, B. 1958. Die Alte Südflora in Brasilien. *Pesquisas Bot.* 2:177–198.

Reitz, R. 1965. Plano de coleção. *Flora Ilustrada Catarinense* IV(A): 1–71.

Reitz, R., ed. 1965–1989. *Flora Ilustrada Catarinense*. Itajaí: Herbário Barbosa Rodrigues. 161 fasc.

Rizzini, C.T. 1979. *Tratado de fitogeografia do Brasil.* v. 2. São Paulo: HUCITEC and Editora da Universidade de São Paulo.

Romariz, D.A. 1964. A vegetação. In *Brasil. A terra e o homem*, ed. A. Azevedo, v.1, pp. 485–536. São Paulo: Companhia Editora Nacional.

Saunders, D.A., R.J. Hobbs, and C.R. Margules. 1991. Biological consequences of ecosystem fragmentation. *Conservation Biology* 5(1): 18–32.

Schnell, R. 1987. *La flore et la végétation de l'Amérique tropicale.* v.1. Paris: Masson.

Sehnem, A. 1979. Semelhanças e diferenças nas formações florestais do Sul do Brasil. *Acta Biológica Leopoldensia* 1(1): 111–135.

Silveira, R.M.B., and R.T. Guerrero. 1991. Aphyllophorales poliporóides (Basidiomycetes) do Parque Nacional de Aparados da Serra, Rio Grande do Sul. *Boletim do Instituto de Biociências* 48:3–121.

Smith, L.B. 1962. Origins of the flora of Southern Brazil. *Contributions from the U.S. National Herbarium* 35(3): 215–249.

Takhtajan, A. 1986. *Floristic regions of the world.* Berkeley: University of California Press.

Teixeira, M.B., and A.B. Coura Neto. 1986. 4—Vegetação. Folhas Porto Alegre e Lagoa Mirim. In *Levantamento de Recursos Naturais*, v. 33, pp. 565–574. Rio de Janeiro: IBGE.

Veloso, H.P., and L. Góes-Filho. 1982. Fitogeografia brasileira. *Boletim Técnico do Projeto RADAMBRASIL. Série Vegetação* 1:3–79.

Veloso, H.P., A.L.R. Rangel Filho, and J.C.A. Lima. 1991. *Classificação da vegetação brasileira, adaptada a um sistema universal.* Rio de Janeiro: IBGE.

Veloso, H.P., and R.M. Klein. 1968. As comunidades e associações vegetais da mata pluvial do Sul do Brasil. VI. Agrupamentos arbóreos dos contra-fortes da Serra Geral situados ao sul da costa catarinense e ao norte da costa sul-riograndense. *Sellowia* 20(20): 127–180.

Wettstein, R.R.v. 1904. *Vegetationsbilder aus Südbrasilien.* Leipzig: Franz Deuticke.

9. Cloud Forest, the Massenerhebung Effect, and Ultraviolet Insolation

J. R. Flenley

As usually defined, the Massenerhebung or mountain mass elevation effect means the occurrence of physiognomically and sometimes floristically similar vegetation types at higher altitudes on large mountain masses than on small isolated peaks, especially those in or near the sea. Although the effect was first reported in the European Alps (Schroeter 1908) and in North America (where it is known as the Merriam Effect; Martin 1963), it is best known in the tropics. Perhaps its clearest expression is the occurrence of tropical mountain cloud forest (TMCF) (upper montane rain forest) at lower altitudes on isolated peaks than on the main mountain masses, which are taken as the norm (Figure 1).

Some explanations of this phenomenon have involved mean temperatures and cloud formation. Cloud formation is often observed on isolated peaks at quite low altitudes, and it was shown by Hastenrath (1968) using radiosonde balloons in Mexico that lapse rates were somewhat steeper over lowlands than over large mountain masses. Similarly, a steep lapse rate (0.81° C per 100 m) was recorded on the 800 m high island of Krakatau in Indonesia (Forster 1982); the regional average is 0.61° C per 100 m (Walker and Flenley 1979). Presumably the afternoon clouding seen on isolated peaks is related to this temperature regime, and the clouding does have pronounced ecological effects. For instance, it increases humidity to 100 percent and reduces total insolation received by about 30 percent compared with unclouded sites in Sabah, Malaysia (Bruijnzeel et al. 1993). The difficulty comes in re-

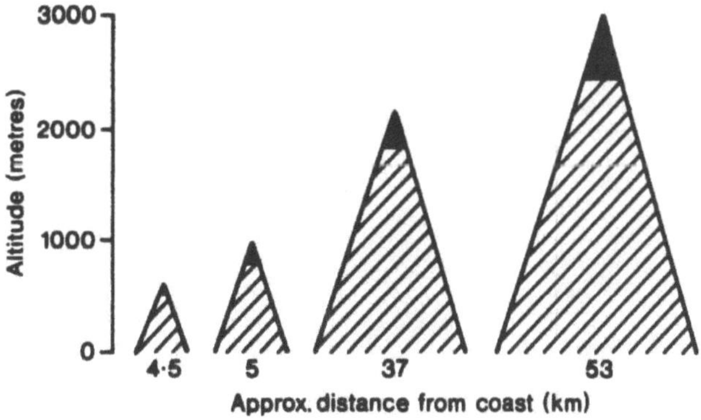

Figure 1. The Massenerhebung (mass elevation) effect illustrated by the occurrence of cloud forest on mountains in Indonesia. From left to right: Mount Tinggi (Bawean), Mount Ranai (Natuna Island), Mount Salak (W. Java), and Mount Pangerango (W. Java). The small mountains near the sea have cloud forest at lower altitudes than the large mountains inland *(after van Steenis 1972).*

lating these changes to the morphological peculiarities of the vegetation. Usually, if plants are grown in high humidity and low insolation, they become etiolated (i.e., they are tall, have long internodes, and large, thin, pale green leaves). This is the exact opposite of the attributes of tropical mountain cloud forest, which are stunted growth, short internodes, small, thick leaves (with a hypodermis), and often extra pigments (anthocyanins or flavonoids).

As I have pointed out elsewhere (Flenley 1992), these attributes have been produced in a number of species by the application of extra UV-B light. Since high altitudes on tropical mountains receive extra UV-B (Caldwell, Robberecht, and Billings 1980), this seems a plausible hypothesis for explaining, at least partially, the morphology of TMCF on high mountains in the tropics. It receives additional support because TMCF was almost absent during the Late Quaternary, according to the palynological record. This may be explained by the absence of a habitat with a suitable combination of temperature and UV-B insolation at that time (Figure 2) (Flenley 1991, 1993).

Perhaps this mechanism might also be relevant in explaining the Massenerhebung effect. UV-B light, like visible light, may experience total reflection at water or cloud surfaces. Total insolation may thus be increased by up to 70 percent through reflection from clouds (Figure 3). In the early morning, low peaks are surrounded by a sea of clouds (Figure 4), which will reflect sunlight strongly up onto vegetation. Islands receive similar reflection from the sea surface. Rayleigh scattering of light from the air molecules in the sky will also be particularly effective in the clear morning and favors UV above all other wavelengths (Dave and Halpern 1976).

There have been several attempts to relate the Massenerhebung effect to

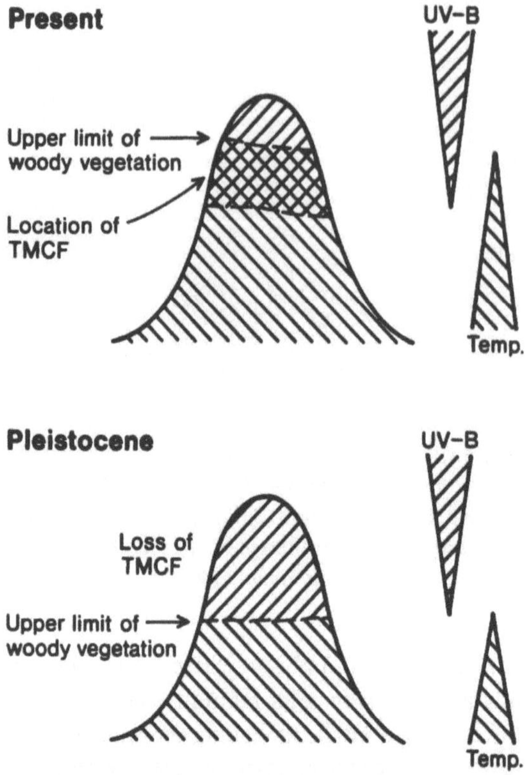

Figure 2. The hypothesis that lack of tropical mountain cloud forest in the Pleistocene may be explained by absence of a habitat with a suitable combination of mean annual temperature and UV-B insolation *(diagram by G. Rapson, unpublished)*.

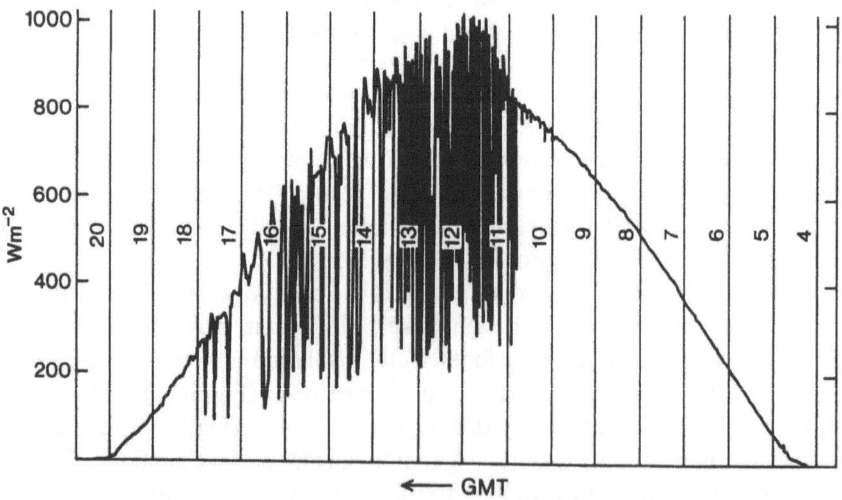

Figure 3. Solar radiation on a day of broken cloud (11 June 1969) at Rothamsted (52° N, 0° W) taken directly from recorder charts. Note very high values of irradiance immediately before and after occlusion of the sun by cloud and the regular succession of minimum values when the sun is completely obscured *(after Monteith 1973)*.

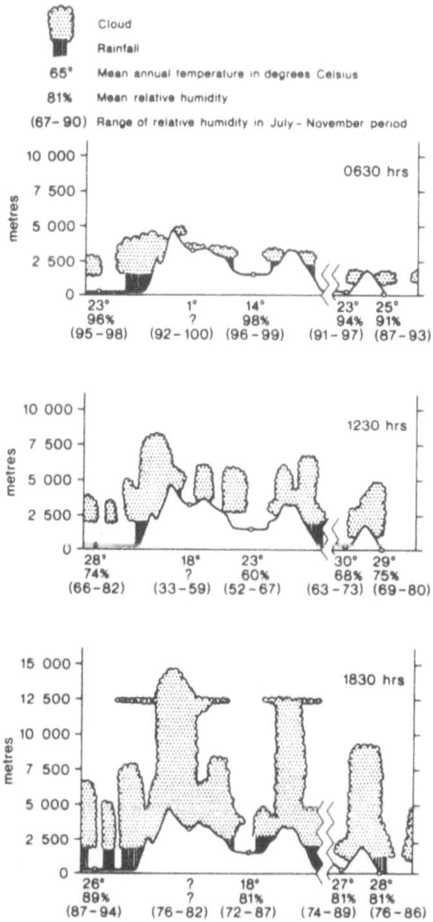

Figure 4. The daily weather regime in the New Guinea Highlands *(modified after Brookfield 1964)*. Note especially the cloud below the peaks at 6:30 A.M., followed by heavy clouding at 12:30 P.M. and later.

soil attributes. Among the more successful of these was that by Bruijnzeel et al. (1993), who found a correlation between the occurrence of stunted forest and phenolic compounds in leaf litter. The latter were thought to cause stunting by harmful effects on plant physiology. This theory is complementary to the UV theory, however. The usual response of plants to excess UV-B is to produce protective compounds that absorb UV-B. These are usually flavonoids or alkaloids (Caldwell 1981), or anthocyanins (Lee and Lowry 1980). Many of these compounds are phenolic or are likely to break down into phenolic compounds in litter. They could provide a reinforcement mechanism, exaggerating the original stunting caused by the UV-B (Bruijnzeel, this volume).

I conclude, therefore, with the tentative hypothesis that the Massenerhebung effect could be partly the result of a high dose of UV-B due to reflection from clouds or the sea in the mornings. Obviously further research is needed to test this idea, although there is little doubt that the full explanation will be a multivariate one. Some factors may be more important in one location, while other factors may dominate elsewhere.

Acknowledgments. I am grateful to Dr. L. A. Bruijnzeel and Dr. R. Hofstede for helpful discussion, and to the East-West Center for making it possible for me to attend the conference.

References

Brookfield, H. C. (1964). The ecology of highland settlement: Some suggestions. *American Anthropologist 66*, 20-38.

Bruijnzeel, L. A., M. J. Waterloo, J. Proctor, A. T. Kuiters, and B. Kotterink. (1993). Hydrological observations in montane forests on Gunung Silam, Sabah, Malaysia, with special reference to the "*Massenerhebung*" effect. *J. Ecol. 81*, 145-167.

Caldwell, M. M. (1981). Plant response to solar ultraviolet radiation. Chapter 6, pp. 169-197 in *Physiological Plant Ecology I, Encyclopedia of Plant Physiology, New Series, Vol. 12A*, Berlin, Springer.

Caldwell, M. M., R. Robberecht, and W.D. Billings. (1980). A steep latitudinal gradient of solar ultraviolet-B radiation in the arctic-alpine life zone. *Ecology 61*, 600-611.

Dave, J. V., and P. Halpern. (1976). Effect of changes in ozone amount on the ultraviolet radiation received at sea level of a model atmosphere. *Atmospheric Environment 10*, 547-555.

Flenley, J. R. (1991). Ultraviolet light and the altitudinal zonation of tropical rain forests, pp. 27-33 in *Aspects of Environmental Change*, ed. T. R. R. Johnston and J. R. Flenley, Department of Geography, Massey University, Miscellaneous Series 91/1.

Flenley, J. R. (1992). UV-B insolation and the altitudinal forest limit, pp. 273-282 in *Nature and Dynamics of Forest-Savanna Boundaries*, ed. P. A. Furley, J. Proctor, and J. A. Ratter. London, Chapman & Hall, 616 pp.

Flenley, J. R. (1993). Two problems of the Late Quaternary in the Sunda-Sahul region and a possible solution. *Proc. International Symposium on Global Change (IGBP), Waseda University, Shinjuku, Tokyo, Japan, 27-29 March 1992.*

Forster, R. M. (1982). Chapter 5: A study of the spatial distribution of bryophytes on Rakata, pp. 103-126 in *The Krakatoa Centenary Expedition, Final Report*. University of Hull, Geography Department, Miscellaneous Series No. 25, 196 pp.

Hastenrath, S. (1968). Certain aspects of the three-dimensional distribution of climate and vegetation belts in the mountains of central America and southern Mexico. *Colloquium Geogr. 9*, 122–130.

Lee, D. W., and J. B. Lowry. (1980). Young leaf anthocyanin and solar ultraviolet. *Biotropica 12*, 75-76.

Martin, P. S. (1963). *The last 10,000 years: A fossil pollen record of the American Southwest*. University of Arizona Press, Tucson, Arizona, 87 pp.

Monteith, J. L. (1973). *Principles of Environmental Physics*. London, Arnold, 241 pp.

Schroeter, C. (1908). Das Pflanzenleben der Alpen. Eine Schilderung der Hochgebirgsflora. 806 pp. Zurich. Verlag von Albert Raustein.
van Steenis, C. G. G. J. (1972). *The mountain flora of Java*, illustrated by Amir Hamzah and Moehamad Toha, 90 pp., Leiden, E. J. Brill.
Walker, D., and J. R. Flenley. (1979). Late Quaternary vegetational history of the Enga District of upland Papua New Guinea. *Phil. Trans. R. Soc. B, 286*, 265-344.

10. Some Low Elevation Fog Forests of Dry Environments: Applications to African Paleoenvironments

Alain Gioda, Jean Maley, Roberto Espejo Guasp,
and Andrés Acosta Baladón

Macaronesian Elfin Cloud Forests: Canarian Examples

We present some examples of cloud water and fog precipitation on vegetation following Kerfoot's pioneer work (1968). High fog frequency allows for the existence of forest communities in otherwise arid (low rainfall) conditions. Using the El Hierro fountain tree as an example, Macaronesian elfin forests are described, with special attention to the Garajonay National Park (La Gomera, Canary Islands) where very significant stands of fog/cloud forest remain.

The Macaronesian region includes the Azores, Madeira, the Selvagens, Canary, and the Cape Verde islands. Of these, the Canary Islands harbor the richest botanical diversity and most extensive remnant stands of fog/cloud forest.

El Hierro is the smallest island (280 km^2) of the Canarian archipelago (28° N; 13–18° W) (Figure 1). Under an arid climate, "fountain" trees (laurels, junipers, pine trees) have been used for centuries to provide freshwater. Water is naturally collected by such trees from the intense fog (*niebla*) that occurs often without any rainfall. From the leaves, the water drops on the soil where it can be collected and stored. One laurel tree (*Ocotea foetens*), the so-called *Garoé,* was used by the native pre-Hispanic populations until 1610, when it was uprooted by a hurricane (Figure 2). The following is an extract from George Glas's *History of the Discovery and Conquest of the Canary Islands* (1764):

> On the top of this rock [»1,000 m] grows a tree called, in the language of the ancient inhabitants, *Garoé* (sacred or holy tree), which for many years has been preserved sound, entire and fresh. Its leaves constantly distill such a quantity of water, as is sufficient to furnish drink to every living creature in Hierro; nature having provided this remedy for the drought of the island. It is distinct from other trees, and stands by itself. The circumference of its trunk is about twelve spans [1 span ≅ 22.8 cm], the diameter four, and in height from the ground to the top of the highest branch forty spans: the circumference of all the branches together is one and twenty feet [1 foot ≅ 30.5 cm]. The branches are thick and extended; the lowest commence about an ell [1 ell ≅ 120 cm] from the ground. Its fruit resembles the acorn, and tastes something like the kernel of a pineapple, but it is softer and more aromatic. The leaves of this tree resemble those of the laurel, but are larger, wider, and more curved; they come forth in perpetual succession, so that the tree always remains green.

In memory of the *Garoé*, another laurel (*Ocotea foetens*; Figure 3) was planted in 1945 at the same place as the old one by Don Zósimo Hernández Martin, who worked with ICONA (Instituto Nacional para la Conservación de la Naturaleza). By 1993, 48 years later, this laurel had become a new fountain tree producing high quantities of fog water like the ancient *Garoé* (Gioda et al. 1992; Gioda, Hernández Martin, and Gonzáles 1993). The average annual rainfall is 800 mm in this El Hierro region, according to data from *Consejeria de Obras Publicas*.

Figure 1. Canary Islands, Spain, including the location of the Garoé and Garajonay National Park. *Drafted by M. Suavin.*

Another famous fountain tree, a juniper (*Juniperus phoenicea*) located at Cruz de los Reyes (1,353 m), was destroyed by a bush fire in 1990, but both the concrete collector under the plant and the water storage system, built on behalf of Don Zósimo Hernández Martin, are still in place (Acosta Baladón 1992). Plans have been made to transplant a new juniper tree in this location (Gioda 1993) with the concurrence of local authorities (island government, ICONA), who in the 1970s and 1980s supported construction of the concrete collector and fog water storage tanks. Today, cattle drink fog water collected from pine trees (*Pinus canariensis*) on Mirador de los Bascos (650 m) and on El Hierro (Gioda, Hernández Martin, and Gonzáles 1993).

Besides these Canarian examples, farmers have used fog water on Brava (Cape Verde Islands) since 1942, mainly collected from agaves (*Furcraea gigantea*) and palm trees (*Phoenix* sp.) (Reis F. Cunha 1964; Acosta Baladón and Gioda 1991, 1992).

Garajonay National Park (La Gomera Island): A Canarian Elfin Cloud Forest

The elfin cloud forest of Mount Garajonay on La Gomera Island has been classified a Spanish national park since 1981 (total area: 3,985 ha). Since 1986, it has been registered as a UNESCO World Heritage site (Pérez de Paz 1990).

Figure 2. El Hierro coat of arms. The fountain tree, the so-called *Garoé,* is at the center of the escutcheon. This tree, which was uprooted in 1610 by a hurricane, is still visible symbolically, especially on taxi cabs that always carry stickers with El Hierro's arms on their doors.

Climate and Hydrology

The Canary Islands are essentially influenced by the subtropical high pressure systems of the north Atlantic Ocean, trade winds being the factor driving the atmospheric circulation. These winds from the NNE are humid (relative humidity 75–90 percent) and collide with the insular blocks, producing a "sea of clouds," the altitude of which usually ranges from 600 to 1,500 m. In the vicinity of Tenerife island, the average frequency of occurrence of the "sea of clouds" is 199 days per year. These clouds do not result in abundant rainfall because an upper-level temperature inversion restricts their vertical growth and the formation of a deep cloud layer necessary for raindrop formation. Mount Garajonay National Park (elevation 700–1,487 m) has an evergreen forest sustained by direct interception of cloud water. The mean annual rainfall in the area ranges from 600 to 800 mm. It is estimated that forest fog/cloud-water interception is more than double this amount (Santana Pérez 1990; Gischler 1991).

Figure 3. *Ocotea foetens*, a Macaronesian laurel. The *Garoé* was an *O. foetens*. *Drawn by M. Djellouli.*

Vegetation and Flora

The laurel forest, the so-called *laurisilva*, is a luxuriant subtropical vegetation with a diverse structure of trees, shrubs, herbs, and climbers so dense that they prevent light penetration into the forest interior. The evergreen forest zone is dominated by four species of laurel (bay): *Laurus azorica, Appollonias barbujana, Persea indica*, and *Ocotea foetens*. These are all Macaronesian endemics and relicts of a now virtually extinct tertiary Mediterranean flora that occupied southern Europe and North Africa about 15–40 million years ago.

Many other Canarian endemic trees and shrubs occur in this forest, including *Appollonias barbujana* spp. *ceballosi, Ilex perado* spp. *lopez-lilloi* (specific to La Gomera), *Arbutus canariensis, Erica scoparia* spp. *platycodon, Gesnouina arborea, Ilex perado* spp. *platyphilla, Juniperus cedrus, Maytenus canariensis, Sambucus palmensis, Teline stenopetala* var. *microphylla, Viburnum tinus* spp. *rigidum*, and *Myrica rivas-martinezii* (Pérez de Paz 1990).

In the Canary Islands, the evergreen forest has vanished on Gran Canaria, where only 1 percent of the pre-Hispanic forest community still exists; and it is endangered on Tenerife where less than 10 percent of this forest type remains, concentrated in the Montes de las Mercedes region (White 1986; García Morales 1989).

South American Cloud Community: The Lachay National Reserve, Peru

In 1977, the Lachay National Reserve, a woodland in a very dry and foggy formation, was declared by the Peruvian authorities (total area 5,070 ha). It is located 105 km north of Lima near the Panamerican Road (11° S; 77° W). It is representative of *lomas* (hill) formations located between ± 8° S (Peru) and ± 20° S (Chile). *Lomas* also refers to botanic formations. In central and southern Peru, not far away from the Pacific coast, many *lomas* such as Lachay are found, with ancient *pueblos* (pre-Colombian settlement ruins [Engel 1987]). Their inhabitants had preserved natural resources without any destructive pressure by grazing and exploitation of wood.

Climate and Hydrology

Based on 44 years of observation, the climate of Lachay is characterized by the following parameters: (1) rainfall, 173 mm; (2) fog (*garúa*) precipitation under the canopy: depth of rainfall multiplied by 3–8; (3) monthly average temperatures: 12.8°–20.8° C; (4) monthly average sunshine periods: 36–218 hours.

SENAMHI (Servicio Nacional de Meteorología e Hidrologia) has been working since the 1960s in the following *lomas*: Lachay, Pasamayo, Cerro Campana, Atiquipa, Cerro Orara (Ventanilla-Ancón), Cerro Colorado (Villa

María de Triunfo), and Parque Recreacional de Cahuide (Ate-Vitarte). The aim of SENAMHI is to develop the use of fog catchers on *lomas* at altitudes between 400 and 1,000 m. SENAMHI obtains an average fog water production of up to 10 liters/m^2/day using a mesh proposed by ESTRATUS Company.

Vegetation and Flora

The vegetation is characteristic of the central Peruvian hills (*lomas*) along the desertic Pacific coast (Weberbauer 1911). The most representative trees are two species: *tara* (*Caesalpina tinctoria*) and *pallilo* (*Capparis prisca*), growing at altitudes ranging from 500 to 700 m. Among the trees and shrubs, the following species are vanishing in the *lomas: Eupatorium sternbergianum, Cassia biflora, Carica candicans, Lobelia decurrens, Capparis prisca,* and *Caesalpina tinctoria* (Dourojeanni and Ponce 1978). The woodland contains few succulents. The trees are scattered, gnarled, and stunted. They are located on often skeletal soils between rocks where Bromeliaceae are growing. The most abundant are the genera *Pitcairnia* and *Puya*.

Some Aspects of African Quaternary Paleoenvironments

Tropical cloud forests help us to understand some paleoenvironmental paradoxes, such as lowland extensions of mountain ecosystems during Quaternary Ice Ages. The mountain ecosystems generally appear above 1,000 m. However, in tropical Africa we find today several lowland spots with mountain ecosystems, for example, on the southern border of the Zaire-Congo basin, in the Mayombe, Chaillu, and Monts de Cristal hills (White 1981; Maley Caballé, and Sita 1988) or in other areas such as the Freetown hills (Liberia), Mount Cameroon, the Angola plateau, and Mount Usambara (Maley 1989). Outside Africa, there are many examples of this phenomenon in the Antilles and around the Caribbean Sea (Baynton 1968; Howard 1970). In the northern tip of Colombia, Cavelier and Goldstein (1989) described the elfin forest of the Serrania de Macuira, which occurs at a maximum elevation of 865 m.

In fact, such communities can be found at elevations as low as 500 m, or even 300 m (Angola). They can occur on lower slopes of high mountains (Mount Cameroon, 4,070 m) (Serle 1964; Thomas 1986), on plateau escarpments (Angola, Sudan), or on hilltops as a result of crest effects (Peru, Chile). These communities are particularly numerous near the sea (Moreau 1966; Kerfoot 1968; Whitmore 1975; Blot 1991). However, cloud phenomena may be present at distances up to 800–1,000 km from the Atlantic Ocean in Gabon and the Congo (Maley 1991).

All studies concerning the history of the vegetation in Africa have concluded that during climate cooling, mountain vegetal species extended to relatively low elevations. Pollen analyses carried out in eastern Africa

(Bonnefille et al. 1992) and in central and western Africa (Maley 1989, 1991) show that temperatures dropped by 3°–8° C during the last Ice Age maximum about 18,000 years BP. On the high mountains of east Africa, the observational evidence of the descent of glaciers by about 1,000 m has led to similar conclusions (Hamilton 1982). To understand the effect of lower temperatures, we can examine the composition of flora and fauna at elevations in the middle altitude range, between 1,000 and 3,000 m. For example, the percentages of plant species common to Mount Cameroon and the mountains of eastern Africa are 53 percent for mountain forests and 49 percent for mountain meadows. Similar percentages occur for mammals and birds (Moreau 1966). To explain these similarities, many authors have ventured that during the coldest periods of the Quaternary, mountain flora and fauna have spread to low altitudes, thus allowing migrations between mountain ranges. Indeed, pollen analyses for location sites far away from high mountains have supported this assumption. For example, a small depression at 600 m (Bois de Bilanko, about 35 km north of Brazzaville, Congo) has yielded a section with pollen types dominated (before the beginning of the Holocene) by mountain species: *Podocarpus latifolius* (50 percent), *Ilex mitis*, and *Olea capensis,* constituting 60 percent of the pollen spectra. Moreover, between the mountain ranges of eastern Africa and Cameroon, isolated occurrences of species typical of mountain areas still exist today (Maley, Caballé, and Sita 1988). These are benchmarks on a preferential path of migration, which probably operated intermittently during the coldest periods of the Quaternary, or before the Quaternary. This path joins eastern Africa to Angola, and through the hills of Mayombe, Chaillu, and Monts de Cristal, it reaches the mountains of Cameroon (Maley 1989, 1991).

Acknowledgments. We thank the Spanish ICONA staff, the Chilean CONAF (Corporación Nacional Forestal) team, Dr. Luis Acosta Alvarez (head of the Peruvian SENAMHI), and Pierre Pourrut (Orstom-Antofagasta) for their contributions. Comments by Pierre Fontanel, Edouard Lefloc'h (CNRS, and Réseau Zones Arides— MAB France-Montpellier), and Yann L'Hôte (Orstom-Montpellier) helped in the preparation of the paper. The Cameroonian program on lowland and mountain forests results from collaboration of several Cameroonian scientific institutions (MESIRES, ISH, CGM, CRH, IRGM, University of Yaoundé). Also, the EEC helped with research on Spanish organisms (Comett-II: Prof. André Van der Beken). The fountain tree project of Cruz de los Reyes (El Hierro, Canary Islands) is supported by the 1993 Rolex Awards for Enterprise (Selected Projects in Environment).

References

Acosta Baladón, A. N. 1992. Niebla potable. *Conciencia Planetaria* 12:50-55.

Acosta Baladón, A. N., and A. Gioda. 1991. L'importance des précipitations occultes sous les tropiques secs. *Sécheresse* 2(2): 132-135.

Acosta Baladón, A. N., and A. Gioda. 1992. Las precipitaciones ocultas y la desertificación. In *Encuentro Meteo 92*, II:370–374. Madrid: Instituto de Meteorología Española.

Baynton, H.W. 1968. The ecology of an elfin forest in Puerto Rico. 2. The microclimate of Pico del Oeste. *Journal of the Arnold Arboretum* 49(4): 419-430.

Blot, J. 1991. Bilan écologique des forêts à *Juniperus procera* de la corne orientale de l'Afrique et de la péninsule arabique. Thèse, Université de Bordeaux III, France.

Bonnefille, R., F. Chalié, J. Guiot, and A. Vincens. 1992. Quantitative estimates of full glacial temperatures in equatorial Africa from palynological data. *Climatic Dynamics* 6:251-257.

Cavelier, J., and G. Goldstein. 1989. Mist and fog interception in elfin cloud forests in Colombia and Venezuela. *Journal of Tropical Ecology* 5:309-322.

Dourojeanni, M.J., and C.F. Ponce. 1978. *Los Parques Nacionales del Perú*. Madrid: INCAFO.

Engel, F.E. 1987. *De las begonias al maiz. Vida y produción en el Perú antiguo.* Lausanne: CIZA.

García Morales, M. 1989. *El bosque de laurisilva en la economía guanche.* S/C. Tenerife: ACT, 12.

Gioda, A. 1993. A new fountain tree for Hierro (Canary Islands.). In *The Spirit for Enterprise: The 1993 Rolex Awards*. Bern: Buri International.

Gioda, A., A. Acosta Baladón, P. Fontanel, Z. Hernández Martin, and A. Santos. 1992. L'arbre fontaine. *La Recherche* 23(249): 1400-1408 (El árbol fuente. *Mundo Científico* 1993, 13[132]: 126-134).

Gioda, A., Z. Hernández Martin, and E. Gonzáles. 1993. Observatoires, brouillards et arbres fontaines aux Canaries. Submitted *Veille Climatique Satellitaire* 46: 38–49.

Gischler, Ch. 1991. *The missing link in a production chain.* Montevideo, Uruguay: Unesco-Rostlac.

Glas, G. 1764. *Descripción de las Islas Canarias 1764.* S/C. Tenerife: Instituto de Estudios Canarios-Goya Ediciones, 2nd ed., 1982 (in Spanish from the British original version).

Hamilton, A.C. 1982. *Environmental history of East Africa. A study of the Quaternary.* London: Academic Press.

Howard, R.A. 1970. The "Alpine" plants of the Antilles. *Biotropica* 2:24-28.

Kerfoot, O. 1968. Mist precipitation on vegetation. *Forestry Abstracts* 29:8-20.

Maley, J. 1989. Late Quaternary climatic changes in the African rain forest: Forest refugia and the major role of sea surface temperature variations. In M. Leinen and M. Sarnthe, editors, *Paleoclimatology and Paleometeorology: Modern and past patterns of global atmospheric transport*, pages 585-616. Dordrecht, the Netherlands: Kluver, NATO Adv. Sc. Inst., Series C, Math. and Phys. Sc., 282.

Maley, J. 1991. The African rain forest vegetation and paleoenvironments during late Quaternary. *Climatic Change* 19:79-98.

Maley, J., G. Caballé, and P. Sita. 1988. Etude d'un peuplement résiduel à basse altitude de *Podocarpus latifolius* sur le flanc congolais du massif du Chaillu. In R. Lafranchi and D. Schwartz, editors, *Paysages quaternaires de l'Afrique centrale atlantique*, pages 336-352. Paris: Orstom, série Didactiques.

Moreau, R.E. 1966. *The bird faunas of Africa and its islands.* New York: Academic Press.

Pérez de Paz, P.L., editor. 1990. *Parque Nacional de Garajonay*. Madrid: ICONA.

Reis F. Cunha. 1964. O problema da captacão de água do nevoeiro em Cabo Verde. *Garcia de Orta (Lisboa)* 12(4): 719-756.

Santana Pérez, L. 1990. La importancia hidrológica de las nieblas en las cumbres del Parque Nacional de Garajonay. In P. L. Pérez de Paz, editor, *Parque Nacional de Garajonay,* pages 67-71. Madrid: ICONA.

Serle, W. 1964. The lower altitudinal limit of the montane forest birds in the Cameroon mountains, West Africa. *Bulletin of the British Ornithological Club* 84:87-91.

Thomas, D.W. 1986. Vegetation in the montane forests of Cameroon. In S.N. Stuart, editor, *Conservation of Cameroon montane forests,* pages 20-27. Cambridge: Report of Intern. Council Bird Preserve.

Weberbauer, A. 1911. *Die Pflanzenwelt der peruanischen Anden.* XII, Leipzig: Wilhelm Engelmann (2nd ed., 1976. Vaduz, Leichtenstein: Gantner).
White, F. 1981. The history of the Afromontane archipelago and the scientific needs of conservation. *African Journal of Ecology* 19:33-54.
White, F. 1986. *La végétation en Afrique.* Paris: Orstom-Unesco.
Whitmore, T.C. 1975. *Tropical rain forests in the Far East.* Oxford: Clarendon.

11. Relationships Between Rainfall, Cloud–Water Interception, and Canopy Throughfall in a Hawaiian Montane Forest

James O. Juvik and Dennis Nullet

Within the topo-climatically complex tropical mountains of Hawai'i, there are four distinctive montane forest zones where the direct canopy interception of wind-driven cloud water (also referred to as "fog drip") plays a significant role in forest ecology and hydrology. Not all of these four zones can be characterized as classic tropical montane cloud forest (TMCF), but to varying degrees each is influenced by ground-level, orographic cloud.

Zone 1. Exposed Windward Mountain Ridges and Peaks at Elevations Between 750 and 1,750 m

Mountain ridges directly exposed to the prevailing northeast trade winds support dwarfed (elfin) montane cloud forest with abundant epiphytes. Although canopy cloud-water interception in such areas is presumed substantial, orographic rainfall is also very high (typically 3,000–10,000 mm/yr). Thus, the moisture augmentation from cloud-water interception in such ridge-forest environments is of more ecological than hydrological importance, since these localized ridge forests are not areally extensive within the mountain watersheds.

No thorough, long-term studies on the climate and hydrology of these TMCF have yet been conducted. The senior author undertook a very preliminary, short-term study of precipitation and throughfall parameters in the dwarfed cloud forest at Alakahi (elevation 1,160 m), an exposed windward site in the Kohala Mountains on the Island of Hawai'i (Figure 1). The site is characterized by a low stature (canopy height 3–5 m) TMCF dominated by the tree species *Metrosideros polymorpha* and *Cheirodendron trigynum*, both Hawaiian endemics.

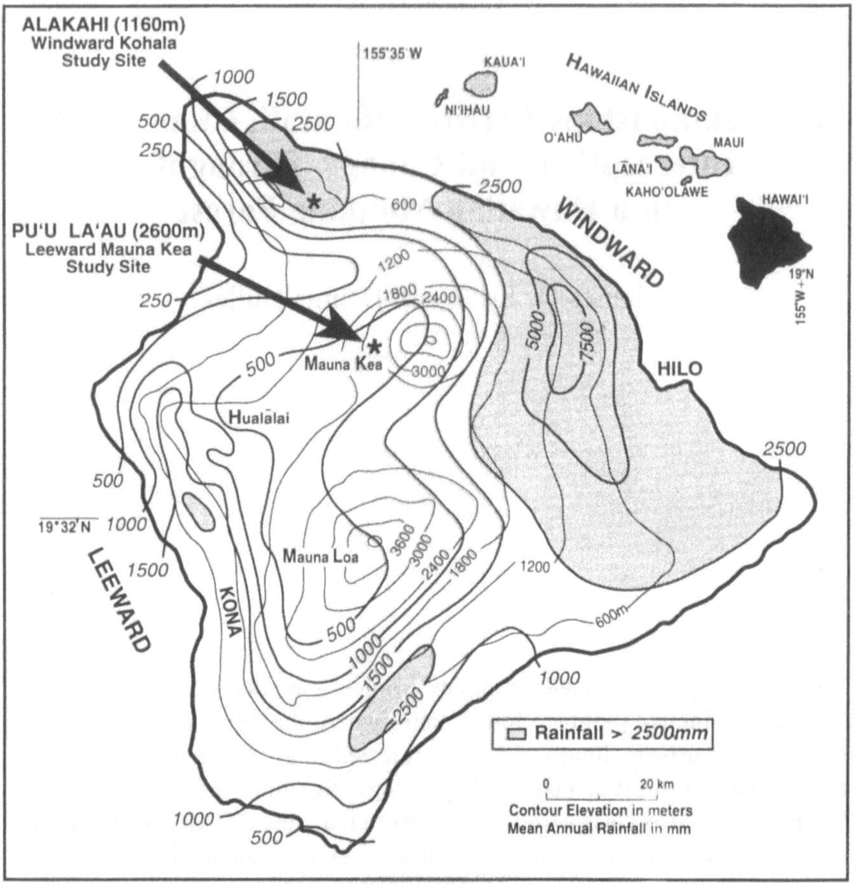

Figure 1. Site location map, island of Hawai'i.

The long-term average annual rainfall at Alakahi is 4,400 mm/yr (Giambelluca, Nullet, and Schroeder 1986). Precipitation, cloud-water interception, and throughfall were measured for a 100-day period (August–November 1984) using tipping-bucket rain gauges. A cylindrical, louvered-screen cloud-water collector (mounted at a height of 3 m) was used to intercept horizontally moving cloud droplets (see methodology section below and McKnight and Juvik [1975], for details of gauge design). A scaler was employed to convert water capture by the vertical silhouette of the louvered screen to the orifice area equivalent for the standard rain gauge. Forest throughfall was measured using multiple trough gauges beneath the canopy. Over the sampling period, measured rainfall (in open) at the site was 336 mm. For the same period, the cloud-water collector recorded 1,365 mm, or 406 percent of rainfall. Canopy throughfall for the corresponding period was 606 mm, or 180 percent of rainfall.

These limited results indicate the potential magnitude of cloud-water interception at exposed windward cloud-forest sites. The high ratio of throughfall to rain during the sampling period was somewhat more exaggerated than would normally be expected because of unusually dry conditions. The 336 mm of measured rainfall represented only about 30 percent of the long-term normal for the period. Juvik and Perreira (1974) have shown that during drought periods in the Hawaiian montane zone, cloud-water interception (and hence canopy throughfall) still occurs and gains relative importance as a moisture source.

Much more detailed work is needed to clarify the relative and absolute contribution of intercepted cloud water in the hydrologic balance of this TMCF zone. In addition to the Kohala Mountains example presented here, other windward cloud forest localities within the Hawaiian archipelago include on O'ahu, ridges and peaks (at 750–950 m) of the Ko'olau Mountains; the Alaka'i Plateau (1,200–1,600 m) on Kaua'i; the Kamakou area (1,300–1,500 m) of Moloka'i; and the West Maui Mountains (1,000–1,750 m).

Zone 2. "Rainshadow" Mountain Ridges and Peaks at Elevations Between 900 and 1,225 m

At a few selected sites within the Hawaiian Islands, mountain ranges and their associated montane cloud forests lie in the lee or rainshadow of other mountains or islands (with respect to the prevailing northeast trade winds). At such locations orographic rainfall is much reduced (1,000–2,000 mm/yr) compared to the windward localities in zone 1, and moisture derived from cloud-water interception becomes relatively more important both ecologically and hydrologically. Cloud forests in the Wai'anae Mountains (1,100–1,225 m) on O'ahu (in the lee of the Ko'olau Mountains) and at Lana'ihale (900–1,025 m) on Lana'i (the entire island lies in the lee of west Maui) are the best examples of "rainshadow" cloud forest.

Some of the earliest and most detailed quantitative studies of cloud-water interception were initiated at Lana'ihale, Lana'i, during the 1950s by the Pineapple Research Institute (Mordy and Hurdis 1956; Ekern 1958, 1964). On Lana'i, high elevation groundwater (confined by volcanic dikes) was essential for pineapple irrigation, but groundwater recharge had been compromised by a history of deforestation. Afforestation with tree species possessing a stature and crown architecture favorable to "cloud catching" or "cloud stripping" was seen as essential to the maintenance and expansion of agriculture. The Norfolk Island Pine *Araucaria excelsa*—a tall, subtropical conifer of the Southern Hemisphere—proved ideal for the task.

Detailed throughfall studies at Lana'ihale (Ekern 1964) showed that isolated pines collected a net increase of 760 mm/yr of water (compared with rainfall in open areas) from cloud-water interception. During the dry summer months, throughfall gauges under the pine canopy yielded 13 times more water than rain gauges in the open (394 mm and 29.5 mm, respectively).

Zone 3. Exposed, Windward Mountain Slopes at Elevations Between 1,200 and 1,800 m

On the geologically younger Hawaiian volcanoes (Haleakala, Maui; and Mauna Kea and Mauna Loa on the island of Hawai'i), fluvial erosion and dissection are comparatively reduced, and the mountains possess broad and gently sloping flanks. Compared with the steep slopes and sharp ridges characterizing zone 1, the gently sloping windward aspects of these larger volcanic mountains experience lower wind speeds on longer overland trajectories and, as a consequence, reduced cloud-water flux. The forest canopies on these slopes typically do not exhibit the dwarfed stature and epiphyte load of zones 1 and 2. These windward slope forests are also characterized by high annual rainfall (4,000–7,000 mm/yr). Although the relative contribution of cloud-water is reduced here (compared with windward ridge forests), the vast areal extent of windward slope forests results in a greater absolute interception contribution to watershed hydrology, irrespective of reduced structural or floristic cloud-forest manifestations.

Juvik and Ekern (1978) used louvered-screen cloud-water collectors to examine altitudinal gradients and seasonal variation in cloud-water interception on the windward slopes of Mauna Loa, on the island of Hawai'i (see Figure 1). In Figure 2 a generalized windward slope model of seasonal cloud-water contribution (relative to rainfall) is presented. Relative cloud-water contribution is greatest in summer when absolute rainfall is at a minimum. During the

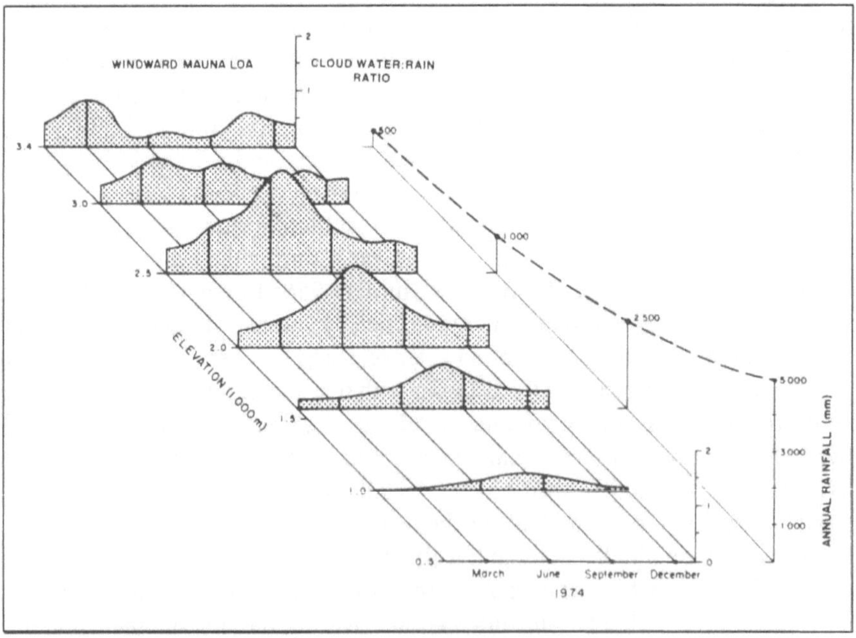

Figure 2. Generalized windward slope model for seasonal cloud-water contribution (relative to rainfall) on Mauna Loa, Hawai'i *(modified from Juvik and Ekern 1978).*

summer the trade winds are stronger (greater upslope vapor flux), and there is also a strong upper level temperature inversion (modal elevation 1,800–2,000 m). The upslope flux of condensing air is thus trapped between the rising slope and the inversion. Cloud depth is restricted, a condition favoring ground-level cloud rather than raindrop formation at mid-slope elevations (1,500–2,500 m).

Although this work (Juvik and Ekern 1978) did not include a companion study of canopy throughfall, the present paper confirms that mechanical cloud-water collectors can be used successfully to model canopy throughfall in cloud-shrouded areas.

Zone 4. Windward and Leeward High Mountain Slopes Above the Trade-Wind Inversion (1,800–3,000 m)

The upper slopes of the highest Hawaiian volcanoes lie above the trade-wind inversion and are generally influenced by dry descending air (Hadley Cell). Annual rainfall is typically 400–1,000 mm/yr at these elevations. Afternoon upslope winds driven by surface heating frequently allow thin ground-level clouds to penetrate the inversion to reach these upper mountain slopes.

Typical vegetation at these elevations on Mauna Kea and Haleakala consists of open "dry-forest," with trees (height to 10 m) lacking a significant epiphyte load. Because of low annual rainfall, canopy interception of cloud water plays a particularly important role in the water balance of these high altitude montane forests.

In the present study, precipitation parameters including rainfall, cloud-water collection, and canopy throughfall were monitored over a 1-year period at a high altitude, montane dry forest site (zone 4) on the island of Hawai‘i. Both empirical and physical models are developed for predicting canopy throughfall from open-site precipitation measurements.

Study Site

A study site was selected above Pu‘u La‘au in open, montane dry forest near tree line on the southwest facing (leeward) slope of Mauna Kea at 2,600 m (see Figure 1). Forest cover at the site is dominated by a low stature (height 5–8 m) endemic tree, the mamane (*Sophora chrysophylla*). Trees are scattered or clumped, and canopy coverage is discontinuous. This open canopy character has resulted at least in part from a century or more of feral ungulate impact (primarily sheep, cattle, and goats) that has only recently been brought under control (Juvik and Juvik 1984).

The prevailing climate of the area is dry and cool. Juvik et al. (in press) have recently described the climatology of the site based on 1 year of monitoring. Mean monthly air temperature averaged 11° C, with little seasonal variation (± 1.5° C). Mean annual rainfall in the area is approximately 500 mm based on long-term records (58 years) from the nearby Pu‘u La‘au ranger station.

The upper level trade-wind inversion is typically present just below the site in the free atmosphere between 1,800 and 2,400 m. This inversion separates a warm, humid, tropical maritime air layer from dry, cool upper air. Because this leeward site is protected from the prevailing northeast trade winds, local circulation is driven largely by diurnal heating and cooling of the mountain slopes. An onshore sea breeze flows upslope during the day and is replaced by a weaker offshore land breeze at night. The upslope, daytime winds frequently penetrate the inversion near the ground and carry a thin layer of moist surface air, ground-level clouds, and occasional light precipitation events into the study area, particularly during the warmer summer months. During winter months, large-scale synoptic disturbances (eastward-moving frontal passages) may occasionally bring heavy rainfall to the site.

Materials and Methods

An automated meteorological station (Campbell Scientific Inc.) was erected at the site in an area with clear upslope and downslope exposure. Two 3-m meteorological masts supported sensor arrays and data storage components (Figure 3). To determine the importance of cloud-water interception, two cloud-water collectors and a rain gauge were mounted at 3 m above the ground (Figure 3). Rain and cloud water drained into Campbell Scientific Model

Figure 3. Study site in open mamane forest at Pu'u La'au, Hawai'i (elevation 2,600 m). Open-site rain gauge and cloud-water collectors on tripod at left. Throughfall sampling under large mamane *(Sophora chrysophylla)* tree in background. *(Photo by J. Juvik.)*

TE525 tipping-bucket rain gauges (resolution 0.254 mm). The cloud-water collectors included a "Grunow" type, wire-mesh cylinder (Grunow 1952) and a louvered aluminum screen collector of larger size (see McKnight and Juvik [1975] for details of gauge design and calibration). The louvered-screen cloud-water collector presents a vertical silhouette interception area of 515.6 cm^2, or 2.83 times the cross-sectional orifice area of the standard (15.2 cm diameter) rain gauges used in the study. After subtracting episode rainfall from the louvered-screen gauge output (since this gauge collects rain as well as intercepting horizontally moving cloud water), the above scaler (2.83) was used to convert the cloud-water values to a comparable "unit vertical catch." Over the 1-year sampling period (October 1991–September 1992), the louvered-screen fog gauge demonstrated slightly greater efficiency at cloud-water collection than the Grunow gauge and thus only data for the louvered-screen gauge are presented here to characterize open-site cloud-water collection.

A mature mamane tree (height 7.6 m, extreme canopy dimensions 16.7 m and 11.6 m) located near the open-site meteorological station (see Figure 3) was instrumented with two parallel throughfall troughs transecting the canopy from the windward to leeward side of the tree (Figure 4). These "v"-shaped aluminum troughs (3.3 cm wide x 380.8 cm long) had a catchment area of 1256.6 cm^2, or 4 times the cross-sectional orifice area of the (20 cm diameter) covered Weathertronics Model 6011–A tipping-bucket rain gauges into which they drained. Calibration of the troughs under open rainfall conditions indicated an average 5 percent splash loss from the shallow troughs as compared to the standard rain gauge. During the study period, throughfall collection by the two parallel troughs typically varied by less than 25 percent. In the analysis that follows, throughfall is expressed as the average of the two troughs. Data were recorded as hourly totals for all precipitation measurements.

Results and Discussion

Observations

Over the 1-year observation period, gross rainfall (open site) at Pu‘u La‘au totaled 257.1 mm; cloud-water collection by the louvered-screen gauge was 98.2 mm, or 38.2 percent of rainfall; and canopy throughfall (average of two troughs) equaled 193.6 mm, or 75.3 percent of rainfall. This precipitation was distributed over 61 events. Figure 5 illustrates the frequency distribution of precipitation by rainfall event size classes. Note from Figure 5 that more than one-half of the precipitation events were comprised of very light episodes totaling less than 2 mm of precipitation. Only three rainfall (and throughfall) events during the year exceeded 12 mm, with the largest being a 7-hour rainfall episode of 45.2 mm (throughfall 32.4 mm) on 5 September 1992. Figure 5 also indicates a total of ten precipitation events with no associated cloud-water collection (rain-only events).

Figure 4. Throughfall precipitation trough gauges (each 380.8 cm long) under mamane *(Sophora chrysophylla)* tree at the Pu'u La'au study site. *(Photo by J. Juvik.)*

The diurnal distribution of all precipitation parameters for the year of record is plotted in Figure 6. Rainfall, cloud-water collection, and canopy throughfall all exhibit strong mid-afternoon maxima (167.3 mm, or 65 percent of the total annual rainfall of 257.1 mm fell between 1200 and 1600 hr). Roughly similar proportions of cloud-water collection (56 percent) and throughfall (65 percent) were also concentrated during this same 4-hour afternoon period. These results clearly indicate that both rainfall and cloud-water interception episodes are associated with the afternoon upslope air flow. Depending on the wind speed and drop size of upslope afternoon mist, either rain- or cloud-water interception may dominate in individual precipitation events.

The seasonal distribution of precipitation at the study site is presented in Figure 7, along with a plot of the long-term median monthly rainfall from the nearby Pu'u La'au ranger station. Total rainfall (257.1 mm) for the sampling period was substantially below the long-term normal of 500 mm, and the first

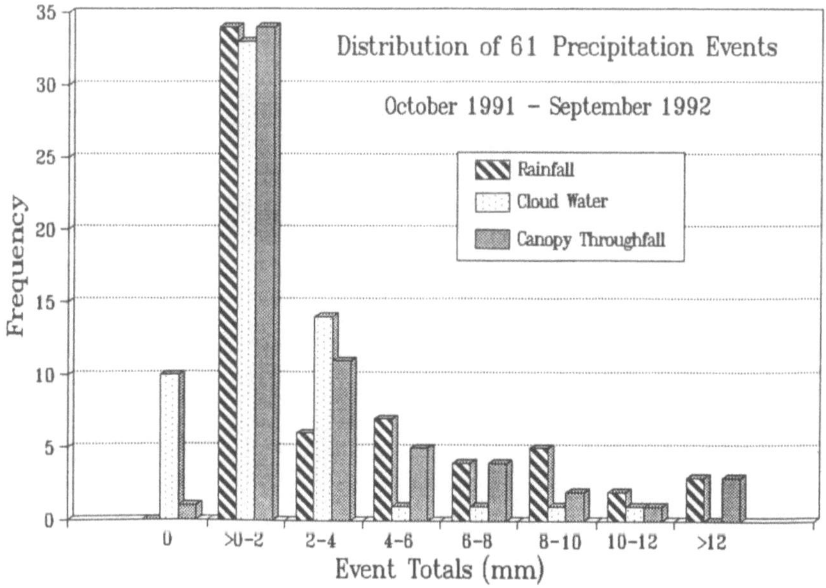

Figure 5. Frequency distribution of 61 precipitation events over the 1-year sampling period plotted by rainfall event size classes.

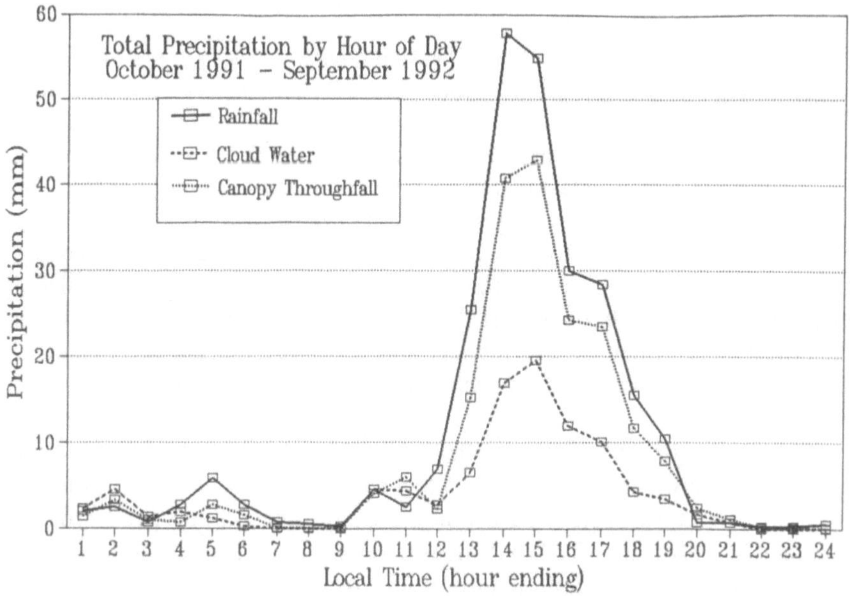

Figure 6. Diurnal distribution of precipitation parameters for the 1-year sampling period.

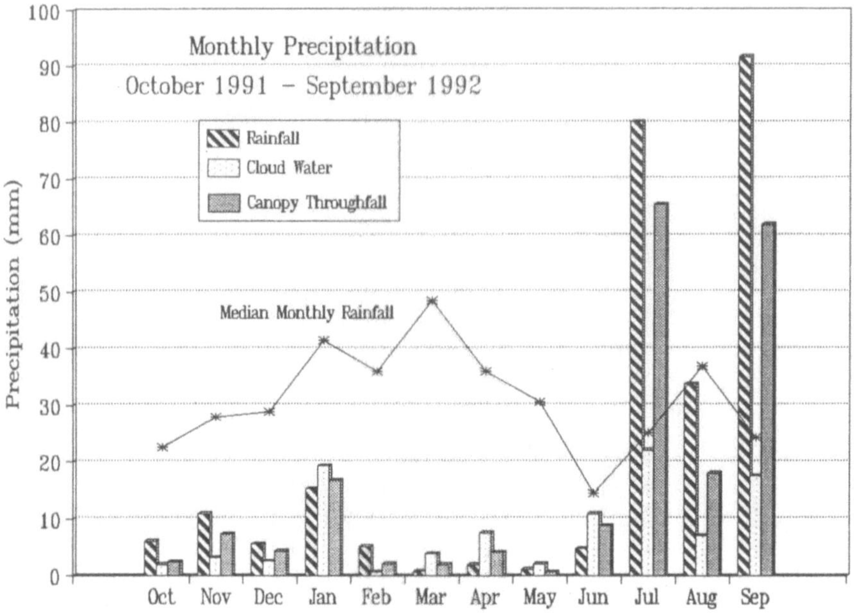

Figure 7. Seasonal distribution of precipitation parameters for the 1-year sampling period at Pu'u La'au. Long-term median monthly rainfall for the nearby Pu'u La'au ranger station is also plotted for comparison.

9 months of the study were extremely dry. Only 17.5 mm of rain was recorded during the driest five months (February though June 1992). With respect to the ecological and hydrological role of cloud-water interception, note that during the same 5-month drought period, canopy throughfall totaled 25.8 mm, or 147 percent of the recorded rainfall. This throughfall gain was clearly a function of the cloud-water interception potential during these months. The louvered-screen collector recorded 34.4 mm (197 percent of rainfall) during the same period. The throughfall results are particularly significant because the very light rainfall events that characterized the period never reached a magnitude sufficient to overcome the tree canopy storage capacity. Without the added cloud water, canopy interception losses would have been substantial (typically 50–90 percent for such light rainfall events with no enveloping cloud).

The results generally indicate that for areas such as Pu'u La'au where throughfall represents a mixture of direct rainfall and cloud-water interception, open-site rainfall measurements alone are a poor predictor of throughfall. This fact is further demonstrated in Figure 8 where the anatomy of three individual precipitation episodes are compared. The events were chosen as examples because each had the same total open-site rainfall (1.27 mm) and exhibited the wide range of conditions present in precipitation episodes at the site. In addition, all three events were characterized by low

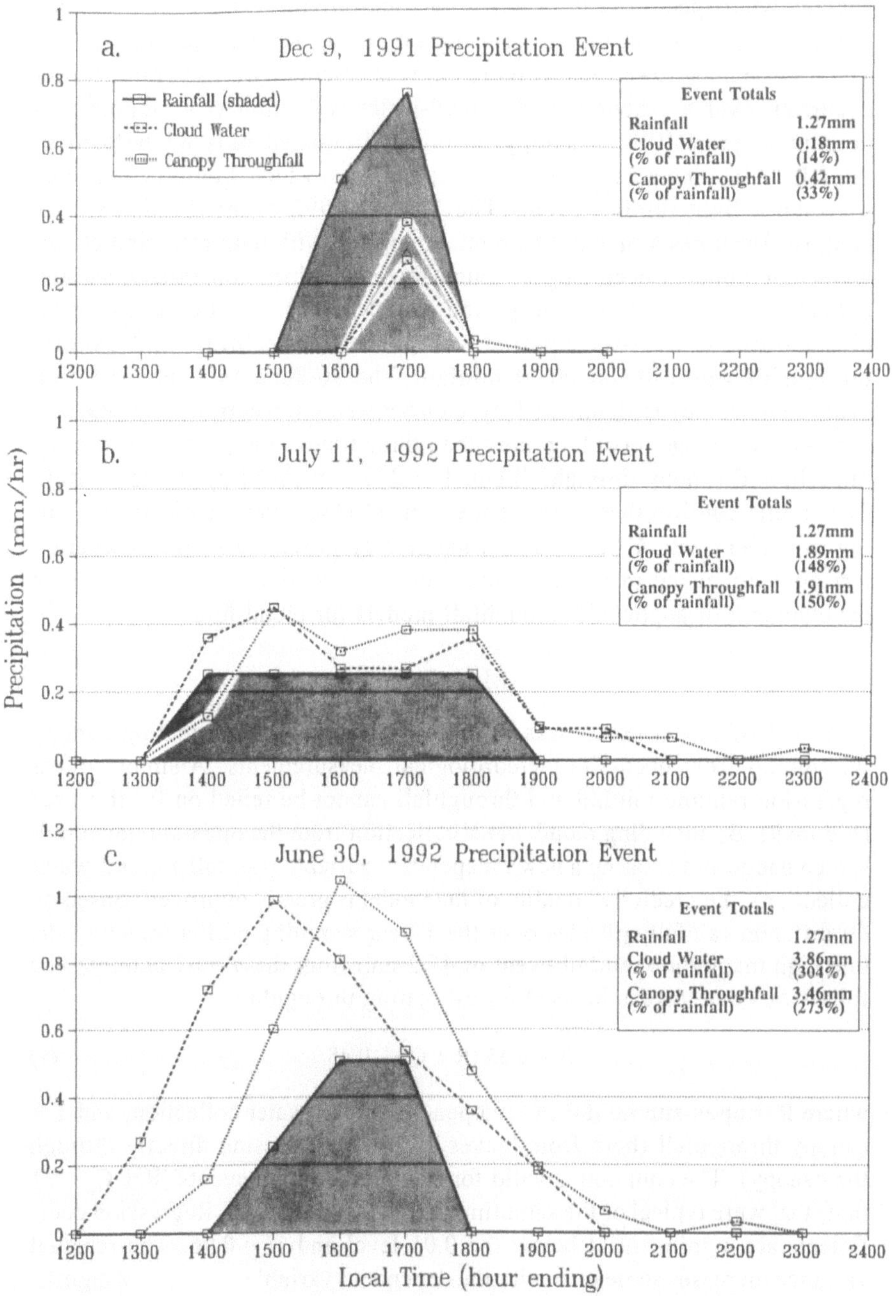

Figure 8. Relationships between rainfall, cloud-water collection, and canopy throughfall for three different precipitation events at the Pu'u La'au sampling site: (a) a rainfall dominant precipitation event; (b) a mixed rainfall and cloud-water event; (c) a cloud-water dominant precipitation event.

average wind speeds (<2.0 m/sec) so the possible influence of non-vertical rainfall (impacting on the cloud-water collecting screen) can be ignored. The 9 December 1991 event (Figure 8a) was essentially a light rain shower occurring over a 2-hour period. Cloud-water collection was minimal (14 percent of rainfall) and canopy throughfall equaled only 33 percent of rainfall, indicating a substantial interception loss, which would be expected for such a light rainfall event. The 11 July 1992 event (Figure 8b) by contrast illustrates a mixed precipitation episode with both rain- and cloud-water components occurring at much lower precipitation intensities and extending over a longer time period (8–10 hours). For this event, total cloud-water collection equaled 148 percent of rainfall and canopy throughfall was 150 percent of rainfall. The 30 June 1992 precipitation event (Figure 8c) is dominated by cloud-water interception with the louvered-screen gauge recording more than three times (304 percent) measured rainfall, and canopy throughfall equal to 273 percent of open-site rainfall. The results confirm that a simple mechanical cloud-water collector operating in tandem with a standard rain gauge may provide a sufficient characterization of mixed rain- and cloud-water interception events to facilitate development of applicable throughfall models for cloud-forest sites.

Modeling

One goal of this research was to develop a model for estimating canopy throughfall from open-site meteorological measurements. A simple linear regression relating rainfall and throughfall cannot be relied on for this area (Figure 9). By including cloud-water collection from the open-site louvered-screen gauge and creating a new independent variable (rainfall + cloud-water collection), the predictive quality of the model is greatly improved. Based on 60 different rainfall episodes over the 1-year sampling period (deleting the one high magnitude rainfall event of 45.2 mm from the $n = 61$ dataset), the following equation can be used for estimating throughfall:

$$T = 0.63\ (R + C_c) - 0.48 \tag{1}$$

where R = open-site rainfall, C_c = open-site cloud-water collection, and T = canopy throughfall (both from leaves and rainfall passing directly through the canopy). The equation is valid for low precipitation events ($R + C_c < 21$ mm) that were typical of the sampling period (see Figure 5). Regression coefficients are significant at below the 0.01 level and $r^2 = 0.96$. The residual variance increases somewhat with the dependent variable, though not significantly enough to warrant a more complex model based on a data transformation.

The above empirical equation can be used to estimate throughfall on Pu‘u La‘au based on measurements of open-site rainfall and cloud-water collection. Figure 9 shows the change in explained variation (r^2) in throughfall over

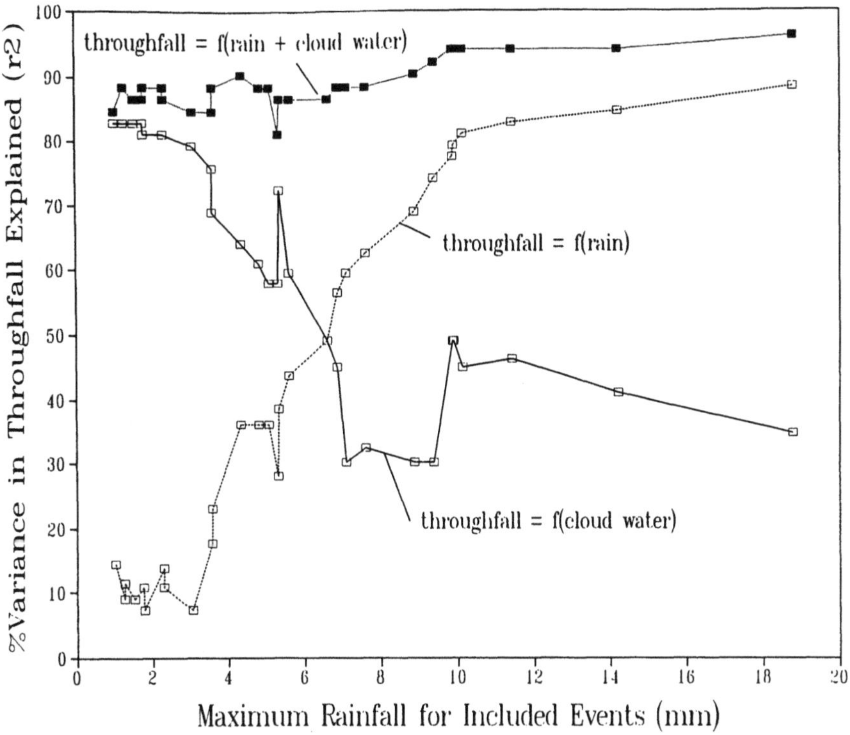

Figure 9. Change in percent explained variation (r^2) for linear regression models estimating throughfall from rainfall, cloud-water collection, and rainfall + cloud water as a function of maximum measured rainfall.

a range of rainfall totals. The coefficients for a predictive linear regression model, however, vary over the range of maximum rainfall values shown on the graph. In general, the coefficient for cloud-water collection decreases and the coefficient for rainfall increases with increasing maximum rainfall.

A more general model might be developed by exploring the physical relationships between rainfall, cloud-water interception by the canopy, and throughfall. The general water balance for the canopy can be given by:

$$R + C_i = S + E + T + Sf \tag{2}$$

where C_i = canopy cloud-water interception, S = canopy storage, E = evaporation from the wet canopy, and Sf = stemflow.

Several assumptions can be applied to simplify the modeling effort. First, stemflow for broadleaf evergreens has been reported as only a minor fraction of rainfall (Jackson 1975; Clarke 1986; Lloyd and Marques 1988; Bruijnzeel 1990; Hutjes, Weirda, and Veen 1990; Kellman and Roulet 1990) and is considered negligible for this study during light precipitation events. Evaporation is estimated using the Penman-Monteith equation (Penman 1948; Monteith

1965) with a low stomatal resistance to simulate evaporation from wetted leaves. Aerodynamic resistance is estimated using the equation given by Thom and Oliver (1977). Estimated evaporation totals for the precipitation episodes range from 0.01 to 1.24 mm. In applying the evaporation estimates, it is assumed that the evaporation rate is limited by the available radiant energy (which is reasonable considering that humidity during the rainfall episodes always exceeds 90 percent). Therefore, assuming the canopy intercepts the bulk of solar radiation, the evaporation rate is independent of total leaf area and can be used as calculated in millimeters without further correction based on canopy geometry. Also, as suggested by Gash (1979), the canopy is assumed to dry between rainfall episodes.

The maximum canopy storage can be estimated using episodes in which it is possible to calculate actual canopy storage as a residual term, that is, when there is little or no cloud-water interception. Canopy storage then becomes

$$S = R - E - T \tag{3}$$

Storage is plotted against rainfall for no cloud-water and low cloud-water collection events in Figure 10. The resulting curve exhibits a typical decaying exponential shape that can be modeled adequately using

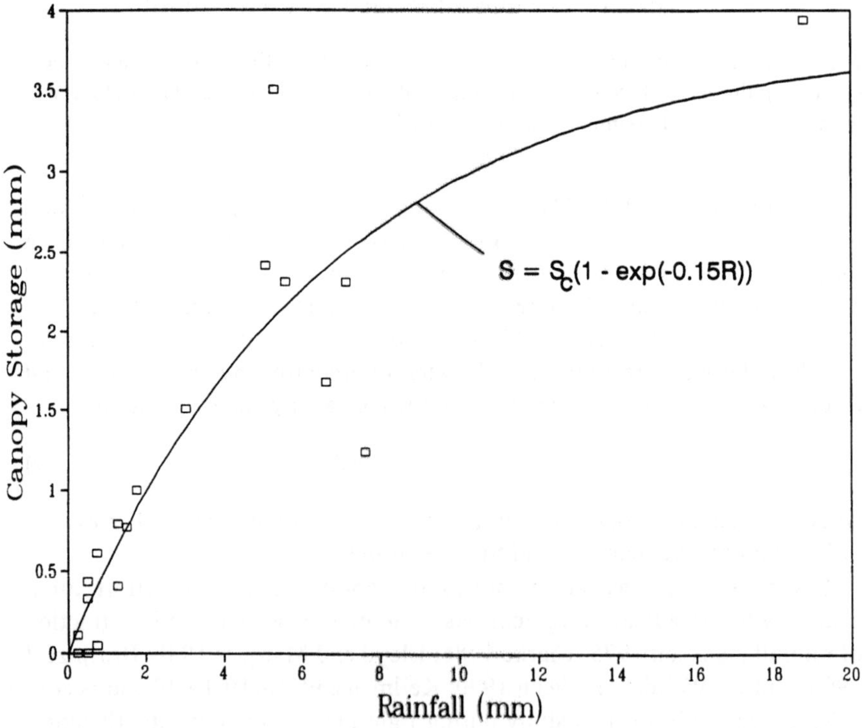

Figure 10. Relationship between rainfall and canopy storage at Pu‘u La‘au for precipitation events with little ($C_c/R<0.15$) or no accompanying cloud-water collection ($n = 19$).

$$S = S_c(1 - e^{-cR}) \tag{4}$$

where c is a constant. S_c can be estimated using the above formula by linear regression of S versus $(1 - e^{-cR})$, with the resulting slope equaling the maximum canopy storage value. Optimizing the fit by maximizing the r^2 value yields c = 0.15, $S_c = 3.8$, and $r^2 = 0.80$ for all events (19 total) in which $C_c = 0$, or $C_c/R<0.15$. The latter events were included to increase the size of the sample as few events were recorded without at least minor cloud-water collection. The limiting fraction $C_c/R = 0.15$ was an arbitrary choice. The derived maximum canopy storage (3.8 mm) is unrealistically high. Since storage was derived as a residual term, this implies unaccounted-for-water in the system. The possibility exists that stemflow (not considered in this analysis) may occur during larger rainfall events (>10 mm).

Knowing the maximum canopy storage value and the relationship between rainfall and canopy storage, it should be possible to determine the relationship between cloud-water collection and canopy cloud-water interception. A comparison of cloud-water collection with wind speed revealed a positive relationship with the highest wind speeds, a reflection of the tendency of the gauge to catch nonvertical rainfall (McKnight and Juvik 1975). At wind speeds below 2 $m\text{-}s^{-1}$, the correlation between wind speed and cloud-water collection was nearly zero. For the remainder of the analysis, only rainfall episodes with average wind speeds less than 2 $m\text{-}s^{-1}$ are considered, a total of 53 events. This limitation greatly reduces variability in the results.

If the relationship between canopy-water interception, canopy storage, and throughfall is similar for both rainfall and cloud water, and if canopy interception of cloud-water (C_i) is directly proportional to mechanical cloud-water collection (C_c), then it follows that S can be estimated by

$$S = 3.8(1 - e^{-0.15\,(R + C_i)}) \tag{5}$$

where

$$C_i = \mu C_c \tag{6}$$

and μ is a constant of proportionality.

By iteration, the value of μ can be converged upon by assuming an initial value of μ in (5), solving for C_i using (2) and (5), and calculating μ by linear regression of C_i versus C_c. This process yields a value of $\mu = 1.06$ ($r^2 = 0.76$). The complete model relating C_c, T, and R then becomes

$$R + 1.06C_c = 3.8(1 - e^{-0.15\,(R + 1.06C_c)}) + E + T \tag{7}$$

Figure 11 shows the results of a comparison of estimated throughfall using (7) versus measured throughfall. The r^2 value for estimated versus measured throughfall is 0.97, a slightly better correspondence than the simple linear regression model suggested in (1). As shown on Figure 11, the model somewhat overestimates throughfall during the heaviest rainfall and cloud-water

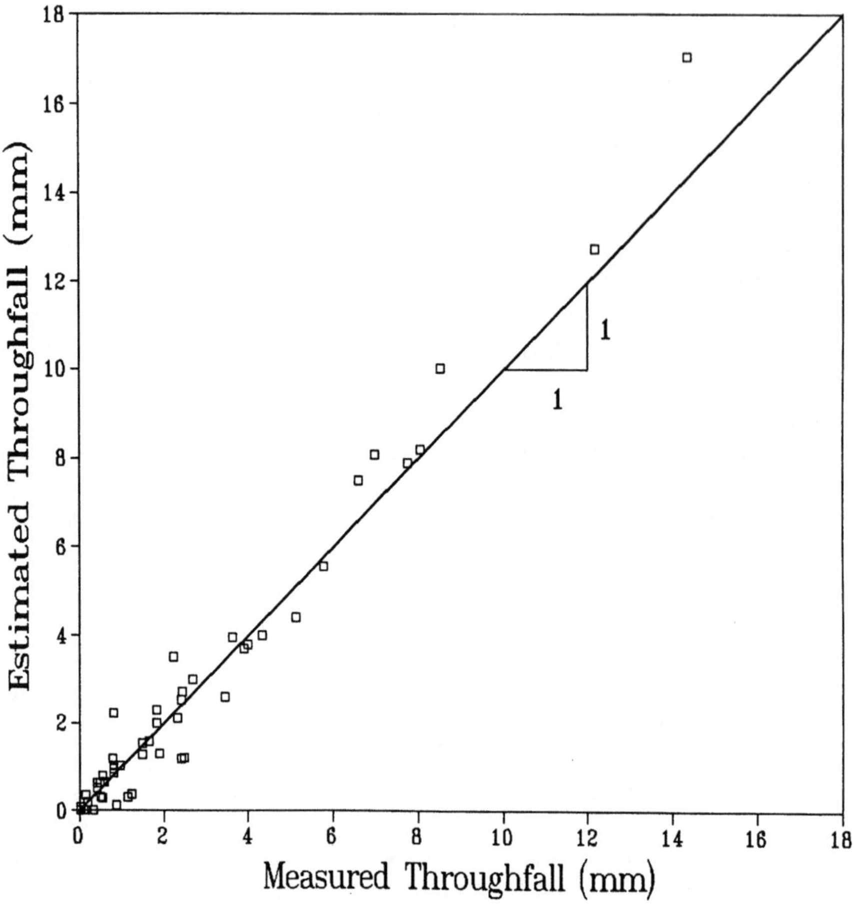

Figure 11. Relationship between measured and estimated (see text equation 7) canopy throughfall for 53 precipitation events ($r^2 = 0.97$).

collection episodes. Use of the model should be limited to $(R + C_c) < 21$ mm, and average wind speed less than 2 m-s^{-1}. This includes 53 (or 87 percent) of the rainfall episodes that occurred during the year of data collection.

Conclusion

The results of this study indicate that cloud-water interception in the high elevation dry forests of Hawai'i can provide a significant moisture source, particularly during dry summer months. Both empirical- and physical-based models have been developed, which successfully predict canopy throughfall based on open-site rainfall and the horizontal cloud-water flux measured with simple screen collectors.

Acknowledgments. This research was funded in part under a contract from the U.S. Fish and Wildlife Service, Hawai'i Field Station. We would like to thank the field station staff and Drs. James Jacobi and Paul Banko for their support of this project. Additional financial support was provided by the National Science Foundation to the University of Hawaii (DEB-89-18526; Dr. D. Mueller-Dombois, principal investigator). Thanks are also extended to Kathy Hughes and Gordon Segal for assistance in data collection and analysis. We dedicate this paper to Paul Ekern, Emeritus Professor of Agronomy and Soil Science, University of Hawaii, Manoa.

References

Bruijnzeel, L.A., 1990. *Hydrology of Moist Tropical Forests and Effects of Conversion: A State of Knowledge Review*. UNESCO International Hydrological Programme. 224 pp.

Clarke, R.T., 1986. The interception process in tropical rainforests: a literature review and critique. *Acta Amazonica*, 16/17, 225-237.

Ekern, P.C., 1958. Direct interception of cloud water on Lana'ihale. *Pineapple Research Institute News*, 6:129-133.

Ekern, P.C., 1964. Direct interception of cloud water on Lana'ihale, Hawai'i. *Soil Sci. Soc. Amer. Proc.*, 28(3): 419-421.

Gash, J.H.C., 1979. An analytical model of rainfall interception by forests. *Q. J. R. Meteor. Soc.*, 105: 43-55.

Giambelluca, T.W., M.A. Nullet, and T.A. Schroeder, 1986. *Rainfall Atlas of Hawaii*. Report R76, Department of Land and Natural Resources, State of Hawai'i, Honolulu, Hawai'i.

Grunow, J., 1952. *Nebelniederschlag: Bedeutung und Erfassung einer Zusatzkomponente des Niederschlags*. Berlin Deutsch Wetterd., US-Zone, 7(42): 30-34.

Hutjes, R.W.A., A. Weirda, and A.W.L. Veen, 1990. Rainfall interception in the Tai Forest, Ivory Coast: application of two simulation models to a humid tropical system. *J. of Hydrol.*, 114: 259-275.

Jackson, I.J., 1975. Relationships between rainfall parameters and interception by tropical forest. *J. Hydrol.*, 24: 215-238.

Juvik, J.O., and P.C. Ekern, 1978. *A climatology of mountain fog on Mauna Loa, Hawai'i*. Tech. Rpt. #118, Water Resour. Res. Ctr., U. Hawai'i, Honolulu. 63 pp.

Juvik, J.O., and S.P. Juvik, 1984. Mauna Kea and the myth of multiple use. *Mountain Res. and Dev.*, 4(3): 191-202.

Juvik, J.O., D. Nullet, P. Banko, and K. Hughes (in press). Forest climatology near the treeline in Hawai'i. *Agric. and Forest Meteor.*

Juvik, J.O., and D.J. Perreira, 1974. Fog interception on Mauna Loa, Hawai'i. *Proc. Assoc. Amer. Geog.*, 6: 22-24.

Kellman, M., and N. Roulet, 1990. Stemflow and throughfall in a tropical dry forest. *Earth Surf. Proc. and Landforms*, 15, 55-61.

Lloyd, C.R., and A. De O. Marques-Filho, 1988. Spatial variability of throughfall and stemflow measurements in Amazonian rainforest. *Agric. and Forest Meteor.*, 42: 63-73.

McKnight, J.H., and J.O. Juvik, 1975. *Methodological approaches in Hawaiian fog research*. Tech. Rpt. #85, Water Resour. Res. Ctr., U. Hawai'i, Honolulu.

Monteith, J.L., 1965. Evaporation and environment. In: *The State and Movement of*

Water in Living Organisms, 8-12 September 1964. Swansea Symp. Soc. Exp. Biol., 19, 205-234.

Mordy, W.A., and J.W. Hurdis, 1956. Fog Drip. *Pineapple Research Institute News,* 4(Mar): 26-35.

Penman, H.L., 1948. Natural evaporation from open water, bare soil, and grass. *Proc. R. Soc. London, Ser. A.*, 193, 120-146.

Thom, A.S., and H.R. Oliver, 1977. On Penman's equation for estimating regional evaporation. *Q. J. R. Meteor. Soc.*, 103: 345-357.

12. Biophysical Conditions of the Montane Cloud Forests of Mount Kinabalu, Sabah, Malaysia

Kanehiro Kitayama

Stadtmüller (1987) defined the cloud forest of the humid tropics as those forests that frequently are covered with cloud or mist in the aseasonal humid tropics. He considered "cloud forest" as synonymous with "montane rain forest." He also discussed the validity of various terminologies that are used to describe tropical montane cloud forest (TMCF).

According to his very broad definition, the cloud forest in Malesia can include an entire montane vegetation zone extending from as low as 900 m to 3,200 m asl on high mountains (e.g., Mount Wilhelm, Papua New Guinea; see Wade and MacVean 1969; Johns 1982). Consequently, there are great variations in the structure and floristic composition in what is considered a "cloud forest" or a "montane forest." For instance, van Steenis (1984) floristically differentiated the submontane from the montane forest in Malesia. Wade and MacVean (1969) and Johns (1982) recognized variations among the lower, middle, and upper montane forests in Papua. Kitayama (1987, 1992) recognized differences in the lower and upper montane forests on Mount Kinabalu, Sabah. Notably, the canopy height of the cloud forest (sensu Stadtmüller 1987) ranges from more than 30 m in the lower montane forest to merely 10 m in the upper montane forest (Whitmore 1975; Kitayama 1992).

In contrast, Wade and MacVean (1969) consider the cloud forest a facies of an upper montane zone; according to them the cloud forest occurs at 3,000 to 3,350 m on Mount Wilhelm, Papua New Guinea. Their much narrower definition agrees with the views of other researchers in Malesia (Whitmore

1975; Okutomi 1977; Frahm et al. 1990; Kitayama 1987, 1992). In this more restricted sense, the cloud forest is generally synonymous with the "mossy forest" of upper montane forests, though some mossy forests are known from much lower altitudes (e.g., the Solomons and the Santa Cruz Islands; see Whitmore 1975).

This discrepancy in the definition probably arises from the question whether one should include forests at altitudes of intermittent cloud occurrence. Alternatively, the definition can be restricted only to forests at altitudes of maximum air-saturation. In relation to the definition, only a few studies (Kitayama and Mueller-Dombois 1992 and in press, a, b) have attempted to answer the following questions: (1) can the presence of cloud cover actually control plant species assemblages that define a distinct floristic TMCF zone on mountain slopes; and (2) if so, at what altitudes?

In this paper, the attempt is made to delineate the TMCF floristically, based on a transect analysis and related studies conducted on Mount Kinabalu (summit elevation 4,101 m), Sabah. Then, the biophysical condition of the so-defined TMCF on Mount Kinabalu and other areas in Sabah is described.

Sabah is an East Malaysian state in north Borneo, and one of the most mountainous states in humid Southeast Asia. The geology of Sabah is almost entirely sedimentary and intrusive rocks. Consequently, the mountains in Sabah have not been influenced by large-scale volcanic disturbances, as have the mountains of the surrounding Sunda Islands.

Mount Kinabalu is located at 6°5' N, 116°33' E and is the highest mountain in Southeast Asia. Its geology consists largely of tertiary sedimentary rock below ca. 2,800 m and adamelite above. The climate is humid-tropical at low elevations with weak influences from the monsoon moisture regime. The mountain is incorporated in a popular national park that attracts more than 30,000 visitors per year.

Delineation of the Cloud Forest on Mount Kinabalu

The following descriptions are based on the results of the studies conducted on Mount Kinabalu in two phases from 1983 to 1990 (Kitayama 1987, 1991, 1992). A study on altitudinal sequence of floristic assemblages of canopy trees (10 cm dbh), and their relationships to edaphic and bioclimatic zones, was undertaken on the south slope of Mount Kinabalu (see Kitayama 1992). Fourteen sampling plots were established at 200 m altitudinal intervals between 600 m asl (the low limit of the natural forest) and 3,400 m asl (the closed-forest line). The two-way indicator species analysis applied to 425 canopy taxa from the plots yielded four distinct floristic vegetation zones by means of indicator species groups.

Figure 1 depicts the altitudinal ranges of diagnostic tree canopy species which are selected and grouped for characterizing the floristic vegetation zones. It also shows bioclimatic zones based on threshold values of absolute

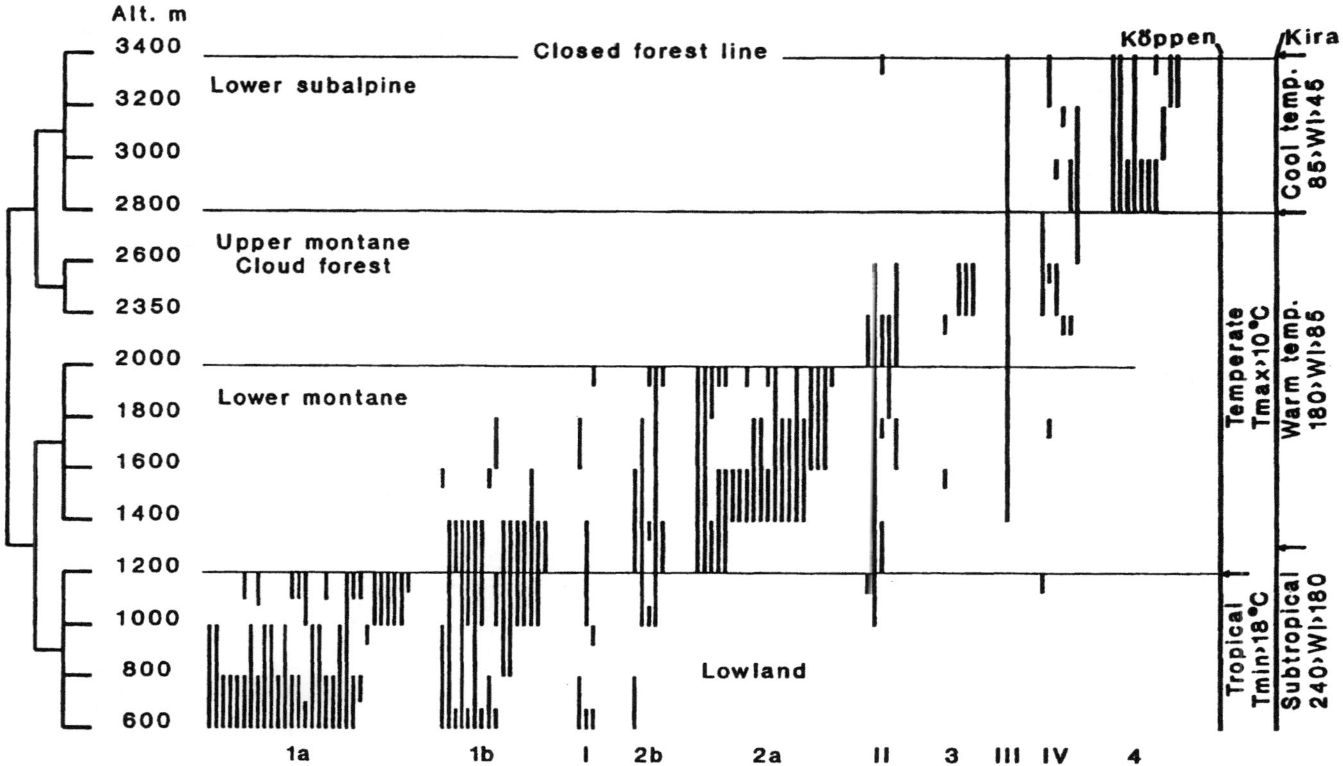

Figure 1. Bioclimate and vegetation zones defined by indicator species below the closed forest line (3,400 m). The altitudinal ranges of indicator species are shown with lines. Threshold altitudes of bioclimatic zones are indicated with arrows.

(Köppen's scheme) and cumulative (Kira's scheme) temperatures. These bioclimatic zones have originally been correlated empirically with the equator-to-pole vegetation zonation by Köppen (1936) and Kira (1976). They are applied to aseasonal Mount Kinabalu by using measured temperature values along Kinabalu's slope. Therefore, it should be noted that the "temperate climate" on the mid-slope of Kinabalu is different from those of highly seasonal mid-latitudes, though the same name is used. Figure 2 shows altitudinal changes in annual potential pan-evaporation which is extrapolated from short-term measurement, and annual rainfall and water budget (rainfall minus potential pan-evaporation) on the same slope where the vegetation analysis was done (Kitayama 1992).

The four zones defined are, in upslope sequence, the lowland zone (defined by the presence of species group 1a), the lower montane zone (group 2a), the upper montane zone (group 3), and the lower subalpine (group 4). Note the definite coincidence of the upper boundaries of the lowland, upper montane, and lower subalpine zones with thermal threshold values (Figure 1). At these threshold altitudes, groups of diagnostic species are jointly displaced by other groups. Thus, the coincidence suggests a hypothesis that the upper boundaries of these zones are thermally controlled. The lower and up-

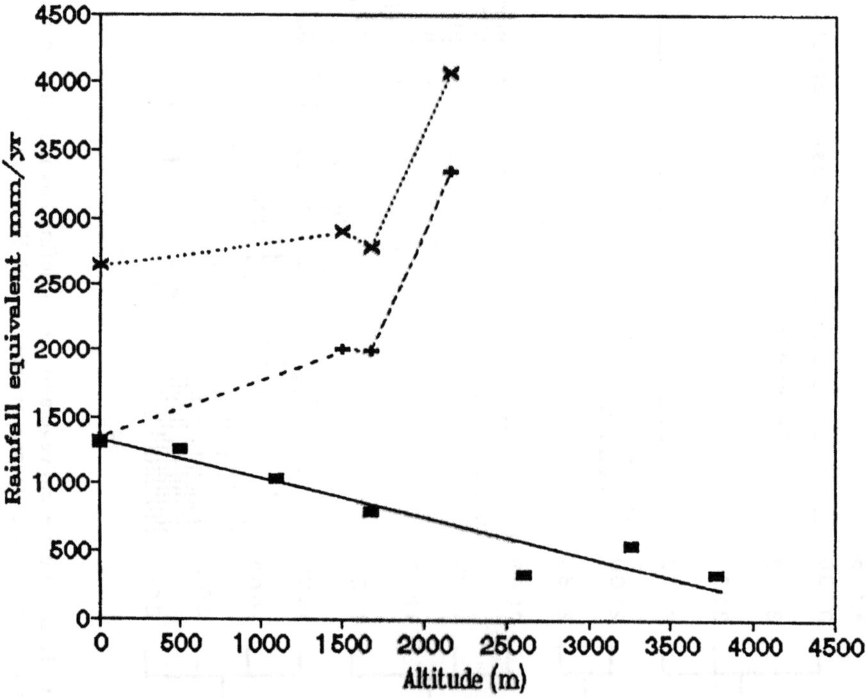

Figure 2. Altitudinal changes in annual potential pan-evaporation (■), rainfall (**x**), and water surplus *(*+, rainfall minus potential pan-evaporation*)* on slopes of Mount Kinabalu. Potential pan-evaporation values were estimated based on a short-term measurement.

per montane zones on Mount Kinabalu are thermally equivalent to a single warm-temperate climate zone (sensu Kira 1976), and there is no threshold value that coincides with the upper limit of the lower montane zone. Instead, the limit of the lower montane zone seems to be related to a sharp increase of water surplus as indicated in Figure 2.

According to Burgess (1969), the lifting condensation level occurs at 1,200 m asl in Malaya, and probably at similar altitudes in other areas of humid equatorial Malesia. However, observations suggest that the lower limit of cloud cover on Mount Kinabalu appears most persistently above 2,000 m asl. Indeed, a lowered potential annual pan-evaporation value, which is below the regression, appears above 2,000 m asl (Figure 2). The sharp water surplus in Figure 2 seems to be a direct consequence of the persistent cloud cover. This increase in water surplus in turn results in the joint displacement of species groups 2a, 2b, and I (a group of species indicating the lowland and lower montane zones) by species group 3 (Figure 1). The results imply that cloud cover can indeed control floristic assemblages of canopy species. In addition to rainfall, a substantial amount of precipitation may be added by horizontal rain in the cloud zone.

Thus results of the multivariate analysis support the view by Wade and MacVean (1969) and Whitmore (1975) that the cloud forest is a facies of the upper montane zone. Then, is it safe to include the entire upper montane zone in the cloud forest? No study has been conducted to determine what regulates the frequent upper cloud limit on Mount Kinabalu. McVean (1974) reported a synoptic air subsidence, which can suppress the cloud cover, at high altitudes on Mount Wilhelm, Papua New Guinea. Kitayama (1987) recorded very irregular and rapid oscillations in relative air humidity within a day at 3,260 m asl on Mount Kinabalu. If air subsidence is a relatively persistent event, the cloud forest is then restricted to at least below 3,260 m asl on Mount Kinabalu. Such persistent air subsidence and consequent temperature inversion occur at much lower altitudes (about 1,900–2,000 m asl) in Hawai'i and coincide with the upper limit of TMCF (Kitayama and Mueller-Dombois 1992).

Biophysical Conditions of the Cloud Forest on Mount Kinabalu

Forest Structure, Diversity, and Flora

Altitudinal changes in the forest layer structure are indicated for forest stands from 600 to 3,400 m (closed forest line) along the transect of Mount Kinabalu in Table 1. The canopy height of the upper montane cloud forest is reduced to ca. 20 m. The structure of the cloud forest is characterized by a dense herb layer (near 100 percent) and sparse shrub and subtree layers.

Aspects of the floristic diversity of the TMCF on Mount Kinabalu are shown in Table 2, and Figures 3 and 4. The data are taken from Kitayama (1991). The numbers of canopy tree species (10 cm dbh) and alpha diversity values

Table 1. Altitudinal changes in the layer structure of the forest stands on Kinabalu's slopes. Height (H) and coverage (C) of each layer are shown: E, emergent trees; T1, first tree layer; T2, second tree layer; T3, third tree layer; S, shrub layer; and Hb, herb layer.

Altitude	Hb		S		T3		T2		T1		E	
(m)	H m	(C%)	H m	(C%)	H m	(C%)	H m	(C%)	H m	(C%)	H m	(C%)
Lowland												
600	0.5	(5)	2	(20)	6	(20)	15	(70)	30	(60)	50	(30)
800	0.5	(40)	3	(20)	7	(25)	15	(60)	25	(70)	45	(60)
1,000	0.5	(5)	2.5	(20)	7	(30)	15	(70)	30	(60)	40	(40)
1,200	0.5	(20)	2	(60)	7	(20)	17	(40)	30	(70)		
Lower montane												
1,400	0.5	(20)	2	(30)	6	(40)	15	(80)	25	(20)		
1,600	0.5	(30)	2	(40)	6	(30)	20	(70)	30	(50)		
1,800	0.5	(20)	2	(30)	5	(25)	15	(70)	25	(60)		
2,000	1	(25)	2.5	(30)	8	(80)	20	(70)	30	(30)		
Upper montane cloud												
2,350	1	(80)	3	(25)			10	(10)	20	(70)		
2,600	1.5	(95)	4	(10)			10	(5)	20	(60)		
Lower subalpine												
2,800	1	(70)	3	(30)			10	(90)				
3,000	1	(40)	4	(25)			10	(80)				
3,200	1	(40)	2.5	(30)			15	(75)				
3,400	0.5	(20)	2	(20)			6	(100)				

Source: Kitayama 1991.

of the upper montane cloud forest are much poorer than that of the lower montane forest (Table 2). The species-area relationships (Figure 3) indicate that a minimal area for canopy tree species (10 cm dbh) in the upper montane cloud forest is reached at ca. 1,500 m^2.

Richness of all vascular species present in a given stand decreases substantially from lower to upper elevations in the cloud forest zone (Figure 4). The number of vascular species per minimal area is 110 (1,600 m^2 plot) at 2,380 m asl and decreases to ca. 70 (400–900 m^2 plot) above 2,500 m asl.

Table 3 shows the floristic composition and abundance of canopy trees for two representative stands in the upper montane cloud forest on Mount Kinabalu (Kitayama, unpubl.). On the south face, the upper montane cloud forest is characterized by the presence of four species of canopy trees (*Ascarina philippinensis, Eugenia punctilimba, Olea rubrovenia*, and *Ilex zygophylla*). The canopy on middle, modest slopes is dominated by the broad-leaved trees *Magnolia carsonii* (Magnoliaceae), *Ascarina philippinensis* (Chloranthaceae), *Horsfieldia* sp. (Myristicaceae), *Eugenia punctilimba* (Myrtaceae), *Elaeocarpus* sp. (Elaeocarpaceae), *Lithocarpus turbinatus* (Fagaceae), *Clethra pachyphylla* (Clethraceae), *Diplycosia* sp. (Ericaceae), *Ilex zygophylla* (Aquifoliaceae), *Olea rubrovenia* (Oleaceae), and Lauraceae sp., and by conifer *Phyllocladus hypophyllus* (Phyllocladaceae). The canopy on steeper

Table 2. Altitudinal changes in the number of canopy tree species (^{3}10 cm dbh) and alpha species diversity (Fisher's Index) on Mount Kinabalu's slopes.

Altitude (m)	Plot area* (m^2)	Sp. no. per plot	Sp. no. (0.1ha^{-1})	Alpha index
Lowland				
600	11,188	153	27	97.3
800	8,063	102	27	54.5
1,000	5,636	93	24	64.6
1,200	4,027	79	33	53.7
Lower montane				
1,400	2,477	70	39	40.5
1,600	3,707	58	29	26.4
1,800	3,441	41	24	15.5
2,000	4,907	50	24	19.0
Upper montane cloud				
2,350	1,439	26	25	10.6
2,600	1,700	13	13	3.8
Lower subalpine				
2,800	1,034	16	17	5.2
3,000	574	17	17	5.6
3,200	932	14	14	4.3
3,400	542	11	11	3.2

Source: Kitayama (unpubl.).
*Plot sizes in the lowland are smaller than a minimal area. Those of the high-altitude forests are close to or exceed a minimal area.

slopes or ridge tops (not shown in Table 3) is predominated by conifer *Dacrydium beccarii* (Podocarpaceae).

The herb and low shrub layers are frequently dominated by a tangle of rattan palm *Calamus gibbsianus*, and a prostrate bamboo species of *Racemobambos*, with solitary shrubs of *Lasianthus, Ardisia, Drymis, Embelia, Cyathea, Schefflera, Rubus, Prunus, Rhododendron, Psychotria*, and *Hedyotis*. Terrestrial ferns are also abundant and include the genera *Dryopteris, Diacalpe*, and *Plagiogyria*. The abundant vascular epiphytes below the canopy include the genera *Hymenophyllum, Trichomanes, Grammitis, Calymmodon, Elaphoglossum*, and *Asplenium*; those in the canopy are species of Polypodiaceae, Ericaceae, and Orchidaceae.

Frahm et al. (1990) concluded that the abundance of epiphytic mosses and liverworts reaches a maximum in this zone (1,900–2,600 m asl) on Mount Kinabalu. According to their account, typical species that are part of dense moss balls include *Lepicolea rara, Lepidozia cladorrhiza, L. trichodes, L. wallichiana, Bazzania harpago, B. indica, B. praerupta, B. spiralis*, and *B. unicigera*.

In summary, Mount Kinabalu's cloud forest is much impoverished in canopy tree species compared to lower-altitude forests. The canopy has a high floristic affinity at genus or family level with cloud forests of continental tropical

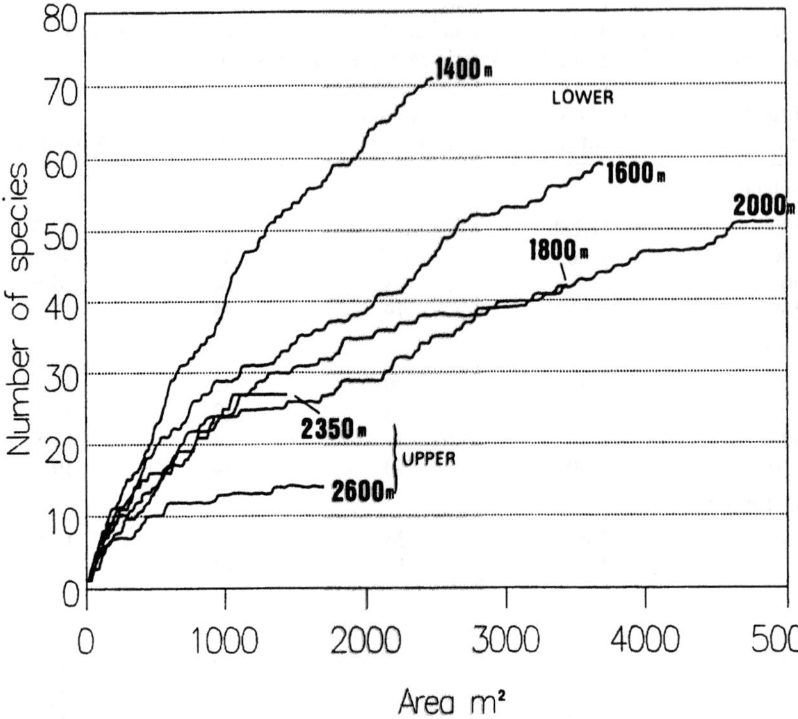

Figure 3. Species-area curves of the lower montane and the upper montane cloud forests on the slopes of Mount Kinabalu.

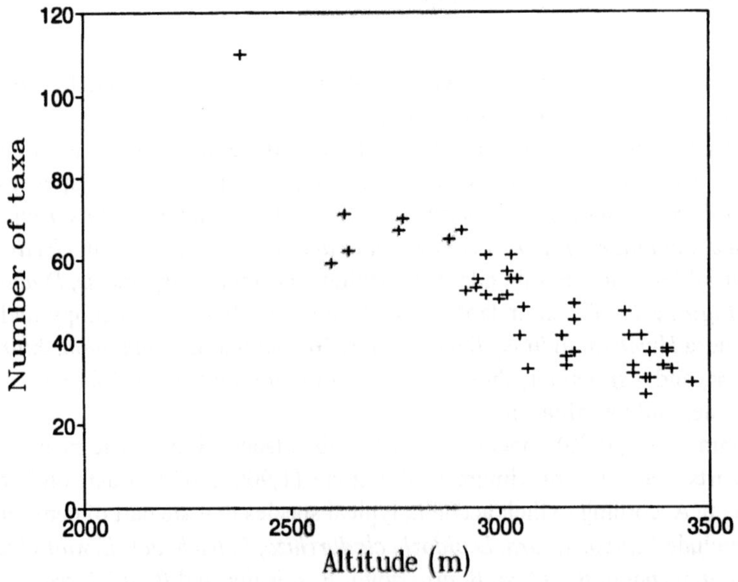

Figure 4. Altitudinal changes in species richness in and above the upper montane cloud forest zone on Mount Kinabalu. Numbers of vascular taxa per minimal area are plotted against altitudes.

mountains elsewhere (see Stadtmüller 1987). The forest is, however, enriched with many species of other life forms such as shrub, herb, and epiphyte, and the total number of vascular species becomes relatively high despite the small plot size. The floristic research is as yet incomplete and little can be concluded about the floristic endemism of the cloud forest.

Soils of the Upper Montane Cloud Forest on Mount Kinabalu

Figure 5 shows altitudinal changes in soil chemical profiles along the south and east slopes of Mount Kinabalu (Kitayama, unpubl.). There is a marked increase in organic carbon content in the topsoils of the lower and upper mon-

Table 3. Composition and abundance of canopy tree species (10 cm dbh) in the upper montane cloud forest at 2,350 and 2,600 m asl on Mount Kinabalu: BA, basal area; RD, relative dominance based on BA; and D, density

	2,350 m			2,600 m		
Species	BA (m^2 ha^{-1})	RD (%)	D (ha^{-1})	BA (m^2 ha^{-1})	RD (%)	D (ha^{-1})
Magnolia carsonii	17.8	29.9	63	4.1	8.4	88
Ascarina philippinensis	6.5	10.8	132			
Unknown	5.2	8.8	69			
Horsfieldia sp.	4.2	7.1	14			
Eugenia punctilimba	3.7	6.3	56	24.9	50.5	235
Elaeocarpus sp.	3.1	5.2	21	0.7	1.4	12
Lithocarpus turbinatus	2.6	4.3	14			
Clethra pachyphylla	2.5	4.2	104	0.4	0.8	18
Diplycosia sp.	2.2	3.8	35			
LAURACEAE sp.	2.2	3.7	21			
Cyathea havilandii	1.2	2.0	14			
Eurya trichocarpa	1.1	1.9	21			
Prunus mirabilis	1.1	1.8	7			
Polyosma sp.	0.8	1.4	7	2.0	4.1	18
Nothaphoebe sp.	0.7	1.3	14			
Olea rubrovenia	0.7	1.1	49	4.0	8.1	141
Helicia sp.	0.7	1.1	49			
Rapanea affinis	0.7	1.1	14	0.1	0.2	6
Ilex zygophylla	0.6	1.0	7	6.5	13.1	53
Eugenia houttuynii	0.5	0.9	21			
Cyathea sp.	0.4	0.7	14			
Phyllocladus hypophyllus	0.4	0.6	14	4.2	8.6	47
Litsea sp.	0.2	0.4	7			
THEACEAE sp.	0.2	0.3	7			
Unknown	0.1	0.2	7			
Lithocarpus havilandii	0.1	0.2	7	1.1	2.3	29
Symplocos pendula	0.6	1.2	12			
Prunus arborea	0.5	1.0	24			
Microtropis wallichiana	0.2	0.4	12			
Total	59.5	100	788	49.2	100	695

Source: Kitayama (unpubl.).

tane forests. Associated with the increase in organic matter, total nitrogen, soluble phosphorus, and exchangeable magnesium and potassium (extracted by N ammonium acetate at pH 7) are higher in the topsoils of the lower and upper montane zones than those of the lowland. Thus, mineral nutrients potentially are not poor in soils of the cloud forest as compared to soils of the lowland. However, it is not known from the data how readily these elements are available to plants. For instance, a large portion of nitrogen may not be available as indicated by high C/N ratios of the montane forest soils. Also, actual nutrient uptake can be lowered by reduced photosynthetically active

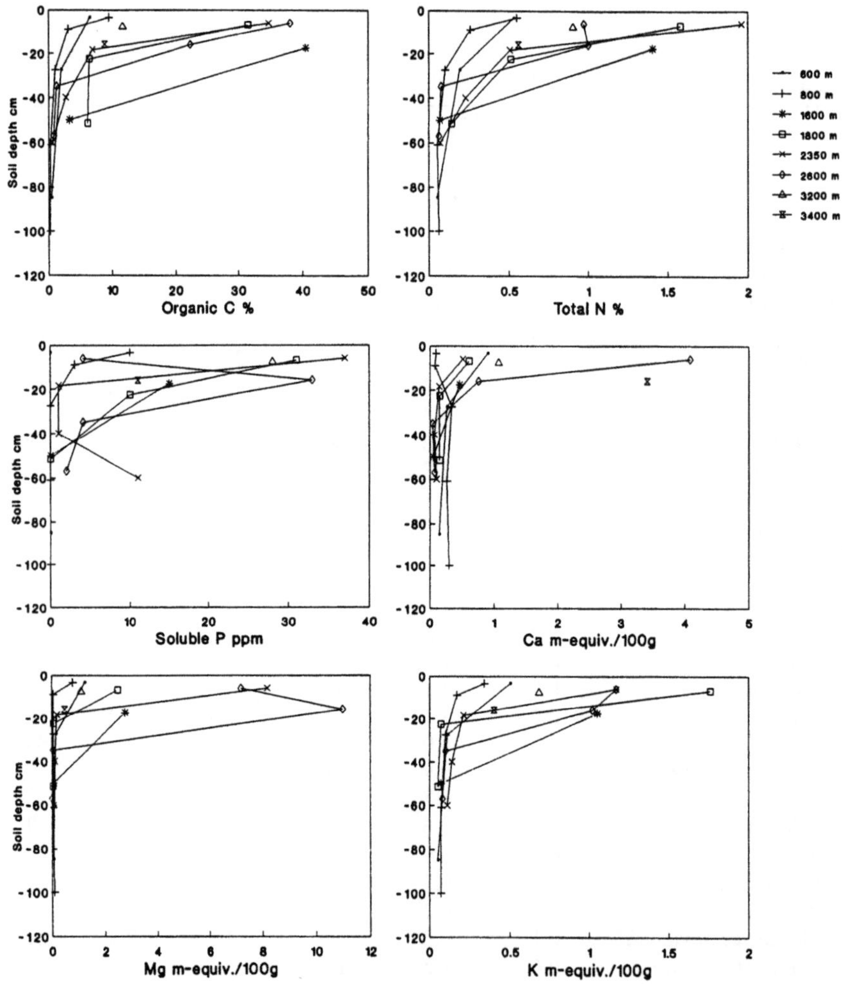

Figure 5. Altitudinal changes in soil chemical properties on the slopes of Mount Kinabalu. Lowland, 600 and 800 m; lower montane, 1,600 and 1,800 m; upper montane cloud forest, 2,350 and 2,600 m; lower subalpine, 3,200 and 3,400 m.

radiation because of cloud cover (Grubb 1977; Körner et al. 1983), and by reduced transpiration because of low saturation deficits of the cloud zone (Leigh 1975; Ash 1987).

Soils of humid tropical mountains change zonally along slopes, and a characteristic zone of peaty soils is thought to appear associated with the "cloud forest" and the "moss forest" facies of upper montane rain forest (Burnham 1975). On Mount Kinabalu, however, peaty soils occur not only in the cloud forest but as low as the lower limit of the lower montane rain forest zone. Soil chemical properties shown in Figure 5 support the broad occurrence of peaty soils. The pedogenesis of the cloud forest is thought to be affected strongly by a high amount of water leading to podzolization, element reduction, acidification, and waterlogging (Askew 1964; Whitmore and Burnham 1969; Burnham 1974, 1975). This process on Mount Kinabalu, as suggested by the occurrence of peaty soils, appears to be present even at elevations with modest water surplus below the maximum condensation zone (Figure 5). Thus, the "cloud forest" zone is not well definable by edaphic characteristics alone.

Microclimate of the Upper Montane Cloud Forest on Mount Kinabalu

Limited information from the transect analysis suggests that this zone's climate is characterized by daily persistent, very low saturation deficits due to the combination of orographic cloud and lowered air temperatures in normal years. A mean daily saturation deficit during wet periods of a normal year is around 3 hPa (Kitayama 1992) and the air is nearly saturated with water. The mean annual air temperature drops to 16.5° C at 2,000 m asl and 12.1° C at 2,800 m asl. The temperature difference between daily maxima and minima is smaller in the cloud zone than in the lowland.

Because of cloud cover the incidence of global and photosynthetically active radiation is expected to be low in the cloud forest. Nakashizuka et al. (1991) reported lowered global radiation values because of cloud cover in Malaya; the low values, however, occur from 800 m asl and suggest that this may be a factor for the occurrence of cloud forests on low-lying mountains.

The above descriptions are based on observations in "normal" years. There are, however, episodic drought events associated with the El Niño Southern Oscillation. The length of droughts ranges from a few weeks to several months, as was the case in 1983 (Beaman et al. 1985). The influence of droughts has been studied with an emphasis on large-scale fire and loss of timber resources in the lowland rain forest zone. However, such droughts can have profound effects even in the wet montane forests and cause wilting of epiphytic fern species (Lowry et al. 1973). In addition, water deficits of air and soils in the otherwise saturated cloud zone may trigger rapid mineralization of accumulated elements, root penetration into deeper horizons, nutrient uptake, and affect the physiology of photosynthesis. The microclimate of the cloud forest during droughts is poorly known.

Cloud Forest and Human Influences in Sabah

Figure 6 depicts the areas of potential distribution of cloud forest in Sabah. Although the summits of most mountains in Sabah are lower than 2,000 m (i.e., the lower limit of the cloud forest on Mount Kinabalu), such low-lying mountains are often capped with cloud forests (e.g., Mount Silam, 884 m; Bruijnzeel et al. 1993). This suppressed occurrence is because of the so-called Massenerhebung or "telescope" effect. To include all potential stands, the cloud forest in Figure 6 is estimated conservatively to occur from 900 m asl. Sizable areas of the "upper montane cloud forest" in Sabah are known only from Mounts Kinabalu (Kitayama 1992), Trus Madi (2,624 m; Kitayama et al. 1993), and Tambuyukon (2,579 m; Kitayama 1987). The floristic composition on Mount Tambuyukon differs substantially from that of the other two mountains as a consequence of geological differences (oligotrophic, ultrabasic substrate).

The cool and wet conditions of the TMCF were not readily exploitable by traditional people in Sabah, and many of the areas in Figure 6 have remained intact until recently. The traditional habitations of the native hill tribes in Sabah were restricted to upper lowland terrains. The highest traditional vil-

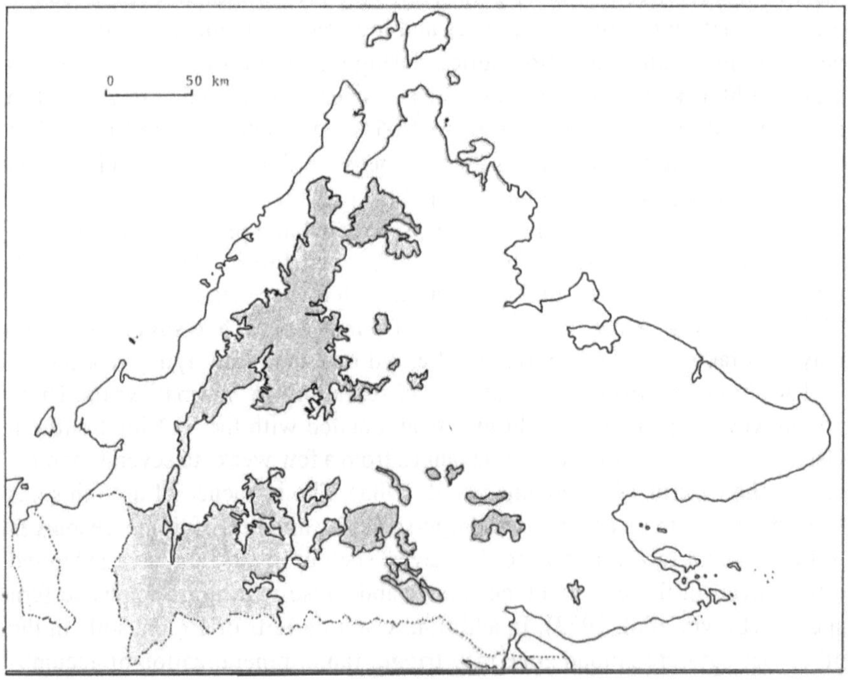

Figure 6. Distribution of potential cloud forest areas (shaded on map) in Sabah, Malaysia.

lage is found at 1,200–1,400 m asl in Sabah (Kitayama 1990). These elevations correspond roughly to the lower limit of the lower montane forest where cloud cover descends daily, and the daily minimum temperature is 18° C. Under this condition, fruits of some important lowland crops abort (e.g., bananas and coconuts); thus harvesting such crops is not possible. The altitudinal range of traditional cultivation coincides with the human habitation range in Sabah (Kitayama, unpubl.).

However, in modern days, cultivation has rapidly invaded the upper montane forest zone (Figure 7). This invasion became possible through the combination of the introduction of temperate crops (e.g., cabbage and asparagus) and modern cultivation skills (a heavy use of pesticide and fertilizer), and road accessibility. The highest elevation of temperate vegetable cultivation reached 2,100 m asl in the cloud forest by 1989 (Kitayama 1990).

In addition, other large-scale, modern exploitative activities (including dairy farming, golf-course development, unregulated tourism, and construction of logging roads) have the potential to directly or indirectly impact TMCF in Sabah. Most areas in the cloud forest zone on Mount Kinabalu are protected by the state government of Sabah. However, the cloud forests on other low-lying, unprotected mountains are increasingly threatened.

Commercial logging operations potentially can be threatening for the cloud forest in Sabah, although current timber extraction is generally practiced be-

Figure 7. Upper montane cloud forest cleared for cultivating temperate vegetables on Mount Kinabalu. Forest alteration proceeded to a border area of Mount Kinabalu Park. *(Photo by author.)*

low the cloud forest zone. At present, useful timber species do not grow in the cloud forest. However, there is a possibility that the continuing dependence of logging operations on highland forest eventually may reach the cloud forest, especially if more forest species become utilized.

Acknowledgments. The unpublished data used in this paper were collected during studies generously supported by The Toyota Foundation, Tokyo (grants 88-I-036 and 90-I-051).

References

Ash, J. 1987. Stunted cloud-forest in Taveuni, Fiji. *Pacific Science* 41:191–199.

Askew, C. P. 1964. The mountain soils of the east ridges of Mount Kinabalu. *Proceedings of the Royal Society London (B)* 161:65–74.

Beaman, R. S., J. H. Beaman, C. Marsh, and P. Woods. 1985. Drought and forest fires in Sabah in 1983. *Sabah Society Journal* 8:10–30.

Bruijnzeel, L. A., M. J. Waterloo, J. Proctor, A. T. Kuiters, and B. Kotterink. 1993. Hydrological observations in montane rain forests on Gunung Silam, Sabah, Malaysia, with special reference to the "Massenerhebung" effect. *Journal of Ecology* 81:145–167.

Burgess, P. F. 1969. Ecological factors in hill and mountain forests of the states of Malaya. *Malayan Nature Journal* 22:119–128.

Burnham, C. P. 1974. The role of the soil forming factors in controlling altitudinal zonation on granite in Malaysia. In J. R. Flenley, ed., *Altitudinal Zonation in Malesia*, pp. 59–74. Transactions of the Third Aberdeen-Hull Symposium on Malesian Ecology, Hull, 1973, Department of Geography, University of Hull, England.

Burnham, C. P. 1975. The forest environment: Soils. In T. C. Whitmore, *Tropical Rain Forests of the Far East*, pp. 103–120. Oxford University Press, London.

Frahm, J. P., W. Frey, H. Kurschner, and M. Menzel. 1990. *Mosses and Liverworts of Mount Kinabalu*. Sabah Parks Publ. 12. Sabah Parks Trustees, Kota Kinabalu.

Grubb, P. J. 1977. Control of forest growth and distribution on wet tropical mountains: With special reference to mineral nutrition. *Annual Review in Ecology and Systematics* 8:83–107.

Johns, R. J. 1982. Plant zonation. In J. L. Gressitt, ed., *Biogeography and Ecology of New Guinea*, vol. 1, pp. 309–330. Junk, The Hague.

Kira, T. 1976. *Terrestrial Ecosystems*. Kyoritsu, Tokyo (in Japanese).

Kitayama, K. 1987. Vegetation of Mount Kinabalu. *Review of Forestry Culture* 8:103–113 (in Japanese).

Kitayama, K. 1990. Threatened montane rain forest. *Greenpower* 133:20–23 (in Japanese).

Kitayama, K. 1991. Vegetation of Mount Kinabalu Park, Sabah, Malaysia; Map of physiognomically classified vegetation. Project Paper, Protected Areas and Biodiversity, the Environment and Policy Institute, East-West Center, Honolulu.

Kitayama, K. 1992. An altitudinal transect study of the vegetation on Mount Kinabalu, Borneo. *Vegetatio* 102:149–171.

Kitayama, K., J. Kulip, J. Nais, and A. Biun. 1993. Vegetation survey on Mont Trus Madi, Borneo, a prospective new mountain park. *Mountain Research and Development* 13:99–105.

Kitayama, K., and D. Mueller-Dombois. 1992. Vegetation of the wet windward slope of Mount Haleakala, Maui, Hawai'i. *Pacific Science* 46:197–220.

Kitayama, K., and D. Mueller-Dombois. In press, a. An altitudinal transect analysis of the windward vegetation on Haleakala, a Hawaiian mountain: (1) climate and soils. *Phytocoenologia* 24.

Kitayama, K., and D. Mueller-Dombois. In press, b. An altitudinal transect analysis of the windward vegetation on Haleakala, a Hawaiian mountain: (2) vegetation zonation. *Phytocoenologia* 24.

Köppen, W. 1936. Das geographische System der Klimate. Handbuch der Klimatologie, Band I, Teil C. Gebruder Borntraeger, Berlin.

Körner, C., A. Allison, and H. Hischer. 1983. Altitudinal variation of leaf diffusive conductance and leaf anatomy in heliophytes of montane New Guinea and their interrelation with microclimate. *Flora* 174:91–135.

Leigh, E. G. 1975. Structure and climate in tropical rainforest. *Annual Review in Ecology and Systematics* 6:67–86.

Lowry, J. B., D. W. Lee, and B. C. Stone. 1973. Effect of drought on Mount Kinabalu. *Malayan Nature Journal* 26:178–179.

McVean, D. N. 1974. Mountain climates of the southwest Pacific. In J. R. Flenley, ed., *Altitudinal Zonation in Malesia*, pp. 47–58. Transactions of the Third Aberdeen-Hull Symposium on Malesian Ecology, Hull, 1973, Department of Geography, University of Hull, England.

Nakashizuka, T., Z. Yusop, and A. R. Nik. 1991. Altitudinal zonation of forest communities in Selangor, Peninsular Malaysia. *Journal of Tropical Forest Science* 4:233–244.

Okutomi, K. 1977. Forest and scrub vegetation in the tropical and subtropical climate zones. In K. Ishizuka, ed., *Distribution of Plant Communities and Their Environments*, pp. 28–71. Asakura, Tokyo (in Japanese).

Stadtmüller, T. 1987. *Cloud Forests in the Humid Tropics*. The United Nations University, Tokyo.

Van Steenis, C. G. G. J. 1984. Floristic altitudinal zones in Malesia. *Botanical Journal of the Linnean Society* 89:289–292.

Wade, L. K., and D. N. McVean. 1969. *The Alpine and Subalpine Vegetation*. Mt. Wilhelm Studies 1. The Australian National University Press, Canberra.

Whitmore, T. C. 1975. *Tropical Rain Forests of the Far East*. Oxford University Press, London.

Whitmore, T. C., and C. P. Burnham. 1969. The altitudinal sequence of forests and soils of granite near Kuala Lumpur. *Malayan Nature Journal* 22:99–118.

13. The Importance of Tropical Montane Cloud Forest for Preserving Vertebrate Endemism in Peru: The Río Abiseo National Park as a Case Study

Mariella Leo

The tropical montane cloud forest (TMCF) is a biome (sensu Whittaker 1972) that is poorly understood. A precise definition based on both quantitative and qualitative characteristics is still lacking, as is a precise definition of its borders with respect to other montane forest types. Thus, different authors have used the term in different ways. TMCF typically occurs at different elevations, in a range of 1,500 to 3,300 m, although regional or local climatic and topographic factors may result in cloud forest occurring as low as 1,000 m (Stadtmüller 1986).

In Peru, TMCF is known as Ceja de Montaña (Weberbauer 1945), or the mountain eyebrow, and has been well differentiated from the more general Montaña (the mountain), which groups other types of montane forest. The Ceja de Montaña is located mainly on the eastern slopes of the Peruvian Andes, along a northwest-southeast axis, but it is also found on the Cordillera Sub-Andina (between the Huallaga and Ucayali rivers) and in the Montañas del Sira range (close to the upper Ucayali River; Figure 1). Smaller areas of TMCF are found along a northeast-southwest axis in the Cordillera del Condor (boundary between Peru and Ecuador) (Weberbauer 1945) and in the western slopes of the Andes along the upper drainage of the Zaña River (Koepcke in Brack 1986), as well as in Piura and Cajamarca (Parker, pers. com.).

According to Weberbauer (1945), most of the Ceja de Montaña is located

Figure 1. General distribution of Peruvian cloud forest.

between 1,800 and 3,600 m, but in central and southern Peru it sometimes may reach 3,900 m. Zamora (unpubl. ms.) proposed a range from 1,500 to 3,500 m for his *bosques nublados* (fog forest) ecological zone. Terborgh (1971), based on his field studies in the Vilcabamba Range, described TMCF located between 1,380 and 2,550 m on the altitudinal gradient, merging with the elfin cloud forest that ranges up to above 3,000 m. Pulgar Vidal (1985) located the Ceja de Montaña between 1,000 and 3,900 m, whereas Peñaherrera (1986) used the name Ceja de Selva (lowland forest eyebrow) for a morphological formation that occurs between 800 and 3,800 m. However, the use of the term by the last two authors seems altitudinally too broad for true TMCF and may include other types of montane forests.

The TMCF in Peru may potentially occur in more than twelve different bioclimatic life zones (sensu Holdridge 1967). In northern Peru, where the Andes have lower elevation, TMCF is usually found at the top altitudinal levels of the cordillera. But in the inter-Andean valleys, it can also occur at the tropical premontane and lower montane moist life zones.

Focus and Methods

This paper presents a preliminary compilation of the species of three vertebrate groups (frogs, birds, and mammals) endemic to Peru, for which the TMCF constitutes primary species habitat or a significant part of it. Within this framework vertebrate endemism is analyzed for the TMCF portion of the Río Abiseo National Park (Figure 1).

For this purpose the author considers TMCF as those areas locally called Ceja de Selva or Ceja de Montaña; habitats described elsewhere as cloud, montane cloud, or elfin forest located in the Cordillera Oriental and adjacent ranges; and those localities studied or known by me, where the environmental conditions and high degree of epiphytal cover justify the use of the name. For the distribution of birds, I use the humid montane forest and the elfin forest habitat designation (in Parker, Parker, and Plenge 1982) as equivalent to TMCF. This paper will not include the TMCF on the western slopes of the Andes since the faunal data are still too scarce. The term montane forest will be used as an equivalent to the Selva Alta geographical region, comprising the Montaña (montane forest other than cloud forest) and the Ceja de Montaña (TMCF).

The main sources of information for the species compilation were the World Conservation Monitoring Centre (unpubl. ms.), Brack (1986), Plenge (unpubl. ms.), Rodriguez, Córdova, and Icochea (in press), Honacki, Kinman, and Koeppl (1982), Nowak (1992), and Parker, Parker, and Plenge (1982). The information species endemism in the Río Abiseo National Park results from fieldwork undertaken by the Peruvian Association for Conservation of Nature (APECO 1988, 1989, 1991a, 1991b) and subsequent lab work (Rodriguez, in press; Leo and Romo 1992; Leo and Handley, in prep.; Leo and Gardner, unpubl. ms.).

Vertebrate Endemism in the Peruvian Montane Cloud Forest

Peru is considered one of twelve megadiversity countries (UNEP 1991) for its highly diverse flora and fauna. However, little attention has been paid to the significance of endemism as an important component of this biodiversity.

The three groups chosen for this analysis of endemism are among the best known for the country, although the information is still incomplete and weak for some families and genera. At present, approximately 450 mammal species (Pacheco et al., in press), 1,702 bird species (O'Neill 1992), and ca. 295 anuran species (Rodriguez, Córdova, and Icochea, in press) are known to occur in Peru. A preliminary compilation of available information shows that 52 mammal species (11.5 percent), 110 bird species (65 percent) (Plenge, unpubl. ms.), and 110 anuran species (37.3 percent) are endemic to the country.

An important proportion of this endemic fauna is distributed mainly along the eastern slopes of the Andes montane forest (including the cloud forest biome): 24 species of mammals, 43 species of birds, and 59 species of anurans. Of these species 17 mammals, 29 birds, and 42 anurans are found in the montane cloud forest.

All the endemic mammals in the montane cloud forest (Table 1) may be considered local endemics (sensu Gentry 1986) due to their relatively restricted geographical distributions. Rodents, the most numerous group among mammals, contain a high number of Peruvian endemics occurring in TMCF. Most of these may be restricted to this biome, as is the case for at least two primate species and the hairy armadillo (*Dasypus pilosus*).

Among birds, just five of the ten families considered most characteristic (O'Neill 1992) of the montane forest zone contain Peruvian endemics: Cotingidae (cotingas), Tyrannidae (tyrant flycatchers), Trochilidae (hummingbirds), Traupidae (tanagers), and Furnariidae (ovenbirds). The last two families and the Formicariidae (antbirds) are represented by the largest number of Peruvian endemics with 5, 4, and 5 species, respectively (Table 2).

Five out of the nine families of anurans existing in Peru are represented by endemics occurring in TMCF. Leptodactylidae, the most species-rich family within the Peruvian anurans, is represented by the largest number of endemic frogs (15 species) found in the TMCF (Table 3); these TMCF endemics represent almost 15 percent of the Leptodactylid species occurring in Peru. The family Centrolenidae is also well represented in TMCF with eight endemic species comprising 36 percent of Peruvian species in this family.

Endemism in the Río Abiseo National Park

Located in northeastern Peru, the 274,520 ha Río Abiseo National Park encompasses an elevational gradient from montane forest to puna (ca. 1,500 m to above 4,000 m). TMCF is found in at least three contiguous life zones: the montane rain forest, montane wet forest, and the lower montane wet forest (Young and León 1988, 1991). Young and León (1990) estimated that the

Table 1. Endemic mammals to Peru which occur in the montane cloud forest

Order/Species	Distribution
Marsupialia	
Lestorus inca	MF(1) CF(12)
Chiroptera	
Vampyresa melissa	MF/CF * (1), MF (2) (3)
Primates	
Lagothrix flavicauda	MF/CF (1), CF (4)
Aotus miconax	MF/CF * (5)
Xenarthra	
Dasypus pilosus	MF/CF (6)
Rodentia	
Sciurus pyrrhinus	MF/CF * (1) (7)
Microsciurus sp. nov. **	CF (8)
Thomasomys incanus	g/CF (9)
Thomasomys notatus	MF/CF (3)
Thomasomys rosalinda	MF/CF * (1)
Thomasomys apeco	CF (10)
Thomasomys sp. nov.1	MF?/CF (8)
Thomasomys sp. nov.2	MF?/CF (8)
Akodon orophyllus	g/MF/CF (11) (9)
Akodon torques	CF (11)
Oxymycteris hiska	CF (11)
Dactylomys peruanus	MF/CF (1)

Sources:

(1) Brack 1986
(2) Thomas 1926
(3) Gardner 1976
(4) Leo 1982
(5) Hershkovitz 1983
(6) Wetzel 1982
(7) Emmons 1990
(8) Leo and Handley (in prep)
(9) Leo and Romo 1992
(10) Leo and Gardner 1993
(11) Patton and Smith 1992
(12) Young, pers. com.

* The localities named by the source comprise montane cloud forest, or the elevation and location suggest cloud forest, or the author has personal knowledge of the locality as being in the montane cloud forest:

g = grassland
MF = montane forest (other than CF)
CF = montane cloud forest (including elfin forest)

** nov. = new to science

montane forest comprises 53 percent of the park's area, but information on TMCF biome in the park is not available. We estimate TMCF at between 20 and 25 percent of the total area.

Thirty-six species of mammals, birds, and frogs endemic to Peru are found in the park, 26 of which occur in the TMCF (Table 4). This figure includes at least 10 species new to science which may occur only within the Río Abiseo National Park. The number reported here for anurans is a conservative estimate that may increase significantly when all collections are identified (Rodriguez, pers. com.).

Table 2. Endemic birds to Peru which occur in the montane cloud forest

Family/Species	Distribution (1)
Stringidae	
Otus marshalli	HT/7
Xenoglaux loweryi	HT, HST/8
Trochilidae	
Loddigesia mirabilis	HT, HST/3,12, MCF (2)
Heliangelus regalis	HST/3,7
Metallura theresiae	HT/9
Metallura empogon	HT/9
Ramphastidae	
Aulacorhynchus huallague	HST/7
Picidae	
Picumnus steindachneri	HST, HUT/7
Furnariidae	
Synallaxis courseni	HT/3,7
Cranioleuca marcapatae	HT/7,8
Schizoeaca helleri	HT/7,8,9
Thripophaga berlepschi	HT/8
Formicariidae	
Grallaria erythroleuca	HST/3,7
Grallaria capitalis	HST/3,7
Grallaria carrikeri	MCF (3)
Grallaria blakei	MCF (3)
Grallaria ochraceifrons	MCF (3)
Rhinocryptidae	
Scytalopus macropus	HT/7,11
Cotingidae	
Ampelion [Doliornis] *sclateri*	HT/8
Tyrannidae	
Anairetes agraphia	HT/7
Hemitriccus cinnamomeipectus	HST/7
Troglodytidae	
Thryothorus eisenmanni	MCF (3)
Henicorhina leucoptera	HST/8
Coerebidae	
Nephelornis oneilli	HT/8,9
Traupidae	
Iridisornis reinhardti	HT/7
Buthraupis aureodorsalis	HT/8
Hemispingus parodii	HT/8
H. rufosuperciliaris	HT/8
Tangara phillipsi	CF (3)

Sources:

(1) Parker, Parker, and Plenge 1982

Life zones		*Habitats:*
humid temperate	=	HT Forest edge 3
humid subtropical	=	HST humid montane forest 7; elfin forest 8;
humid upper tropical	=	HUT paramo 9; montane lakes, streams and their margins, marshes 11; montane scrub 12

(2) Brack 1986

(3) Plenge, pers. com.

MCF = montane cloud forest

CF = montane cloud forest (including elfin forest)

Table 3. Endemic anurans to Peru which occur in the montane cloud forest

Family/Species	Distribution
Bufonidae	
Bufo nesiotes	MCF (1)
Bufo trifolium	g/MCF (2)
Centrolenidae	
Cochranella miriae	MCF (1)
Cochranella truebae	MCF (11)
Cochranella ocellata	MF/MCF (11)
Cochranella phenax	MF (2)/MCF *
Cochranella spiculata	MF/MCF (11)
Centrolene lemniscatum	MCF (4)
Centrolene muelleri	MCF (4)
Centrolenella azulae	MCF (15)
Dendrobatidae	
Colostethus sylvaticus	MCF * (2)
Dendrobates silverstonei	MCF (16)
Epipedobates sp. nov. **	MCF (3)
Hylidae	
Gastrotheca excubitor	HP/EF (5)
Gastrotheca ochoai	HP/UMF? (5)
Gastrotheca lateolata	MCF (13)
Gastrotheca rebeccae	MCF (13)
Gastrotheca sp. nov.1	MCF (3)
Gastrotheca sp. nov.2	MCF (3)
Hyla sp. nov.	MCF (3)
Hyla aperomea	MCF (3)
Hyla elkejungingerae	MCF (3)
Phyllomedusa baltea	MCF (1)
Phyllomedusa coelestis	MCF (3)
Phyllomedusa duellmani	MCF (3)
Leptodactylidae	
Eleutherodactylus mendax	LF/MF/MCF (6) (9)
E. cosnipatae	MCF (7)
E. danae	MCF (7)
E. scitilus	MCF (7)
E. lindae	MCF (7)
E. salaputium	MCF (7)
E. rhabdolaemus	MCF (7)
E. schulti	MCF (12)
Eleutherodactylus sp. nov.	MCF (3)
Phrynopus bracki	MF/MCF (9)
P. cophites	g/MCF/MCF *
P. lucida	MCF
P. montium	MCF (8)
P. peruvianus	MCF (8)
Phrynopus sp. nov.1	MCF (10)
Phrynopus sp. nov.2	MCF (10)
Phyllonastes lynchi	MCF (14)

Sources:

(1) Duellman and Toft 1979
(2) Frost 1985
(3) Rodríguez, pers. com.
(4) Duellman and Schulte 1993
(5) Duellman and Fritts 1972
(6) Duellman 1978a
(7) Duellman 1978b
(8) Brack 1986
(9) Hedges 1991
(10) Rodríguez (in press)
(11) Duellman 1976
(12) Duellman 1990
(13) Duellman and Trueb 1988
(14) Duellman 1991
(15) Flores and McDiarmid 1989
(16) Myers and Daly 1979

* The localities named by the source comprise montane cloud forest, or the elevation and location suggest cloud forest, or the author has personal knowledge of the locality:

g = grassland
HP = humid paramo
LF = lowland forest
MF = montane forest (other than CF)
EF = elfin forest (= montane cloud forest)
UMF = upper montane forest (= montane cloud forest?)
MCF = montane cloud forest

** nov. = new to science

Table 4. Peruvian endemic species found in the cloud forest of the Río Abiseo National Park

Taxa/Species	Life Zones (1)
Mammals	
Lagothrix flavicauda	TM-wf and TLM-wf
Dasypus pilosùs	TM-wf
Microsciurus sp. nov. *	TM-wf and TLM-wf
Thomasomys incanus	TM-rp to TLM-wf/TPM-wf
Thomasomys sp. nov.1	TM-rp
Thomasomys sp. nov.2	TM-rf?
Thomasomys sp. nov.3	TLM-wf
Akodan orophyllus	TM-rp to TLM-wf/TPM-wf
Birds	
Metalura theresia	TM-rf
Aulacorhynchus huallague	TLM-wf to TLM-wf/TPM-wf
Thripophaga berlepschi	TM-rf
Grallaria carrikeri	TM-rf
Heliangelus regalis	TLM-wf to TLM-wf/TPM-wf
Scytalopus macropus	TM-rf
Ampelion [Doliornis] *sclateri*	TM-rf
Nephelornis oneilli	TM-rf
Iridisornis reinhardti	TM-rf
Buthraupis aureodorsalis	TM-rf
Hemispingus rufosuperciliaris	TM-rf
Anurans	
Phrynopus sp. nov.1	TM-rf
Phrynopus sp. nov.2	TM-rf
Hyla sp. nov.	TLM-wf
Gastrotheca sp. nov.1	TM-rf and E/TM-rf
Gastrotheca nov.2	TLM-wf
Eleutherodactylus sp. nov.1	TM-rf to TLM-wf
Eleutherodactylus sp. nov.2	TM-rf to TLM-wf/TPM-wf

Sources: (1) Sensu Holdrige 1967. Local determination of life zones by Young and León 1988:

TM-rf	= tropical montane rain forest
TM-wf	= tropical montane wet forest
TLM-wf	= tropical lower montane wet forest
TM-rp	= tropical montane rain paramo (grasslands)
TLM-wf/TPM-wf	= transition zone from tropical montane wet forest to tropical premontane wet forest (not considered as cloud forest)
E	= ecotone

* nov. = new to science

Detailed information published on these three vertebrate groups is unavailable for comparison from other Peruvian TMCF localities. Patton and Smith (1992) collected small mammals from the slopes of the eastern Andes, which reveal the presence in TMCF of at least three small Peruvian endemic rodents. Other endemic mammals, including *Dactylomys peruanus* and *Lestoros incanus*, may occur in this region (Honacki, Kinman, and Koeppl 1982). The remaining 12 species of Peruvian endemic mammals with distributions in TMCF are restricted to either central or northern Peru. Hedges (1991), who collected anurans from TMCF at the southern end of Cordillera Yanachaga, revealed two endemic species (Table 3) and "about six undescribed species of Eleutherodactylus and at least one undescribed species of bufonid" (not considered in Table 3).

Conclusion

Stadtmüller (1986) has suggested that a high level of endemism is a characteristic of TMCF. Gentry (1986) reported that in the northern Andes and Central America the concentration of plant endemism is "mostly along the moist lower slopes of the Andes and to a lesser extent in the cloud forest of southern Central America." Terborgh and Weske (1975) mentioned the endemic richness of the cloud forest and elfin forest vegetation in Cordillera del Sira (Peru). Available data presented here show that a high percentage of all endemic vertebrate species in Peru is found in the montane cloud forest and, for the three groups analyzed, approximately 30 percent of the endemic species have TMCF as their primary habitat (Table 5).

Mares (1992) has shown that although the South American eastern slope montane forest is only 3.2 percent of the continental area, it holds 63 percent of its endemic mammals. This region has a ratio of 1.5 endemic mammal species per million hectares versus 0.26 species per million hectares in the Amazon forest. The data presented here produced a lower ratio for the montane forest as a whole (0.89–1.09) but higher (2.8) for the TMCF portion (Table 6).

As expected, because of their rapid generation turnover rate and favorable moisture conditions in the cloud forest biome, the anurans present the highest proportion of endemism, with 110 species endemic to Peru (37.3 percent), whereas mammals and birds exhibit much lower levels of endemism (11.5 percent and 6.5 percent, respectively). For all Peruvian endemics in these three groups (total 272 species), 88 species (32 percent) occur within the TMCF habitat (Table 5). Of the endemics distributed in the montane cloud forest of Peru, the Río Abiseo National Park protects 47 percent of the mammals, 38 percent of the birds, and almost 17 percent of the anurans. Although little information is available for comparison, the high number of endemic species found in the park suggests that the area might be part of a "center of endemism" (sensu Lamas 1982), as has already been suggested based on data for birds (Parker, pers. com.).

Table 5. Vertebrate species endemic to Peru and their distribution along the montane forest

Taxa	No. of Endemic Spp. in Peru	% MF *	% CF	% RANP (CF)
Mammals	52	46	33	15
Birds	110	39	26	10
Anurans	110	54	38	6
Combined 3 taxa	272	46	32	9

* The montane forest as a whole, including cloud forest.
RANP = Río Abiseo National Park

Table 6. Peruvian endemic species present in the montane cloud forest (no. of species/area ratio)

	Area (millions of ha)		
Taxa	Montane Forest 27 (1) 22 (2)	Montane Cloud Forest (TMCF) 6 (2)	RANP (0.27)*
Mammals	0.89–1.09 (24)	2.8 (17)	29.63 (08)
Birds	1.59–1.95 (43)	4.8 (29)	37.04 (10)
Anurans	2.18–2.68 (59)	7.0 (42)	25.92 (07)
Combined 3 taxa	4.6–5.7 (126)	14.1 (88)	77.78 (25)

Sources:
(1) Alva and Huamani 1991
(2) Zamora 1988
* Total area of the park montane. Cloud forest might be less than one-fourth of the area.
Note: The figures in parentheses for each taxonomic group are the actual number of Peruvian endemic species in the biome, or in the park.
RANP = Río Abiseo National Park

As Gentry (1986) has noted, "One important criterion for determining the optimal design for conservation units is the degree of biotic endemism. . . . Even if local endemism is high, however, concentration of locally distributed species in one or two centers of endemism might make possible preservation of most of a region's species within a few reserves chosen to include the high-endemism areas." To apply this criterion to the conservation of TMCF biodiversity, a detailed geographical analysis of the distribution of the endemic species present in this biome should be undertaken and other centers of endemism identified. For this purpose, the six physiographic provinces proposed for Peru by Young (1992) may provide an adequate framework.

Endemic species should be of special interest for conservation purposes. They are a major fraction of the South American fauna and often because of limited range are unusually vulnerable to habitat destruction. Such endemics are important in conservation efforts as they can be made a point of national pride as a unique part of the country's natural patrimony (Terborgh and Winter 1983). Endemic species are also important in analyzing biogeographic patterns, which in turn are useful for selecting more efficient protected areas.

A preliminary analysis shows that even within a relatively small area, the Peruvian TMCF supports a large number of endemic species warranting concerted national protection efforts. Centers of endemism that include TMCF, such as in the Río Abiseo National Park, should be identified so that protection for additional areas can be established. A network of such montane reserves could preserve a majority of Peru's endemic species.

Acknowledgments. I thank J. Córdova, Lily Rodriguez, Víctor Pacheco, Carlos Zamora, and Manuel Plenge for allowing me to use information from unpublished manuscripts, and Dr. Charles O. Handley and Dr. Alfred Gardner for their continuous support and involvement in studying the small mammals collection from the Río Abesio National Park.

Research in the park was supported by the University of Colorado (Boulder), David and Lucile Packard Foundation, and the J.N. Pew, Jr., Charitable Trust and Conservation International. I thank the Smithsonian Institution Short-Term-Visits Program, the former Direción General Forestal y de Fauna, and park personnel, as well as the inhabitants of Pataz and Los Alisos for their field support. Appreciation is extended to my friends and colleagues Monica Romo, Enrique Ortiz, Mirko Fernández, J. Aldave, and Lily Rodriguez for their dedication to research in the park; and to Patti Moore, Abby Bravo, and Daniel La Torre for their continuous support in different aspects of the inventory. Dr. Lawrence Hamilton of the East-West Center encouraged me to write this paper, which has also benefited from comments by L. Rodriguez, M. Plenge, T. Parker, and V. Merino.

Apenheul Zoo (Holland) provided us with a stereo microscope, and the Peruvian Association for Conservation of Nature assisted with logistical support for the fieldwork and the facilities to produce this paper.

References

Alva, C., and G. Huamani. 1991. Sistema de información geográfico de la selva alta Pemaná. Resultados de la evaluación de las áreas intervenidas acumuladas en las décades del 60' y del 80' (manuscript).

Brack, A. 1986. Ecología de un país complejo. In *Gran Geografía del Perú*, ed. Manfer and J. Mejía Baca, 2:177-319. España.

Duellman, W. 1976. Centrolenid frogs from Peru. *Occasional Papers of the Museum of Natural History*, 52:1-11.

Duellman, W. 1978a. New species of Leptodactylidae frogs of the genus

Eleutherodactylus from the Cosñipata Valley, Peru. *Proceedings Biological Society of Washington* 91(2): 418-430.
Duellman, W. 1978b. Three new species of Eleutherodactylus from Amazonia, Peru (Amphibia: Anura: Leptodactylidae). *Herpetologica* 34(3): 264-270.
Duellman, W. 1990. A new species of Eleutherodactylus from the Andes of northern Peru (Anura: Leptodactylidae). *Journal of Herpetology* 24(4): 348-350.
Duellman, W. 1991. A new species of Leptodactylid frog genus Phyllonastes, from Peru. *Herpetologica* 47(1): 9-13.
Duellman, W. E., and T. H. Fritts. 1972. A taxonomic review of the Southern Andean marsupial frogs—Hylidae: *Gastrotheca.* Occasional Paper Mus. Nap. Hisp. Univ. Kansas 9: 1-37.
Duellman, W., and R. Schulte. 1993. New species of Centrolenid frogs from northern Peru. *Occasional Papers of the Museum of Natural History*, 155:1-33.
Duellman, W., and C. Toft. 1979. Anurans from Serrania de Sira, Amazonian Peru: Taxonomy and biogeography. *Herpetologica* 35(1): 60-70.
Duellman, W., and L. Trueb. 1988. Cryptic species of Hylid marsupial frogs in Peru. *Journal of Herpetology* 22(2): 159-179.
Emmons, L. 1990. *Neotropical Rainforest Mammals: A Field Guide*. Chicago and London: The University of Chicago Press.
Flores, G., and R. McDiarmid. 1989. Two new species of South American Centrolenella (anura: Centrolenidae) related to C. mariae. *Herpetologica* 45(4): 401-411.
Frost, D. 1985. *Amphibian Species of the World*. Lawrence, KS: Allen Press and the Association of Systematics Collections.
Gardner, A. 1976. The distributional status of some Peruvian mammals. *Occasional Papers of the Museum of Zoology*, 48:1-18, Louisiana State University.
Gentry, A. 1986. Endemism in tropical versus temperate plant communities. In *Conservation Biology*, ed. M. Soul, 8:153-181. Sunderland, MA: Sinauer Press.
Hedges, B. 1991. A new species of Phrynopus (anura: Leptodactylidae) from Peru. *Copeia* (1): 108-112.
Hershkovitz, P. 1983. Two new species of night monkey, genus Aotus (Cebidae, Platyrrhini): A preliminary report on Aotus taxonomy. *American Journal of Primatology* 4:209-243.
Holdridge, L. 1967. *Life Zones Ecology*. San José de Costa Rica: Tropical Science Center.
Honacki, J., K. Kinman, and J. Koeppl. 1982. *Mammal Species of the World*. Lawrence, KS: Allen Press and the Association of Systematics Collections.
Lamas, G. 1982. A preliminary zoogeographical division of Peru, based on butterfly distribution (Lepidoptera, Papilionoidea). In *Biological Diversification in the Tropics*, ed. G. Prance, 18:336-357. New York: Columbia University Press.
Leo, M. 1982. Conservation of the yellow-tailed wooly monkey (*Lagothrix flavicauda*) in Peru. In *International Zoological Yearbook*, 22:47-52.
Leo, M., and A. Gardner. 1993. A new species of a giant Thomasomys (Mammalia, Muridae, Sigmodontinae) from the Andes of north-central Peru. *Proceedings of the Biological Society of Washington* 106(3): 417-428.
Leo, M., and Ch. Handley. In prep. Andean rats, Genus Thomasomy Coues, and other rodents of Parque Nacional Río Abiseo.
Leo, M., and M. Romo, 1992. Distribución altitudinal de roedores sigmodontinos (Cricetidae) en el Parque Nacional Río Abiseo, San Martín, Perú. In *Biogeografía, Ecología y Conservación del Bosque Montano*, ed. K. Young and N. Valencia. Lima: Memorias del Museo de Historia Natural, Universidad Nacional Mayor de San Marcos, 21:105-118.
Mares, M. 1992. Neotropical mammals and the myth of Amazonian biodiversity. *Science* 225:967-979.
Myers, C., and J. Daly. 1979. A name for the poison frog of Cordillera Azul, eastern

Peru, with notes on its biology and skin toxin (Dendrobatidae). *American Museum Novitatus* 2674:1-24.

Nowak, R. 1992. *Walker's Mammals of the World*. Baltimore: Johns Hopkins University Press.

O'Neill, J. 1992. A general overview of the montane avifauna of Peru. In *Biogeografía, Ecología y Conservación del Bosque Montano*, ed. K. Young and N. Valencia. Lima: Memorias del Museo de Historia Natural, Universidad Nacional Mayor de San Marcos, 21:47-55.

Pacheco, V., H. de Macedo, E. Vivar, C. Ascorra, R. Arana-cardo, and S. Solari. In press. Lista anotada de los mamferos Peruanos. *Conservation International*.

Parker, T., Susan Allen Parker, and M. Plenge. 1982. *An Annotated Checklist of Peruvian Birds*. Vermillion, SD: Buteo Books.

Patton, J., and M. Smith. 1992. Evolution and systematics of Akodontine rodents (Muridae:Sigmodontinae) of Peru, with emphasis on the genus Akodon. In *Biogeografía, Ecología y Conservación del Bosque Montano*, ed. K. Young and N. Valencia. Lima: Memorias del Museo de Historia Natural, Universidad Nacional Mayor de San Marcos, 21:83-103.

Peñaherrera, C. 1986. Geografía física del Perú. In *Gran Geografía del Perú*, ed. Manfer and J. Mejía Baca, 1:1-323. España.

Peruvian Association for Conservation of Nature—APECO. 1988. Results of 1987 fieldwork faunal inventory, Río Abiseo National Park, Peru. A report to the David and Lucile Packard Foundation.

Peruvian Association for Conservation of Nature—APECO. 1989. Results of 1988 fieldwork faunal inventory, Río Abiseo National Park, Peru. A report to the David and Lucile Packard Foundation.

Peruvian Association for Conservation of Nature—APECO. 1991a. Resultado del trabajo de campo 1990, inventario de fauna del Parque Nacional Río Abiseo, Perú. Informe para Conservación Internacional.

Peruvian Association for Conservation of Nature—APECO. 1991b. Results of 1989 fieldwork faunal inventory, Río Abiseo National Park, Peru. A report to the David and Lucile Packard Foundation.

Plenge, M. Unpubl. ms. List of the birds of Peru—Lista de la aves del Perú.

Pulgar Vidal, J. 1985. Las tres selvas del Antisuyo. *Boletín de Lima* 39(7): 59-72.

Rodriguez, L. In press. Five new species of frogs genus *Phrynopus* (Leptodactylidae) from the Río Abiseo National Park, northeastern Peru. *American Museum Novitates*.

Rodriguez, L., J. Córdova, and J. Icochea. In press. Lista de los anfibios del Perú. *Publicaciones del Museo de Historia Natural de la UMSM Ser. Zoologia*.

Stadtmüller, T. 1986. *Los Bosques Nublados en el Trópico Húmedo*. Tokyo and Turrialba: Universidad de las Naciones Unidas and Centro Agrónomico Tropical de Investigacion y Ensenanza,.

Terborgh, J. 1971. Distribution on environmental gradients: Theory and a preliminary interpretation of distributional patterns in the avifauna of the Cordillera Vilcabamba, Perú. *Ecology* 52(1): 23-40.

Terborgh, J., and J. Weske. 1975. The role of competition in the distribution of Andean birds. *Ecology* 56:562-576.

Terborgh, J., and B. Winter. 1983. A method for siting parks and reserves with special reference to Colombia and Ecuador. *Biological Conservation* 27:45-58.

Thomas, O. 1926. The Godman-Thomas expedition to Peru—III, On mammals collected by Mr. R. W. Hendee in the Chachapoyas Region of North Peru. *Annals and Magazine of Natural History* 18(9): 156-167.

UNEP (United Nations Environment Programme). 1991. Guidelines for the preparation of country studies on cost benefit and unmet needs of biological diversity within the framework of the planned Convention on Biological Diversity. Nairobi.

Weberbauer, A. 1945. *El Mundo Vegetal de los Andes Peruanos*. Lima: Ministerio de Agricultura.

Wetzel, E. 1982. Systematics, distribution, ecology and conservation of South American edentates. In *Mammalian Biology in South America*, ed. M. Mares and H. Genoways. Linesville, PA: University of Pittsburgh.

Whittaker, R. 1972. *Communities and Ecosystems*. New York: Macmillan.

World Conservation Monitoring Centre. Unpubl. ms. *Conservation Status Listing: Animals of Peru*. Cambridge, UK.

Young, K. 1992. Biogeography of the montane forest zone of the eastern slopes of Peru. In *Biogeografía, Ecología y Conservación del Bosque Montano*, ed. K. Young and N. Valencia. Lima: Memorias del Museo de Historia Natural, Universidad Nacional Mayor de San Marcos, 21:119-140.

Young, K., and B. León. 1988. Vegetación de la zona alta del Parque Nacional Río Abiseo, San Martín. *Revista Forestal del Perú* 15(1): 3-20.

Young, K., and B. León. 1990. Catálogo de las plantas de la zona alta del Parque Nacional Río Abiseo. *Publicaciones del Museo de Historia Natural*. Lima: Universidad Nacional Mayor de San Marcos (B) 34:1-37.

Young, K., and B. León. 1991. Diversity, ecology and distribution of high-elevation Pteridophytes within the Río Abiseo National Park, north-central Peru. *Fern Gazette* 14(1): 25-39.

Zamora, C. 1988. Las regiones ecológicas del Perú.

Zamora, C. Unpubl. ms. Las regiones ecológicas del Perú (Bases para una planificación de desarrollo).

14. Cloud Forests of the Rwenzori Mountains, Uganda: Research and Management Possibilities

Cosmo Lush

> In the rare sunshine one may fancy oneself in the scenery of a Russian pantomime . . . you may be familiar with the Alps and the Caucasus, the Himalaya and the Rockies, but if you have not explored Rwenzori you still have something wonderful to see
>
> —D. W. Freshfield (1906)

The Rwenzori Mountains—known romantically as the Mountains of the Moon—present an interesting and valuable biogeographical environment. However, due to problems of access and political turmoil in the region, the Rwenzoris are significantly under-researched. Increasing population pressure and a growing tourist industry are now placing the mountains under threat from expanding agricultural and subsistence land uses and from environmental damage caused by excessive visitors to the mountains. Scientific research is therefore needed not only for its own sake, but also to draw attention to the lack of a comprehensive mountain management plan in the Rwenzoris.

This paper summarizes the research potential in the Rwenzori cloud forests and discusses the possibilities for a management plan in the mountains, as well as the difficulties of implementing such a plan in the Ugandan context.

The Rwenzori Range

Rising to a height of 5,109 m, the Rwenzori range includes Africa's third highest peak, Mount Stanley; yet, it is better known for its unusual vegetation and its inhospitable climate. Geographically, the mountains are located 30° E of Greenwich and less than 1° N of the equator (see Figure 1). They are situated near the convergence zone of the two weather systems that have the greatest effect in the region—namely, the southeast monsoon and the northeast wind. Thus the already considerable orographic effect of the mountain range is supplemented further, and precipitation levels are high: annual rainfall averages 2,500 mm, and during the two wet seasons (March to May and August to December) monthly totals can be as high as 375 mm.

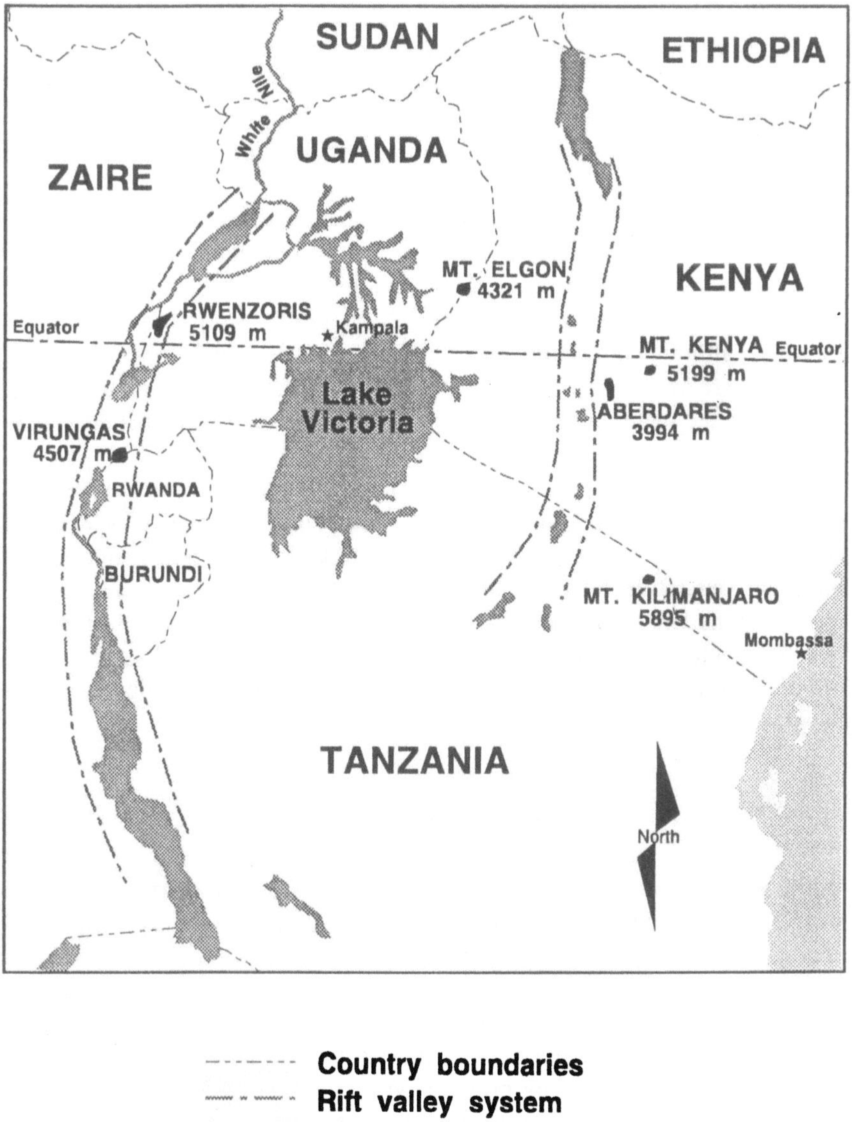

Figure 1. East Africa, showing the Rwenzoris and other principal mountains with cloud forests.

The east African region contains several areas of high altitude—namely, Mounts Kenya, Kilimanjaro, and Elgon, the Virunga volcanoes, the Aberdare Mountains, as well as the Rwenzoris (Figure 1). These are collectively known (in biogeographical terms) as the afroalpine region (Hedberg 1951). This ecological zone is characterized by high levels of plant endemism and specialization among the flora and fauna.[1] Of the 278 woody plant taxa found in the afroalpine zone, 81 percent are endemic to east Africa and 19 percent are found only in the afroalpine belt (Hedberg 1961). The development of this unusual flora is hard to trace. Many of the species share physiological characteristics with Andean paramo species; yet patterns of endemism indicate that they have evolved in isolation (Hedberg 1986). Tertiary-era climate change and tectonic upheaval created the biogeographical pattern that we see today—afroalpine areas as islands of high-altitude vegetation in a sea of lowland forest and savannah.

East African mountain vegetation is also characterized by marked altitudinal zonation, with terms like "ericaceous" and "afromontane" commonly used to describe the various ecological communities found at different altitudes (Hedberg 1951, 1964; Vuilleumier and Monasterio 1986; Hamilton 1982). However, although the communities in these areas clearly depend on the altitudinal range of individual species, to categorize them into altitudinal zones can be misleading because other factors affecting vegetation distribution may be ignored.

The Rwenzoris are unique in the afroalpine context because higher precipitation rates and other meteorological conditions such as extensive cloud cover and a comparatively low diurnal temperature range allow much more luxuriant vegetative growth at high elevation (to 4,500 m). Another characteristic of the Rwenzori range not shared by other afroalpine areas is the extensive development of tropical montane cloud forests (TMCF) at an altitude of between 3,200 and 4,000 m, and these forests are the most relevant to this paper.

The Rwenzori Cloud Forests

While sharing many characteristics with cloud forests in other parts of the world, such as physical structure, floral composition, and the meteorological conditions in which they exist, the Rwenzori TMCF also displays certain unique traits that place them apart from other similar communities and prevent generalizations from being made. The two features that stand out most clearly are (1) the relatively high altitude of the forests above sea level and (2) the height to which they grow above the ground. Although cloud forests on most tropical mountains and islands are found below an altitude of 2,500 m, in the Rwenzoris they are found above 3,200 m. And while a majority of cloud forest tree species on other tropical mountains are characteristically stunted, on Rwenzori this is by no means the case. In fact quite the opposite is true; the forests grow to heights of up to 20 m above ground level.

The TMCF here is dominated by tree species of the genus *Philippia* (*P. philippia* and *P. johnstonii*). Although these trees are found elsewhere in east Africa, the Rwenzoris provide the right conditions for extensive growth, with high precipitation, moist air, and steep slopes. The *Philippia* trees provide the canopy structure for a well-developed cloud forest in which mosses, herbs, and lichens all benefit from the near-constant temperature, humidity, and light levels. The *Philippia* trees can grow up to 20 m tall; bryophytes and pteridophytes grow in great profusion on the trunks and branches of the trees as well as cover the forest floor with a mossy carpet. *Usnea*—Spanish moss—is commonly found throughout the cloud forest, draped over the upper branches, giving the forests their characteristic bedraggled look. At the ground level other woody species such as *Rapanea rhododendroides* and *Hypericum lanceolatum* struggle upward, as do the giant *Lobelia lanuriensis* and *L. gibberoa*. Other plants, such as the terrestrial orchid *Disa stairsii*, contribute to the floristic diversity.

The overall character of the TMCF is similar to those found on other tropical mountains as described in this symposium, but in terms of biomass the Rwenzori forests excel. The height to which they grow above ground level, their wide altitudinal range (3,200–4,000 m), and the density of their growth at these heights are all factors contributing to a floral phenomenon not characteristic of other such communities. Yet hydrologically speaking the Rwenzori forests play a similar role to other cloud forests, regulating fluvial discharge and the runoff of precipitation to the lower slopes, and thus protecting the lower areas from erosional and flood damage.

The cloud forests clearly rely on large quantities of precipitation, both direct and indirect, and their structure is indicative of the amount of moisture they retain. While the mosses and lycopodiums are almost permanently saturated with water, the soil beneath the mossy carpet is crisscrossed with small streams and rivulets draining the forest. The primary economic role of these moss forests in the hydrological system of the Rwenzoris is to regulate the fluvial discharge from the mountains. The extensive area of cloud forest is indicative of the amount of water that is contained within the alpine system. As Yeoman (1989) wrote, "The bog-moss forest ecosystem . . . lies at the very heart of the Rwenzoris' role as the most permanent source of the Nile." As water resources come under increasing demand in subtropical areas of the world (including the lower stretches of the river Nile in Egypt and Sudan), the hydrological characteristics and role of the Rwenzori cloud forests appear increasingly important. On a local scale, much of the lowland areas around the Rwenzori range receive less than 1,000 mm of rainfall annually (Downer 1967), and irrigation schemes using Rwenzori runoff are vital to the local agricultural economies.

Potential Scientific Research in the Rwenzori Mountains

In spite of the above-mentioned value, the cloud forests remain significantly under-researched. Twenty years of civil strife in Uganda have meant that biogeographical and ecological works have been diverted to other TMCF areas, notably in the New World. Now that political stability in Uganda is being maintained and access to the mountains is less difficult, there is a chance for the unique characteristics of the Rwenzori forests to be studied.

There are possibilities for research on a variety of scales. At the regional level there is scope for exploring the Rwenzoris' role as an important source of the Nile; while on the microscale, one could investigate the hydrological and ecological relationships between the cloud forests and the peat bogs. These two perspectives will be addressed in more detail.

The Rwenzori range is drained by 11 rivers. Five flow westward into the Semliki River and then northward into Lake Albert; six drain eastward into the Lake George–Lake Edward complex and then into the Semliki. From Lake Albert the water flows into the Albert Nile. The perennial flowing streams and rivers of the Rwenzoris are therefore a permanent source of Nile headwater, and as such play a crucial role in the regional drainage system.

Hydrographic data for most Rwenzori rivers were recorded in the past by the Uganda Water Development Department (Osmaston and Pasteur 1972); whether this is still the case is extremely doubtful. Monitoring fluvial discharge would therefore be a necessary precondition for future studies of the regional hydrology. A long-term project could assess the effect of fluctuations in Rwenzori rainfall upon the river flows and even upon the ecological dynamics of the surrounding lakes. This is especially relevant at present because large areas of Uganda have suffered from a lack of rain for two consecutive years (1992 and 1993), and this is affecting not only agricultural production but also fish stocks in the lakes, an important food source (*New Vision* 1993).

Work more directly relevant to cloud forest research should be directed at investigations into the role of the *Philippia* forests in controlling rainwater discharge. The role of organic forest soils, and the *Carex* and peat in the surrounding bogs in retaining groundwater and retarding its progress downslope warrants detailed study. The cloud forests cover approximately 282 km^2 (Howard 1988) and must therefore play an important role in the fluvial dynamics of the entire mountain range. A comparison of precipitation rates with fluvial discharge at a range of locations and altitudes would shed some light on this role. This could also have important implications for small-scale hydropower generation both in the mountains and in the more densely inhabited foothills.

The Rwenzoris are blessed with 2.5 m of precipitation each year. A large proportion of this flows through the *Philippia* forests and the *Carex* bogs, having drained relatively rapidly off the upper slopes due to the thinner vegetation cover and the porous screes and moraines. The dense vegetation and organic soils between 3,000 and 4,000 m acts as a sponge, absorbing the throughflow from above. The effect of the vegetation on the speed of throughflow and on the eventual fluvial discharge lower down should be studied, as well as the effect on the hydrology of the biomass/unit area ratio. How much water is stored in the cloud forests, and how much is stored in the peat bogs? How rapidly can water pass through this ecosystem and into the main river channels? Answers to these questions would shed light on the functions of the cloud forest structure, on the effect of the bryophytes and pteridophytes that blanket the trees, and on the slow process of organic decay that takes place on the forest floor.

Other research possibilities include forest-atmosphere interactions and the relative importance of direct and indirect precipitation in the forest zone. Clouds envelop the mountains quite regularly at altitudes above 2,500 m and can remain for 2–3 days at a time. Clearly the indirect precipitation from clouds and mist plays an important role in the hydrology of the *Philippia* forests, helping to keep them moist and cool. The rates of direct cloud-water interception by TMCF, and the methods by which it is brought into the hydrological system of the forest and larger watershed are all worthy of future study.

In the cloud forest zone there is a wealth of research possibilities that would be relevant not only to studies of the Rwenzoris but also to mountain hydrology in the wider context. This paper is merely an outline of the Rwenzori forests, intended to give an impression of what that environment is like and to offer suggestions for further research.

Management Concerns in the Rwenzoris

Scientific research in the Rwenzoris is necessary not just for its own sake but also to give substance to the calls from conservationists and foresters for the protection and management of the mountains. There is a great need for concerted environmental management, as well as scientific research. The mountain range is under increased threat because of two primary reasons: (1) population pressure and agricultural expansion and (2) the growing tourist industry.

Uganda's population is currently growing at a rate of 1.7 percent (Economist Intelligence Unit 1992), and it has been growing even more rapidly in previous years. The effect of this on local land use has been considerable: agricultural expansion up the lower slopes of the Rwenzoris has led to land degradation and loss of valuable topsoil. As the population continues to grow, its activities encroach on land at increasingly higher elevations and steeper slopes.

In the last 5 years Uganda's tourist industry has burgeoned immensely as the country has returned to relative political stability. The Rwenzori range is one of the country's primary tourist attractions, and the mountains are coming under increasing threat from the trampling of the trekker's boot. In 1987 there were 217 visitors to the mountains; in 1989, 690; and in 1992 more than 1,500 (pers. com., Rwenzori Mountaineering Services 1993). These are unsustainable levels, and the problems that they cause are exacerbated by lack of a coordinated management plan.

Trekkers visiting the mountains all follow the same route, and all stay in the same huts on each night of their 7-day circuit. Thus the damage they cause is concentrated along an increasingly treacherous and degraded path, and around the camps, which are rapidly being denuded of their surrounding forests. The number departing on a trek each day is not controlled, since the organization that manages the tourism is more concerned with maximizing its turnover than with protecting the environment. As a result 30–40 trekkers, and an equivalent number of porters, can descend on a hut equipped for 15. The facilities and the natural surroundings cannot take this sort of strain.

The local communities are aware of the need to protect the mountain environment, but lack of funds and an underdeveloped infrastructure hamper all activities. A certain amount of effort, however, is being put into conservation and environmental management by various government departments, private organizations, and overseas aid projects. The Rwenzori Mountaineering Services is a local, private organization set up in 1986 to manage tourism in the area. Its aim is to provide trekkers with all the services they require (porters, huts, trail maintenance, and rescue facilities) and to reinvest tourist dollars in the local community into social welfare projects. How far these worthy aims have been adhered to is questionable. There is little evidence of development projects in the community apart from those established by USAID at the time the Rwenzori Mountaineering Services was originally founded. Thus, the organization's understanding of conservation and mountain management issues appears to be somewhat lacking, judging from the state-of-the-tourist infrastructure in the mountains.

The government also shows an interest in protecting the Rwenzoris, but bureaucracy and a lack of communication between the various ministries responsible for the area prevent much good from being achieved. Government authority hampers the work of the Rwenzori Mountaineering Services, especially since the creation of a national park in the Rwenzoris in 1991 (see below), since the organization's responsibility for the tourist industry has been taken away. What is made clear by the situation in the Rwenzoris is that no matter how many concerned parties there are in a conservation situation, very little is actually achieved without a focused and enforced management strategy that unites the various players.

Conservation in the Rwenzoris

The region desperately needs a consistent policy for environmental protection and conservation. But first one must ask, why does the area need conserving? There are several reasons. First, the central Rwenzori is one of the last places on earth that is little touched by human activity. Although members of the local community and tourists pass through the mountains, they are forced to keep to well-worn paths because of the impenetrable vegetation; the vast majority of the mountain range is never visited. Steep slopes, soil infertility, and the inhospitable climate have constrained human settlement in the mountains, and so the natural ecosystems have remained largely undisturbed. But these are fragile ecosystems, ones that are coming under increasing attack as the number of visitors to the mountains grows.

Second, the numerous endemic species of plants and animals need to be protected in their own right. *Ruwenzorornis johnstoni johnstoni* (the Rwenzori Turaco), *Colobus angolensis ruwenzori* (the Rwenzori colobus monkey), and *Dendrohyrax arboreus ruwenzorii* (the Rwenzori hyrax) are all indigenous species of animal. Many floral species including *Lobelia* (*L. bequaertii*, *L. wollastonii*), *Senecio* (*S. adnivalis*, *S. erici-rosenii*), and *Alchemilla* (*A. subnivalis*, *A. stuhlmanii*)—to name but a few—are endemic to the range.

Third, the Rwenzoris represent an island of alpine habitat "in a continent-wide archipelago of mountains" (Howard 1988) and as such are valuable for the study of isolated biogeographical communities. The subject is becoming increasingly important as protected areas of the world shrink in size and number.

Fourth, there is a practical need to protect the water catchment area. According to the 1980 population census (Census Office 1982), 460,000 people rely on the Rwenzori catchment for drinking water and on the vegetation in the mountains to prevent flooding. Two hydroelectric schemes provide low-cost power to local communities. Damage to the upper catchment area would prove costly to these local communities, just as in the lower catchment area the use of marginal land for agriculture to support the growing population has caused extensive damage and loss of land through erosion of the topsoil.

Clearly there is a need to protect these mountains. From 1941 to 1991, the area above 2,100 m was protected by the Forest Department as a Central Forest Reserve. Several proposals for national parks were drawn up during this time (Wheater 1971; Yeoman 1985; Howard 1988), and the area was eventually gazetted in 1989. In 1991 a new national park was established, but without many of the guidelines advised in recent proposals. Howard (1988) and Yeoman (1985) had proposed the establishment of a park divided into two zones: a central "core" area and an outer "traditional use zone." Surrounding the park would be a third buffer zone to enhance the direct benefits of the national park for the local population. The traditional use zone would enable local communities to continue making use of the forest resources (for activities like bamboo harvesting, beekeeping, firewood gathering, and transit) but under sustainable management policies. Preventing these traditional activities is not only impossible, but also inappropriate; prevention would be unenforceable and create friction between the national park staff and the local population, whom the park should be benefiting anyway.

Actually, the park has been established as a single zone but is not being patrolled, so that infringements of the boundaries are common. Moreover, establishment of the park has provided little benefit to the mountains' environment, the local population, or indeed the tourists. Bureaucracy, mismanagement, and a lack of staff and resources prevent concerted action from being taken. The Rwenzori Mountaineering Services was responsible for upkeeping the huts and trails in the mountains; even these activities are now bound up in red tape so that very little is being done. Financial constraints are problematic in Uganda, but they need not be exacerbated by bureaucracy.

Meanwhile the number of visitors to the mountains is rapidly rising. Benefits from the tourist boom should be filtering into the local economy, but predictably only a few people appear to be profiting. While tourist dollars should be spent on much-needed social infrastructure projects (school and medical equipment, improved roads) and improvements to the huts and trails in the mountains, a top-heavy organizational structure in which all decisions are referred to the capital, Kampala, hinders such progress. While those in charge quibble over who is responsible for what, the cloud forests, which people come to marvel at up in the mountains, are being cut down for firewood, the peat bogs are being seriously damaged by ever-widening trails, and the amount of litter left on the mountains is increasing dramatically.

An Action Plan for Rwenzori Tourism

An action plan for Rwenzori tourism is called for and the following steps are proposed:

1. The number of visitors must be limited. At the moment there is unrestricted access to the mountain trails, and during the peak summer months as many as 40 tourists and even more porters travel the trekking circuit each day. In the evenings they all stay in the same huts, since there is only one route around the mountains. These huts are designed to sleep 15 people, excluding porters. There is neither room nor the resources for more, and the damage to the surrounding vegetation caused by firewood collection and trampling has reached unsustainable levels. Tourist numbers must be limited, and a daily quota of 15 is the most suitable balance. Tourists should be asked to book in advance, a system that is employed for visiting the mountain gorillas in nearby areas.

2. Equipment must be provided and maintained. More porters than tourists pass through the Rwenzoris, and their equipment is negligible. Not only do they lack protection against the cold and wet, but they also rely on firewood cut locally for cooking and warmth in the evenings. Providing kerosene stoves and sleeping bags would be the simplest way of improving this situation. Solar power or small in-stream hydroelectric projects are also a possibility. The older huts should be renovated or replaced to increase their warmth and provide adequate shelter. All huts should be equipped with two-way radios for safety reasons. More equipment in the camps would create the need for a resident guard or a housekeeper who would be responsible for the upkeep and cleanliness of the camp.

3. Litter and latrines must be controlled. All nonbiodegradable and noncombustible litter must be taken down the mountain and not left in the garbage pits. Long-drop latrines must be constructed properly and maintained regularly. Environmental awareness is high among porters, and sometimes abysmally low among tourists.

4. Path management must be improved. Currently there is virtually no trail management. In the peat bogs the damage caused is considerable, as well as discomforting to trekkers who suddenly find themselves up to their waists in mud. As more visitors pass through, the paths widen, destroying the fragile bog ecosystem. The paths must be better maintained or even rerouted to pass through drier land. With the current number of visitors, the lack of trail maintenance is simply not sustainable.

A wider choice of trekking routes must also be established. The Rwenzoris are crisscrossed with the paths of hunters and traders—indeed the trekking route followed nowadays was traditionally a network of such trails. Widening tourist access to the mountains would be a difficult task because the trails would have to be improved and connected. Nevertheless, spreading the effect of visitors over a wider area should be a long-term goal of any Rwenzori management plan.

5. Prices must be raised. All these changes and improvements do, of course, have to be paid for, and the most direct means of raising money is by increasing the price for visiting the mountains. Unfortunately, experience shows that in these circumstances the extra dollars often end up in the pockets of a small number of people, or are simply taken out of the local area and sent back to the central government departments. Before prices are raised, it is important to ensure that there is some means of monitoring how the income will be used, and this is the hardest task of all. Many projects established along these lines eventually come to nothing, and it is important to learn from experiences.

There is still time to save the Rwenzoris, but it requires a concerted effort by conservationists, foresters, and the scientific community. The Rwenzoris are a valuable scientific and tourist resource. At the moment the potential prosperity that the tourist industry could bring to this area is largely being squandered due to the lack of a long-term management plan. The cloud forests, peat bogs, and other alpine ecosystems—containing a wealth of biogeographical and hydrological information—are also under threat for the very same reasons. The various agencies and government bodies that are working in the area must be coordinated to provide (1) protection for the natural environment and (2) prosperity for the local population. These two goals can only be achieved in tandem: the natural environment can never be protected without the cooperation, involvement, and understanding of the local people, just as their communities will never be able to prosper unless they protect and manage their natural surroundings. Only when these fundamental concepts are fully understood and implemented can the Rwenzoris ever be saved.

Note

1. Animal endemism is also high in this zone but fauna were beyond the scope of this study.

Bibliography and References

Census Office. 1982. Report on the 1980 Population Census, Volume 1: the provisional results by administrative areas. Census Office, Ministry of Planning and Economic Development, Kampala.

Downer, E.R. 1967. *Atlas of Uganda.* Dept. of Lands and Surveys. Entebbe: Government Printer.

Economist Intelligence Unit. 1992. Uganda Country Report, Kampala.

Fishlock, C.W.L., and Hancock; 1933. Notes on the flora and fauna of Ruwenzori with special reference to Bujuku Valley. *J. of E. African & Ugandan Nat. Hist. Soc.* 44:205-229.

Freshfield, D.W. 1906. Ruwenzori. *Alpine Journal* 23(171): 45-50.

Hamilton, A.C. 1982. *Environmental history of East Africa: a study of the Quaternary.* New York: Academic Press.

Hedberg, O. 1951. Vegetation belts of the East African mountains. *Botany no. 1.*

Hedberg, O. 1961. The phytogeographical position of the afroalpine flora. *Rec. Adv. Botany.* 1:914-919.

Hedberg, O. 1964. Features of afroalpine plant ecology. *Acta phytogeogr. suecica* 49:1-144.

Hedberg, O. 1986. Origins of the afroalpine flora. In *High altitude tropical biogeography,* eds. P. Vuilleumier and M. Monasterio. Oxford University Press.

Howard P. 1988. Proposal for the establishment of a new national park in the Rwenzori Mountains, Uganda. World Wildlife Fund International and The New York Zoological Society.

Lush, C. 1991. The effects of altitude and other geographical factors on sub-alpine vegetation patterns in the Rwenzori Mountains, Uganda, and the human role in protecting this environment. B.A. Hons. Thesis, Oxon.

New Vision. 25 March 1993. *Fish shortage in Kasese district.*

Osmaston, H.A., and Pasteur, D. 1972. *Guide to the Rwenzori.* Reading: West Col Productions.

Rwenzori Mountaineering Services, pers. com. 1993. P. O. Box 33, Kasese, Uganda.

Vuilleumier, P., and Monasterio, M., eds. 1986. *High altitude tropical biogeography.* Oxford University Press.

Wheater, R.J. 1971. New national parks and additions to existing ones. Mimeographed report pursuant to Board Minutes 27/71, Uganda National Parks.

Woosnam, R.B. 1907. Ruwenzori and its life zones. *Geog. Journal* 30:630-642.

Yeoman, G. 1985. Can the Rwenzori be saved? *SWARA—J. of E. Af. Wildlife Soc.* 8(3):8-12.

Yeoman, G. 1989. *Africa's Mountains of the Moon.* London: Elm Tree Books.

15. Conservation of Cloud Forests in Maui County (Maui, Moloka'i, and Lana'i), Hawaiian Islands

Arthur C. Medeiros, Lloyd L. Loope, and Robert W. Hobdy

Introduction

Cloud forests of the Hawaiian Islands are crucially important as watershed and as reservoirs of endemic biological diversity. This paper (1) sketches the environment and biology of the cloud forests of Maui County, Hawai'i, with emphasis on upper Haleakalā volcano, East Maui, and (2) describes threats to these forests and management efforts under way to maintain biologically intact cloud forests. The islands of Maui County and the approximate location of the cloud forests are shown in Figure 1.

The remarkable evolution in isolation of the Hawaiian biota has resulted in high endemism, textbook cases of evolutionary adaptive radiation, and vulnerability to biological invasions (Howarth, Sohmer, and Duckworth 1988; Loope and Mueller-Dombois 1989). Having evolved in long isolation from many of the evolutionary forces (most notably hooved mammals) that shape continental organisms, island organisms are highly vulnerable when such forces are introduced into their environments. Evolutionary studies on Hawaii's native biota have been extremely important in the development of modern theories of speciation and evolution, with many practical spinoffs (molecular genetics, genetic engineering, and other biotechnology). It has been argued that Hawaii's native biota provides the world's best laboratory for evolutionary studies and is the state's single most important natural resource (Kaneshiro 1989).

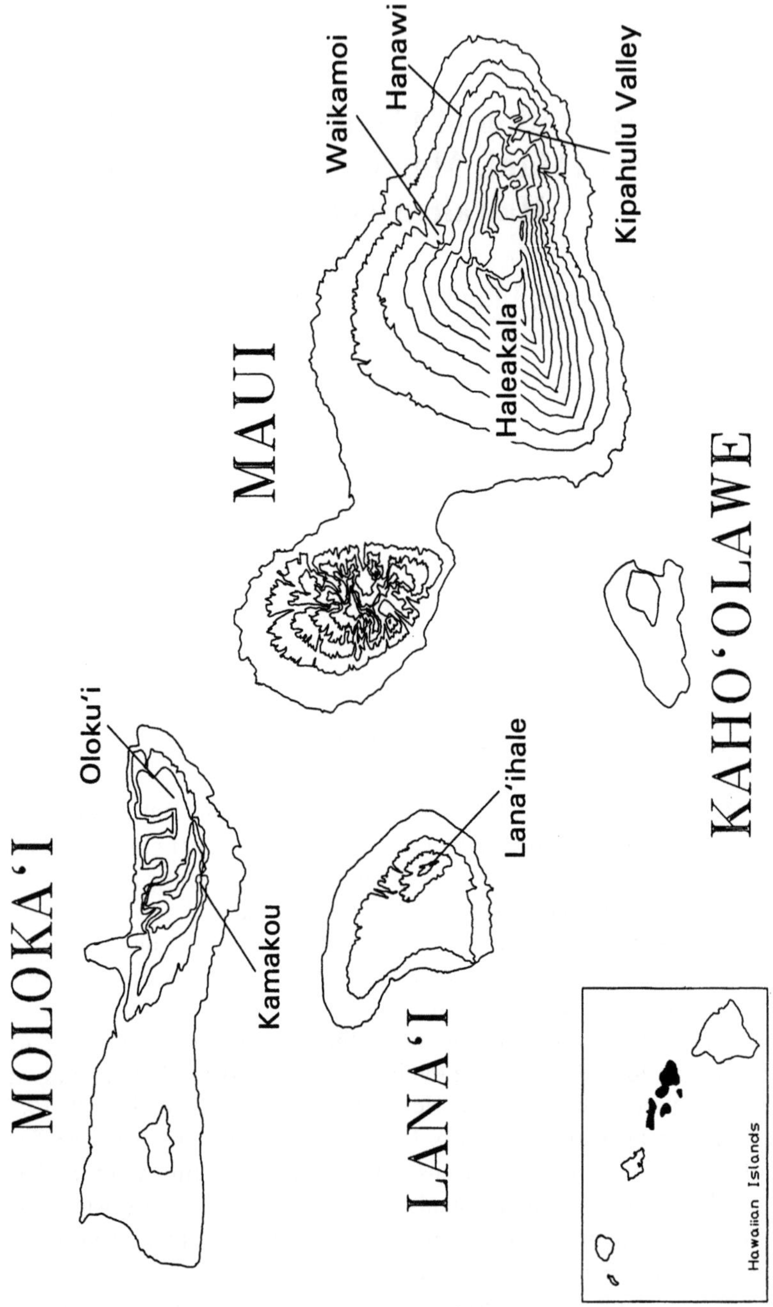

Figure 1. The islands of Maui County (State of Hawai‘i) and the approximate location of the cloud forests.

In the period between the arrival of Polynesians around the fourth century A.D. and Western contact in the late eighteenth century, lowland landscapes were profoundly altered through forest clearance for agriculture and the use of fire (Kirch 1982). The rate of habitat destruction and introduction of invasive species accelerated in the nineteenth and twentieth centuries, so that human impacts became pervasive in the lowlands, but high-elevation ecosystems remained relatively intact (Cuddihy and Stone 1990). In windward tropical montane cloud forests (TMCF) of Maui County, the major deterioration began as recently as the 1970s (Diong 1983; Loope, Medeiros, and Gagne 1991); these ecosystems are still among the most intact remaining reservoirs of biological diversity in the Hawaiian Islands.

Prime TMCF sites lie within Haleakalā National Park, Hanawi (State) Natural Area Reserve, and Waikamoi Preserve (managed by The Nature Conservancy). The primary threat to these forests is rooting and trampling by feral pigs, resulting in loss of native plant and animal species and subsequent invasion of nonnative invasive plants. A secondary threat involves loss of endemic bird species through habitat alteration, predation, competition, and disease. Cloud forests of West Maui and Moloka'i are also relatively intact, but subjected to essentially the same threats. The small cloud forest area of Lana'i has lost all but one native forest bird species and is currently being dramatically impacted by populations of axis deer. This cloud forest has suffered a marked loss of native biodiversity accompanied by massive invasions of nonnative plant species.

Climate

Over the open ocean near Hawai'i, annual rainfall averages 60–75 cm. Yet certain locations in the islands, such as windward summits and valley confluences, receive up to 15 times this amount. These high precipitation rates are a result of orographic rain that forms within the northeasterly moist trade wind air as it moves inland from the sea and overrides the steep and high terrain of the islands (Price 1983). The TMCF that occupy the northern and northeastern exposures of Haleakalā receive cloud cover and moisture in this fashion. We accept the delineation of the cloud forest zone on East Maui posed by Kitayama and Mueller-Dombois (in press)—the zone between the lower cloud limit (i.e., the lifting condensation level) at ca. 1,000 m elevation and the upper cloud limit set by the trade wind inversion at ca. 1,900 m elevation. Within this zone, fog may add significant amounts of precipitation to a very wet environment (Juvik and Ekern 1978). Since the highest peaks on West Maui and Moloka'i reach only 1,765 m and 1,515 m, respectively, cloud forests there are less extensive and somewhat less diverse, possibly because they are narrower and more limited in extent. The summit of Lana'ihale (elevation 1,027 m) on Lana'i is occupied by a very small but nevertheless interesting patch of marginal cloud forest vegetation; because of the lack of mountain mass there, mean annual precipitation is only ca. 100 cm, yet prolonged cloud cover and fog drip are characteristic (Ekern 1964).

The upper cloud forest limit on Haleakalā at ca. 1,900 m marks a major microclimatic discontinuity, the zone of the trade wind inversion (Leuschner and Schulte 1991; Giambelluca and Nullet 1991). The inversion zone normally limits upward movement of clouds; above the inversion, the climate quickly becomes substantially drier because of lower rainfall, lower humidity, and higher solar radiation.

Based on limited data, annual rainfall is believed to reach a maximum of ca. 6,500 mm at ca. 1,000 m elevation on windward Haleakalā (Giambelluca, Nullet, and Schroeder 1986). At that elevation, mean monthly rainfall reaches a maximum (>600 mm) in March–April and minima (ca. 350 mm) in June and from September to October. Yet in spite of mean annual precipitation totals among the highest recorded in the world and very high rainfall means in even the driest months, Hawaiian cloud forests are periodically subjected to severe drought, often associated with El Niño/Southern Oscillation conditions (Lyons 1982; Chu 1989). As an extreme example, rainfall during the first half of 1983 (during the exceptionally strong 1982–83 El Niño event) at Haleakalā National Park headquarters was 9 percent of normal; consequently epiphytes and bryophytes in the normally luxuriant cloud forests of windward Haleakalā were powdery dry.

Soils of cloud forests in Maui County are developed on Pleistocene volcanic rocks and layered volcanic ashes. These soils are better drained at high elevations, becoming less so with decreasing elevation. Kitayama and Mueller-Dombois (in press) found strong acidification and low Ca/Al ratios in the mineral horizons at the lower edge of TMCF (1,000 m elevation) of Waikamoi on East Maui, suggesting toxic effects of soluble aluminum upon vegetation.

Biota and Vegetation

The native biota of the cloud forest zone of East Maui includes about 240 species of flowering plants (with ca. 95 percent endemic to Hawai'i), 100 species of ferns (ca. 50 percent endemic), 600–1,000 species of native invertebrates (ca. 65 percent endemic), one endemic mammal taxon (subspecies of hoary bat), and nine endemic forest bird species in the family Fringillidae, subfamily Drepanidinae (Hawaiian honeycreepers). Cloud forests of West Maui, Moloka'i, and Lana'i have similarly high endemism but somewhat lower species diversity.

The vegetation of undisturbed Hawaiian TMCF is dominated by two tree species, *Metrosideros polymorpha* (Myrtaceae) and *Acacia koa* (Fabaceae). *Metrosideros* is the overwhelming dominant in most cloud forests, whereas *Acacia* replaces *Metrosideros* as the dominant at lower elevations (below ca. 1,400 m) or in slightly drier sites in the transition between windward and leeward slopes. The height of the *Metrosideros* canopy in the middle cloud forest zone reaches 10–18 m; near the upper forest line the trees are gnarled and low (ca. 5 m) in stature; at the lower limit of cloud forest, *Metrosideros* may reach 25–30 m in height. *Acacia* is a taller tree, reaching 35 m.

Metrosideros is bird- and insect-pollinated; it flowers sporadically throughout the year, and provides a major food source (nectar) for Hawaiian honeycreepers. *Acacia*, like most leguminous plants, has nitrogen-fixing nodules associated with its roots; it is believed to provide an important nitrogen source available to associated species on generally impoverished soils.

Other tree species common in Maui County cloud forests include *Cheirodendron trigynum* (Araliaceae), *Ilex anomala* (Aquifoliaceae), *Myrsine lessertiana* (Myrsinaceae), *Coprosma* spp. (Rubiaceae), and *Melicope* spp. (Rutaceae). These species contribute primarily to a subcanopy layer, along with the arborescent shrub *Styphelia tameiameiae* and tree ferns (*Cibotium* spp.).

Ferns overwhelmingly dominate the forest floor. Mat-forming ferns, including *Diplazium sandwichianum, Microlepia setosa, Ctenitis rubiginosa, Pseudophegopteris keraudreniana*, and *Hicriopteris pinnata*, are among the dominants. These species typically provide dense cover in TMCF at 1,200–1,700 m elevations if they are undisturbed by feral pigs. Near the upper cloud forest limits (1,700–1,900 m), mat-forming ferns are replaced by scattered large "shuttlecock"-shaped *Dryopteris* spp. (to 1.5 m in diameter). In the lower part of the cloud forest zone, *Dicranopteris pinnata* commonly forms thick mats, especially on steep slopes. Fallen logs provide habitat for mosses, liverworts, small fern species, and small herbs (*Astelia, Peperomia* spp.).

Presumably as a result of abundant moisture and relatively high light levels in Hawaiian cloud forests, most species grow epiphytically as well as terrestrially, with as many as 15 vascular species commonly occupying the trunk and branches of a single tree. Native tree ferns (*Cibotium* spp.) are the prime host species for epiphytes. Populations of facultative epiphytes appear to serve as important "reservoirs" of propagules to hasten recovery of biodiversity after feral pigs have denuded the forest floor (personal observation).

Management and Conservation of Cloud Forest in Maui County

Native Cloud Forest Vegetation, Feral Pigs, and Alien Plant Species

Hawaiian cloud forests, especially at higher elevations, are more resistant to displacement than most other native Hawaiian plant communities. Because of high rainfall regimes, effects of fire, though not entirely absent, have never been consequential. Because of dense vegetation and suboptimum habitat, feral goats (*Capra hircus*) are a relatively minor threat. Axis deer (*Axis axis*) have degraded TMCF on both the islands of Lana'i and Moloka'i. Though present on Maui, Axis deer are thus far absent in that island's cloud forests.

The greatest threat to the long-term integrity of Hawaiian TMCF is the feral pig. There are three primary impacts of feral pigs: (1) direct selective consumption of native plant species, (2) the creation of frequent, often widespread ground disturbance through rooting and trampling, and (3) distribution of propagules of alien plants on their manes (e.g., grass and sedge seeds) and in their droppings (*Psidium cattleianum*).

Although pigs were brought by the early Hawaiians, this form of pig (*Sus scrofa* ssp. *vittatus*) rarely, if ever, escaped into the wild (Diong 1983). There are no records of pigs causing damage in forests until the twentieth century. Prior to the 1930s, pigs were absent from upper elevations of Haleakalā. Since the 1930s, pigs have become increasingly common and progressively more invasive into previously pristine forests. In the early 1900s, the damage caused by feral breeds of European pigs in native rain forests was recognized when the Hawai'i Territorial Board of Agriculture and Forestry started a feral pig eradication project that lasted until 1958, removing nearly 170,000 pigs during that period (Diong 1983). Pigs were first reported in the upper areas of East Maui's Kipahulu Valley in 1945; by 1967 they occupied most of the valley (Diong 1983). By the 1970s, the damage had become dramatic. Pigs invaded the northeast rift zone of Haleakalā in the 1970s and 1980s (Loope, Medeiros, and Gagne 1991). This invasion is now spatially complete; there is currently no accessible area on Haleakalā where pigs have not reached (although some areas are pig-free due to fencing).

Control of feral pigs, especially in remote mountain areas, is achievable by isolating areas with woven wire fencing and natural barriers. Once an area is protected from ingress, the standing population is removed by snaring (Anderson and Stone 1993). Recently, snaring of feral pigs for conservation purposes in Hawai'i has come under criticism by a coalition of local hunters and an animal rights group. This has created a dilemma for conservation agencies since snaring is the only available and effective pig removal strategy in remote areas.

Increasingly, proactive management in monitoring and selective control of alien plant species is regarded as essential in the long-term protection of Hawaiian cloud forests. Most of the alien plant species in Maui County forests can be classified as pioneer species that thrive on disturbance (animal-induced or natural). Control of feral pigs in a natural area with intact native vegetation results in a sharp reduction in degree and frequency of ground disturbance. If control of feral pigs in cloud forests of Maui County is sustained, the cover of alien plant species may be diminished with new invasions largely confined to areas of frequent natural disturbance (e.g., trailsides, river courses, flood areas, and landslides).

However, the threat to native cloud forests posed by certain low-light-tolerant weed species is not entirely mitigated by removal of feral pigs. In Kipahulu Valley (Haleakalā National Park) where protection from feral pigs in some areas is approaching a decade, recovery of native vegetation is occurring on a large scale. The cover of alien species has diminished sharply

Table 1. Serious[a] invasive weeds of cloud forests in Maui County, Hawaiian Islands

	Life Form	Family	Elev. Limit (m)
Clidemia hirta	subshrub	Melastomataceae	1,100
Cyathea cooperi	tree fern	Cyatheaceae	1,040
Hedychium gardnerianum	large herb	Zingiberaceae	1,100
Paspalum conjugatum	grass	Poaceae	1,980
Psidium cattleianum	small tree	Myrtaceae	1,500
Rhynchospora caduca	sedge	Cyperaceae	1,435
Tibouchina herbacea	subshrub	Melastomataceae	1,765

[a] We define a weed as a serious threat to nature areas when it has demonstrated the ability to occupy significant cover, thereby causing reduction of native species by competition and displacement.

Other invasive alien plant species in Maui County cloud forests:

Ferns: *Adiantum raddianum, Athyriopsis japonica, Blechnum occidentale, Christella dentata, Christella parasitica*

Grasses, sedges, and rushes: *Andropogon virginicus, Anthoxanthum odoratum, Axonopus affinis, Carex longii, Cyperus halpans, Ehrharta stipoides, Elaeochoeris obtusa, Holcus Ianatus, Juncus planifolius, Kyllinga brevifolia, Melinis minutiflora, Oplismenus hirtellus, Paspalum dilatatum, Pennisetum clandestinum, Sacciolepis indica*

Dicot herbs and subshrubs: *Ageratina adenophora, A. riparia, Ageratum conyzoides, Cuphea carthagenensis, Drymaria cordata, Epilobium billardierianum, Erechtities valerianaefolia, Hydrocotyle verticillata, Hypochoeris radicata, Lapsana communis, Ludwigia octovalvis, Lythrum maritimum, Polygonum punctatum, Prunella vulgaris, Rubus argutus, R. rosiifolius, Rumex acetosella, Stachytarpheta urticifolia, Youngia japonica*

Dicot trees: *Psidium guajava, Spathodea campanulata*

in area; yet certain alien species are persistent, especially at the lower elevation fringe of TMCF. In the cloud forests of Maui County, 7 alien plant species have been identified as serious threats to native vegetation, and 42 additional alien species are identified as present but either not serious, or not yet serious, threats (see Table 1). The greatest impacts of these alien species are in the lower regions of cloud forests. Thus far, above 1,500 m, few weeds pose a severe threat to windward Hawaiian cloud forests in Maui County (Loope, Nagata, and Medeiros 1992).

The subshrub *Clidemia hirta* (Melastomataceae) has spread rapidly on Moloka'i, East Maui, and West Maui; if it reaches the invasive potential on these islands that it has exhibited on O'ahu, it may come to dominate much windward vegetation, potentially including lower TMCF. *Cyathea cooperi* is an alien tree fern that escaped from horticultural plantings and is invasive in low elevation cloud forests on northern to eastern flanks of Haleakalā volcano (Medeiros et al. 1992). It appears to impoverish understory vegetation in areas it dominates, usurping native vegetation. *Hedychium gardnerianum* is an invasive ginger (Zingiberaceae) with bird-dispersed seeds that has the potential (based on experience elsewhere in Hawai'i) to dominate partially open cloud forest understory and stream courses. *Paspalum conjugatum* (Poaceae) and *Rhynchospora caduca* (Cyperaceae) are mat-forming graminoids that quickly colonize and are persistent in sites of pig disturbance. *Psidium cattleianum* (Myrtaceae) is a small tree that develops dense, nearly monospecific stands, invading and replacing smaller statured but somewhat open-canopied lower cloud forests. The bright red fruits are eaten and seeds dispersed by pigs and fruit-eating alien birds. Wind-dispersed *Tibouchina herbacea* (Melastomataceae) develops dense thickets in former pig rootings and in open areas in lower elevation cloud forests, especially on West Maui.

Several other species that are absent or with incipient populations thus far in Maui County may be counted as serious threats in the future (e.g., banana poka [*Passiflora mollissima*] and *Miconia calvescens*). *Miconia calvescens* is a soft-wooded tree native to montane forests (to 1,800 m elevation in Ecuador) of Central and South America and is highly tolerant of low light conditions (Gagne et al. 1992). The massive invasion of similar forests in the Society Islands by *M. calvescens* following Hurricanes Reva and Veena in 1983 may be one of the most dramatic of documented plant invasions (Birnbaum 1991). The incipient but substantial population of *M. calvescens* on East Maui (~15,000 individuals destroyed, 75 percent less than 2 m tall; C. Chimera, unpubl. data) has been largely controlled with excellent community cooperation. The potential modification of native forests by invasive plant species in the absence of feral pigs is probably the greatest "unknown" factor in their conservation.

Native Forest Birds, Alien Birds, and Disease

Of the estimated 48 species of Hawaiian honeycreepers, 28 species (58 percent) are extinct, 12 species (25 percent) are listed by the U.S. Fish and Wildlife Service as endangered, and 8 species (17 percent) are not listed and are considered either rare, localized, or relatively common (James and Olson 1991; Scott et al. 1988; J. Jacobi, pers. com.). Also, 10 of 22 endemic genera of Hawaiian honeycreepers are extinct. Of 6 species of native honeyeater (Meliphagidae), only 1 or 2 species may possibly survive. Their low population numbers (< 5 surviving individuals combined for both species) suggest their imminent demise, hence eliminating all native members of the honeyeater family. Fossil evidence has demonstrated that native forest birds currently confined to montane windward cloud forests were, prior to human contact, widely distributed in leeward and lower elevation areas (e.g., *Melamprosops phaeosoma*) (James et al. 1987).

The catastrophic losses of species diversity of Hawaiian forest birds may be largely due to dramatic reduction of native habitat and the influence of introduced predators (i.e., mongoose [*Herpestes auropunctatus*], cats, and black rats [*Rattus rattus*]). A complete understanding of other factors such as disease and competition with alien passerine species that may be important components of their decline is lacking (Scott et al. 1986).

In Maui County, forest bird conservation is little more than a moot point on four of the five montane areas, where native forest birds are either already extinct or at catastrophically low population levels. The only exception is on the northern and eastern slopes of Haleakalā volcano where five endangered species (*'akepa*, crested honeycreeper, *nukupu'u*, Maui parrotbill, and *po'o-uli*) are protected and four other still-common native Hawaiian honeycreepers thrive (*'apapane, 'amakihi, 'i'iwi*, and Maui creeper).

A sixth endangered species may possibly be present, the presumed-extinct black-and-gold honeyeater—the Bishop's 'o'o—(*Moho bishopi*) first reported on Maui just over a decade ago but not since (Sabo 1982). Of the five federally endangered forest birds, only two have been sighted recently (J. Jacobi, pers. com.; USFWS unpubl. data). These two, the crested honeycreeper and Maui parrotbill, are apparently at comparatively "safe" population levels, and first records for nesting in both species have recently been made (E. van Gelder, unpubl. data).

New invasive passerine species become established regularly in the Hawaiian Islands. In stations throughout Haleakalā in 1980, only a single unmistakable vocalization of the Japanese bush warbler (*Cettia diphone*) was recorded. Just over a decade later (1992), the call was one of the most common bird calls in Haleakalā's forest below 1,500 m elevation (J. Jacobi, pers. com.; USFWS unpubl. data).

The thrust of current management efforts to protect remaining forest bird populations on East Maui can be described as an ecosystem approach. The immediate objective is to improve habitat quality of their host cloud forests, as previously described, by removing ungulates and controlling key invasive alien plant species. Monitoring of forest birds is conducted by the U.S. Fish and Wildlife Service in cooperation with land management agencies. The primary method uses 8-minute counts at fixed sites at regular intervals (134 m) along transects through forest habitats (approximately 1.6 km interval between transects) (Scott et al. 1986).

Conclusions

Despite provisos, the prognosis for long-term conservation of cloud forest is perhaps more favorable than that of any other native Hawaiian plant community. Because of the destruction of most Hawaiian ecosystems at lower elevations and on leeward exposures, cloud forests contain much of the islands' remaining natural diversity. With active management and prevention of new invasive introductions, long-term maintenance of this diversity appears possible.

Exclusion of feral pigs from cloud forests is the most important management action to promote conservation of Hawaiian cloud forests. After feral pigs are removed, recovery of native species can be rapid and extensive, especially at higher elevations and/or in areas that largely lack alien weed species. Of the 49 species of invasive weeds found in Maui County cloud forests, none is as yet aggressive above 1,500 m elevation. At the lower elevation fringe of cloud forests, where alien species have already become well established, recovery of native vegetation after removal of feral pigs is less likely.

The prospects for conservation of Hawaiian forest birds are less optimistic, however. Since their first contact with man, many native Hawaiian passerine birds have become extinct (James and Olson 1991). Once occupying extensive ranges, most surviving birds are now extirpated at lower elevations and on leeward exposures and are now nearly entirely confined to relatively small tracts of higher elevation cloud forests. Despite recent protection of forest bird habitat by fencing and animal removal, numbers of native forest birds, most notably the Hawaiian honeycreepers (Fringillidae, endemic subfamily Drepanidinae), have continued to decline. Without an unexpected reversal of current trends, extinction of several species of still-extant honeycreepers seems reasonably certain; prospects for survival of other less endangered but still rare species may be little better. With the active, recent (<10 years) protection of significant tracts of montane cloud forests on Haleakalā, the resultant increase in native vegetation may stabilize populations of endangered species and/or may stem the decline of some of the more common birds, but this is not assured by any means.

References

Anderson, S. J., and C. P. Stone. 1993. Snaring to control feral pigs *Sus scrofa* in a remote Hawaiian rain forest. *Biological Conservation* 63:195–201.

Birnbaum, P. 1991. Exigences et tolerances de *Miconia calvescens* a Tahiti. Papeete, Tahiti: Centre ORSTOM de Tahiti.

Chu, P. 1989. Hawaiian drought and the Southern Oscillation. *Intl. Jour. Climatology* 9:619–631.

Cuddihy, L. W., and C. P. Stone. 1990. *Alteration of native Hawaiian vegetation. Effects of humans, their activities and introductions*. Honolulu: Univ. of Hawai'i, Cooperative National Park Resources Studies Unit.

Diong, C. H. 1983. Population biology and management of the feral pig *Sus scrofa* L. in Kipahulu Valley, Maui. Ph.D. thesis, University of Hawai'i at Manoa, Honolulu.

Ekern, P. C. 1964. Direct interception of cloud water on Lanaihale, Hawaii. Soil Soc. Am. Proceedings 28:419–421.

Gagne, B. H., L. L. Loope, A. C. Medeiros, and S. J. Anderson. 1992. *Miconia calvescens*: A threat to native forests of the Hawaiian Islands. *Pacific Science* 46(3): 390–391 (abstract).

Giambelluca, T. W., and D. Nullet. 1991. Influence of the trade-wind inversion on the climate of a leeward mountain slope in Hawaii. *Climate Research* 1:207–216.

Giambelluca, T. W., M. A. Nullet, and T. A. Schroeder. 1986. Rainfall atlas of Hawai'i. Report R76. Honolulu: Hawai'i State Department of Land and Natural Resources, Division of Water and Land Development.

Howarth, F. G., S. H. Sohmer, and W. D. Duckworth. 1988. Hawaiian natural history and conservation efforts. *BioScience* 38(4): 232–237.

James, H. F., and S. L. Olson. 1991. Descriptions of thirty-two new species of birds from the Hawaiian Islands: Part II: Passeriformes. *Ornithological Monographs* No. 46. Washington, D.C.: The American Ornithologists Union.

James, H. F., T. W. Stafford, Jr., D. W. Steadman, S. L. Olson, P. S. Martin, A. J. T. Jull, and P. C. McCoy. 1987. Radiocarbon dates on bones of extinct birds from Hawaii. *Proc. Ntl. Acad. Sci. USA* 84:2350–2354.

Juvik, J. O., and P. C. Ekern. 1978. A climatology of mountain fog on Mauna Loa, Hawaii Island. Technical Report 118. Honolulu: Univ. of Hawai'i, Water Resources Research Center.

Kaneshiro, K. 1989. Uniqueness of Hawai'i's biota. In *Conservation Biology in Hawai'i*, ed. C. P. Stone and D. Stone. Honolulu: Univ. of Hawai'i, Cooperative National Park Resources Studies Unit.

Kirch, P. V. 1982. The impact of prehistoric Polynesians on the Hawaiian ecosystem. *Pacific Science* 36:1–14.

Kitayama, K., and D. Mueller-Dombois. In press. An altitudinal transect analysis of the windward vegetation on Haleakalā, a Hawaiian island mountain: (1) climate and soils. *Phytocoenologia.*

Leuschner, C., and M. Schulte. 1991. Microclimatological investigations in the tropical alpine scrub of Maui, Hawaii: Evidence for a drought-induced alpine timberline. *Pacific Science* 45(2): 152–68.

Loope, L. L., A. C. Medeiros, and B. H. Gagne. 1991. *Aspects of the history and biology of the montane bogs of Haleakalā National Park.* Technical Report No. 76. Honolulu: Univ. of Hawai'i, Cooperative National Park Resources Studies Unit.

Loope, L. L., and D. Mueller-Dombois. 1989. Characteristics of invaded islands. In *Ecology of biological invasions: A global synthesis*, ed. H. A. Mooney et al., 257–280. Chichester, U.K.: John Wiley & Sons.

Loope, L. L., R. G. Nagata, and A. C. Medeiros. 1992. Introduced plants in Haleakalā National Park. In *Alien plant invasions in native ecosystems of Hawaii: Management and research*, ed. C. P. Stone, C. W. Smith, and J. T. Tunison, 551–576. Honolulu: Univ. of Hawai'i, Cooperative National Park Resources Studies Unit.

Lyons, S. W. 1982. Empirical orthogonal function analysis of Hawaiian rainfall. *Journal of Applied Meteorology* 21:1713–1729.

Medeiros, A. C., L. L. Loope, T. Flynn, L. Cuddihy, K. A. Wilson, and S. Anderson. 1992. The naturalization of an Australian tree fern (*Cyathea cooperi*) in Hawaiian rain forests. *American Fern Journal* 82(1): 27–33.

Price, S. 1983. Climate. In *Atlas of Hawaii*, ed. R. W. Armstrong, 59–66. 2d ed. Honolulu: Univ. of Hawai'i Press.

Sabo, S. R. 1982. The rediscovery of Bishop's *'o'o* on Maui. *'Elepaio* 42:69–70.

Scott, J. M., C. B. Cameron, C. van Viper III, and S. I. Fefer. 1988. Conservation of Hawaii's vanishing avifauna. *BioScience* 38(4): 238–253.

Scott, J. M., S. Mountainspring, F. L. Ramsey, and C. B. Kepler. 1986. *Forest bird communities of the Hawaiian Islands: Their dynamics, ecology, and conservation.* Studies in Avian Biology 9. Berkeley, CA: Cooper Ornithological Society.

16. Montane Cloud Forest in the Tropical Pacific: Some Aspects of Their Floristics, Biogeography, Ecology, and Conservation

Mark D. Merlin and James O. Juvik

In the Pacific region, tropical montane cloud forest (TMCF) typically occurs as small and isolated patches on the rugged upland ridges and peaks of high volcanic islands. In addition to copious amounts of orographic rainfall, these forests receive substantial "horizontal precipitation" through direct canopy interception of wind-driven cloud water. Pacific Island cloud forests are also known as mossy, dwarf, or elfin forests because of the plethora of herbaceous epiphytes that festoon the gnarled and stunted trunks and branches of the woody vegetation.

Within the vast region of Oceania, these "islands within islands" of restricted TMCF habitat occur over a diverse altitudinal range in response to a combination of atmospheric and topographic variables. In general the cloud forest belt occurs at higher elevations on larger islands and reaches its lowest altitudinal expression on islands near the equator, where the dual effects of precipitous mountains near the sea and the extremely humid tropical air (very low lifting condensation level) combine to create TMCF at elevations as low as 450–500 m (e.g., Kosrae, Federated States of Micronesia, see Figure 1; Rarotonga, Cook Islands; and for a low altitude Fijian example, see Watling and Gillison, this volume). This "telescoping" phenomenon with respect to vegetation zonation and mountain size (the Massenerhebung effect) is further explored by Bruijnzeel et al. (1993) and Bruijnzeel and Procter (this volume).

Tropical Pacific cloud forests often contain floristic elements with either north or south temperate biogeographic affinities, thus distinguishing them from the low elevation rain forest flora that is dominated by tropical Malesian

Figure 1. On Kosrae Island (Federated States of Micronesia), Mount Mutunte (589 m), in the background, is one of five mountain peaks that support 70 ha of TMCF at comparatively low elevation *(photo by Mark Merlin).*

species. In many cases, TMCF species in the Pacific are endemic to a specific island, or even to a particular mountain range or peak. In general, this interesting and little-studied endemic TMCF biota (including plants, invertebrates, and bird species) is highly vulnerable to disturbance or extirpation simply by virtue of the very small dimensions of extant TMCF habitat. In spite of this inherent vulnerability, Fosberg (1963) pointed out that among the diverse ecosystems occurring on tropical Pacific Islands, by comparison, the TMCFs remain little disturbed by anthropogenic impacts because of their isolated occurrence on precipitous mountain ridges and peaks.

This paper reviews literature on Pacific Island TMCF and presents results from the authors' fieldwork in the cloud forests of Micronesia and Polynesia. The regional focus of this paper is especially intriguing since the lowest elevations at which TMCF can be found are on small, remote volcanic islands in the tropical Pacific. In the following discussion we summarize various aspects of TMCF vegetation (including current conservation threats and protection status) for selected areas of insular Southeast Asia (Mount Maquiling, Philippines), New Guinea (Arfak Mountains, Irian Jaya), insular Melanesia (Solomon Islands and Fiji), Micronesia (Pohnpei and Kosrae), and Polynesia (the Cook Islands and French Polynesia). In addition, we offer notes on selected genera of important cloud forest plants with respect to their possible origins, dispersal, and biogeography.

Philippines

In an early, now classic, study of altitudinal forest zonation and ecology in the Philippines, Brown (1919) described in detail the cloud forest of Mount Maquiling, characterizing it as "typical of mossy forests in the Philippines." Occurring at elevations between 900 m and the mountain summit (1,140 m), Brown referred to this forest as a "*Cyathea-Astronia* association," since the tree fern *Cyathea caudata* and the melastome *Astronia lagunensis* are the two dominant species in the single story, low-stature forest (6–10 m canopy height). Of comparative floristic interest, with respect to other Pacific Island TMCF, is the common presence of *Elaeocarpus argenteus*, *Melastoma polyanthum*, *Freycinetia* spp., and *Medinilla* spp. in the Mount Maquiling cloud forest flora. Filmy ferns (family Hymenophyllaceae), especially *Trichomanes* spp. and the much larger epiphytic "bird nest fern" *(Asplenium nidus)*, festoon tree trunks and branches in this cloud forest zone.

As noted, tree ferns (*Cyathea caudata*) are the most abundant canopy species in the Mount Maquiling cloud forest. Stadtmüller (1987) noted that the pantropical distribution of the family Cyatheaceae is largely congruent with tropical cloud forest occurrence worldwide. This is certainly true for the Asia-Pacific region. They are widespread in the upper montane rain forests of Sri Lanka (Werner, this volume) and are a dominant genus in the TMCF of Mount Kinabalu, Borneo (Kitayama, this volume), as well as in most of the Pacific Islands discussed below. Furthermore, Holttum (1964) observed that the *Cyathea* species "of Australasia and the Pacific are closely related to those of Malesia."

Brown (1919) makes an important point in his closing remarks regarding the TMCF on Mount Maquiling; although this forest may be regarded as fairly typical of "mossy forest" in the Philippines, "the character of these forests varies greatly in different places. . . . In some the mosslike covering is composed very largely of true mosses; in others, of liverworts; and in still others, of mosslike filmy ferns." Both regional and local variation in cloud forest composition, and interaction with climatic and topographic variables have not yet received serious investigation in the Philippines.

At present approximately 1.1 million ha of TMCF (officially termed mossy forest) are estimated to remain in the Philippines, constituting 18 percent of all currently forested land (Forest Management Bureau 1991). The majority of remaining cloud forest is concentrated on the large islands of Luzon and Mindanao. All cloud forest areas receive official government protection under a variety of legislation, including the National Integrated Protected Area System program implemented in 1990. However, actual protection has remained an elusive goal. In Mount Data National Park in the central Cordillera of Luzon, most of the native cloud forest has been cleared in recent years for temperate vegetable cultivation (see Penafiel, this volume). On Mount Maquiling, in spite of official protection and administration of the forest by the nearby University of the Philippines (Los Baños), forest clearing by landless peasant farmers now extends to the 600 m elevation, within a few kilometers of the summit cloud forest. In addition, current harvesting of native plant material within the cloud forest (particularly orchids and tree fern for the horticultural trade) threatens the integrity of remaining forest areas not yet cleared for agriculture (P. Sajise, pers. com., August 1993).

Melanesia

Just below the equator, 1,000 km to the southeast of the Philippines, is New Guinea and its associated Melanesian island groups extending outward into the South Pacific through the Solomon Islands, Vanuatu, New Caledonia, and Fiji.

New Guinea

New Guinea, with its large size (900,000 km^2) and topographic complexity (extensive upland areas above 3,000 m and numerous peaks reaching 4,000–5,000 m), harbors a diverse and largely endemic flora containing an estimated 9,000 angiosperm species (Good 1953). In the most recent general review of New Guinea vegetation, Paijmans (1976) associates the occurrence of distinctive cloud or mossy forest with the broad category of "lower montane forest" occurring generally between 1,000 and 3,000 m asl: "The frequency and duration of low cloud cover in particular appear to have a marked if not overriding effect on the physiognomy of lower montane forest. In regions frequently covered by low cloud, the forest begins to present a lower montane appearance well below the normal lower limit for this type, whereas above a strongly monsoonal lowland zone lower montane features begin to show at an altitude much higher than normal."

Gibbs (1917) has provided a valuable description of cloud forest vegetation and flora in western New Guinea (Irian Jaya) based on surveys of a number of TMCF areas in the Arfak Mountains (generally above 2,300 m). Among the more important woody species found in the TMCF of the Arfak Mountains are members of the following common Pacific genera: *Pandanus* (many species), *Cyathea*, *Vaccinium* (many species), *Dendrobium* (several species of this orchid genus occur in Pacific TMCF), *Dicranopteris* (*D. linearis*), *Freycinetia* (many species), *Spiraeanthemum*, *Psychotria*, and *Medinilla* (both very abundant), and *Melastoma (M. malabathricum)*. *Metrosideros arfakensis*, a species belonging to an important cloud forest genus (found throughout the Pacific Islands), was also recorded by Gibbs from the "intermediate mossy forests" and higher mossy forests of the Arfak Mountains. She also reported an abundance of associated epiphytes, including the fern genera *Hymenophyllum*, *Trichomanes*, *Polypodium*, and *Dryopteris*.

Brass (1941) also describes "mossy forest" in western New Guinea occurring on a ridge at only 900 m elevation and dominated by trees species of

Nothofagus, *Phyllocladus*, and *Astronia* (a common woody melastome genus in both Asian and western Pacific TMCF).

Although no specific information is available on the current distribution and protection status of New Guinea cloud forest, the widespread and ongoing anthropogenic transformation of montane vegetation (particularly through the use of fire) by highland subsistence agriculturalists in New Guinea is well documented (e.g., see Paijmans 1976; Robbins 1972) and doubtless the survival of many remaining cloud forest areas must be considered at risk.

Solomon Islands

The Solomon Islands lie to the east of New Guinea, forming an extension of the Bismarck Archipelago. They are almost completely covered by tropical rain forest (Whitmore 1969), although much of this forest is in various stages of ecological succession following repeated natural disturbances and the increasing impact of traditional shifting cultivation because of rapid human population growth. The Solomon Islands are located at the northern edge of the southwest Pacific tropical cyclone belt, and consequently large areas of forest are periodically devastated by typhoon winds. The influence of tropical cyclones forms a significant aspect of forest ecology in tropical East Asia and the western Pacific, and is of particular significance for TMCFs that occupy exposed mountain ridges and summits.

As elsewhere in the tropical Pacific, rain-forest composition in the Solomons changes significantly with relatively moderate increases in elevation. Whereas TMCF is not typically found below 2,000–2,400 m on the major mountains of New Guinea, Whitmore (1969) observed "stunted elfin woodland facies of upper montane rain forest as low as 690 m on the crater rim of Vangunu, and at similar elevation on Mount Gallego, and on north-east San Cristobal in the Pagato area. . . . The combined result of compression of vegetation zones and absence of certain floristic groups is that no clear lower montane rain forest zone can be distinguished in the Solomons. . . . The only sharp boundary with increasing elevation is the upper montane rain forest. In the Solomons this formation type is very 'mossy' with the ground and tree trunks and limbs usually swathed in bryophytes to 30 cm thick. A thick layer of peat lies over a shallow mineral soil."

Whitmore (1969) also noted that "the flora of the Solomon Islands is in main a poor relation to the Malesian. The diminution in flora is more strongly marked in montane than lowland groups." Ericaceae, a family with numerous species in montane New Guinea, is also poorly represented in the montane forests of the Solomon Islands (including the TMCF) with only four endemic epiphytic rhododendrons and a single shrubby *Vaccinium*.

According to Henderson and Hancock (1988), the upper montane rain forest (TMCF) in the Solomon Islands is characterized by stunted trees (maximum height 7–9 m) and exhibits a very broken canopy. The dominant species appear to be in the genera *Dacrydium* and *Eugenia* (*Syzygium*). Several species of *Weinmannia* and climbing *Freycinetia* also occur in this forest zone, along with a variety of epiphytic ferns, including several in the genera of *Gleichenia*, *Dipteris*, *Hymenophyllum*, and *Trichomanes*.

Based on limited field observations by the senior author (in 1993), the TMCFs in the Solomon Islands appear to be comparatively undisturbed at present by direct human activity since agricultural and forestry activity are concentrated in coastal and low- to mid-altitude locations.

Fiji

Centered on 18° S, the islands of Fiji represent the eastern-most extension of Melanesia. The highest mountain elevations occur on Viti Levu (1,323 m) and Taveuni (1,241 m) while four other islands include peaks rising to more than 600 m.

Detailed studies of Fijian cloud forest are limited to the recent work by Ash (1987, 1992) on the island of Taveuni. He noted that both species diversity and forest canopy height decrease on the slopes of Mount Koroturanga with increasing elevation. The orographic cloud base is reached at between 600 and 800 m near the coast, increasing to 900–1,100 m inland, leaving the island's summit ridge typically shrouded in clouds and supporting dwarfed TMCF (canopy height 3–7 m).

Some endemic trees abundant in the Taveuni TMCF, such as *Ascarina swamyana* and the melastome *Medinilla waterhousei*, are only rarely recorded in other forest zones on the island. On the other hand, climbing *Freycinetia* spp., which dominate in the TMCF canopy, also occur in other forest zones. "At the generic level the cloud-forest flora has strong affinities to that on other neighboring Pacific archipelagoes, suggesting direct dispersal rather than evolution from local rainforest taxa" (Ash 1987).

Listed below are some selected TMCF genera species from Mount Koroturanga, Taveuni (from three altitudinal samples at 800, 1,000, and 1,200 m), as identified by Ash (1987): *Ascarina swamyana* (only on Taveuni and on Mount Voma, Viti Levu), *Cyathea* spp., *Cyrtandra tempestii*, *Dysoxylum* spp., *Fagraea vitiensis*, *Freycinetia urvilleana*, *Medinilla waterhousei* (only on Taveuni and Mount Seatura, Vanua Levu), *M. spectabilis* (only on Taveuni), *Weinmannia* sp., *Astronidium parviflorum*, *Clinostigma exorrhizum* (endemic palm, same genus in the eastern Carolines), *Spiraeanthemum serratum*, *Elaeocarpus chelonimorphus*, *Calophyllum vitiense* (high percentages in 800 and 1,000 m elevational transects), and *Paphia vitiensis* (also on peaks above 870 m on Viti Levu). With respect to the general floristics of Fijian cloud forests, Ash (1987) concluded: "As yet many of these cloud-forest localities have not been thoroughly investigated by botanists and the distribution of most cloud-forest species is inadequately known."

According to Watling and Gillison (this volume), the TMCF on Ngau Island, Fiji (with summit peaks reaching 715 and 705 m), is comprised of "stunted ridge thickets." They suggest that the "ecological signature" for the Fijian TMCF is the "super abundance" of climbing *Freycinetia* spp. and large ferns such as *Cyathea* spp. and *Leptopteris* spp. resulting from an "ecological release" of such groups within a relatively impoverished flora (the flora of the Fiji group includes approximately 2,900 native species of vascular plants).

Micronesia

To the north of Melanesia, lying relatively close to the equator in the Northern Hemisphere, are the Caroline Islands of Micronesia. Although predominately low coralline atolls, the Carolines also include some raised reef islands and high volcanic islands with sufficient elevation to support TMCF. The montane cloud forests of Micronesia are confined to a few small summit sites on the two rugged volcanic islands of Pohnpei and Kosrae in the eastern Carolines (both islands are politically affiliated with the Federated States of Micronesia).

Pohnpei

Pohnpei Island (7° N, 158° E) is roughly circular with 335 km^2 of land. Cloud forest on the island is limited to only a few hectares at the island's mountain summit of Nahnalaud (elevation 772 m). Summit rainfall is estimated to exceed 5,000 mm/year and ground-level orographic clouds typically envelop the area daily beginning in the late morning hours.

Even with its restricted size, this Nahnalaud TMCF ecosystem harbors a distinctive flora including endemic species. The forest canopy is 5–6 m high. Hosokawa (1952) divided the Pohnpei TMCF into two associations: (1) the *Maesa carolinensis* (shrub) association—with *Clinostigma ponapensis* palms forming the upper tree layer (canopy emergents reaching up to 25 m high) and *Cyathea* spp. tree ferns comprising the lower tree layer; and (2) the *Cyathea ponapeana/Pandanus patina* association. Hosokawa further divided the latter association into three "consociations," which are distributed according to the substrate conditions of deeper soil (*Cyathea ponapeana/Gynotroches axillaris* consociation), shallow humus on rocky terrain (*Garcinia ponapensis/Astronidium ponapensis* consociation), and swampy, flat ridge spots (*Pandanus patina* consociation).

Glassman (1952) estimated that the summit cloud forest on Pohnpei included about 20 pure stands of *Pandanus patina*, but concluded that this species should be lumped into an association with *Cyathea*. Figure 2 illustrates this endemic montane pandan shrouded by *Freycinetia ponapensis*, another member of the Pandanaceae.

Other common plants in the Pohnpei TMCF include *Elaeocarpus* (probably *E. carolinensis*) and, to a lesser extent, *Campnosperma brevipetiolata*, both of which may grow to a height of 40 m or more in the less humid mountain forest environments below the summit. A number of other woody species endemic to Pohnpei, or the Caroline Islands, are found in the Pohnpei TMCF, including *Astronidium ponapense*, *Garcinia ponapensis*, and *Syzygium carolinese*. *Cyrtandra urvillei*, another woody plant found in the Pohnpei

Figure 2. The prop-rooted tree in the foreground is *Pandanus patina*, an endemic species found only in the TMCF of Pohnpei Island (Federated States of Micronesia). The liana wrapped around this tree is another member of the family Pandanaceae, *Freycinetia ponapensis*. It is also endemic to the Pohnpei cloud forest *(photo by Mark Merlin)*.

TMCF, belongs to a genus commonly forming endemic species in many Pacific TMCF. As in other TMCF, this forest type on Pohnpei is exceedingly mossy, with many vascular epiphytes such as the conspicuous iridescent filmy fern *Trichomanes pallidum* (Figure 3). Merlin et al. (1992) provide additional descriptive information on the Nahnalaud cloud forest. The Pohnpei cloud forest must be considered highly vulnerable to possible human disturbance because of its restricted distribution. Raynor (this volume) provides a thorough discussion of current issues relating to montane forest protection in Pohnpei.

Kosrae

Kosrae Island is located at the extreme eastern end of the Caroline Islands (5° N, 163° E). The island is roughly triangular (10–15 km across) and has an area of 109 km^2. The highest elevations of interior ridges and peaks are all near 600 m (asl) at places such as Mount Finkol (629 m) and Mount Mutunte (589 m). Kosrae supports considerably more TMCF than Pohnpei, with approximately 70 hectares distributed among five separate mountain peaks and ridges in the island's rugged interior (Whitesell et al. 1986). This cloud forest habitat, which ranges down to about 480 m elevation from the summit of Mount Finkol, represents the lowest elevation in the world at which TMCF is known to occur.

The existence of cloud forest at such low elevation is certainly related to very high equatorial humidity and resulting low-lifting condensation level. However, because of continuing scientific debate over the relative contribution of different environmental factors (wind, temperature, cloudiness, humidity, waterlogged soils) in shaping the unique dwarf stature of tropical cloud forests, the low altitude cloud forests of Kosrae may hold as yet unstudied clues to a global understanding of cloud forest–atmospheric interaction and linkage.

The trees and larger plants of the Kosrae TMCF "tend to be stunted and bent or slanting, rarely reaching a height greater than 5 m" (Maxwell 1982). These trees often support an abundance of nonwoody epiphytic species including orchid species in the genera of *Moerenhoutia*, *Phreatia*, and *Oberonia*, ferns primarily in Hymenophyllaceae (e.g., *Trichomanes pallidum*) and Polypodiaceae, and mosses. Similar epiphytes are also found in the cloud forest on Pohnpei. On some summit ridges of Kosrae, only a few scattered trees have huge moss "cushions." Maxwell observed the TMCF of Kosrae generally "to be more common and more developed on the leeward sides of the highest mountains, possibly indicating a low tolerance for direct wind and driven rain."

The most common woody species in the Kosrae TMCF include *Elaeocarpus carolinensis*, *Astronidium kusaianum* (endemic to Kosrae), and *Cyathea ponapeana*, and to a lesser extent, *Campnosperma brevipetiolata*, *Cinnamomum carolinense*, *Eugenia (Syzygium) stelechanthum*, *Angiopteris evecta*, *Cyrtandra* spp., and *Ptychosperma kusaiensis* (also endemic to Kosrae).

Figure 3. Filmy ferns in the genera *Hymenophyllum* and *Trichomanes* (seen here in the cloud forest of Pohnpei Island, Federated States of Micronesia) are common components of the epiphytic vegetation that often festoons the trunks and limbs of the woody plants in the Pacific TMCF *(photo by Mark Merlin)*.

Freycinetia sp. is the most common liana both in the TMCF and the less humid forests below. In a floristic survey of the Mount Mutunte cloud forest in 1993 (Merlin, Taulung, and Juvik 1993), the authors also collected specimens of undescribed species of *Psychotria* and *Medinilla*.

Perched on precipitous mountain ridges, the cloud forest areas of Kosrae appear relatively undisturbed by anthropogenic forces at present. However, with an increasing human population and the possibility of an expansion of ecotourism to the more remote wilderness areas of the island, the TMCF of Kosrae remains vulnerable to future disturbance.

Polynesia

The many remote island groups of Polynesia include numerous high volcanic mountains supporting cloud forest habitat and often harbor a distinctive endemic flora. In this volume a number of papers treat the comparatively well-known cloud forests of Hawai'i and Samoa (see also Kitayama and Mueller-Dombois 1992 for additional comments on Hawaiian cloud forest composition). We therefore limit our discussion in this section to a review of cloud forest environments in the Cook Islands and French Polynesia.

Cook Islands

Rarotonga is the largest (64 km^2) and most mountainous (summit elevation 652 m) of the Cook Islands. This high volcanic island is located near the southern end of the archipelago at 21°12' S latitude and 158°46' W longitude. TMCF in the Cook Islands is found only on Rarotonga where it is restricted to the relatively small areas of interior mountain summits (total area approximately 1.5 km^2, or about 3 percent of the inland forest). This type of forest on Rarotonga reaches its lowest altitudinal expression on exposed windward ridges down to about 500 m (e.g., on ridges surrounding the Avana Valley; see Merlin 1985). This TMCF on Rarotonga is normally less than 8 m tall and dominated by gnarled and wind-deformed *Metrosideros collina* festooned with epiphytic mosses and ferns. The indigenous liana, *Freycinetia arborea* (= *F. wilderi*), is also common in this cloud forest.

Some of the other, more common native species found in the TMCF on Rarotonga include *Weinmannia samoensis*, *Elaeocarpus tonganus*, *Cyathea* sp., *Ascarina diffusa*, *Fagraea berteroana*, *Xylosma suaveolens*, *Vaccinium cereum*, *Elaphoglossum savaiiense*, and *Hymenophyllum involucratum*. *Pittosporum arborescens*, a medium-sized, indigenous tree, is an occasional or rare component in the TMCF. Among the endemic woody species found in the TMCF are the following: *Fitchia speciosa* (dominant understory woody species with prop roots, genus restricted to French Polynesia and Rarotonga), *Geniostoma rarotongensis* (occasional to rare), *Sclerotheca viridiflora* (very rare, genus found only in French Polynesia and Rarotonga), *Cyrtandra rarotongensis* (rare), and *Cyrtandra lillianae* (very rare, first collected in 1929, then not again until 1988). There are nine flowering plant species only found in the TMCF of Rarotonga.

Merlin and Juvik (1992) have noted the scarcity of alien plant species in the Rarotonga TMCF. They attributed this absence to the relative inaccessibility by people and an insignificant feral ungulate population on the island.

In 1969, the Te Manga Nature Reserve was proposed to the Cook Islands government (Figure 4). The 118 ha contained in this proposed preserve include the three highest peaks on the island and comprises 80 percent of the total TMCF on Rarotonga. As of September 1993, this proposed reserve had not yet been officially sanctioned (Gerald McCormack, Prime Minister's Department, Rarotonga, Cook Islands; pers. com. 1993).

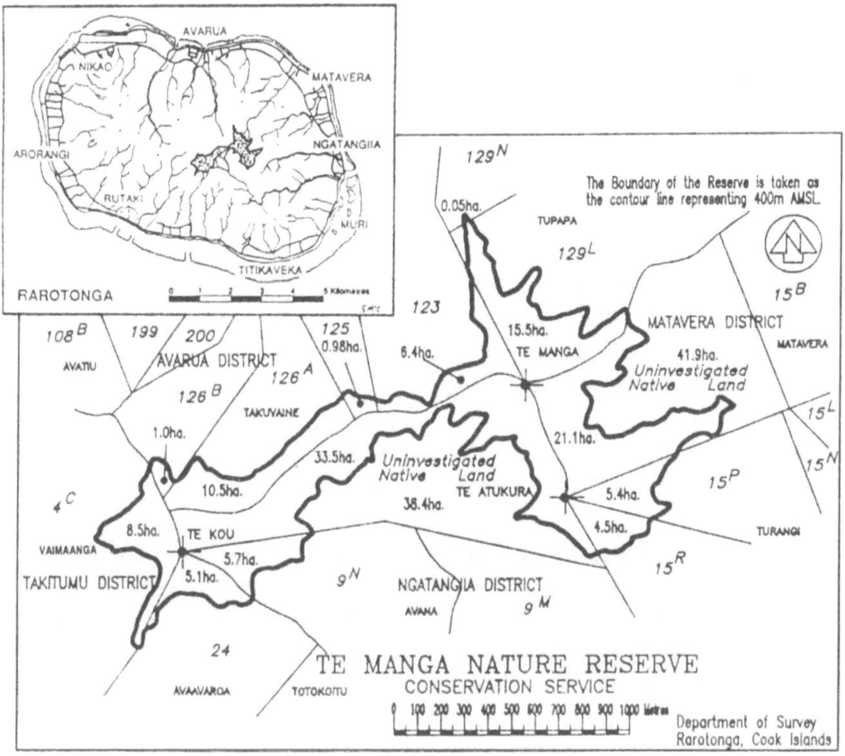

Figure 4. Proposed Te Manga Nature Reserve, containing 118 ha of TMCF on Rarotonga, Cook Islands. *(Larger map courtesy of Cook Islands government, Department of Survey; smaller map courtesy of Gerald McCormack, Prime Minister's Department. Reprinted with permission.)*

Society Islands

In the Society Islands, Tahiti, Moorea, Huahine, and Raiatea all possess steep mountainous interiors of sufficient elevation to support limited areas of TMCF. Fosberg (1992) has recently characterized the cloud forest flora of these islands (based on his fieldwork there in the 1930s), and the following species lists and comments are drawn from that research.

At higher elevations in the Society Islands, TMCF occurs where there is almost continuous cloud or fog. This forest, like other TMCF elsewhere, is of lower stature than the montane rain forest below it. In the Society Islands the TMCF is somewhat similar in composition to the lower montane forests, but more sclerophyllous. The "outstanding characteristic" is the great abundance, both on the ground and on tree trunks and branches, and even on leaves, of bryophytes (hepatics and mosses), as well as some lichens. Epiphytic ferns and orchids are also frequently found in these forests. Common woody plants in the TMCF of Tahiti include dwarf forms of *Metrosideros* and *Myrsine* and shrubs of genera such as *Vaccinium*, *Ascarina*, and *Styphelia*. Forests on ridges above 1,000 m are often dominated by *Weinmannia parviflora*. Tree ferns (*Cyathea* spp.) tend to be abundant and give the vegetation a spectacular character.

Fosberg (1992) also reported that considerable local variation in cloud forest floristic composition and species dominance occurs within the islands, the ecological basis of which is poorly understood. He notes, for example, that on Huahine, the climbing vine *Freycinetia impavida* "becomes so prominent and tangled as to make forest penetration difficult."

The Temehani Plateau, on the Island of Ra'iatea, also supports a distinctive dwarfed cloud forest association, including *Alstonia costata*, *Metrosideros collina*, *Descaspermum fruticosum*, *Weinmannia*, *Myrsine*, *Crossostylis*, *Alyxia*, *Glochidion*, *Astronia*, and *Pandanus*. Fosberg (1992) described this forest in some detail: "The plateau is a gently sloping surface, undulating between about 600 and 650 m, with hills and crest to 700 m or more. The soil is thin peat or mor, lying on a weathered rock surface, desiccated when I examined it [in 1934], but obviously saturated at most times. The shallowness of this soil may be the cause of the general low stature of the woody vegetation which varies between a few centimeters and 2–3 m, but except for patches of *Pandanus raiateensis* is usually less than 1 m, mostly about 0.5 m. Genera making up this low scrub are *Decaspermum*, *Alyxia*, *Weinmannia*, *Canthium*, *Wikstroemia*, *Alstonia*, *Astronia*, *Glochidion*, *Styphelia*, *Bidens*, *Xylosma*, *Metrosideros*, *Garnotia*, *Machaerina*, *Gahnia*, *Sphenomeris*, *Lycopodium*, and *Cassytha*, plus undoubtedly others not noted or collected. . . . In slightly protected areas, the same component of shrubs becomes taller. *Pandanus temehaniensis*, a relatively dwarf species of *P. tectorius* affinity, but amply distinct, occurs scattered in the scrub."

The cloud forests of the Society Islands today remain relatively undisturbed by anthropogenic forces because of their generally restricted occurrence on inaccessible mountain ridges and summits. However, the interaction of periodic natural disturbance and forest recovery in the presence of alien species is well illustrated in Tahiti (Birnbaum 1991). In the wake of two successive devastating hurricanes in 1983 (see Figure 5), a fast-growing South American melastome tree, *Miconia calvescens,* rapidly invaded montane cloud forest habitat. More time will be required to assess the ultimate impact of this alien species on natural, post-hurricane vegetation recovery in the TMCF of Tahiti.

Marquesas Islands

In contrast to the Society Islands, the Marquesas are much drier, and fire and feral ungulates (e.g., goats) have had a more destructive impact on TMCF. Nevertheless, some remnant cloud forest still remains on the more inaccessible high ridges and precipitous volcanic spires (dikes) in the Marquesas.

Five islands in the remote Marquesas have sufficient elevation (>900 m) to support cloud forest habitat: Nuku Hiva (1,185 m), Fatu Hiva (960 m), Hiva Oa (1,190 m), Tahuata (1,000 m), and Ua Pu (1,232 m). Where TMCF still exists in the Marquesas, the dominant trees, according to Adamson (1936), include *Metrosideros collina*, *Weinmannia*, *Crossostylis biflora*, *Cyrtandra*, *Sclerotheca*, *Cheirodendron*, and *Reynoldsia*. Adamson also noted that *Cyathea* tree ferns are a common large plant in the TMCF. He described "undergrowth in sheltered places" as normally quite dense and dominated by *Freycinetia* and "a great many species of ferns." In addition, Adamson pointed out that tree limbs "are almost everywhere covered by a thick epiphytic growth of mosses, ferns—especially the filmy ferns, *Hymenophyllum* and *Trichomanes*, and the bird's nest fern (*Asplenium nidus*)—and here and there *Lycopodium phlegmaria*, *Selaginella*, and *Procris pedunculata*."

On wind-swept ridges and slopes, Adamson (1936) observed low, dense scrub. Among the main woody species in this scrub are stunted *Metrosideros*, *Cyathea*, *Freycinetia*, *Sclerotheca*, and *Vaccinium*; and the fern *Dicranopteris linearis* is dominant in these areas.

Selected Cloud Forest Taxa in the Pacific

The following annotated list of selected genera includes some of the more important, common, and/or unique woody plants found in Pacific Island TMCF. In many cases, geographical origin, distribution, and dispersal agency of the taxa are suggested, and where endemic species are known, these are noted; much of the information has been drawn from Balgooy (1971).

Weinmannia (Cunoniaceae)

There are approximately 160 species in this genus, which has a pan-subantarctic distribution (e.g., extending to the southern tip of South Island, New

Figure 5. The devastating impact of 1983 hurricane damage to tree ferns in the Pacific TMCF is showr here—summit region of Mount Marau, Tahiti *(photo by W. Arthur Whistler).*

Zealand), with some species found north into tropical mountains, where they are often a common member of the TMCF. The dispersal agency of the small "diaspore" is uncertain, but Carlquist (1974) noted "That seed size and viability are prime factors in absence of trees from oceanic islands is well illustrated by the exceptions: *Metrosideros* has reached virtually all the high islands of the Pacific because its seeds are small and resistant to cold, desiccation and even seawater. *Weinmannia*, the chief forest tree on Tahiti, has similar seeds. These two genera are virtually the only tree species occurring on reasonably remote oceanic islands for which wind dispersal is possible or probable."

Species of *Weinmannia* are reported from Madagascar, Malesia, Philippines, New Guinea, Bismarck Archipelago, Solomon Islands, Vanuatu, New Caledonia, Fiji, Samoa, Cook Islands, Society Islands, Rapa, Marquesas, Central America, South America, Mexico, and the West Indies. Good (1953) notes that *Weinmannia* is one of a number of plants that "seem to be specially characterized by the proportion of their species which occur on relatively small islands."

Elaeocarpus (Elaeocarpaceae)

This genus comprising about 200 species belongs to a relatively small, discontinuous family of tropical and subtropical trees and shrubs (Good 1953). *Elaeocarpus* is found in the Mascarene Islands, Eastern Asia, Malesia, Australia, and the Pacific area. In fact, the genus occurs throughout Malesia and is common on many tropical high islands of the Pacific. New Guinea appears to be the main center with more than 100 species (Whitmore 1966). There are at least nine species in the Solomons, but just one on Rarotonga. The attractive, dark-colored fruits probably indicate bird dispersal, possibly by fruit doves or pigeons. It is often a major component in the Pacific area TMCF (e.g., Mount Kinabalu, Borneo, and the eastern Caroline Islands of Pohnpei and Kosrae).

Cyathea (Cyatheaceae)

The tree fern family Cyatheaceae has a pantropical distribution; and the genus *Cyathea* is a major, if not dominant, component in most TMCF of tropical Asia and the Pacific region. The genus is missing from Hawai‘i, where it is replaced by a related endemic genus (*Cibotium*) in the same family. Wind is the obvious agency for dispersal of *Cyathea*.

Ascarina (Chloranthaceae)

This genus belongs to a small family of herbs, shrubs, and trees native to the tropics and southern temperate regions. Some members of this genus occur as far south as New Zealand, but the center of species development is clearly situated in the Pacific. There is one species in eastern Malesia and six in the

Pacific. In the Kermadec Islands, *Ascarina* occurs as a subcanopy dominant below *Metrosideros kermadecensis* in much of the wet forest; and on some of the highest (>500 m), narrow ridge tops on Raoul Island (29° S) a type of "wet moss forest" occurs where *Ascarina* is the main canopy genera (Sykes 1977). In Fiji, Ash (1987) indicates that *A. swamyana* occurs only in the cloud forests of Taveuni and on Mount Voma in Viti Levu. On Rarotonga in the Cook Islands, *A. diffusa* is one of the common small trees in the TMCF (Merlin 1985). The somewhat succulent drupe with copious endosperm surrounding the tiny embryo in the seed suggests birds as the dispersal agents.

Metrosideros (Myrtaceae)

Although the great majority of the species in this genus occur in the Malay Archipelago, Australia, New Caledonia, and New Zealand, there is a species in South Africa and several, widespread species in Melanesia, Polynesia, and the Bonin Islands, but not in Micronesia. The Pacific Island members of the genus are most closely related to temperate zone New Zealand species, and this may explain their general preference for montane habitats in the islands of the tropical Pacific. *Metrosideros* is very common or dominant in native lowland and montane rain forest of Hawai'i. The tiny nonsucculent seeds are believed to be wind dispersed.

Freycinetia (Pandanaceae)

This genus includes about 100 species of climbing perennials with slender stems that produce penetrating roots to give them support on their host plants. The genus is widespread, occurring from Sri Lanka in south Asia, north to Taiwan, south to Australia and New Zealand, and on most high islands in the tropical Pacific eastward to Hawai'i. The greatest number of described species are found in the Philippines and New Guinea. *Freycinetia* includes many important endemic climbers in the rain forests and TMCF of the Pacific. Gibbs (1917) suggested that the "very wide range" of *Freycinetia* spp. in New Guinea indicates this as the origin area. The colorful spadix with brightly colored spathes produces fleshy fruits attractive to birds and bats, which probably act as the dispersal agents of the genus.

Medinilla (Melastomaceae)

Many endemic, low-lying or epiphytic species in this genus are found primarily in Asia and Malesia, but not in Australia; some species are also found in Africa. More specifically, *Medinilla* spp. are found in mainland and insular Southeast Asia, the Philippines, New Guinea, the Bismarck Archipelago, the Solomon Islands, Santa Cruz, Vanuatu, Fiji, Samoa, the Mariana Islands, and the Caroline Islands. Endemic species of the genus also occur in the TMCF of remote Pacific islands (e.g., see earlier discussion of Fiji). A probable new species of *Medinilla* was recently collected by the authors in the cloud forests of Kosrae, Micronesia (Merlin, Taulung, and Juvik 1993). The genus is probably bird dispersed.

Vaccinium (Ericaceae)

Belonging to a family found in almost all parts of the world, this large genus of more than 300 species is widespread, occurring in Africa, Eurasia, Asia, Southeast Asia, Malesia, the Philippines, New Guinea (but not in Australia), and in the Pacific, including the rain forests or TMCF of the following: Solomon Islands, Vanuatu, New Caledonia, Bonins, Samoa, Cook Islands, Society Islands, Tubuai, Rapa, Marquesas, and Hawai'i. It is also represented by species in South America, Central America, and North America. The fleshy fruits are bird dispersed; and since the seeds are tiny, they may be transported externally as well as internally. For example, Guppy (1906) referred to the *Vaccinium* fruit as a typical food of the Hawaiian goose (*Nesochen sandvicensis*).

Ptychosperma (Arecaceae)

The Australia and eastern Malesia regions are probably the general area of origin for this palm genus, but it is not distinctly centered in either of these regions. There are endemic species in the Philippines, Australia, New Guinea, Bismarck Archipelago, Solomon Islands, and both the western and eastern Caroline Islands. The genus is probably bird dispersed.

Conclusions

The ecology of TMCFs in the Pacific remains largely unstudied. In many Pacific Island settings where anthropogenic landscape transformation has extensively modified native ecosystems, the TMCF habitat, restricted to inaccessible mountain ridges and peaks, often retains a comparatively pristine composition with respect to native species. Thus, remnant TMCF habitats can, in many cases, be viewed as biodiversity "hot spots" and refugia for insular endemic species. In addition, the TMCF habitat often occurs at different elevations, and with varying aspect, on different mountains and islands providing unique opportunities for comparative research on basic cloud forest ecology and atmosphere-vegetation relationships.

At the same time that some Pacific Island TMCF habitat can still be characterized as relatively undisturbed, these ecosystems are intrinsically vulnerable simply by virtue of their extremely limited spatial extent on most Pacific Islands. Other papers in this volume draw attention to some of the problems facing the integrity of cloud forest ecosystems in Oceania. Forest disturbance caused by fire and feral ungulates, as well as the altitudinal expansion of various agricultural activities (e.g., kava cultivation), poses significant threats to specific cloud forests around the Pacific.

In addition to their biodiversity value, in many cases TMCF anchor the mountain watersheds of small Pacific Islands, where limited water resources are critical constraints on downslope agriculture and economic development. For both these reasons, TMCF protection, conservation, and active management should be given a very high priority within the Pacific region.

Acknowledgments. The authors express their appreciation to Lawrence Hamilton for funding support for the authors' fieldwork in Micronesia, provided under a grant to the East-West Center, Honolulu, Hawai'i, from the John D. and Catherine T. MacArthur Foundation. We dedicate this paper to the memory of F. Raymond Fosberg (1908–1993) who, for more than 60 years, had been a leading force in Pacific Island botany and biogeography. The authors feel particularly privileged to have shared in Dr. Fosberg's final Pacific Island botanical fieldwork on the island of Kosrae, where he returned in 1992 (after an absence of 46 years) to assist in our ethnobotanical survey of the island.

References

Adamson, A. M. 1936. *Marquesan Insects: Environment.* Bull. 139. Bishop Museum Press, Honolulu.

Ash, J. 1987. Stunted Cloud-Forest in Taveuni, Fiji. *Pacific Science* 41(1–4): 191–199.

Ash, J. 1992. Vegetation Ecology of Fiji: Past, Present, and Future Perspectives. *Pacific Science* 46(2): 111–127.

Balgooy, M. van. 1971. Plant-Geography of the Pacific. *Blumea*, Supplement VI.7.

Birnbaum, P. 1991. *Exigences et tolerances de* Miconia calvescens *a Tahiti.* Centre ORSTOM de Tahiti, Papeete.

Brass, L. J. 1941. The 1938–39 Expedition to the Snow Mountains, Netherlands New Guinea. *Journal of the Arnold Arboretum* 22:271–342.

Brown, W. H. 1919. *Vegetation of Philippine Mountains.* Bureau of Printing, Manila.

Bruijnzeel, L. A., Waterloo, M. J., Proctor, J., Kuiters, A. T., and Kotterink, B. 1993. Hydrological Observations in Montane Rain Forests on Gunung Silam, Sabah, Malaysia, With Special Reference to the 'Massenerhebung' Effect. *Journal of Ecology* 81:145–167.

Carlquist, S. 1974. *Island Biology.* Columbia University Press, New York.

Forest Management Bureau. 1991. *Philippine Forestry Statistics.* Department of Environment and Natural Resources, Quezon City, pp. 11–12.

Fosberg, F. R. 1963. Disturbance in Island Ecosystems. In *Pacific Basin Biogeography* (ed. J. L. Gressit). Bishop Museum Press, Honolulu, pp. 557–561.

Fosberg, F. R. 1992. Vegetation of the Society Islands. *Pacific Science* 46(2): 232–250.

Gibbs, L. S. 1917. *Dutch N.W. New Guinea: A Contribution to the Phytogeography and Flora of the Mt. Arfak Mountains.* Taylor and Francis, London.

Glassman, S. 1952. *The Flora of Ponape.* Bull. 209. Bishop Museum Press, Honolulu.

Good, R. 1953. *The Geography of the Flowering Plants.* Longmans, London. 2nd ed.

Guppy, H. B. 1906. *Observations of a Naturalist in the Pacific Between 1896 and 1899, Vol. 2: Plant Dispersal.* MacMillan & Co., London.

Henderson, C. P., and Hancock, I. R. 1988. *A Guide to the Useful Plants of Solomon Islands.* Research Department, Ministry of Agriculture and Lands, Honiara.

Holttum, R. E. 1964. The Tree-Ferns of the Genus *Cyathea* in Australasia and the Pacific. *Blumea* 12(2): 241–274.

Hosokawa, T. 1952. A Plant-Sociological Study in the Mossy Forests of Micronesian Islands. *Memoirs of the Faculty of Science, Kyushu University,* Series E. 1, pp. 65–82.
Kitayama K., and Mueller-Dombois, D. 1992. Vegetation of the Wet Windward Slope of Haleakala, Maui, Hawaii. *Pacific Science* 46(2): 197–220.
Maxwell, B. D. 1982. Floristic Description of Native Upland Forests on Kosrae, Eastern Caroline Islands. *Micronesica* 18(2): 109–120.
Merlin, M. 1985. Woody Vegetation in the Upland Region of Rarotonga, Cook Islands. *Pacific Science* 39(1): 81–99.
Merlin, M., Jano, D., Raynor, W., Keene, T., Juvik, J., and Sebastian, B. 1992. *Tuhke en Pohnpei: Plants of Pohnpei.* East-West Center, Honolulu.
Merlin, M., and Juvik, J. 1992. Relationships Among Native and Alien Plants on Pacific Islands With and Without Significant Human Disturbance and Feral Ungulates. In *Alien Plant Invasions in Native Ecosystems of Hawaii* (eds. C. P. Stone, C. W. Smith, and J. T. Tunison). University of Hawaii Cooperative National Park Resources Studies Unit, Honolulu.
Merlin, M., Taulung, R., and Juvik, J. 1993. *Sahk Kap Ac Kain In Acn Kosrae: Plants and Environments of Kosrae.* East-West Center, Honolulu.
Paijmans, K. 1976. Vegetation. In *New Guinea Vegetation* (ed. K. Paijmans). Australian National University Press, Canberra, pp. 23–105.
Robbins, R. G. 1972. Vegetation and Man in the South-West Pacific and New Guinea. In *Man in the Pacific Islands* (ed. R. G. Ward). Oxford University Press, London.
Stadtmüller, T. 1987. *Cloud Forests in the Humid Tropics.* The United Nations University/CATIE, Turrialba, Costa Rica.
Sykes, W. R. 1977. *Kermadec Islands Flora: An Annotated Check List.* Bull. 219. New Zealand Department of Scientific and Industrial Research, Wellington.
Whitesell, C., Maclean, C., Falanruw, M., Cole, T., and Ambacher, A. 1986. *Vegetation Survey of Kosrae, Federated States of Micronesia.* Resource Bull. PSW–17. USDA Forest Service, Pacific Southwest Forest and Range Experiment Station, Honolulu.
Whitmore, T. C. 1966. *Guide to the Forests of the British Solomon Islands.* Oxford University Press, London.
Whitmore, T. C. 1969. The Vegetation of the Solomon Islands. *Philosophical Transactions of the Royal Society* 255:259–270.

17. The Montane Cloud Forest and Its Gradational Changes in Southeast Asia

Masahiko Ohsawa

Tropical and subtropical montane forests, which include cloud forests (TMCF), provide us with excellent opportunities to study biological and ecological adaptation and evolution along the dramatic environmental gradients found on tropical mountains. The ecological complexity of the tropical montane habitat manifests itself in:

- Distinctive altitudinal thermal zones (little temperature overlap) due to an aseasonal temperature regime in the tropics (Janzen 1967).
- Stark differences in floristic composition or even physiognomy of vegetation on windward versus leeward mountain slopes in trade-wind dominated regions. This "aspect" effect is well illustrated in the Canaries (Walter 1985; Kunkel 1987; Ceballos and Ortuno 1976), Mesoamerica (Howard 1968; Sugden 1982), and southeastern Brazil (Monteiro and Kaz 1991–92; Por 1992).
- Regional contrast in vegetation on mountains influenced by seasonally reversing monsoon wind systems such as occur in the Himalaya (Schweinfurth 1957; Ohsawa 1987) and other areas of Southeast Asia (Whitmore 1984).
- The occurrence of tropical montane forests as isolated high-altitude "habitat islands" in a sea of different vegetation types (i.e., surrounding lowland rain forest) of seasonal forest (Sugden 1982), savanna, or even desert. This isolation sometimes promotes an explosive speciation of biota and results in high endemism in montane forest habitat (Gentry 1992).

Regarding the human factor, TMCF is often exploited as a source of timber, fuelwood, and charcoal (in some places even the forest litter and humus are extracted for agricultural use). However, the most important role of the TMCF may be the watershed value for the surrounding dry lowlands, particularly important on small tropical islands (Stadtmüller 1987) and even in some continental situations (e.g., in the Atlantic rain forest of Brazil; see Por 1992). Today, throughout the world, the relentless human exploitation of cloud forest has led to a drastic reduction of this forest area (e.g., in Gran Canaria only 2 percent of the original laurel forest remains [Walter 1985]; and in the Atlantic rain forest of Brazil, only about 5 percent has survived [Por 1992]). There is an urgent need to conserve this distinctive forest ecosystem on a global scale for sustainable human development, as well as for the conservation of what are often biodiversity "hot spots." Furthermore it may even be necessary to restore this important ecosystem in some locations, as has already been done in the Tijuca forest reserve in Rio de Janeiro in the 1860s—an area that was once cleared and used as a coffee plantation (Monteiro and Kaz 1991–92).

The present study intends to clarify the distribution, and the ecological mechanisms and environmental constraints for differentiation of TMCF in Southeast Asia.

Definition and Distribution of Tropical Montane Cloud Forest in Southeast Asia

The characteristic features of TMCF include a thick coat of bryophytes (Frahm and Gradstein 1991) and vascular epiphytes such as filmy ferns, bromeliads, crassuraceans, gingibers, orchids, and ericaceous shrubs. The trees are more or less stunted and gnarled, and canopy surfaces are relatively smooth (Leigh 1975; Stadtmüller 1987). However, in humid climates such as in Southeast Asia, the TMCF cannot easily be identified solely by such physiognomic or structural features, because the surrounding upland forests are also relatively humid from heavy precipitation throughout the year (Whitmore 1984). The "true" cloud forest, which is mainly sustained by cloud humidity, is easily recognizable when surrounded by totally different arid areas such as savanna or desert. This type of cloud forest can be clearly distinguished from the surrounding dry area because its altitudinal position results in significant moisture augmentation from fog and cloud. This situation is sometimes referred to as an isolated cloud forest (Sugden 1982) or an altitudinal oasis (Cloudsley-Thompson 1977). Such montane cloud forest often forms a narrow belt with an upper and lower treeline, and in most cases is confined to a windward exposure.

Botanists have observed that cloud forests on the tropical mountains of Southeast Asia roughly correspond to the upper montane rain forests dominated by microphyllous trees (Figure 1). This is also true for the Neotropical

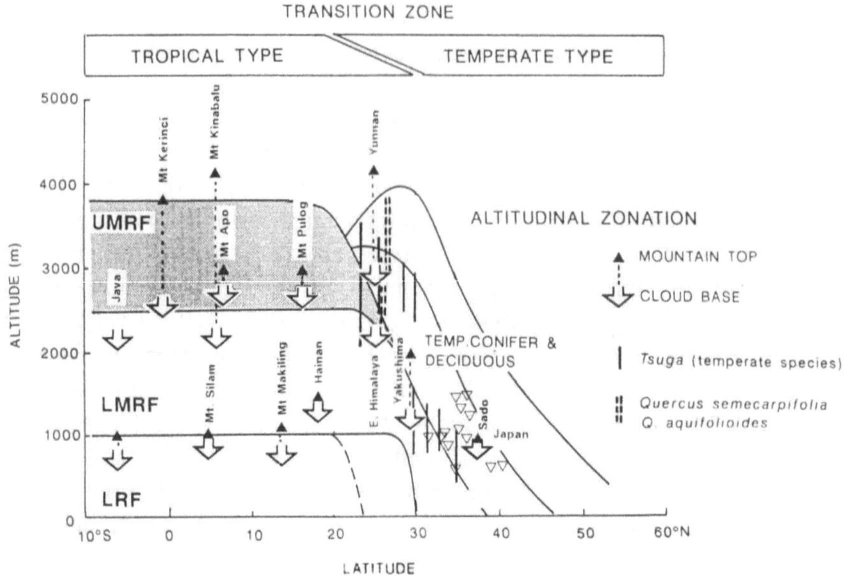

Figure 1. Several distributional features that are related to cloud forests in South and East Asia. The solid demarcation lines are the boundary of vegetation zones in humid South and East Asia *(based on Ohsawa 1993b).* The shaded area is tropical upper montane rain forest (UMRF including cloud forest). Cloud bases are indicated by arrow heads. Inverted triangles for Japan are the altitude of the estimated cloud base in summer *(after Okanoue and Ohtani 1981). Other sources: Mount Kinabalu, Kitayama (1992); Mounts Apo, Pulog, and Makiling, Anon. (1971).* Altitudinal ranges of Tsuga, Quercus semecarpifolia, and Q. aquifolioides are based on various sources. (LMRF = lower montane rain forest and LRF = lowland rain forest.)

region (Sugden 1982). In tropical Southeast Asia below 20° N, upper montane forests occupy an altitudinal zone from 2,500 to 3,600 m, where they occur in and grade into a narrow subalpine zone consisting mainly of ericaceous scrub above the treeline. *Dicranopteris* and *Gleichenia* are common indicators of an ever wet mossy forest or TMCF belt in Southeast Asia (van Steenis 1972). A clear indication of this is found on Mount Kerinci (3,801 m), Sumatra, where a continuous belt of *Dicranopteris-Gleichenia* occurs between 2,350 and 2,720 m on the south slope. This belt quite accurately coincides with the nearly daily passage of clouds from west to east in July (Ohsawa et al. 1985). That the *Gleichenia-Dicranopteris* vegetation occurs in a similar altitudinal belt on the surrounding mountains in Sumatra (Jacobs 1958) suggests its coincidence with the mountain cloud zone.

On isolated peaks near the sea or on small islands around Java, structurally similar cloud forests, though in some cases having different floristics, appear at much lower altitudes of 550–800 m (van Steenis 1972). A lowering of the cloud base near the sea and increased atmospheric humidity appear to cause an altitudinal contraction of vegetation zones (so-called Massenerhebung effect) on many small coastal mountains in the humid tropics (e.g., stunted montane forest at 400 m on the 500 m high Gunung Payung, Java; Bruijnzeel et al. 1993). Similar types of cloud forest may, however, also appear at low altitudes under different environmental influences (e.g., ridge topography or poor soils), which may complicate the altitude/climate relationship to vegetation zones (Whitmore 1984).

Above 20° N, the TMCF zone disappears because of low winter temperature (see Figure 2a; Ohsawa 1991). Most of the floristic elements (e.g., *Symplocos, Rapanea, Myrica, Ilex, Eurya*) become minor forest components in narrow ridge habitats, or as the understory plants in various forest types such as evergreen notophyllous forests or temperate coniferous and deciduous forests. These species are also early successional colonizers at disturbed sites. In the humid eastern Himalaya, several species of *Symplocos*, a typical TMCF genus, often become important elements of tall forests in a transition zone between the lower elevation evergreen notophyllous forests dominated by *Quercus/Castanopsis* and the *Tsuga/Lithocarpus* forests at higher elevations above 2,400 m (Ohsawa 1983).

In the eastern Himalaya, a distinctive moss cover on tree trunks begins to appear in *Castanopsis tribuloides* forests at about 1,500 m elevation. The zone of most frequent cloud cover (summer) begins at about 2,000 m, where *Castanopsis* forest is replaced by *Quercus* forest, and the epiphytic moss layer around tree trunks may reach up to 5 cm thick. This cloud base elevation also imposes a upper limit on maize and upland rice cultivation.

In southwest China the cloud level in Yunnan (July and August) is high at above 2,750 m on the western slope of the Diancang Shan range, west of Dali, and 2,800–3,500 m on the eastern side of the range. The level becomes still higher, above 3,400 m, on the south slope of the Yulongxue Shan near Lijiang at 27° N (Shimizu 1991), and this may be from the dry continental climate. The lowest cloud level coincides with the humid forests of *Lithocarpus variolosus* and *Tsuga dumosa*, with a thick moss cover on the trunks.

Differentiation of Upper and Lower Montane Forest

Evidence suggests that the forests of the upper montane zone may be largely derivative from those of the lower montane zone (Brown 1919; Whitmore 1984; Ohsawa 1991). Thus the canopy of the upper montane forests, to some degree, is equivalent in floristics to the understory of the lower montane forests (Ohsawa et al. 1985). There are several explanations such as mineral shortage, frequent fog/cloud incidence (Grubb 1977), temperature effects (Ohsawa 1991), and UV-B radiation (Flenley 1992) for this replacment phenomenon along altitudinal gradients on tropical mountains.

The upper montane forest is dominated by microphyllous, small trees (less than 10–15 m high) of *Eurya, Ilex, Symplocaceae, Myrsinaceae*, and others;

whereas the lower montane forest is dominated by notophyllous, large trees of lauro-fagaceous origin (e.g., *Schima, Hamamelidaceae*) (Figure 2). The two formations differ in several other structural and functional characteristics (Grubb 1977; Whitmore 1984; Ohsawa 1991). These differences may suggest further keys elucidating the causes of the differentiation of the two formation types along these complex altitudinal gradients. A similar differentiation pattern of evergreen broad-leaved forests comparable to upper and lower montane zones can be found along several different environmental gradients of various scales such as at the top of isolated small peaks (Grubb 1977; Whitmore 1984) and on small mountainous islands in the sea exposed to strong winds and/or capped by clouds (Whitmore 1984; Ohsawa 1993a).

A detailed structural analysis of these small-scale differentiations of micro- and notophyllous forests indicated that in a favorable habitat, the notophyllous trees have more rapid initial growth and taller maximum tree heights at maturity than microphyllous trees. In an unfavorable, stressful habitat, the notophyllous trees were more strongly suppressed in their growth than the microphyllous trees, and thus the latter take the dominant position over the former in TMCF (Ohsawa 1993a). This finding is compatible with the statement made by Grubb (1977) that the changing pattern of tropical montane zonation, lower montane, upper montane, and subalpine forests "may result from reduced competition being provided by the plants of the next lower formation." Many factors hinder notophyllous trees from reaching higher altitudes (e.g., cloud cover, low temperature, reduced growing season, adverse soil conditions, UV-B), but in any case, microphyllous species replace notophyllous species in the upper montane tropical zone.

Change of Upper Montane Cloud Forest from Tropical to Temperate Type

In East Asia, tropical upper montane forests reach their latitudinal limit between 20° and 30° N. They are replaced in their altitudinal ranges by three different formation types: sclerophyllous, coniferous, and less frequently deciduous forests as indicated in Figure 1 (Ohsawa 1991). The main causes of this replacement are decrease of minimum winter temperature and/or seasonal drought, both of which may control the leaf forms of canopy trees. This latitudinal change climatically coincides with the location of the subtropical high-pressure cells (Barry and Chorley 1987). Figure 3 shows a profile of free-air temperature isotherms for summer and winter seasons in relation to latitude. The tropical air masses at lower latitude are rather homogeneous throughout the year, and the annual temperature range is less than 10° C. This condition results in an altitudinal vegetation zonation that is horizontally tiered below 20°–30° N. The critical air temperature isotherm of 0° C in winter (the cold limit for subtropical evergreen broad-leaved trees of both microphyllous and notophyllous types) lies at about 3,600 m at around 25° N, and then drops

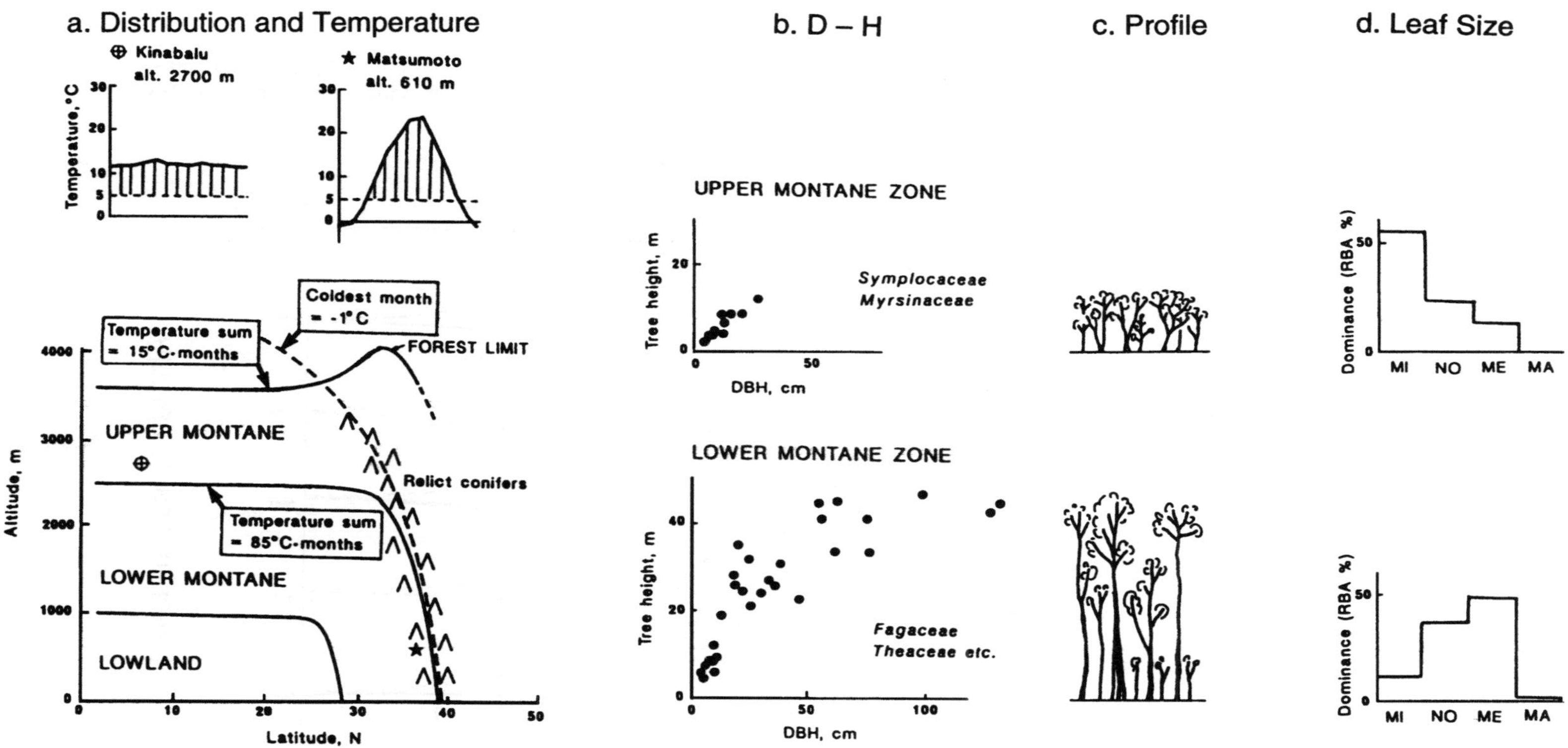

Figure 2. Schematic illustration of temperature constraints (a) and three structural features (b–d), for tropical upper and lower montane zones *(data are based on Ohsawa et al. 1985; Ohsawa 1990)*. (a) coldest month temperature of –1° C controls the northern and upper limit of evergreen broad-leaved forests, while temperature sum conditions limit the forest limit (15° C months) and the upper limit of the lower montane forests (85° C months). The coincidence of the coldest month temperature of –1° C and temperature sum of 85° C months due to seasonality of temperature exclude the thermosphere for the upper montane forests above 30° N *(see Ohsawa 1991)*. The other three sets of illustrations are structural characteristics of the upper and lower montane forests as an example of Mount Kerinci. (b) tree diameter at breast height (DBH)–tree height (H) relationship for all individuals of the tree layer. (c) forest profile. (d) leaf-size spectrum as the species dominance basis (MI, microphyll; NO, notophyll; ME, mesophyll; MA, macrophyll).

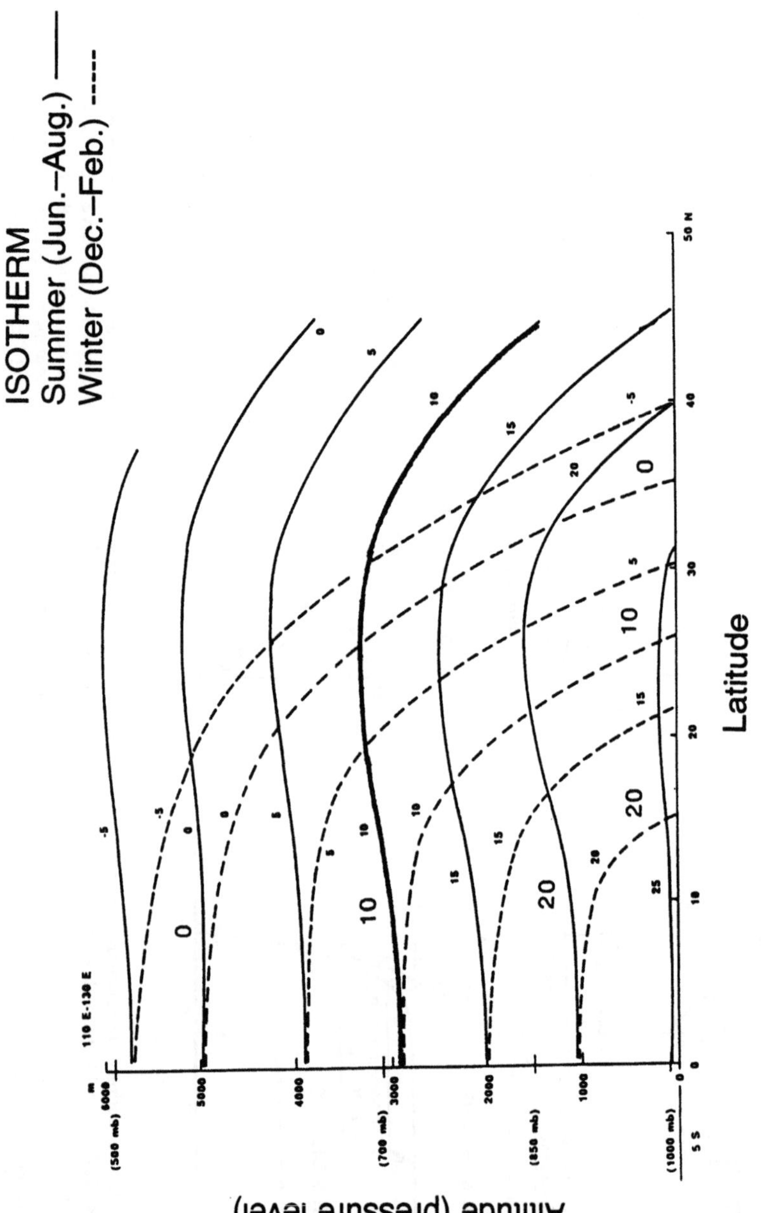

Figure 3. Free-air temperature isotherms for summer and winter seasons in South and East Asia *(data are based on Newell et al. 1972)*.

drastically to sea level at around 37° N (Figure 3). These inclined isotherms of free-air temperature in winter explain the replacement of tropical evergreen broad-leaved forests by temperate coniferous/deciduous forests (see Figure 1). At around 30° N, although summer temperatures are still equal to or even a little higher than in equatorial regions, and the isotherms are horizontally parallel up to about 40° N, the winter cold limits the northward distribution of tropical plants.

Tropical upper montane microphyllous forests are replaced by temperate coniferous forests (*Abies, Tsuga, Cryptomeria, Picea, Chamaecyparis, Cupressus, Taiwania, Cunninghamia,* and *Taxus*) with increasing latitude (Ohsawa 1991). These coniferous forests constitute a zone between subtropical evergreen notophyllous forests and cool-temperate deciduous forests, ranging from the humid eastern Himalaya through Taiwan to Japan (Su 1984; Ohsawa 1987). On southern to southeastern peripheral slopes of the Tibetan plateau and the dry western Himalaya, tropical upper montane forests are replaced by high altitudinal sclerophyllous oak forests dominated by *Quercus semecarpifolia, Q. aquifolioides,* and related species, together with the temperate conifers mentioned earlier (Jin 1981; Yang 1990). A similar replacement also can be observed in montane forests of Mesoamerica, above 2,000–2,500 m, in south central Mexico, western Guatemala, and Costa Rica. Here the tropical montane forests change into a subtropical montane belt, which is a low-latitude center of diversity for conifers such as *Abies, Cupressus, Juniperus, Pinus, Podocarpus, Taxodium,* and *Taxus* (Hartshorn 1988). Interestingly, "temperate conifer" forests rather than deciduous broad-leaved forests replace the tropical upper montane evergreen microphyllous forests, because both leaf forms seem to be effective for "fog-stripping" (Kerfoot 1968).

The climatic conditions that characterize the boundary between the coniferous forest zone and the tropical upper/lower montane forest zone were examined using Holdridge's life zones illustrated in Figure 4 (Holdridge 1967; Ohsawa and Ozaki 1993). The evergreen notophyllous forests are located in the moist forest realm and are replaced by various coniferous forests depending on environmental gradients— *Pinus* forest in very dry, subhumid, or dry forest sites; *Abies* forest at cold sites; and *Cryptomeria* forest at superhumid or rain forest sites. This last occurrence of *Cryptomeria* rain forest is equivalent to the cloud forest zone in temperate mountains. On Yakushima at about 30° N in southwestern Japan, the cloud belt is located above 1,000 m (with a peak at 1,935 m), and this level coincides with the boundary between the *Cryptomeria japonica* forest above and evergreen notophyllous forest below. On Sado island at about 38° N in northwestern Japan, *Cryptomeria* forest occurs above 700 m, and this altitude coincides with the summer cloud level. Okanoue and Ohtani (1981) estimated the cloud base levels for the Japanese archipelago and concluded that the coniferous forest or upper part of evergreen notophyllous forest zone dominated by *Quercus acuta* coincided with the summer cloud base (e.g., 700 m in Miyazaki near Yakushima, and 600 m in Niigata on the mainland coast opposite Sado island).

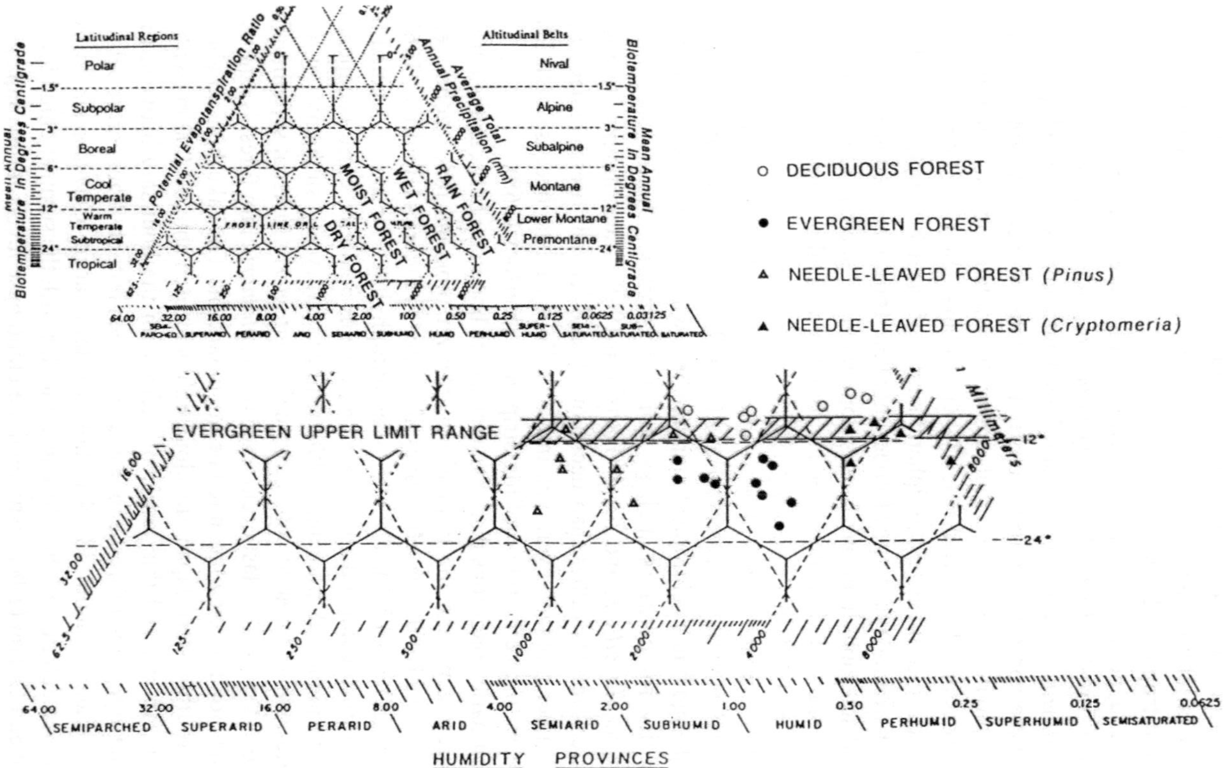

Figure 4. Holdridge's life zone illustrates habitat segregation among subtropical/warm-temperate evergreen notophyllous forests and coniferous forests in South and East Asia *(based on Ohsawa and Ozaki 1993)*.

Management and Conservation Implications

In conclusion, in Southeast Asia the cloud forest or tropical upper montane forest borders the upper margin of the tropical evergreen forest. At the northern margin of tropical evergreen forests, this role is taken up by temperate conifer or sclerophyllous forests. The constituent taxa of tropical upper montane forests are thought to be derived from the taxa of lower zones, though some widely distributed taxa may suggest possibly wider biogeographic affinities. The common condition of high species endemism necessitates the conservation of the tropical montane ecosystems at the various sites.

The microphyllous trees of TMCF can tolerate adverse conditions such as an early successional survival or extreme habitat conditions. Detailed studies on ecophysiological characteristics of microphyllous trees are much needed in Southeast Asia to form a scientific basis for conservation action. Furthermore, few studies deal with population dynamics, patch dynamics, or succession of cloud-determined forest formations. Mueller-Dombois (1988) has reported mass dieback of monospecific *Metrosideros* stands in Hawai'i and concluded that cohort-senescence, interacting with abiotic stresses, was the main cause of this dieback. Werner (1988) has also reported canopy dieback in the upper montane rain forests of Sri Lanka, which occurred after an exceptionally dry year but recovered to some degree after high rainfall. Byer and Weaver (1977) studied the 6-year secondary succession of TMCF in Puerto Rico and concluded that the recovery rate is the slowest among other tropical forest types. These results suggest that the cloud forest ecosystem is vulnerable to human disturbances as well as certain natural hazards. For a rational conservation of this fragile ecosystem, long-term studies of forest dynamics and maintenance mechanisms are essential.

References

Anon. 1971. *Plants of the Philippines.* Quezon: The Univ. of the Philippines Press.

Barry, R.G., and Chorley, R.J. 1987. *Atmosphere, weather and climate.* 5th ed. London: Methuen.

Brown, W.H. 1919. *Vegetation of Philippine mountains.* Manila: Bureau of Science.

Bruijnzeel, L.A., Waterloo, M.J., Proctor, J., Kuiters, A.T., and Kotterink, B. 1993. Hydrological observations in montane rain forests on Gunung Silam, Sabah, Malaysia, with special reference to the "Massenerhebung" effect. *Journal of Ecology* 81:145-167.

Byer, M.D., and Weaver, P.L. 1977. Early secondary succession in an elfin woodland in the Luquillo mountains of Puerto Rico. *Biotropica* 9:35-47.

Ceballos, L., and Ortuno, F. 1976. *Estudio sobre la vegetacion y flora forestal de las Canarias occidentales.* 2nd ed. Santa Cruz de Tenerife: Excmo. Cabildo Insular.

Cloudsley-Thompson, J.L. 1977. *Man and the biology of arid zones.* London: Edward Arnold.

Flenley, J.R. 1992. Ultraviolet V-B insolation and the altitudinal forest limit. In *Nature and dynamics of forest-savanna boundaries*, eds. P.A. Furley, J. Proctor, and J.A. Ratter, 273-282. London: Chapman and Hall.

Frahm, J.-P., and Gradstein, S.R. 1991. An altitudinal zonation of tropical rain forests using bryophytes. *Journal of Biogeography* 18:669-678.

Gentry, A.H. 1992. Tropical forest biodiversity: Distributional patterns and their conservational significance. *Oikos* 63:19-28.

Grubb, P.J. 1977. Control of forest growth and distribution on wet tropical mountains: With special reference to mineral nutrition. *Ann. Rev. Ecol. Syst.* 8:83-107.

Hartshorn, G.S. 1988. Tropical and subtropical vegetation of Meso-America. In *North American terrestrial vegetation,* eds. M.G. Barbour and W.D. Billings, 365-390. Cambridge: Cambridge Univ. Press.

Holdridge, L.R. 1967. *Life zone ecology*. San Jose: Tropical Science Center.

Howard, R.A. 1968. The ecology of an elfin forest in Puerto Rico, 1. Introduction and composition studies. *Journal of the Arnold Arboretum* 49:381-418.

Jacobs, M. 1958. Contribution to the botany of Mount Kerintji and adjacent area in west central Sumatra-I. *Annales Bogorienses* 3:45-104.

Janzen, D.H. 1967. Why mountain passes are higher in the tropics. *The American Naturalist* 101:233-249.

Jin Zhen-zhou. 1981. The sclerophyllous forests in China. *Journal of Yunnan University* 2:13-20.

Kerfoot, O. 1968. Mist precipitation on vegetation. *Forestry Abstracts* 29:8-20.

Kitayama, K. 1992. An altitudinal transect study of the vegetation on Mount Kinabalu, Borneo. *Vegetatio* 102:149-171.

Kunkel, G. 1987. *Die Kanarischen Inseln und ihre Pflanzenwelt.* 2. Auflage. Stuttgart: Gustav Fischer Verlag.

Leigh, E.G., Jr. 1975. Structure and climate in tropical rain forest. *Ann. Rev. Ecol. and Syst.* 6:67-86.

Monteiro, S., and Kaz, L. 1991–92. *Atlantic rain forest.* Rio de Janeiro: Livroarte Editora.

Mueller-Dombois, D. 1988. Towards a unifying theory for stand-level dieback. *GeoJournal* 17:249-251.

Newell, R.E., Kidson, J.W., Vincent, D.G., and Boer, G.J. 1972. *The general circulation of the tropical atmosphere and interactions with extratropical latitudes.* Vol. 1. The Massachusetts Institute of Technology.

Ohsawa, M. 1983. Distribution, structure and regeneration of forest communities in eastern Nepal. In *Structure and dynamics of vegetation in eastern Nepal,* ed. M. Numata, 89-120. Chiba: Chiba Univ.

Ohsawa, M. 1987. Vegetation zones in the Bhutan Himalaya. In *Life zone ecology of the Bhutan Himalaya,* ed. M. Ohsawa, 1-71. Chiba: Chiba Univ.

Ohsawa, M. 1990. An interpretation of latitudinal patterns of forest limits in south and east Asian mountains. *Journal of Ecology* 78:326-339.

Ohsawa, M. 1991. Structural comparison of tropical montane rain forests along latitudinal and altitudinal gradients in south and east Asia. *Vegetatio* 97:1-10.

Ohsawa, M. 1993a. Forest pattern along hierarchical habitat gradients in east Asia. *Biology International* (special issue, in press).

Ohsawa, M. 1993b. Latitudinal pattern of mountain vegetation zonation in southern and eastern Asia. *Journal of Vegetation Science* 4:13-18.

Ohsawa, M., Nainggolan, P.H.J., Tanaka, N., and Anwar, C. 1985. Altitudinal zonation of forest vegetation on Mount Kerinci, Sumatra: With comparisons to zonation in the temperate region of east Asia. *Journal of Tropical Ecology* 1:193-216.

Ohsawa, M., and Ozaki, K. 1993. Hierarchical analysis of vegetation/environment pattern for east Asian extra-tropical evergreen broad-leaved forests. *Journal of Japanese Biometeorology* 40 (in press).

Okanoue, M., and Ohtani, Y. 1981. On estimation for altitude of cloud belt. *Shinrin Ritchi* 23:31-34.

Por, F.D. 1992. *Sooretama, the Atlantic rain forest of Brazil.* The Hague: SPB Academic Publishing.
Schweinfurth, U. 1957. Die horizontale und vertikale Verbreitung der Vegetation im Himalaya. *Bonner Geographische Abhandlungen Heft* 20.
Shimizu, Y: 1991. Forest types and vegetation zones of Yunnan, China. *Journal of the Faculty of Science* (Univ. of Tokyo), Sect. 3, 15:1-71.
Stadtmüller, T. 1987. *Cloud forests in the humid tropics: A bibliographical review.* Tokyo and Turrialba: The United Nations University and Centro Agronomico Tropical de Investigacion y Ensenanza.
Su, H.J. 1984. Studies on the climate and vegetation types of the natural forests in Taiwan (2): Altitudinal vegetation zones in relation to temperature gradient. *Quarterly Journal of Chinese Forestry* 17:57-73.
Sugden, A.M. 1982. The vegetation of the Serrania de Macuira, Guajira, Colombia: a contrast of arid lowlands and an isolated cloud forest. *Journal of the Arnold Arboretum* 63:1-30.
van Steenis, C.G.G.J. 1972. *The mountain flora of Java.* Leiden: E.J. Brill.
Walter, H. 1985. *Vegetation of the earth and ecological systems of the geo-biosphere.* 3rd ed. Berlin: Springer-Verlag.
Werner, W. 1988. Canopy dieback in the upper montane rain forests of Sri Lanka. *GeoJournal* 17:245-248.
Whitmore, T.C. 1984. *Tropical rain forests of the Far East.* 2nd ed. Oxford: Oxford Univ. Press.
Yang, Q.Z. 1990. The characteristics and classification of oak durisilvae in the Himalayan region of China. *Acta Phytoecologica et Geobotanica Sinica* 14:197-211.

18. The Biological and Hydrological Values of the Mossy Forests in the Central Cordillera Mountains, Philippines

Samuel R. Penafiel

The Philippines has a total land area of 30 million ha, of which the remaining forests today cover only 6.7 million ha, compared with about 17 million ha in 1934. Remaining forest cover includes about 984,000 ha of old-growth commercial forest and 3,455.8 million ha of second-growth forest. The majority of the remaining noncommercial native forest is represented by tropical montane cloud forest (TMCF), which occurs in the upper mountain slopes and peaks above 900 m elevation. This forest type is referred to as "mossy forest" in the Philippines.

Distribution

Among the three largest islands in the Philippine archipelago, Luzon has the most extensive TMCF. These are located in the Central Cordillera Mountains and cover about 79,800 ha. Leyte Mountains and Panay Mountains in the Visayas and Mount Apo in Mindanao also have large areas of mossy forest (Table 1 and Figure 1). Remnant TMCFs remain unexploited because of steep slopes and the presence of crooked and stunted trees that lack commercial value.

The Mossy Forests of the Central Cordillera Mountains

Among the most important TMCF are those occurring in the Central Cordillera Mountains of Luzon, which cover about 15,000 km^2 and consist of three

Table 1. Area of mossy forests in relation to total regional forest cover

Region	Mossy Forest (ha)	Total Forest (ha)	% Mossy
1. Ilocos and Cordillera	79,800 *	421,000	18.50
2. Cagayan Valley	290,000 **	1,521,400	19.06
3. Central Luzon	41,600 *	221,200	18.80
4. Southern Tagalog	149,200 **	1,088,300	13.70
5. Bicol	7,700 *	59,100	13.02
6. Iloilo-Panay	29,000 *	81,400	35.62
7. Central Visayas	8,600 *	31,000	27.74
8. Eastern Mindanao	5,500 *	326,300	1.68
9. Western Mindanao	—	222,000	—
10. Northern Mindanao	221,300 **	886,100	24.97
11. Central Mindanao	139,400 **	808,600	17.23
12. Southern Mindanao	130,300 **	348,900	37.34
TOTAL	1,102,400	6,015,300	18.32

Source: Forest Management Bureau (1991)
* Mossy Forest.
** Includes lowland forest below 1,000 m occurring in ultra-basic soils where trees are stunted.

parallel north-to-south trenching mountain ranges: the Malayan range on the west, and the Central and the Polis ranges in the east. Summit altitudes in these mountains range from 2,400 to 2,900 m.

Physiognomic Characteristics

The Philippine TMCF is generally two-tiered with trees reaching an average height of 10 m and having a diameter of not more that 40 cm. The dominant tree species in the mossy forests of the Central Cordillera Mountains are *Lithocarpus jordanae, L. scleriana, L. coopertus, Vicinium* spp., *Melastoma topingii, Eurya acuminata,* and a few gymnosperms represented by *Taxus sumatrana, Phyllocladus stugii,* and *Dacrycarpus cumingii.* These species form the canopy story, whereas the subcanopy is made up by *Clethrea canescens, Cyathea* spp., *Smilax* spp., *Decaspermum* spp., *Medenilla* spp., *Melastoma topingii*, and *Duplocosia* spp. (Penafiel 1980).

The lower elevation limit of TMCF corresponds to the mountain cloudline or condensation level. The cloud forest environment is characterized by low light intensities, low temperature, high rainfall, and direct cloud-water interception. These conditions promote slower forest litter decomposition, allowing for accumulation of decomposed and undecomposed organic matter. Various epiphytes, liverworts, lichens, and moss species also envelop the trunks and branches of TMCF trees.

Biodiversity

Past botanical surveys of Philippine TMCF show that there are about 57 bryophyte species, 91 fern species, 18 gymnosperms, and 377 angiosperm species.

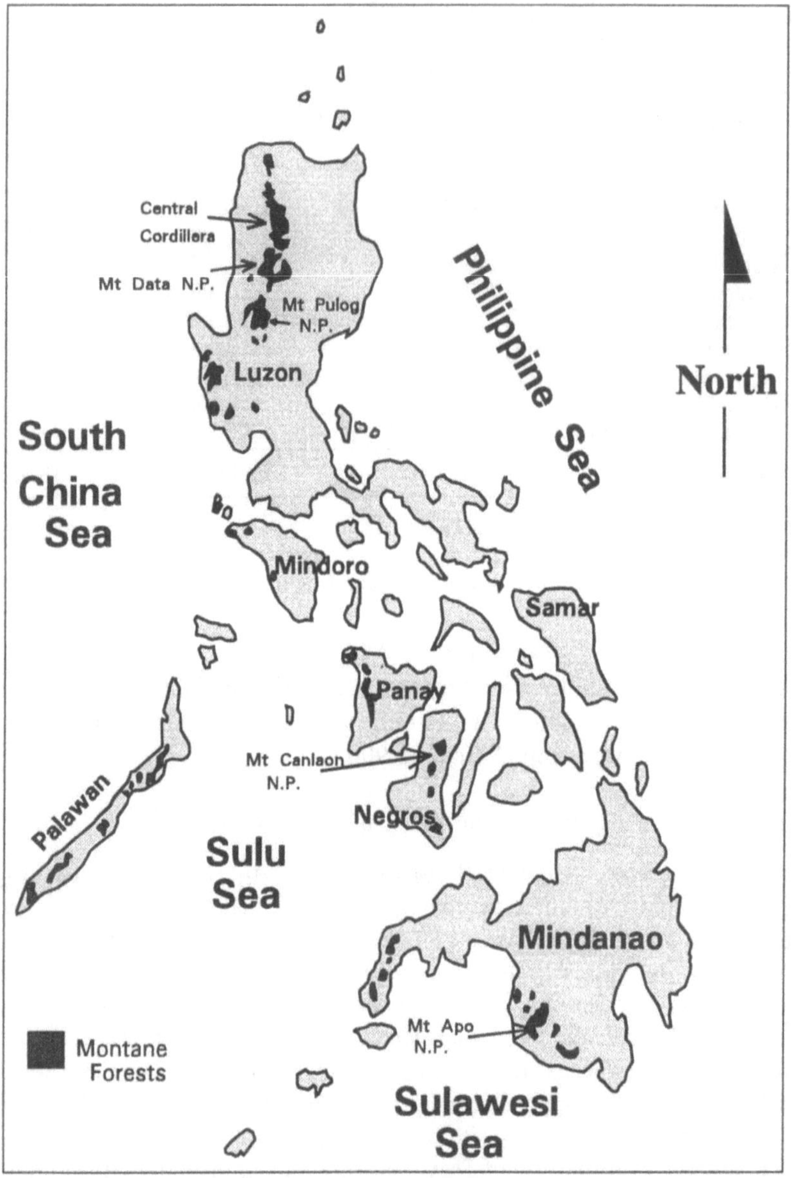

Figure 1. The Philippines, showing its major montane forests and the regions.

The number of endemic genera and species in Mount Pulog National Park is quite high: 19 genera and 44 species of bryophytes, 2 genera and 34 species of pteridophytes, 60 genera and 173 species of angiosperms (DENR 1992). The mossy forest is microphyll dominated, and the trees do not possess buttresses. The leaves of most species occurring along cloudlines are dark green, leathery, and hairy. Pinnate leaf arrangements are rarely found among plants growing in the area. On the other hand, the trees growing above cloudlines have leaves that are small, shiny, and thick. The mossy forest is therefore a gene bank for many species occurring at higher elevations. The proclamation of two protected areas—the Mount Data National Park and the Mount Pulog National Park in the Cordillera Mountains—ensures an in-situ conservation area for the species occurring in the Cordillera biogeographic zone.

Fauna

The Central Cordillera TMCF also harbors a distinctive fauna. The International Council for Bird Protection (Jensen et al. 1991) conducted a faunal survey in Mount Pulog and discovered 77 bird species, of which 33 were found in Luzon, 14 migratory and 9 endemic in Mount Pulog, and 8 endangered. The prominent mammals in the area are the Philippine deer (*Cervus philippinensis*), *Crocidura gravi*, *Poteropus vanypyrus*, *Otopterus catilagonodous*, and bushy-tailed cloud rats (*Crateromys schadenbergi*).

Watershed Values

The role of TMCFs in water production is critical because they form the vegetative cover of the upper portion of the mountain watersheds. The Central Cordillera range is valuable because it is the headwater area of the Upper Agno River Basin, the Chico River Basin, and the Amburayan River Basin. The watershed in the Upper Agno River Basin supplies water to two hydroelectric power plants. The ability of watersheds to store water during the monsoon season, which starts in May and extends to October, minimizes the occurrence of flash floods. During the dry months, water captured and stored by the mossy forest watersheds is released slowly, supplying water to reservoirs for power generation and irrigation.

Rainfall records in a TMCF watershed located at Monamon Sur in Mount Data, Mount Province, over a 3-year period showed an average annual rainfall of 3,770 mm at an elevation of 2,352 m (Forest Research Institute 1980). As an example of the variability of rainfall in this mossy forest, the October rainfall for 1978, 1979, and 1980 was 954, 343, and 442 mm, respectively (Table 2). Dry season precipitation is much lower, with January rainfall being only 7, 9, and 66 mm, respectively. This is "vertical" precipitation and does not include cloud water which is intercepted.

The mean daily streamflow recorded over a 3-year period in a mossy forest showed that the lowest average daily streamflow occurred in March. The streamflow in a gauged mossy forest watershed located at Monamon, Mount Data Bontoc, Mount Province, showed that there is a dry-season streamflow. The total streamflow between May and October (the rainy season) constitutes 73 percent of the annual flow.

Table 2. Total and mean monthly rainfall (mm) within the right fork watershed, Monamon, from 1978 to 1980

Month	1978	1979	1980	Mean
January	6.95	8.70	66.35	27.34
February	16.55	18.25	27.55	20.78
March	68.90	26.00	143.55	79.49
April	337.60	333.60	91.50	254.23
May	157.85	414.90	430.10	334.28
June	325.50	470.60	150.75	315.62
July	440.15	450.35	782.90	557.80
August	695.20	534.35	313.35	514.30
September	840.50	423.40	429.10	564.33
October	954.20	343.35	442.10	579.88
November	169.20	203.15	575.55	315.97
December	270.30	153.50	132.05	185.28
TOTAL	4,282.90	3,380.15	3,584.85	3,749.30

Source: Forest Research Institute (1980)

The mean daily temperatures at this site averaged over the year (at 8:00 A.M. and 4:00 P.M.) were 12° C and 14.4° C, respectively. The absolute minimum temperature of 30° C was recorded in January at the Mount Data mossy forest stream-gauging stations. Mossy forests are generally wet; thus, the average annual relative humidity measurements taken in the morning were 79 percent.

Under rainfall conditions in the Central Cordillera Mountains, the water yield at the Mount Data mossy forest was estimated at 945.0 m^3/ha/day. In a dipterocarp forest the variation in streamflow as influenced by rainfall duration and intensity is about 79 percent compared to 68 percent in a mossy forest (Ufano 1990), suggesting that the mossy forest has an influencing effect in sustaining and moderating streamflow.

The forest floor in mossy forests consists of a layer of organic material about 70–100 cm thick, which allows for high rainfall infiltration. The soil in mossy forests are generally acidic (pH 4), and the organic matter is high (47 percent) with bulk density averaging 0.97 g/cm^3.

Threats to Mossy Forests

The greatest threat to mossy forests largely exists in the Central Cordillera Mountains. Here the mountain people have accepted modern technology and are growing temperate and semi-temperate vegetable crops such as potato, cabbage, cauliflower, broccoli, cucumber, and carrot. The low temperature and availability of upland farming technology, plus a steady market, created an environment conducive to growing these semi-temperate crops. Because of limited flatlands for agriculture, agricultural technology using bench terracing was adopted. This led to clearing of the mountain forest and converting the lands into vegetable gardens. The Mount Province portion of the Mount Data National Park, which was previously covered with mossy forests located on the Mount Data plateau, has been transformed into a Mount Data vegetable garden with about three-fourths of its mossy forest already cleared in favor of growing semi-temperate vegetable crops (DENR 1992). With a profitable vegetable industry, communities have also started to flourish, and more mossy forest will be lost as settlements increase in size and number. Since most of the tribes still depend on fuelwood for heating and cooking, cutting of trees in the forest for fuelwood has also increased alarmingly.

The TMCF also harbors various plants such as *Medenilla* and tree ferns *Cyathea* spp., which are now collected as ornamentals. Many of these plants are illegally collected from the mossy forest and are sold as far away as Metro Manila (300 km) to the orchid and ornamental plant industry.

The few trees such as *Phyllocladus*, *Taxus* spp., and *Dacrycarpus* spp. are also cut and used for house construction. And the few wildlife species that occur are also hunted for their fur or skin, such as those of the cloud rats.

The situation described here is no different from the TMCF that supplies the water needs of the rice terraces in the Banawe area. The relatively low upland rice yields forced many of the tribes to expand their rice terraces by clearing mossy forests on the slopes of the mountain.

However, the situation is not yet widespread in other mossy forests of Mindanao and the Visayas, except in Mount Apo National Park and the Mount Canlaon National Park, where potato and cabbage gardening are starting to expand.

Conservation Efforts

Conservation of TMCF has been attempted by establishing national parks such as the Mount Data National Park, Mount Apo National Park, and Mount Canlaon National Park. However, conservation of mossy forests was considered only coincidental because the upper sections of these mountains were already covered by mossy forests. The government attempted to conserve mossy forests by taking the following action:

- Issuance of Letter of Instruction 917

In recognition of the value of mossy forest as "protection forest," LOI 917, which was issued in 1979, mandated that all mossy forests, whether inside or outside a Timber License Agreement, be made wilderness areas under a protected status and were to be properly identified and demarcated. Under this policy forest occupancy, use of, or the cutting, gathering, and use of timber and other forest products are prohibited.

- Proclamation of Mount Pulog as a National Park

In 1978, the Mount Pulog mountain ranges, containing 5,821 ha of mossy forest, were proclaimed a National Park for preserving the biodiversity of the area. The national park soon received funds from a Debt for Nature Swap negotiated with the World Wide Fund for Nature, USA. The Mount Pulog National Park represents the Cordillera biogeographic zone under the National Integrated Protected Areas System.

- Implementation of the National Integrated Protected Areas System

The Department of Environment and Natural Resources, in its commitment to save certain representative samples of ecosystems in the different biogeographic zones, implemented a program of Integrated Protected Areas System (IPAS) in 1990. To carry out and support the IPAS program, the Philippine Congress passed Republic Act 7856, known as the National Protected Areas System Act of the Philippines (NIPAS). The IPAS program and NIPAS law now provide the major impetus in conserving the remaining mossy forests of the Philippines.

- Ban on cutting of trees in areas above 1,000 m elevation

In its further effort to save its remaining native forests, the Department of Environment and Natural Resources also banned logging operations in areas with slopes greater than 50 percent and in areas above 1,000 m elevation, as well as in commercial old-growth forests. The protection of commercial old-growth forests means that the mossy forests will be eventually protected because they are located above the former.

Recommendations

Considering the role of mossy forests (TMCF) in the conservation of biological diversity and its protective watershed values, there is an urgent need to address the proper protection of what remains of this important Philippine mountain ecosystem. The Department of Environment and Natural Resources, the agency responsible for conserving mossy forests, should address the following:

- Undertake documentation of the biological diversity in other mossy forests of the Philippines where floral and faunal surveys have not been conducted.
- Enforce proper protection of the remaining forest with assistance from local people and non-governmental organizations, including demarcation of appropriate boundaries.
- Implement a strong information, education, and communication program about the value of mossy forests, and direct it to local people who are recipients of the benefits from mossy forests.
- Establish buffer zones and initiate livelihood projects that shift the activities of forest users away from the forest resources of the core protected areas.

- Develop and apply ecological guidelines and conservation plans for mossy forest protected areas (e.g., Mount Data National Park and the Mount Pulog National Park).
- Carry out restoration ecology projects in the degraded mossy forest lands.

References

DENR (Department of Environment and Natural Resources), 1992. Mt. Pulog National Park Development Plan. DENR-CAR, Baguio City, Philippines.

Forest Management Bureau, 1991. Philippine Forestry Statistics, Department of Environment and Natural Resources, Diliman, Quezon City, Philippines, pp. 11-12.

Forest Research Institute, 1980. Annual Report on Hydrology of Mossy Forests Project. WCLFRC, Baguio City, Philippines.

Jensen, A., et. al., 1991. Bird Preservation and Integrated Conservation of Mt. Pulog National Park, Philippines. International Council for Bird Protection Report, Cambridge.

Penafiel, S. R., 1980. The Mossy Forests of the Central Cordillera Ranges and the Philippines. *Canopy International* 6(8): 6-7.

Ufano, G., 1990. Determination of Rainfall and Streamflow Patterns in a Gauged Mossy Forest Watershed. M.S. thesis, Benguet State University, La Trinidad, Benguet, Philippines.

19. Montane Cloud Forests in Micronesia: Status and Future Management

Bill Raynor

Tropical montane cloud forest (TMCF) is found on only two high islands in the Micronesian region of the western Pacific Ocean, Pohnpei and Kosrae (Figure 1). The cloud forests of these two relatively young volcanic islands are unique in that they are among the lowest occurring in the world, starting at just above 450 m altitude. The reason for this is generally attributed to the location of the two islands (in the wettest part of the Pacific Ocean), their proximity to the Intertropical Convergence Zone (ITCZ), and their steep, volcanic, high-island topography with its related orographic effects.

Because the islands are relatively close to biologically rich Southeast Asia, and are isolated, their forests are richly biodiverse with a high incidence of endemism. The islands' small size, however, has made the forests important traditional sources of food, building materials, medicines, and various other material products (Anson and Raynor 1991; Merlin et al. 1992; Merlin et al. 1993) throughout the more than 2,500 years of human habitation (Haun 1984). The cloud forests isolated on steep ridgetops were exploited less but nevertheless several products, especially medicine and specialty items, were traditionally extracted from them. In the past, a few areas of lower cloud forest on Pohnpei were actually inhabited as a result of warfare and perhaps population pressure. However, depopulation due to disease after the arrival of the Europeans (circa 1826) greatly reduced human pressure on the islands' upland forests. With the exception of limited Japanese military use during World War II, Pohnpei's and Kosrae's TMCFs have been left virtually alone up to the 1970s.

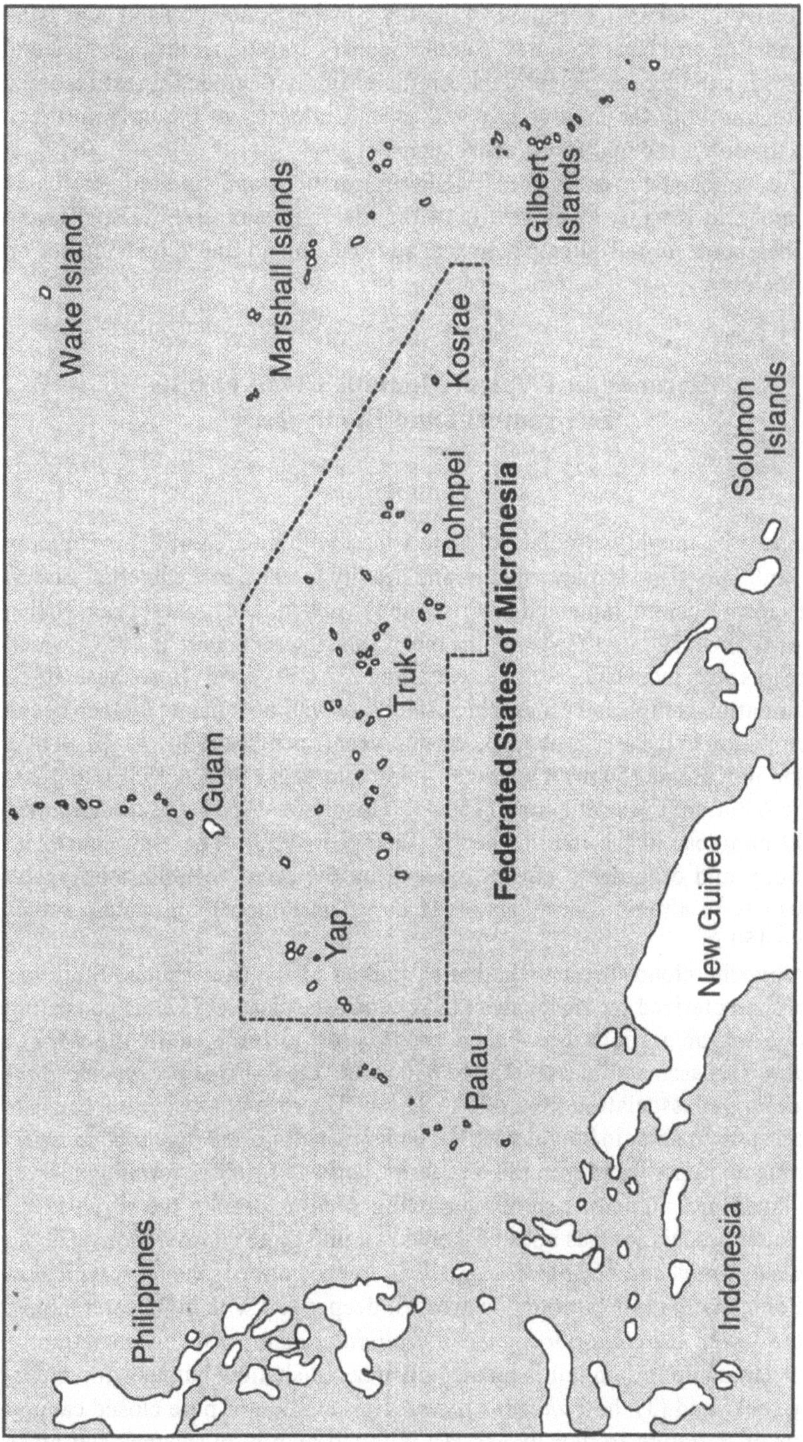

Figure 1. Location of Pohnpei and Kosrae islands in the Western Pacific. *(Source: MacLean et al. 1988. Reprinted with permission.)*

Lately, a resurgence of forest exploitation on both islands has occurred, especially on Pohnpei, because of a rapidly growing population and increasing dependence on a cash economy. Simultaneously, traditional management and respect for the forest have eroded. An increasingly degraded forest resource is affecting both the biodiversity and overall integrity of the upland forest, even the summit cloud forest ecosystems.

This paper summarizes the TMCF environment and its biodiversity on Pohnpei and Kosrae, gives details of the major threats to the TMCFs, and outlines some initial attempts to manage the upland and cloud forests on Pohnpei.

Micronesian Tropical Montane Cloud Forests—Environment and Biodiversity

Pohnpei

Pohnpei is a roughly circular volcanic island with an area of approximately 355 km^2. The island is mountainous, and heavily forested in the interior. Eleven peaks form a central tableland that rises above 600 m. The highest peak, Nahna Laud, attains 772 m. At sea level, the mean annual temperature is 27° C, which decreases with increasing altitude, averaging 23° C at 450 m (Hosokawa 1952). Precipitation is high and orographic effects are well documented. Open ocean precipitation in the area is about 3,650 mm/year (Spengler 1990). At the NOAA station in Kolonia (50 m) it averages 4,850 mm/year (NOAA 1987), increasing to 6,200 mm/year at Nihpit (450 m) (Hosokawa 1952), and to more than 9,000 mm/year in the interior peaks (Spangler 1990). The high equatorial humidity and orographic effects induced by the steep volcanic topography contribute to almost continuous cloud cover on Pohnpei's mountain ridges above 450 m.

Pohnpei's cloud forest is the better studied of the two islands. Flora has been characterized by Hosokawa (1952) and Glassman (1952) as consisting of two major associations based roughly on altitude, with the *Maesa carolinensis* association from 450 to 680 m and the *Cyathea ponapeana/Pandanus patina* association from 600 to 780 m. The former association consists of forest up to 25 m in height with the endemic palm *Clinostigma ponapensis* forming an upper tree layer, the tree fern *Cyathea nigricans* forming a lower tree layer, and numerous shrubs including *Maesa* forming the shrub layer. The latter association has been broken down into three consociations: (1) the *Cyathea ponapeana/Gynotroches axillaris* consociation is the most typical in the Pohnpei cloud forest found in areas of deeper soil with high water tables; (2) the lower *Garcinia ponapensis/Astronidium ponapensis* consociation of small scrubs up to 1-m tall is located on areas of shallow humus with underlying rock; and (3) the *Pandanus patina* consociation of pure closed canopy stands of *P. patina* from 3 to 5 m tall is found mainly in swampy flat areas along the ridges. In a few areas, open marshes dominated by *Thoracostachyum pandanophyllum* exist. Epiphytes and bryophytes abound in the upper cloud forests, the moss line being located relatively high at about 3.0–4.5 m above the ground (Hosokawa 1952). On and near ridges, trees are characteristically stunted and bent away from the prevailing wind direction. Plant leaves are reduced in size and more glabrous, compared to the same species at lower elevations, because of exposure to severe wind and other factors.

The cloud forest's fauna has been less completely studied. Of the fauna inhabiting Pohnpei's forest, only the avifauna has been extensively surveyed. Of the 29 bird species recorded on Pohnpei, 25 are known to make extensive use of the island's upland forest. Five bird species (and 8 subspecies) are endemic to the island. Species of special concern include the Pohnpei Mountain starling (*Aplonis pelzelni*), believed to be extinct; the Pohnpei Greater white-eye (*Rukia longirostris*), included on the U.S. Endangered Species List; the Micronesian pigeon (*Ducula oceanica*), a popular game bird; and the common Pohnpei lory (*Trichoglossus rubiginosus*), the only endemic parrot in Micronesia. All of these species utilize upland forest habitat, including cloud forest.

The TMCF also provides the habitat for a number of other fauna species including endemic tree snails, which comprise 74 percent (26 species) of the total land snail fauna reported from Pohnpei (Smith, pers. com.), and numerous insects and arthropods.

Kosrae

Kosrae is a roughly triangular volcanic island with an area of approximately 116 km^2. The island is topographically more rugged than Pohnpei, with several peaks exceeding 600 m. The highest, Mount Finkol, reaches 654 m. The mean annual temperature at Lelu is 27° C. Precipitation is high and orographic effects exist. For example, Hosokawa (1952) reports average precipitation at Lelu (<10 m) at 4,495 mm/year, while that at Mwot (100 m) averages 6,472 mm/year. Combined with the steep volcanic topography, these factors contribute to a favorable climate for the development of cloud forest at elevations as low as 480 m.

The area under cloud forest on Kosrae is estimated at 6 ha (Whitesell et al. 1986). Kosrae's montane cloud forest has been characterized as a single association, the *Cyathea ponapeana* association at 480–654 m altitude (Hosokawa 1952). While similar to Pohnpei cloud forest in physiognomy, it appears that *C. ponapeana* is the only species that is a major component of both forests. *C. ponapeana* is the common upper tree layer, and *Polyscias subcapitata* and other species form a lower shrub layer, with an undergrowth of the fern *Marattia fraxinea* and associated ground cover. Maxwell (1982) reports that cloud forest is restricted mostly to the sides of summit areas facing leeward (west), eastern exposed crests being dominated by almost pure patches of the fern *Dicranopteris linearis*.

Kosrae's cloud forest fauna has not been well studied, although reports do show a high rate of endemicity in tree snails (Smith, pers. com.).

Major Threats to the Cloud Forest Ecosystem

Conversion to Agroforestry and Other Agriculture

On Pohnpei, the cultivation of "sakau en Pohnpei" (*Piper methysticum*) has been identified as the most important agricultural threat to the upland forest. *Sakau*, a perennial plant the roots of which have a mild narcotic effect, is an important crop on Pohnpei, both for ceremonial and recreational use. Since World War II, the use of *sakau* by the general populace has been increasing steadily, and its cultivation and marketing have become commercialized. The expanding market and commercial value of *sakau* have produced a significant crop theft problem in lowland agricultural areas. This factor, along with the plant's need for rich organic soils, has led to increased cultivation in more remote upland forest. The clearing of overstory trees during cultivation has contributed to increased erosion and mass wasting on steep slopes. Few data exist on the level of *sakau* cultivation, but indications are that it is substantial.

Settlement

Homesteading has already encroached on the upland forest in some parts of Pohnpei. Anson et al. (1985) reported that settlement patterns tend to be along streams, up to as high as 500 m elevation, and added that several landslides and other mass wastings were noted in or adjacent to clearings associated with burning by settlers. Population growth and unequal land tenure are the major exacerbating factors involved in increased settlement of the island's interior. So far the Pohnpei government has done little to address the issue of squatters in the upland forest, despite its status as public land.

Road Construction

Roads are a threat both in terms of their negative environmental effects and the function they serve in making watershed lands more accessible for agriculture, settlement, and other types of use. Many existing and planned (already funded) secondary and tertiary roads reach the vicinity of the upland forests on both islands. Often these roads are poorly designed and have extreme gradients and virtually no surfacing (Zeimer and Megahan 1991). At present, little coordination on these key issues exists between municipal and state governments and the various agencies carrying out road construction.

Hunting

On Pohnpei, several of the popular game birds, especially the Micronesian pigeon (*Ducula oceanica*) and the Caroline Islands ground dove (*Gallicolumba *kubaryi*), are suffering decline because of overhunting by an increasing rural population, inland settlement, and growing local and off-island markets. The Pohnpei population of the Micronesian pigeon was estimated at only 822 birds in 1983 and the ground dove population was even more limited (Engbring, Ramsey, and Wildman 1990). Currently there is no agency dealing with terrestrial wildlife conservation on either island.

Tourism and Trails

Growth of tourism on Pohnpei and Kosrae has meant that more people desire the experience of visiting the upland and cloud forests. Traditional dependence on the upland forests for various products has led to a system of existing trails, consisting of "feeder trails" and a system of ridge trails in the interior of both islands. According to Griswold (1992), the trails on Pohnpei are little more than unconstructed paths-of-least-resistance evolved over many years of use and sustained without resource damage by the "light" overall use. However, with increased hiking by boot-clad ecotourists, these trails could soon become muddy eroded gullies in the steeper sections, with damage accelerating as "hikers attempt to avoid walking in the eroded sloppy ruts—a typical pattern of resource and trail degradation that could quickly result" (Griswold 1992). Removal of forest resources by tourists and other forest visitors (e.g., cutting of trees for firewood and bedding) can also be locally severe.

Developing a Management Strategy

In 1983, the Pohnpei Division of Forestry requested assistance from the Pacific Islands Forester Office (USDA Forest Service Institute of Pacific Islands Forestry) to delineate and develop legislation to establish a watershed forest reserve area on interior public lands. Utilizing aerial photos, soil survey, and aerial reconnaissance, the watershed boundaries were determined by "carefully mapping, from the air, places [on Public lands] where people have not yet settled on the highly erodible soils" (Anson et al. 1985). As a result, the Pohnpei Legislature enacted "The Pohnpei Watershed Forest Reserve and Mangrove Protection Act of 1987" (SL #1L-128-87) designating some 5,100 ha (13,000 acres) of the central upland forest area and 5,525 ha of coastal mangrove forest of Pohnpei Island as a protected area (Figure 2). It quickly became evident, however, that community awareness of the law was virtually nonexistent. When people heard of the proposed rules and regulations, which failed to recognize traditional Pohnpeian resource use in the forest areas, they were universally rejected. The government boundary survey team was turned back in many areas of the island, sometimes forcefully. This led to the formation in 1990 of the Watershed Steering Committee (WSC) task force made up of representatives from several Pohnpei government agencies and non-governmental organizations (NGOs). Based on municipal meetings and field surveys, the group designated three areas as priorities for watershed education

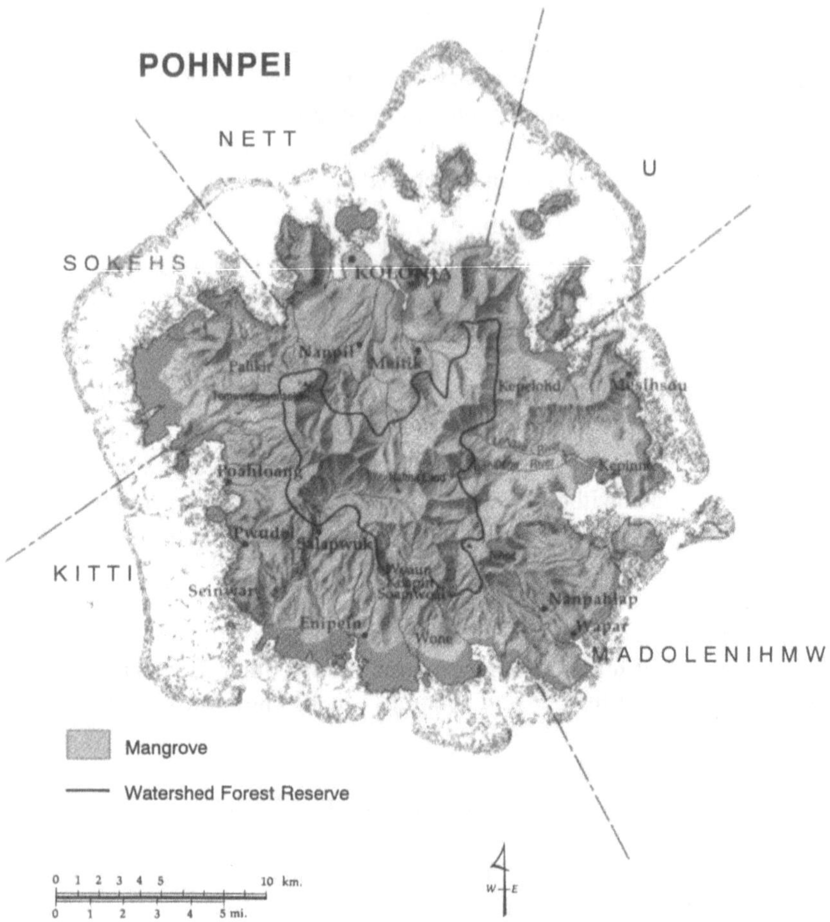

Figure 2. Location of the legally designated Watershed Forest Reserve on Pohnpei Island. *(Copyright Manoa Mapworks Inc. Reprinted with permisssion.)*

and negotiation—the Koapin Soamwoai area of Kitti, the Lehdau-Senipehn area of Madolenihmw, and the Nanpil area of Nett Municipality.

In 1991, the Pacific Islands Forester Office funded a pilot watershed extension project targeting the Koapin Soamwoai area. A local NGO group, Woaun Koapin Soamwoai Board (WKSB), made up of representatives from four villages and their chiefs, was convinced that watershed forest protection was needed in their area but desired more input in watershed management. It agreed to cooperate with the Division of Forestry and the WSC in watershed education and negotiation. The resulting, highly successful program involved educational and negotiation meetings. Since then, this program has been extended to all other municipalities, resulting in the formation of local NGO management groups in each area. The education program will culminate in an islandwide meeting of traditional chiefs in late 1993, at which time rules and regulations and a community-based management structure will be adopted.

A Community-Based Management Approach

In response to public input, a community-based approach to the management of the island's upland forests has been developed. Known locally as "comanagement," this approach recognizes several cultural and social factors:

1. Government resources are inadequate to actively manage, monitor, and enforce the watershed forest reserve;
2. The upland forest is not strictly a common property regime—past exploitation was regulated by traditional authority including various land use designations;
3. Communities have a much greater stake in forest resources and values than do government managers, thus "enlightened self-interest" should be encouraged through participation in management activities;
4. Local community institutions will be more effective than the state or national government in detecting infractions and imposing sanctions to regulate resource exploitation; and
5. The legally protected forest areas represent only 15 percent of Pohnpei's land area and less than half (42 percent) of the remaining forested area (excluding swamp/mangrove forests). Sustainable land use practices must extend to a greater area for management success.

The resulting management program combines local community and traditional institutions with the municipal and state governments. Four entities are proposed to participate in watershed management. The Division of Forestry (DoF) is designated as the lead agency, maintaining the ultimate responsibility and authority to develop and implement the management program and regulate use within the Watershed Forest Reserve. The DoF chief acts as chairperson of the WSC, which serves as the advisory board to the DoF and its parent agency, the Department of Conservation and Resource Surveillance. Municipal watershed protection officers will liaise with the DoF/WSC in matters related to the watershed, particularly infrastructure development. Local autonomous Watershed Area Management Committees (WAMCs) will be formed for each discrete watershed management unit (about 10), consisting of the local *soumas en kousapw* (village chiefs) or their delegates, and will act as comanagers with the DoF.

Following completion of the community education program in each unit, the WSC will work with the WAMCs to develop and implement Watershed Unit Plans. These plans will cover the entire watershed unit, from the cloud forest to the reef. Outside the legally designated areas, the plans will only recommend action—compliance will be largely voluntary—but by involving the communities and their leaders, compliance is expected to receive more support. The plans will also integrate existing and proposed infrastructure, including roads, trails, and water systems. Road construction in the vicinity of the watershed forest reserve will be emphatically discouraged.

Data generated during the planning process will be organized and made available to all management entities and will eventually result in a management system based on zoning and permits. Suitable use areas in the watershed forest reserve and surrounding buffer areas will be designated as agricultural, tourist, wildlife (nonhunting), or historical/cultural zones. The remaining areas will be left unzoned and protected through a permit system.

In addition, realizing that conservation efforts will only be successful if people are able to maintain/improve their standard of living, a number of sustainable development projects will be identified by the various communities during the planning process. These are envisioned to include small-scale ecotourism projects, alternative crop production systems (especially for *sakau*), and other small-scale sustainable resource extractive activities.

Finally, using the Pohnpei watershed area as a model, exchanges of extension and technical personnel between the island and other states in the Pacific region will be promoted, and a nationwide workshop will be held to encourage the delineation of watershed areas, development of legislation, and promotion of forest conservation education.

Conclusions

The cloud forests of Pohnpei and Kosrae are unique biological treasures worthy of protection. The continued health of these important ecosystems under current trends is dependent on involving the traditional leaders and local communities in their management and preservation. Pohnpei's experience demonstrates that while legislation "on paper" is an important first step, real success can be realized only through its adoption by the local community. Modern and traditional resource managers of Pohnpei's forests have been able to develop a mutually agreeable strategy of comanagement, thus effectively incorporating community and landowner input into forest planning and management. Their success has been based largely on effective integration of traditional forest management practices with sound scientific information. With continued local support and appropriate external assistance, the proposed management approach for Micronesia's upland and cloud forests will become a reality.

References

Anson, H., L. Newell, E. Pettys, and M. Falanruw. 1985. A discussion draft for upland watershed forest reserves on Pohnpei, Federated States of Micronesia. 13 pp.

Anson, H., and B. Raynor 1991. Traditional resource management and the conservation of biological diversity on Pohnpei Island, F.S.M. *South Pacific Forestry Newsletter* 2:4–5.

Engbring, J., F. Ramsey, and V. Wildman 1990. *Micronesian forest bird surveys; the*

Federated States: Pohnpei, Kosrae, Chuuk, and Yap. USDOI Fish and Wildlife Service, Honolulu, HI. 312 pp.

Glassman, S. 1952. *The flora of Ponape*. Bulletin no. 209. Bernice P. Bishop Museum. Honolulu, HI. 152 pp.

Griswold, S. 1992. Unpublished trip report—One week in Pohnpei, F.S.M., 22–28 March 1992. In 30 March 1992 letter to Emensio Eperiam, Historic Preservation Officer, Pohnpei State Government. 5 pp.

Haun, A. 1984. Prehistoric subsistence, population, and socio-political evolution on Ponape, Micronesia. PhD dissertation. University of Oregon, Eugene. 311 pp.

Hosokawa, T. 1952. A plant sociological study in the mossy forests of Micronesian islands. *Mem. Fac. Sci. Kyushu Univ. Ser. E. (Biol.)* 1:65–82.

MacLean, C., T. Cole, C. Whitesell, M. Falanruw, and A. Ambacher. 1986. The vegetation of Pohnpei, Federated States of Micronesia. USDA Forest Service, Resource Bulletin PSW-18. Berkeley, CA.

MacLean, C., C. Whitesell, T. Cole, and K. McDuffie. 1988. Timber resources of Kosrae, Pohnpei, Truk, and Yap, Federated States of Micronesia. Pacific Southwest Forest and Range Experiment Station, USDA Forest Service, Resource Bulletin PSW-24. Berkeley, CA.

Manner, H., A. Rinehart, and B. Borelli. 1993. Report on a survey of the upland forest watershed, Kitti Municipality, 20–29 May 1993. Unpublished report, University of Guam, Mangilau. 47 pp.

Maxwell, B. 1982. Floristic description of native upland forests on Kosrae, Eastern Caroline Islands. *Micronesica* 18(2): 109.

Merlin, M., D. Jano, B. Raynor, T. Keene, J. Juvik, and B. Sebastian. 1992. *Tuhke en Pohnpei: Plants of Pohnpei*. East-West Center, Program on Environment, Honolulu, HI.

Merlin, M., R. Taulung, and J. Juvik. 1993. *Sahk Kap Fin Acn Kosrae: Plants and Environments of Kosrae*. East-West Center, Program on Environment, Honolulu, HI.

National Oceanic and Atmospheric Administration (NOAA). 1987. Local climatological data: Annual summary with comparative data: Pohnpei, Eastern Caroline Islands, Pacific. National Climatic Data Center, Asheville, NC.

Smith, B. 1992. Pers. com. Marine Laboratory, University of Guam, Mangilao.

Spengler, S. 1990. Geology and hydrogeology of the island of Pohnpei, Federated States of Micronesia. PhD dissertation, University of Hawai'i, Honolulu.

Whitesell, C., C. Maclean, M. Falanruw, T. Cole, and A. Ambacher. 1986. Vegetation survey of Kosrae, Federated States of Micronesia. USDA Forest Service, Resource Bulletin PSW-17. Berkeley, CA.

Zeimer, R., and W. Megahan. 1991. Erosion and sedimentation control on roads and construction sites in the Federated States of Micronesia. Unpublished trip report to EAPI, East-West Center, Honolulu, HI.

20. Human Impacts on the Cloud Forests of the Upper Guayllabamba River Basin, Ecuador, and Suggested Management Responses

Fausto O. Sarmiento

The fate of tropandean landscapes in Ecuador is of special concern because of the ongoing biological impoverishment and habitat fragmentation. Ecuador hosts a great variety of ecosystems, ranging from the lush Amazonian forests to snow-capped mountains and from the Andean vergent to the semiarid Galapagos islands. The topographic complexity of the Andes mountains produces a large number of bioclimatic zones, each with different land use regimes. Ecuador is one of the most biologically diverse nations per unit area (Steinitz-Kannan, Colinvaux, and Kannan 1983) in the neotropics. In addition to the great biodiversity, Ecuador has a rich variety of indigenous cultures, which have lived there for millennia. Post-Colombian anthropogenic pressures have had severe impacts on the land and biotic resource base, with feedbacks on the traditional lifestyles of indigenous peoples. Currently, Ecuador stands at an important environmental crossroad. Therefore, protection of the fragile and fragmented natural tropandean ecosystems is an immediate concern for Ecuadorian land use planning.

Generally, tropical montane forests were considered in literature to be tree-dominated plant formations in mountainous tropical regions. However, local topographic and orographic conditions require a more precise definition, as in the Equatorial Andes, where Cenozoic tectonic events created a 150-km-wide series of mid-altitude plateaus and transversal ranges between two parallel, north-south trending cordilleras (Figure 1). This configuration defines two distinct domains: (1) the *transandean forests* that occupy the

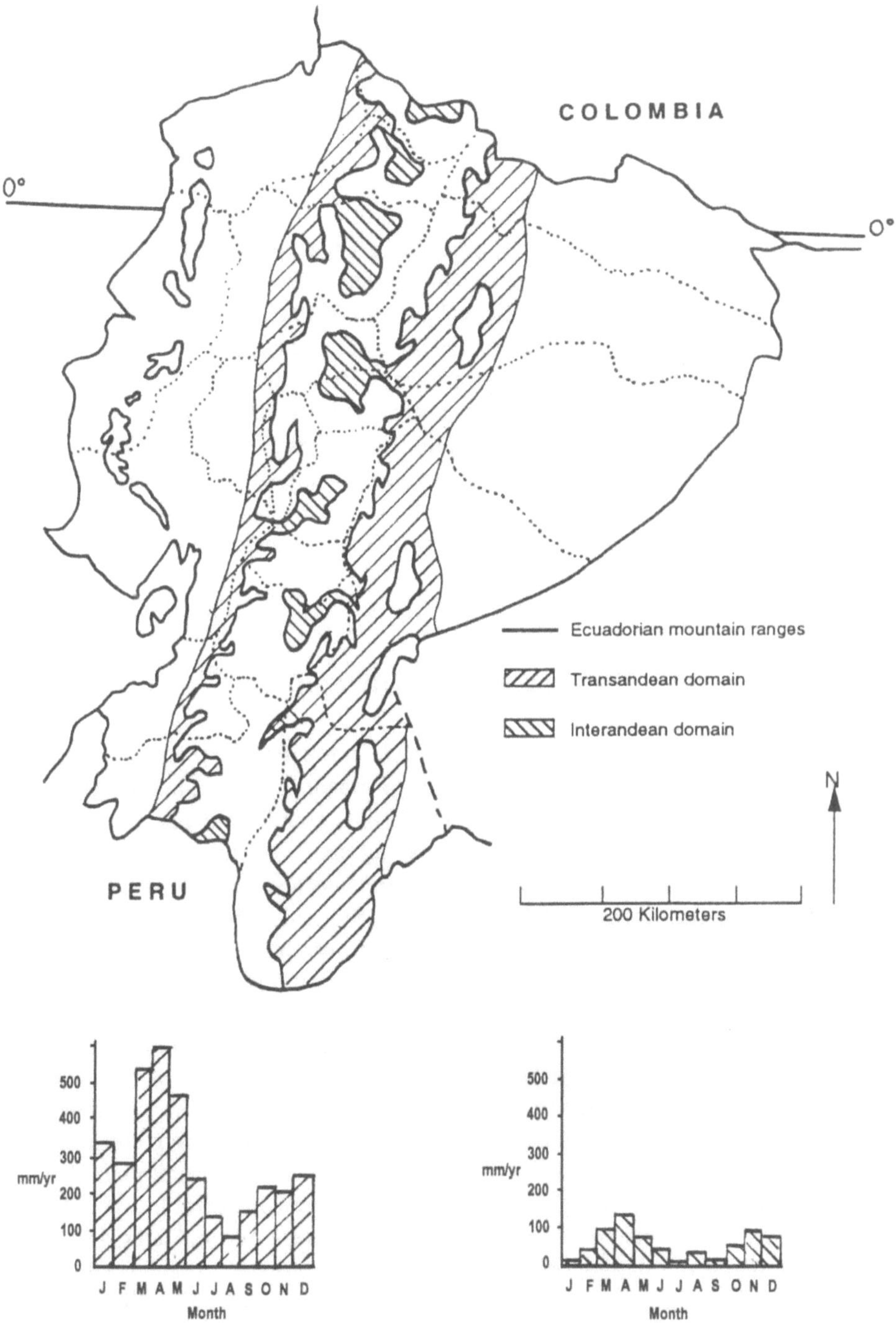

Figure 1. Interandean and transandean domains of the Tropandean Orobioma of Ecuador.

outer slopes of both of the western cordillera facing the Andean Piedmont of the coastal region (Sarmiento 1988) and the eastern cordillera facing the Amazon basin (Sarmiento 1987); and (2) the *interandean forests* that occupy small watersheds called *hoyas* in a climatically variegated array of highland plateaus and transversal ranges, locally known as knots, or *nudos*. These two domains form what Acosta-Solis (1977) called the *tropandean bioma* (Figure 1).

Authors sometimes refer to the tropandean landscape as the northern Andes (i.e., from the Huancabamba depression between Peru and Ecuador, up to the Talamanca cordillera in Costa Rica) (Van Venzen 1992). However, the variations in geomorphology and microclimate in the tropandean region make the interandean valleys and transandean coves of Ecuador unique. Further description of these peculiar landscapes is provided elsewhere (Sarmiento 1987; Sarmiento, in press). Here, a description of a mountain transect along the 0°4'0" N within the Guayllabamba River Basin calls attention to the need to protect remnant forest patches (especially of cloud forests) and to enhance the heterogeneity of the landscape mosaic of the interandean and transandean domains.

The Guayllabamba River Basin

As shown in Figure 2, the Machángara River drains from the western cordillera, catches the runoff from the Illinizas, Corazón, Atacazo, and Pichincha volcanoes, and flows northward through fertile agricultural lands and the industrial and urban areas of Quito, the capital city. The river then continues into the lower valley of Tumbaco where it joins the San Pedro River. The San Pedro drains the catchment of the eastern cordillera, including the Nudo de Tiopullo, Cotopaxi, Pasochoa, Antisana, and Cayambe volcanoes. Land uses in its drainage include isolated residual interandean forest patches and agricultural, industrial, and residential areas in the Los Chillos valley.

The Guayllabamba River proper begins at the confluence of the San Pedro and the Machángara rivers, and already carries a heavy load of pollutants including sewage, industrial effluents, and heavy metals from the interandean region. As the Guayllabamba flows westward through the western cordillera, it creates an undulating canyon that opens up toward the coastal plain. The Pisque River imports into the Guayllabamba River a considerable amount of suspended sediments eroded from the semi-arid areas of the Andes. Downstream, inputs of relatively clear water from numerous streams and small rivers from the pristine windward catchment basins refresh the Guayllabamba, which finally reaches the Andean Piedmont. A few kilometers farther downstream, the Guayllabamba joins the Blanco River and forms the Esmeraldas River, the second largest South American river that drains into the Pacific Ocean.

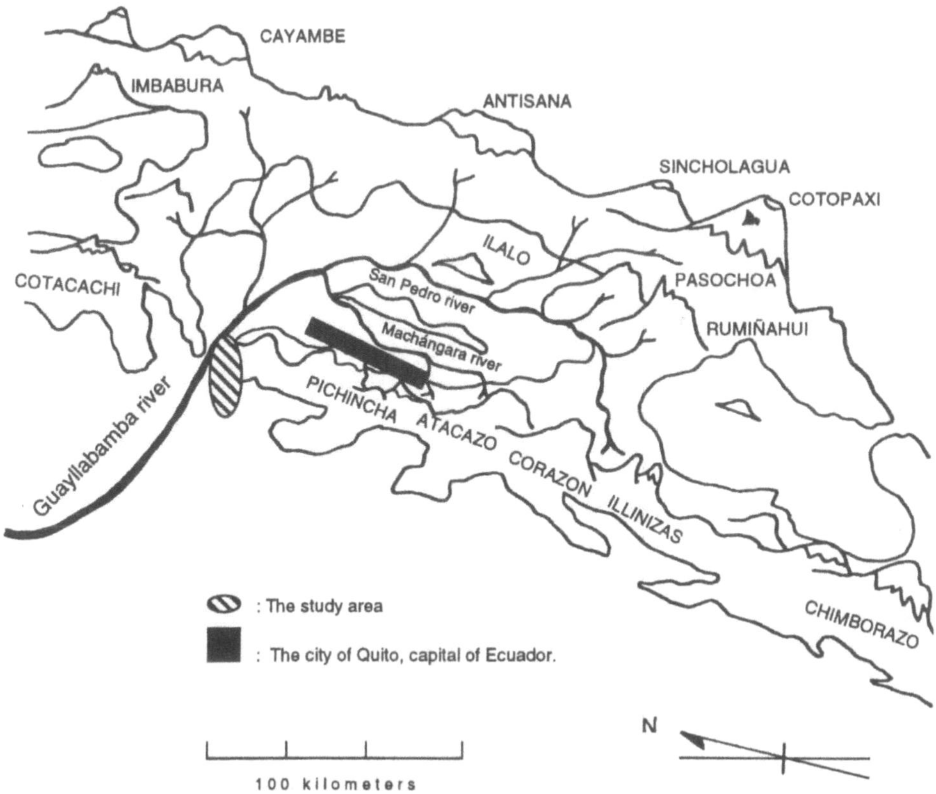

Figure 2. Scheme of the upper Guayllabamba River basin showing the relation of mountains in the western and eastern cordilleras to create the basin, locally known as *hoya*. The study area is shown in the oval at the mountain pass. Transandean valleys located at the Andean Piedmont are open toward the coastal plain, while interandean valleys occupy small highland plateaus.

The Study Area

The study area (Figure 3) is a few minutes to the north of the equator, and encompasses three main sampling zones:

1. The *Pululahua Geobotanical Reserve*, a 3,300-ha governmentally managed reserve including an old caldera with andesitic domes and interandean forests;
2. The *Maquipucuna Reserve*, a 3,000-ha privately managed reserve including pristine montane cloud forests of the tropandean domain; and
3. The *frontier agriculture zone* in between, with active and abandoned farm fields and pastures, remnants of cloud forest patches, and hedgerows and windbreaks that serve as corridors for wildlife (Figure 3).

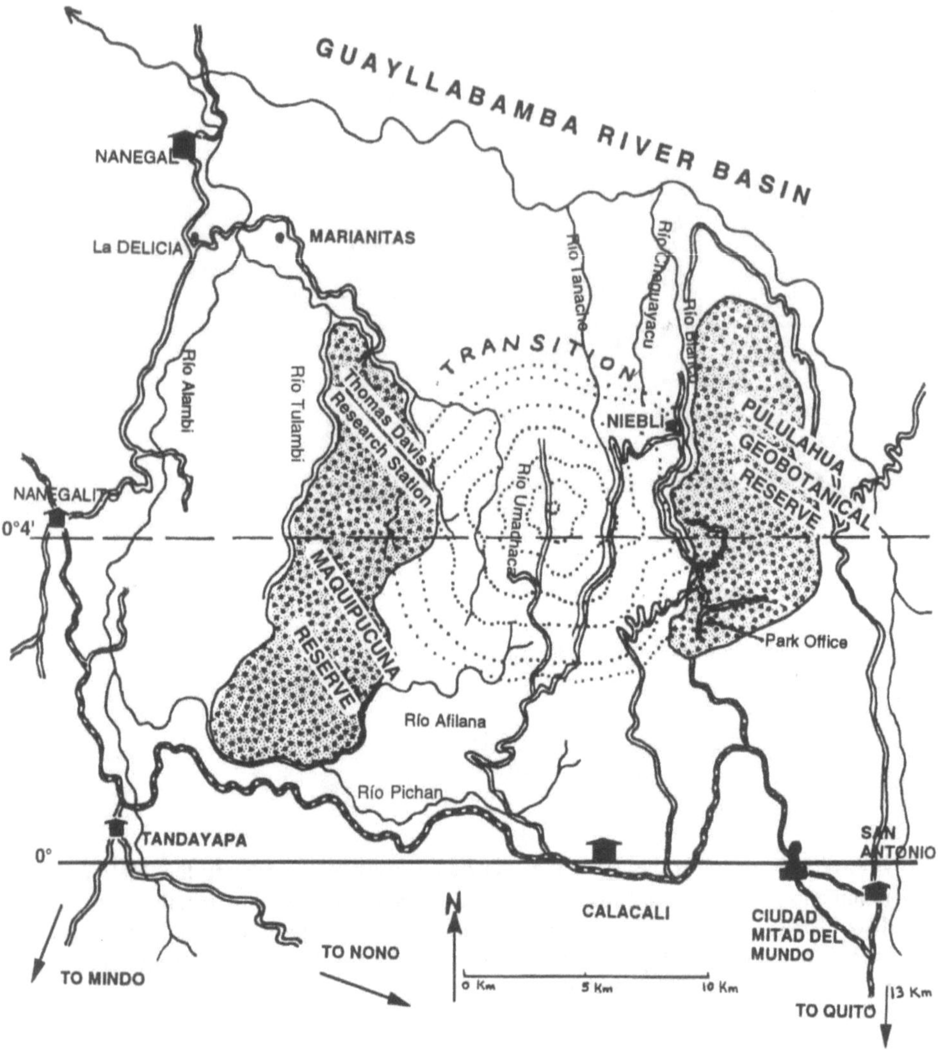

Figure 3. The transect follows the 0°4' N and spans ca. 24 straight-line kilometers. The study area encompasses ca 18,000 ha, where small towns are shown linked by a paved road and dirt roads.

The study area and surrounding region have a rich cultural heritage dating back to a pre-Incan civilization that involved agriculture and ranching of some wild animals (e.g., Cracidae, Cervidae). They also built pyramids—*pucarás,* or stone fortresses—along the Guayllabamba canyon using a local volcanic tuff known as *Cangahua*. Distinctive cultures flourished on opposite sides of the continental divide: the Caranquitus in the interandean domain and the

Yumbos in the transandean domain, where today a small colorful bird, the Yumbo (*Semnornis ramphastinus,* Capitonidae), bears its name as a reminder of this former powerful aboriginal culture of the TMCF of the western Andean slopes.

Increasing amounts of land are being cleared, even of scrublands, to make room for what is at best marginal agriculture (Knapp 1992). The primary limiting factor for biomass productivity is water. The typically low relative humidity, a meteorological trait of the region, is exacerbated by deforestation of the water-capturing TMCF (Cañadas 1983). This aridity promotes water-erosion gullies and high wind flow in intensities such that loess is visible in the air. A new paved road parallels the study area and connects with older roads to the lowlands of northwestern Ecuador. The road construction significantly degraded river quality, since all the debris was thrown downslope to be entrained in the fluvial system. The study area is beginning to receive pressure from the expanding capital city. Now, with reduced travel time to Quito, agricultural products and colonizers come and go more easily. Small towns along the original road have subsisted under precarious conditions, converting the original forest to farm fields and pasture.

The Transect Reflects the Föhn Effect

The study area's 24-km linear transect crosses, from west to east, the tropical rain forest in the Andean Piedmont and the semi-arid scrublands of the leeward rainshadow at the Guayllabamba. There are dramatic changes in species composition and climatic parameters along the transect (Figure 4).

The transect begins in the west in the Maquipucuna Reserve at an elevation of 1,200 m and extends eastward to Cerro de la Marca in the Pululahua Reserve, the highest point being at 3,350 m; then it descends the leeward slope of the Guayllabamba River, which drains at the Tanlagua region at 1,700 m. The rugged terrain makes the journey longer since roads and trails must switchback on the steep slopes, locally known as *kingos*, a common regional practice that dates back to prehistoric times. The relief varies from a ridge-and-valley type in the Piedmont to a ridge-and-abysm in the Guayllabamba. Slopes of 60 percent are a common feature in the region; thus, landslides naturally occur. The richness of the cloud forest flora and epiphytic aroids, bromeliads, and orchids decreases clearly from west to east, giving way to the rather depauperated flora in the drier leeward areas. A list of common native plant species of the tropandean region is presented in Table 1 with reference to their humidity regime of the tessera in the mountainous landscape, with its horizontal preferential location and its vertical stratification.

The windward to leeward gradients in air mass humidity and cloudiness are well developed along the transect (Figure 5). Decreasing air temperature at night helps to condense water vapor. Although not enough for precipitation to occur, vertical structures (i.e., tree poles, electric cables, antennas) collect

Table 1. Common plant species of the tropandean region, as related to the hydrological regime of the site, the spatial configuration of horizontal areal location, and the vertical stratification

Family	Latin Name	Local Name	Humidity	Locale	Strata
Agavacea	*Agave americana*	penco	Us	Ed	Sh
Amaranthaceae	*Alternanthera* spp.		Us	Gp	Un
	Amarantus hybridus	bledo	Ud	Ed	Un
Amarilidaceae	*Bomarea* spp.	campanitas	Ud	Ed	Un
	Fourcroya spp.	cabuya	Us	Ed	Sh
Anacardiaceae	*Schinus molle*	molle	Me	In	Do
	Bistropogon mollis	tipo	Me	In	Do
Araliaceae	*Oreopanax equatorialis*	puma-maqui	Ud	In	Do
Arecaceae	*Ceroxylum andinum*	palma andina	Ud	In	Do
	Euterpe ensiformis	palma	Ud	In	Un
	Chamadorea poepigiana	palma chica	Ud	In	Un
	Parahubaea cocoides	palma sureña	Me	In	Do
Asteraceae	*Bacharis latifolia*	chilca grande	Me	Gp	Sh
	Bacharis polyantha	chilco	Us	Ed	Sh
	Psoralea mutisii	trinitaria	Me	Ed	Un
	Clibadium eggersii		Me	Ed	Un
	Barnadesia spp.	paloespino	Ud	Gp	Do
	Eupatorium hitchcockii		Ud	In	Do
	Gynoxis halli	piquil	Ud	In	Do
Betulaceae	*Alnus acuminata*	aliso	Me	Ed	Do
	Alnus jorulensis	aliso	Me	Gp	Do
Bignonaceae	*Labium hostifolium*		Ud	In	Do
	Tecoma stans	cholán	Me	In	Do
Boraginaceae	*Tournefortia* spp.	sietecueros	Ud	Gp	Do
	Boconia frutescens	uva de monte	Me	In	Do
Bromeliaceae	*Tillandsia recurvata*	guaycundo	Ud	In	Ep
	Tillandisia usneoides	guaycundo	Us	Gp	Ep
	Puya spp.	achupalla	Ud	Ed	Un
Buddleiaceae	*Buddleia incana*	quishuar	Ud	In	Do
Cactaceae	*Borcycactus sepium*	sirio	Us	Ed	Un
	Opuntia cilindrica	cacto	Us	In	Sh
	Opuntia ficus indica	tuno	Us	Ed	Sh
Caesalpinaceae	*Cassia tomentosa*	chanchilvas	Us	In	Do
Caricaceae	*Carica paniculata*	chaupipapaya	Us	In	Do
Cecropiaceae	*Cecropia obtusifolia*	guarumo plateado	Me	Gp	Do
	Cecropia pentandra	guraumo chico	Ud	Gp	Do
Chlorantaceae	*Hedyosmum* spp.	guayusa de monte	Ud	In	Ep
Clusiaceae	*Clusia dixonii*	ducu	Ud	In	Do
	Vismia obtusa	sangre de gallina	Ud	In	Do
Cunonaceae	*Weinmania descendens*	encino	Me	In	Do
Elaeocarpaceae	*Vallea stipularis*	sacha capulí	Me	In	Do
Ericaceae	*Mackleania popenio*	joyapa	Me	Ed	Sh
Euphorbiaceae	*Croton wagneri*	mosquera	Us	In	Do
	Jantropha spp.	pigllo	Me	Ed	Sh
	Euphorbia laurifolia	lechero	Me	Ed	Do
Fabaceae	*Erythrina edulis*	guato	Me	Ed	Sh
	Casuarina equisetifolia	casuarina	Us	In	Do
Flacourtiaceae	*Tetrathylacum nutans*		Me	In	Do
Gutiferae	*Chrysochlanys dependens*		Me	Ed	Sh
Juglandaceae	*Juglans neotropica*	tocte o nogal	Me	In	Do
Labiaceae	*Sypocampylus giganteus*	fucuna	Ud	Ed	Sh
Lauraceae	*Ocotea puberula*	aguacatillo	Ud	In	Do
	Nectandra spp.	canelo	Ud	In	Do
Melastomataceae	*Miconia longiraremosa*	colca	Me	Ed	Sh

continued

Table 1. *continued*

Family	Latin Name	Local Name	Humidity	Locale	Strata
	Miconia dielsii	colca	Me	Gp	Do
	Miconia crocea	colca	Ud	In	Do
	Brachiotum sedifolium	illinchi	Me	Ed	Do
Meliaceae	*Cedrella montana*	cedro de montaña	Ud	In	Do
	Cedrella rosei	cedro rojo	Ud	In	Do
Mimosaceae	*Acacia pellacantha*	algarrobo	Us	In	Do
	Mimosa quitensis	espino	Me	Gp	Sh
	Prosopis sp.	jaique	Us	Ed	Do
	Inga edulis	guaba	Me	In	Do
	Hesperomeles ferruginea	sacha manzana	Me	In	Do
	Tara spinosa	guarango	Us	In	Do
Myricaceae	*Myrica parvifolia*	laurel de cera	Ud	In	Do
	Myrica grandiflora	laurel	Ud	In	Do
Myrsinaceae	*Rapanea andina*	samil	Me	In	Do
Myrtaceae	*Eugenia mirtelloides*	arrayán	Me	In	Do
	Terminalia spp.	arrayancillos	Me	Ed	Do
Onagraceae	*Fuchsia* spp.	flor de pena	Ud	In	Sh
Passifloraceae	*Passiflora pichinchensis*	taxo	Me	Gp	Ep
Piperaceae	*Piper miersinum*	cordon	Ud	In	Do
	Piper aduncun	cordoncillo	Me	In	Do
Poaceae	*Sporobolus poiretii*	hierba	Us	Gp	Un
	Cynodon dactylum	hierba alta	Me	Gp	Un
	Arundo donax	carrizo	Me	Ed	Un
	Chusquea scandens	sural	Ud	Gp	Un
	Chusquea spp.	suros	Me	Gp	Un
	Cortadeira spp.	sigse	Me	Ed	Un
Podocarpaceae	*Podocarpus oleifolious*	sumi	Ud	In	Do
	Podocarpus rospiglossi	pino andino	Ud	In	Do
Polygalaceae	*Monnina obtusifolia*	igualán	Me	In	Do
Proteaceae	*Embothrium grandiflorum*	gañal	Me	Ed	Sh
Rosaceae	*Prunus serotina salicifolia*	capulí	Me	In	Do
	Polylepis incana	yagual o quina	Ud	In	Do
Rubiaceae	*Condaminea corymbosa*		Ud	In	Do
	Cinchona officinalis	cascarilla	Ud	In	Do
Salindaceae	*Salix humboldtiana*	sauce	Ud	Ed	Do
Sapindaceae	*Dononaea viscosa*	chamana	Me	Ed	Sh
Sapotaceae	*Pouteria lucuna*	lucma	Ud	In	Do
Saxifragaceae	*Escalonia myrtilloides*	sacha peral	Me	In	Do
Simarubaceae	*Picramia dolichobotyra*		Me	In	Do
Solanaceae	*Cestrum stubelii*	sauco andino	Me	Ed	Do
	Cestrum aureum	sauco blanco	Me	In	Do
	Cestrum quitense	sauco quiteño	Me	Ed	Do
	Cestrum tomentosum	sauco negro	Me	Ed	Do
	Brugmansia sanguinea	guantug	Me	Ed	Sh
	Brugmansia arborea	floripondio	Me	Ed	Sh
	Solanum quitoense	lulo	Me	Gp	Sh

Source: Sarmiento (1987).

Codes:

Humidity
Ud: Udic (↑ humidity)
Us: Ustic (↓ humidity)
Me: Mesic (intermediate)

Locale
Ed: Edge (ecotone space)
Gp: Gap-phase (open space of the gap-phase)
In: Interior (protected space of the forest interior)

Strata
Sh: Shrub (intermediate level)
Do: Dosel (higher level of big trees in the canopy)
Un: Understory (lower level of scandent and dwarf flora)
Ep: Epiphyte (epiphytic gardens of the overstory)

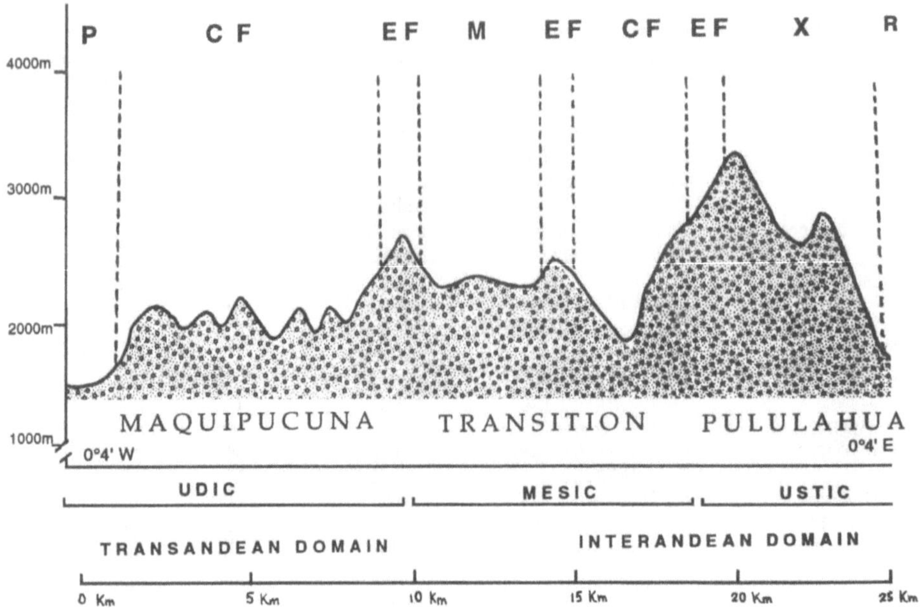

Figure 4. Altitudinal transect along 0°4' N on the equator. Different regions are shown: P: Andean Piedmont, CF: cloud forest, EF: elfin forest, M: mixed forest–agricultural land, X: xerophytic formation, and R: the Guayllabamba River.

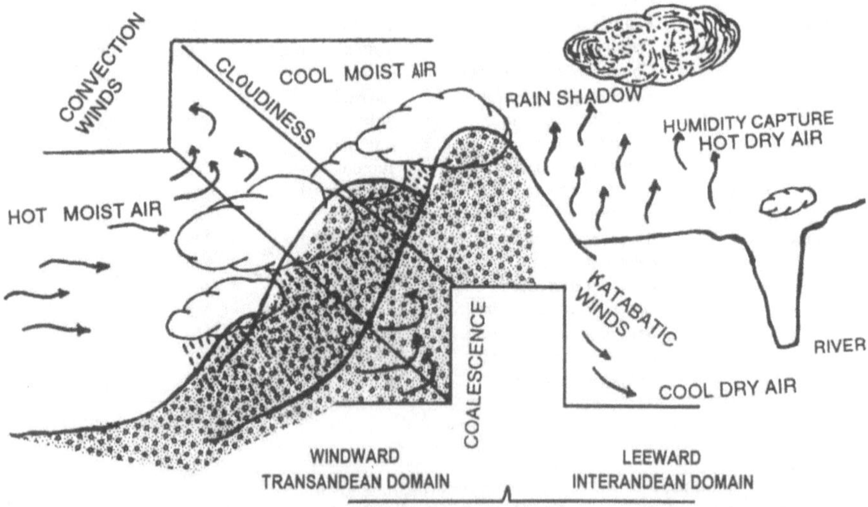

Figure 5. The föhn effect creates the rainshadow area on the leeward of the Equatorial Andes. Evening cooling provides dew and rime to maintain epiphytes in the interandean valley, whereas the morning temperatures generate humidity from the area.

dew or rime; hence, bromeliads, such as *Tillandsia recurvata* and *T. miristicoides*, are seen festooning the older telegraph lines in places such as San Antonio or Pomasqui in the drier basin. In contrast, most of the Acacia trees as well as the tunos (*Opuntia* spp., Cactaceae) are heavily foraged by goats and cattle, in an exotic vista of an African-like savanna in the rainshadow (known as *abrigado*) at Tanlagua. The monthly rainfall regime as climatic parameter between the transandean and interandean domains can be compared in Table 2.

Conclusion

Habitat modification is easily observable in a patchy array of remnants of TMCF and shrubby-woody "islands" in the dry leeward. The patchwork mosaic is influenced heavily by topography, slope, aspect, and soil type, but the current configuration is more strongly influenced by land use practices. The presence of abandoned fields, fallowed marginal agricultural land on steep slopes, secondary forest growth from timber exploitation, and the overall scheme of spatial configuration depend on the relative humidity of the site. The degree of habitat alteration, forest decline, and devastation experienced by the site is correlated with the transportation network, which affects the spatial process of human-induced habitat fragmentation.

Aggressive programs for environmental education at the local level are needed, as well as identification of suitable alternative sources for peasants' income. More basic and applied ecological research of montane ecosystems (Sarmiento 1992) of the tropandean biome is needed to prepare sound management alternatives. Several conservation management ideas can be tested (i.e., agroforestry, ecotourism, pharmacological gardens, nurseries), but all of them require an assumption of the reversibility of environmental damage experienced by the area. Thus, landscape ecological approaches and restoration ecology research can lead the decision-making process toward an amelioration of the tropandean human-caused stress in a sustainable trend for long-term use of resources and maintenance of environmental amenities in the landscape.

Coda

Neotropical conservation has hitherto emphasized the Amazon basin (Redford and Padoch 1992). The dynamics of tropical rain forest fragments, from a biological perspective (Bierregaard et al. 1992), have been developed experimentally for the lowlands. This paper presents descriptive arguments supporting more scientific work in the tropandean region based on the shifted paradigm in ecology (*sensu* Pickett, Parker, and Fiedler 1992), which allows

Table 2. Monthly rainfall regime of selected meteorological stations from INERHI (1987) within the influence of the upper Guayllabamba River basin. Nono is representative of the mesic transition frontier agriculture zone where no records were available. Values of Nanegal in April and September were interpolated

Station Locality	Jan	Feb	Mar	Apr	May	Jun	Jul	Aug	Sep	Oct	Nov	Dec	Total	Zone
San Antonio	33.3	42.5	92.9	88.8	35.7	8.1	2.8	11.3	4.8	47.4	64.5	57.1	489.2 mm/y.	Interandean
Guayllabamba	18.6	43.1	94.6	118.1	56.7	37.5	4.3	27.7	7.9	49.8	80.6	79.6	618.5 mm/y.	Interandean
Nono	115.1	102.2	173.5	207.2	55.8	16.4	1.7	36.1	3.7	92.4	126.9	113	1044.1 mm/y.	Transition
Nanegalito	334.3	233.6	531.1	568.4	450.7	240.3	128.4	63.3	149.1	207.3	204.8	250.3	3361.5 mm/y.	Transandean
Nanegal	215.2	209.3	589.3	479.5	461.6	182.1	162.3	75.7	115.3	184.6	268.6	254.9	3198.4 mm/y.	Transandean

inclusion of humans in the normal landscape dynamics. The biology of remnant montane forest fragments, especially TMCF, still awaits investigation. The management of these fragments still provides an optimistic outlook for restoration practices and still meet some of the needs of the local villagers, which in turn promotes biodiversity conservation, equity, and sustainable development in the tropandean region of the Equatorial Andes.

Acknowledgments. This study was made possible by the endorsement of the Ecuadorian Museum of Natural Sciences and the Maquipucuna Foundation in Ecuador, and research support from the Philadelphia Academy of Natural Sciences and the Institute of Ecology at the University of Georgia. Lawrence Hamilton from the Program on Environment, East-West Center, motivated my participation in the symposium and in the mountain protected area network. The author appreciates comments on earlier versions of the manuscript from Albert Parker, Ronald Carroll, and William Straw. Xavier Viteri has been my great field assistant and friend. Thank you all.

References

Acosta-Solis, M., ed. 1977. *Ecología y Fitoecología.* Casa de la Cultura Ecuatoriana, Quito.

Bierregaard, Jr., R.O., T.E. Lovejoy, V. Kapos, A. dos Santos, and R.W. Hutchings. 1992. The biological dynamics of tropical rainforest fragments. *BioScience* 42 (11): 859-866.

Cañadas, L., ed. 1983. *El Mapa Ecológico y Bioclimático del Ecuador.* Ministerio de Agricultura y Ganaderia, Program de Regionalización Agraria, Quito.

INERHI. 1987. *Reporte hidrológico.* Instituto Ecuatoriano de Recursos Hidraulicos, Quito.

Knapp, G. 1992. *Andean Ecology: Adaptive Dynamics in Ecuador.* Westview Press. Boulder, CO.

Pickett, S.T., T. Parker, and P. Fiedler. 1992. The New Paradigm in Ecology: Implications for Conservation Biology Above the Species Level. In: P. Fiedler and S. Jain (eds.) *Conservation Biology: The Theory and Practice of Nature Conservation, Preservation and Management.* Chapman & Hall. New York.

Redford, K.H., and C. Padoch. 1992. *Conservation of Neotropical Forests: Working from Traditional Resource Use.* Columbia University Press. New York.

Sarmiento, F.O., ed. 1987. *Antología Ecológica del Ecuador. Desde la selva hasta el mar.* Casa de la Cultura Ecuatoriana, Quito.

Sarmiento, F.O. 1988. El Toaza: un valle de esperanza. *Revista Colibri* 2:74-78.

Sarmiento, F.O. 1992. Research in tropandean protected areas of Ecuadorian landscapes. *Forum* 9(3-4): 148-160.

Sarmiento, F.O. In press. Interandean valleys: The most endangered tropandean landscapes. *Mountain Research and Development.*

Steinitz-Kannan, M., P.A. Colinvaux, and R. Kannan. 1983. Limnological studies in Ecuador: 1. A survey of chemical and physical properties of Ecuadorian lakes. *Arch. Hydrobiol.* 65(1): 61-105.

Van Venzen, H. P. 1992. Priorities for conservation of the biodiversity in the Colombian Andes. Novedades Colombianas 4 (Special Issue): 1-32.

21. The Management of Luquillo Elfin Cloud Forest Ecosystems: Irreversible Decisions in a Nonsubstitutable Ecosystem

F. N. Scatena

The uniqueness and importance of tropical montane cloud forest (TMCF) ecosystems have been recognized for several decades (Beard 1944; Daugherty 1973; LaBastille and Pool 1978; Stadtmüller 1987). While their importance has been acknowledged, explicit guidelines for managing these forests are lacking. This paper will review management of elfin cloud forest in the Luquillo Mountains of Puerto Rico to provide insight into the management of similar areas elsewhere (Figure 1). These forests have been managed by the USDA Forest Service since 1917. Forest Service personnel with the Caribbean National Forest (CNF) are responsible for overall management and protection of the forest. Forest Service research scientists with the Luquillo Experimental Forest (LEF) are responsible for conducting and managing research activities within the forest.

Introduction

The elfin cloud forest of the Luquillo Mountains of Puerto Rico is a unique wetland ecosystem that occurs on higher elevation mountain summits. The ecosystem is characterized by short stature trees, endemic rich flora, saturated soils, and relatively slow rates of recovery following disturbance. Since the 1930s, more than 20 buildings, 6.5 km of road and trails, and over $50 million of communication facilities have been built within this ecosystem.

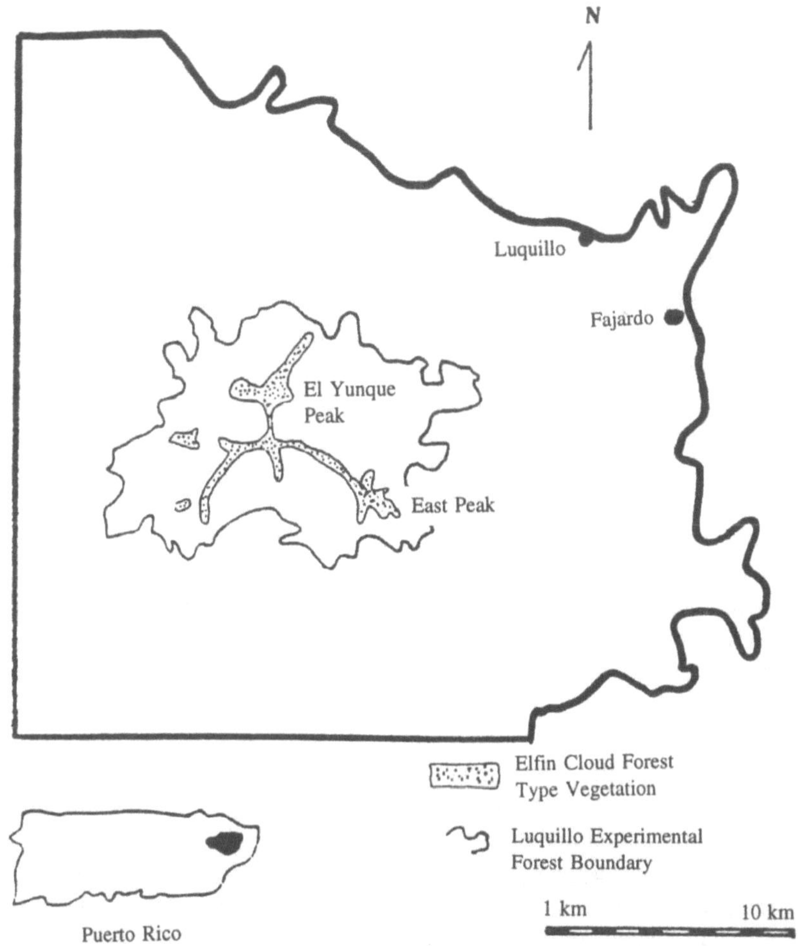

Figure 1. Luquillo National Forest, experimental forest, and extent of elfin cloud forest.

Approximately 83 percent of the original 225 ha of elfin cloud forest remains undisturbed.

Estimates of the total economic and strategic value of this TMCF site range from hundreds of millions to billions of dollars. Two factors contribute to the high importance of these forested lands: (1) the lack of other forests or landform that could be used to accommodate the communication, research, and recreation activities that occur within the ecosystem, and (2) the irreversibility of land use decisions given the slow rate of recovery of these forests following disturbance.

The first section of this paper describes some aspects of the ecology of the

Luquillo elfin cloud forest that directly influence cloud forest management. The second section addresses communication facilities, recreation, water supply, and research activities within the forest. The last section discusses the social and ecological values of this ecosystem.

Site Description

The elfin cloud forest is the smallest of the four primary vegetation types within the LEF/CNF, occupying less than 2 percent of the National Forest (Table 1). It is distinguished here from the Colorado-type forest and palm-type forest that also occur in an upper montane cloud-forest setting. The elfin cloud forest occurs on mountain summits and comprises dense stands of short, small diameter, moss-covered trees and shrubs (Table 2). The windy, cloudy, and rainy characteristic of this forest environment can directly influence management (Table 3). Air humidity is high, saturation deficits and rates of evapotranspiration are low, and plants intercept additional amounts of cloud moisture (Howard 1968; Weaver et al. 1986).

Compared to lower elevation forests, the arborescent vegetation of the elfin cloud forest has a lower stature and lower diversity. Trees are evergreen and have leaves that are typically microphyllous and thick. Epiphytes, mosses, and aerial roots cover nearly all trunks, branches, and twigs (Gill 1969). Most plants are insect- or bird-pollinated (Nevling 1971), and at least 28 endemic plant species exist within this forest type (Weaver 1972b). The closed canopy forest has little undergrowth, but grasses and sedges are found in small gaps, along trails, and in isolated meadows.

The area is underlaid by a clay-rich pedon that has weathered from undisturbed and contact metamorphosed volcanoiclastic and quartz dioritic bedrock (Sieder 1971). On ridgetops the surface soils are organic, root filled, saturated peat that have been classified as loamy, mixed, dysic, tropofolist (Soil

Table 1. Approximate area by land use of the elfin cloud forest ecosystem of the Luquillo Experimental Forest

Land Use	ha	Percent
Undisturbed elfin cloud forest	190	83
Roads and road margins	15	7
Communication facilities	13	6
Research sites	4*	2*
Developed recreation sites	4	2
Total elfin cloud forest	226	100

Areas were measured from 1:20,000 scale topographic maps.

*The Luquillo cloud forests have less than 1 ha of permanent research plots scattered over approximately 4 ha.

Table 2. Some environmental and ecological traits of elfin cloud forests in the Luquillo Experimental Forest

Elevation	900–1,050 m asl
Climate	
Mean annual rainfall	4,500–5,000 mm/year
Air temperature	18°–20°C
Mean relative humidity	99.7%
Windspeed average	18 km/hr
Vegetation	
Total number of woody species	40
Number of woody endemic	18
Tree species/ha	15
Stems/ha	2,651–4,312
Canopy height	3–5 m
Typical dbh range	4–15 cm
Mean dbh growth	0.03 cm/year
Basal area	37–65 m^2/ha
Above ground volume	120 m^3/ha
Above ground woody biomass	80 t/ha
Above ground woody biomass growth	0.45 t/ha/year
Leaf biomass	2.9 t/ha
Leaf area index	3–3.5
Soil organic matter, 0–50 cm	550 t/ha

Additional information can be found in summaries by Howard 1970, Brown et al. 1983, and Weaver and Murphy 1990.

Table 3. Summary of various ecosystem properties and their implication for managing cloud forests in the Luquillo Experimental Forest

Ecosystem Property	Implication for Management
High rainfall and steep slopes	Increases the construction and maintenance cost of roads and trails
Location on exposed summits	Provides unrestricted views which are ideally suited for recreation and communication facilities
Organic-rich soils with numerous macropores	Increases potential for negative impacts associated with soil compaction and changes in subsurface flow patterns
Short stature and contorted shape of canopy trees	Reduces commercial value of timber resources
Slow rate of succession	Dictates that restoration of degraded areas may require decades or centuries
Importance of root sprouts in regeneration	Limits viability of restoring areas with damaged soil or root stocks
High endemism and low species diversity	Increases possibility that management activities will affect rare or endangered species

Conservation Service 1987). These peats thin away from the ridge crest and eventually grade into mineral soils on slopes. Macropores and pipes are common and suggest a significant amount of shallow subsurface flow. Although several major rivers have their headwaters in the elfin cloud forest, there are no permanently flowing streams within the forest type. Landslides are common features along roadways but are rare in the undisturbed forest.

Succession of vegetation following disturbance is extremely slow (Byer and Weaver 1977; Weaver 1990) and reduces the viability of restoring degraded areas (Table 3). The average rate of above-ground biomass accumulation for this forest type was estimated at 0.45 t/ha/year and is slower than any forest type within the LEF (Weaver and Murphy 1990). At this rate it would take approximately 175 years to accumulate the 80 t/ha of above-ground biomass of the mature forest. After 18.5 years of monitoring a site that had been cleared and burned by an airplane crash, it was estimated that it could take two centuries for the site to return to its original biomass (Weaver 1990). The slow rate of above-ground biomass accumulation, limited successional pathways, and the dominance of root sprouting in primary species have been observed in other upper montane forests (Moll 1972; Sugden, Tanner, and Kapos 1985) and may be a general property of these ecosystems.

Current Land Use within the Luquillo Elfin Cloud Forest Ecosystem

Many tropical cloud forests have been directly or indirectly modified by harvesting for timber and nontimber products, grazing, burning, water diversions, and agriculture (Moll 1972; LaBastille and Pool 1978; Stadtmüller 1987). In the Luquillo Mountains, these activities have not significantly influenced the elfin cloud forest ecosystem. Apparently, the forest's low volume of commercial timber, inaccessibility, harsh climate, and strategic value have provided an effective barrier against these land uses. However, communication facilities and recreation and research activities have impacted these forests.

The first major structure within the elfin cloud forest was a recreational lookout tower that was constructed in the 1930s by the Civilian Conservation Corps. During World War II, radar facilities and single-lane access roads were built on an adjacent peak. In the early 1960s additional communication facilities and a two-lane road were constructed on a third peak. Today, more than 20 buildings, 6.5 km of road and trails, and over $50 million of communication facilities exist within the elfin cloud forest ecosystem.

Communication Facilities

Approximately 13 percent of the Luquillo elfin cloud forest has been replaced with communication facilities and their associated roads (Table 1). These facilities carry most of Puerto Rico's overseas telephone communications and a substantial portion of the domestic telephone, television, and radio service. Approximately 200 users, including the Federal Aviation Administration, the military, television, radio stations, and cellular telephone companies, have communication equipment within the cloud forest. Unlike other uses, these facilities are located within the cloud forest because of the elevation and exposure of the mountain summits, not because of any property of the elfin cloud forest ecosystem itself.

The use of the area as a site of modern communication facilities began in World War II. By 1962 electromagnetic congestion became so intense that the Forest Service insisted on forming an association to coordinate use, establish sound design standards, and designate special receiving and transmitting areas. However, due to continued growth and demand for space, the Forest Service placed a moratorium on development in the early 1970s, and the first of several plans for site management and consolidation were developed. The major considerations and operating rules that currently govern the Forest Service's management of these facilities are summarized in Table 4.

Table 4. Summary of considerations and guidelines governing the management of electronic communication facilities in the cloud forest ecosystem of the Luquillo Experimental Forest

1. Communication facilities are an important use of these areas because of the favorable location of the site and the public value of these facilities. Therefore, the Forest Service supports multiple use of the area and manages the site to accommodate communication facilities in a manner consistent with proper land management and the overall uses of the forest.
2. The Forest Service maintains the responsibility for the quality of the lands where the facilities are located and, when necessary, acts as a mediator between users. The Forest Service is not primarily concerned with problems of frequency interference, maintenance, or security of facilities.
3. The cloud forest ecosystem is recognized as a unique and sensitive ecosystem whose climate, steep slopes, and unique vegetation limit the potential for development. Therefore, the Forest Service will manage to minimize adverse environmental impacts of existing users and limit future use to licensees who demonstrate that off-forest sites will not meet their needs and that the LEF is the only feasible site.
4. The Forest Service recognizes that with use of modern technology the amount of area needed for electronic facilities is decreasing. Therefore, Forest Service policy is to promote the modernization and consolidation of existing facilities and the restoration of degraded lands.
5. The summits of the LEF are a dominant landscape feature in the eastern part of Puerto Rico and an important recreation and cultural resource. Therefore, the Forest Service will manage to preserve and restore the natural characteristics of the peaks as a scenic resource by promoting the reduction in the visibility of facilities and the placement of utilities underground.
6. All activities must be in compliance with the National Environmental Act regulations and the Master Plan of the Caribbean National Forest.

Additional details and design specifications can be found in CNF (1991).

Today, all electronic users are members of a user association and are subject to explicit Forest Service guidelines that govern the areas designated for electronic facilities (Electronic Site Plan, CNF 1991). Except for noncommercial purposes of the military and Federal Aviation Administration, all users are required to pay annual user fees based on the amount of space and equipment they have within the forest.

During the past 50 years that communication facilities have been operating in the LEF, there has been considerable improvement in telecommunication technology. The early tube-based equipment required large buildings, substantial cooling systems, and multiple antennas. Modern digital, fiber optic, and ceramic technology is compact and can significantly reduce electromagnetic interference, antenna size, and the amount of space needed for operation (Watson 1991). For example, with current technology as many as 30 systems can use the same antenna compared to the previous 4 systems per antenna limit (Watson 1991). With improvements in technology and emphasis by management on consolidating existing infrastructure, the extensive impact of communication facilities on the forest is expected to decline as existing equipment is replaced. In the future, the present array of 15 buildings and numerous multistoried towers will probably be replaced by one, nearly invisible below-ground complex and a few ground-level antennas (Watson 1991).

The challenge for future management is expected to focus on balancing the desire to consolidate facilities for aesthetic purposes with the need to realize the capital investment of the existing infrastructure. An additional challenge will involve the reforestation of abandoned electronic facilities. However, given the slow rate of regrowth of these forests, this may take decades or centuries to achieve.

Roads

All roads within the Luquillo cloud forest ecosystem are paved and were constructed to provide access to communication facilities. These roads are closed to public traffic but are used for guided tours, access to research sites, and hiking. Because of the difficulties encountered in building and maintaining these roads, they have been cited as examples of the construction problems that can occur in this environment.

Although the Forest Service raised considerable concerns about the project (F. Wadsworth, pers. com.), in 1960 the U.S. military commissioned a private contractor to build a 6 km road through the cloud forest to their new communication facilities (Sowers 1971). The initial design was typical of roads in mountainous areas: the center line maintained a flat grade by following the sides of the mountain, and the material excavated on upslope cuts was to be used as fill on the downhill side of the road. Since the Forest Service objected to having extensive borrow pits within the National Forest, the design envisioned that the volume of material excavated would balance the amount of material needed for fill.

Unfortunately problems were encountered during construction, which rendered the initial design obsolete, and the entire project had to be redesigned. The final road required 3 years instead of 8 months to complete, cost more than twice the original estimate, and resulted in extensive litigation between the contractor and the government. Landslides, some as large as 75,000 m^3, were the major problem. In addition, some of the excavated soils remained so wet and unstable they could not be used for fill; a tractor sank 5 m in soft "jelly-like" saturated silt; and blasting triggered rock falls in unexpected areas.

Eventually the road was completed. Corrective measures included realigning the road to avoid particularly unstable areas, flattening cut slopes, establishing deep-rooted grass on exposed areas, and trimming bedrock slopes to fit the geometry of the bedding plane. However, adjusting the design to account for all the problems was not possible, and slope failures are still a chronic and expensive problem. In addition, areas that were disturbed during construction are still covered with grasses and shrubs and have not regenerated with primary cloud forest species.

It is unlikely that additional roads will be built in the Luquillo elfin cloud forest. The present and future challenge to managers will be to reduce the maintenance costs of these roads and to establish native vegetation along existing road corridors.

Research

Because of their unusual ecosystem properties, TMCFs have received disproportionate study relative to their spatial extent and global distribution. Much of this research has been motivated by the abundance of rare and endemic species and the unique nutrient cycles and physiological traits of these forests. The close association of these forests with global air masses also makes them ideal locations for detecting and monitoring regional and global processes (Lugo and Scatena 1992).

In the LEF, cloud forest research has been limited to a few areas (Table 1). Although most of the research has been observational, several research impacts are recognizable. These include soil compaction and local changes in drainage due to trampling, introduction of grasses and sedges along foot paths, and accumulation of debris. While the spatial extent of these disturbances is small, their impact is compounded by the forests' inherently slow rate of recovery. For example, foot trails that were established during the mid 1960s are still recognizable today, although they have been essentially abandoned for nearly 25 years.

To mitigate and reduce research impacts, Forest Service research managers are consolidating research activities into a few areas and are implementing protocols and guidelines for researchers (Table 5). Due to the sensitive nature of this ecosystem, these guidelines are much more stringent than guidelines for conducting research in other areas of the LEF.

Table 5. Summary of guidelines that govern research activities within the cloud forest ecosystem of the Luquillo Experimental Forest

1. The cloud forest ecosystem is recognized as a unique ecosystem that merits significant research. However, the area's climate, steep slopes, and unique vegetation limit the opportunities for extensive development. Therefore, the policy of Forest Service research is to minimize adverse environmental research activities and to limit research to those who demonstrate that other forest sites will not meet their needs.
2. To obtain a research permit, all studies must have detailed study plans that are reviewed by Forest Service research scientists and outside peers. Projects will be reviewed on their scientific merit, design, and potential for impacting the cloud forest ecosystem.
3. All study plots must be located within designated research areas, be geographically referenced, and be well marked and maintained.
4. All studies must be in compliance with the National Environmental Act regulations and the Master Plan of the Caribbean National Forest.

Recreation

Recreation is an important use of the CNF/LEF. At least 1.75 million people live within a 1.5-hour drive of the forest and more than 1.2 million people pass their daily lives within sight of the mountain's summits. The coolness and mistiness of the elfin cloud forest climate and the stunted, contorted, moss-covered vegetation make them a favorite recreation site within the forest. Approximately 750,000 people per year recreate within sight of this distinct vegetation and 5,000–10,000 people per year actually visit the ecosystem (CNF 1991).

Currently all of the major peaks within the forest experience some recreational use. Most of this use consists of short visits to observe the forest and surrounding landscape. The collecting of bromeliads and wild raspberries also occurs. Heavy rains, strong winds, and lack of level ground limit opportunities for camping, picnics, or other recreational activities. Existing facilities include an observation tower and paved trails.

Regardless of the attraction of the elfin ecosystem, recreation is not without its problems or limitations. For example, trampling of unpaved areas by guided tours became so apparent that in 1989 the Forest Service requested that all groups limit their visit to paved recreation areas. For all forms of recreation, the greatest problem appears to be associated with trampling of the forest floor and the potential damage to regeneration. The presence of grasses along the trails also indicates the possibility of introducing exotic species into the forest. The possibility of litigation resulting from hikers slipping and falling on the slick moss-covered surfaces could also be a potential problem.

Due to environmental and safely problems associated with recreation within the cloud forest, the present trails are paved and have individual drainage ditches to route runoff. Most of these trails were built by hand in the 1930s by workers with the Civilian Conservation Corps. Today the construction of these trails cost approximately 2.4 times as much as unpaved trails in the lower forest (pers. com. 1993, CNF Forest Engineer).

Future challenges for recreation managers will be to protect the aesthetic and ecological values of the cloud forest ecosystem while allowing the public to experience this unusual ecosystem. Present solutions include limiting visits to existing paved trails and roadways.

Water Supply

The Luquillo elfin cloud forests may have the highest water yield per unit area in Puerto Rico and are considered an important source of water (Weaver 1972b). No water intakes occur within the ecosystem, but water intakes do occur at lower elevations. Given that the area receives an average rainfall of 4,500 mm/year and that cloud interception can be equal to transpiration losses (Weaver 1972b), the annual water yield from the cloud forest to the lower elevation forest area may exceed 10 million m^3. If the total exploitable use of this water were limited to 50 percent, to account for instream flow needs and other ecological and hydrological abstractions, at average domestic prices (Nieswiadomy 1992) the exploitable water could have a market value between \$2.5 and \$4.5 million per year. The future challenge to managers will be to maintain the quality of this water as well as the local ecosystem benefits that accrue from having it flow freely to the ocean.

Social and Economic Value of the Luquillo Elfin Cloud Forest Ecosystem

The monetary value of an ecosystem is difficult, if not impossible, to calculate or define. This is particularly true for Luquillo elfin cloud forests whose dominant uses are nonmarket activities related to information transfer, strategic defense, and recreation. Nevertheless, the relative importance of different land use activities can be ascertained from estimates of the present monetary value of these activities (Table 6). Regardless of the uncertainties involved, it is clear that the strategic and economic value of the communication installations is orders of magnitude greater than the economic activity associated with recreation, research, and water supply.

In addition to the monetary value associated with the cloud forest ecosystem, these forests are public goods that have a value to the thousands of people who recreate within sight of their distinct vegetation and to the millions of people who consider them part of their cultural heritage and depend on them for communication and national security. Two additional factors influence the value of this unique ecosystem: the lack of viable substitutes and the irreversibility of land use decisions.

Table 6. Estimated present value of various nonmarket activities in the Luquillo cloud forest and the assumptions used in estimating the nonmarket values (in thousands of 1992 dollars)

Activity	Assumptions	Present value in $1,000
Communication		
Strategic value	The Luquillo cloud forest is the only possible site for the facilities, and the Puerto Rican economy is completely dependent on the facilities	Possibly billions
Investment value	Present value of $50 million investment at a 10% discount rate over 20 years	336,000
Water	50% of 4,500 mm/year at average domestic prices	2,500–4,500
Research	Total annual research expenditures	200–300
Recreation		
Visitors who enter the cloud forest	Transportation cost of $8/person/trip for the 5,000–10,000 persons/year	40–80
Visual recreation	Transportation costs of $8/person/trip for the 750,000 people/year who visit close enough to distinguish the cloud forest from other forest types	6,000

Substitutability

The value of any resource is not only related to supply and demand, but is also dependent on the quantity and quality of other resources that can be used to substitute for a particular commodity (Mansfield 1985). Much of the inherent value of the Luquillo elfin cloud forest land results from its limited supply and the lack of other sites that could be used to accommodate the communication, research, and recreation activities that occur within the ecosystem.

Likewise, recreationists in Puerto Rico do not have readily available substitutes for the cool climate, moss-covered vegetation, or cloud-shrouded views offered by these forests. Finally, scientists cannot substitute study areas in lowland forests for study areas in the endemic-rich, ecologically distinct cloud forests, nor can resources from plantations be used to augment or substitute for the products of the elfin cloud forests.

Irreversibility

In most ecosystems, the consequences of many land use decisions are somewhat reversible given sufficient inputs of energy and time. However, when a decision cannot be readily adjusted to meet some future desire or provide the amenities of the original environment, then the decision can be considered as permanent and irreversible (Krutilla and Fisher 1975). Such irreversible development carries an additional cost associated with the loss of options and opportunity.

Due to the unique gene pools, small size, and slow recovery of the Luquillo elfin cloud forest ecosystem, activities that remove elfin forest cover may be considered irreversible. Moreover, the loss of scientific and genetic information that results from the loss of TMCF is essentially irreversible since no other ecosystem has similar attributes. The loss of recreation opportunities is also irreversible since no other location offers the climatic or visual amenities of the cloud forests, nor the undisturbed, virgin nature of the forest.

Because the recovery of TMCF may take decades or centuries, restoration or reforestation projects cannot be considered as simple remedies for correcting the impact of inappropriate decisions that disturbed the original ecosystem. Even if they can be achieved in socially acceptable time periods, the loss of authenticity associated with being in an undisturbed ecosystem would be permanently lost from the recreational experience.

References

Beard, J.S. 1944. Climax vegetation in tropical America. *Ecology* 25:127-158.

Brown, S., A. El Lugo, S. Silander, and L.L. Liegel. 1983. Research history and opportunities in the Luquillo Experimental Forest. General Technical Report SO-44. USDA Forest Service Southern Forest Experiment Station. New Orleans. 128 pp.

Byer, M.D., and P.L. Weaver. 1977. Early secondary succession in an elfin woodland in the Luquillo Mountains of Puerto Rico. *Biotropical* 9(1): 35-47.

CNF (Caribbean National Forest). 1991. General Report to the Public 1991, Caribbean National Forest. USDA Forest Service, Rio Piedras, Puerto Rico. 29 pp.

Daugherty, H.E. 1973. The Montecristo cloud-forest of El Salvador—A chance for protection. *Biological Conservation* 5(1): 227-230.

Gill, A.M. 1969. The ecology of an elfin forest in Puerto Rico. 5. Aerial roots. *J. Arnold Arboretum* 50(2): 197-209.

Howard, R.A. 1968. The ecology of an elfin forest in Puerto Rico. 1. Introduction and composition studies. *J. Arnold Arboretum* 50:225-267.

Howard, R.A. 1970. The ecology of an elfin forest in Puerto Rico. 10. Notes on two species of Marcgravia. *J. Arnold Arboretum* 51(1): 41-55.

Krutilla, J.V., and A.C. Fisher. 1975. *The economics of natural environments*. John Hopkins Press, Baltimore.

LaBastille, A., and D.J. Pool. 1978. On the need for a system of cloud-forest parks in Middle America and the Caribbean. *Environmental Conservation* 5(3): 183-190.

Lugo, A.E., and F.N. Scatena. 1992. Epiphytes and climate change research in the Caribbean: A proposal. *Selbyana* 13:123-130.

Mansfield, E. 1985. Microeconomics: Theory and applications, 5th ed. W.W. Norton & Company, Inc., New York. 590 pp.

Moll, E.J. 1972. The current status of mistbelt mixed Podocarpus forest in Natal. *Bothalia* 10(4): 595-598.

Nevling, L.I., Jr. 1971. The ecology of an elfin forest in Puerto Rico. 16. The flowering cycle and an interpretation of its seasonality. *J. Arnold Arboretum* 52(4): 586-613.

Nieswiadomy, M.L. 1992. Estimating urban residential water demand: Effects of price structure, conservation, and education. *Water Resources Research* 28(3): 609-616.

Sieder, V.M. 1971. Geologic map of the El Yunque Quadrangle, Puerto Rico. Misc. Geol. Maps I-658, scale 1:20,000.

Soil Conservation Service. 1987. Puerto Rico soil characterization and description report. Unpublished report.

Stadtmüller, T. 1987. *Cloud forests in the humid tropics*. The United Nations University Press, Tokyo, Japan. 81 pp.

Sugden, A.M., E.V.J. Tanner, and V. Kapos. 1985. Regeneration following clearing in a Jamaican montane forest: Results of a ten-year study. *J. Tropical Ecology* 1:329-351.

Watson, J. S. 1991. El Yunque communication and consolidation report. Unpublished report for the Caribbean National Forest.

Weaver, P.L. 1972a. Cloud moisture interception in the Luquillo Mountains of Puerto Rico. *Caribbean J. Science* 12(3-4): 129-144.

Weaver, P.L. 1972b. The dwarf forest of Pico del Este and the Luquillo Mountains. *Revista Interamerican Review* (2): 174-186.

Weaver, P.L. 1990. Succession in the elfin woodland of the Luquillo Mountains of Puerto Rico. *Biotropica* 22(1): 83-89.

Weaver, P.L., E. Medina, D. Pool, K. Dugger, J. Gonzales-Liboy, and E. Cuevas. 1986. Ecological observations in the cloud forest of the Luquillo Mountains in Puerto Rico. *Biotropica* 18(1): 79-85.

Weaver, P.L., and P.G. Murphy. 1990. Forest structure and productivity in Puerto Rico's Luquillo Mountains. *Biotropica* 22(1): 69-82.

22. Lichens as Indicators of Cloud Forest in Hawai'i

Clifford W. Smith

Lichens are extremely sensitive indicators of microclimatic conditions (Rose 1976). Their use as bioindicators of air quality is well established (Huckaby 1993). Several species are powerful indicators of substrate types and conditions (Barkman 1958).

Fog-zone vegetation in many parts of the world (e.g., California, Peru, Namibia) is characterized by a significant biomass of fruticose (bush- or beardlike growth) lichens (Redon and Lange 1983; Rundel 1978). Lichens are ideally suited to cloud- and fog-shrouded environments because:

1. lichens are capable of absorbing water vapor from fog or cloud even though little or no condensation occurs (Lange and Kilian 1985);
2. fruticose lichens have a high surface area to volume ratio that maximizes interception and absorption; and
3. the stability of lichen symbiosis is dependent on, or at least highly tolerant of, repeated dessication/hydration events.

TMCF typically may exhibit a high biomass of epiphytic fruticose lichens except under extremely wet conditions when they are replaced by bryophytes.

This paper describes the cryptogamic flora, with an emphasis on lichens, in three broad habitat types in Hawai'i, where cloud contributes a significant quantity of water to the ecosystems. Such environments are notably different from altitudinally adjacent areas where frequent ground-level cloud is not a significant factor.

The three habitat types examined here include:

1. Seasonal Forests—where moisture deficits occur for at least 3 months each year;
2. Montane Rainforest—where the rainfall exceeds 100 mm in every month; and
3. Montane Mesic Forest—where moisture deficits may occur for short periods, and mean monthly rainfall exceeds 100 mm for only half the year.

Montane Seasonal Forest

The correlation of the abundance of fruticose lichens and the frequency of ground-level cloud can be seen by following an elevation gradient up the south slope of Haleakalā, East Maui, Hawai'i. The area is in the leeward rain shadow of the mountain but subject to afternoon cloud buildup in the altitudinal zone between 1,300 and 2,000 m. Rainfall is concentrated in winter months (December and March), but ground-level cloud occurs throughout the year. Average annual rainfall on this leeward slope ranges from 500 mm at the coast to 1,000 mm at the 1,300 m contour (Giambelluca et al. 1986). The vegetation in this lower zone is scrubland with interspersed herbs and has been modified heavily by human activity (fire, cattle grazing, etc). Dry forest trees are increasingly common above 800 m principally on a'a lava flows (Medeiros et al. 1986). The lichen communities are similiarly poorly developed other than crustose lichens, particularly species of *Buellia, Caloplaca, Dirinaria,* and *Lecidea,* most of which are saxicolous. A few small, corticolous foliose species also occur, their frequency increasing with elevation. Fruticose lichens (e.g., diminutive specimens of *Ramalina ?intermedia*) are occasionally found in highly protected, shaded habitats, particularly in rock clefts. Bryophytes are uncommon and generally ephemeral.

At 1,200 m, the lichen flora changes somewhat abruptly. Over a 100 m increase in elevation, the abundance of the orange, fruticose lichen *Teloschistes flavicans* increases to the point that the color is observable from a distance. The abrupt increase in lichen biomass is almost certainly the result of frequent ground-level cloud at this elevation. Elsewhere in Hawai'i *Teloschistes* is normally found well below this elevation, but along this transect there is insufficient moisture except where the cloud is present. Lichens pay a high energy cost in alternately dehydrating and rehydrating (Smith and Molesworth 1973). There is a point below which the frequency and duration of rehydration events do not compensate for the energy lost during the phase change.

No corresponding boundary exists in the angiosperm vegetation. In fact, it is difficult to delineate vegetation types on this Haleakalā slope other than those associated with substrate differences (i.e., lava flow age and type).

At 1,300 m the stature of the vegetation changes to forest and the number of angiosperm species increases. Hawaiian dry forests have the highest diversity of angiosperms in the islands (Rock 1913). These forests also have a very high number of lichen species, close to 100 in this seasonal cloud forest on Haleakalā.

Above 1,300 m other fruticose species, particularly green and yellow *Usnea* and *Ramalina* spp., become extremely abundant. *Ramalina soraligera, Usnea entoviolata*, and *U. osseoalba*, all endemic, and the indigenous *R. farinacea, U. australis*, and *U. rubicunda* displace *Teloschistes flavicans* to a considerable degree. Foliose species are not common except on exposed rocks which are covered by a variety of species, principally *Rimelia reticulata* and *Dirinaria aegialata.* The cyanobacterial phytobiont (blue-green algal symbionts) species are not a significant component of the community. Their thalli are small, not well developed, and confined to the interior of shrubs in protected gullies or rock crevices. Tussock mosses, plus the carpet moss *Racomitrium lanuginosum*, and liverworts (principally *Frullania* spp.), though present, represent a minor element of the community.

Interception, condensation, and fog-drip are further demonstrated by the abundance of another fruticose lichen, the terricolous *Cladina skottsbergii,* plus the foliose *Rimelia reticulata* under trees, though the association has been obscured by the invasion of the aggressive, matt-forming grass *Pennisetum clandestinum* (Kikuyu grass). The cyanobacterial lichen *Pseudocyphellaria crocata* can be abundant in sunken pockets where, once wetted, it is protected from the drying effects of the wind.

Above the inversion layer at 2,000 m, the annual rainfall decreases with increasing elevation. The lichen biodiversity decreases and fruticose species become a minor element of the vegetation. Bryophytes continue to be poorly represented.

Above the inversion layer on the windward side where clouds move upslope during the day or spill over the crater rim, fruticose lichens are still common though the species composition is very different. Here, dark to light brown species predominate, principally the almost black *Bryoria smithii* and the lighter *B. lanestris* where the cloud is more frequent. Their distribution correlates very well with the path of cloud movement. They are cryptic, their color blending with the cyanobacteria-covered twigs of *Styphelia tameiameia* and *Vaccinium* spp. *Cladina skottsbergii* and *C. leiodea*, two endemic reindeer lichens, are common under shrubs where they receive fog-drip. Again, corticolous bryophytes are scarce and confined to shaded or protected areas, but saxicolous carpet mosses are commonly interspersed with the *Cladina*.

Montane Mesic Forest

The general pattern in mesic forest enveloped frequently in cloud is that fruticose lichens are not abundant. They are replaced by foliose species, particularly *Hypotrachyna* spp., together with carpet mosses. *Rimelia reticulata,* another foliose species, is also abundant and frequently fertile. This represents an unusual condition for this species, which normally reproduces asexually and is the most common lichen in the islands. In one area, the local endemic *Hypotrachyna longiloba* mimics the fruticose habit by hanging from the branches. Cyanobacterial lichens are common though not abundant, principally species of *Leptogium*, *Pseudocyphellaria*, and *Sticta.* Liverworts are represented by numerous species but their biomass is low. Montane mesic cloud forest, therefore, is not characterized by an abundance of any growth form or phytobiont. It differs from areas not enveloped in cloud principally because of the greater biomass of species.

Montane Rain Forest

In rain forest environments liverworts and pleurocarpic mosses dominate. Fruticose species are absent except in certain special situations. The lichen community is dominated by foliose cyanobacterial phytobiont species, one of which, *Pseudocyphellaria crocata,* is large and generally pendent. Corticolous lichen biodiversity is low, less than 20 species. A terricolous fruticose species, *Stereocaulon ramulosum*, occurs in great quantities on exposed rocks of cliffs or landslides. The pseudopodetia are much more well developed than elsewhere, which led Tuckerman (1874) to describe this form as a new species, *S. pilophoroides.* Snags or dead branches are covered with *Usnea australis* to a greater or lesser degree depending apparently on the degree of exposure.

There is, however, considerable variation in the lichen and bryophyte epiphytic flora in the rain forest. A comparison of two rain forest areas on Maui—one on the north face of Haleakalā, west of Olinda, the other on the southeast slope in Kipahulu Valley, both at 1,300 m elevation and receiving an estimated 5,000 mm rainfall every year—illustrates the difficulty of interpreting the floristic data. The Olinda forest, which is exposed directly to the prevailing tradewinds and has an open *Metrosideros* canopy with *Dicranopteris linearis* as a dominant understory species, has 11 species of foliose lichens. The dominant epiphytic species are *Pseudocyphellaria crocata* and the pendent carpet moss *Aerobryopsis longissima.* The sheltered Kipahulu forest with a denser canopy has a more diverse understory of ferns. Only two foliose lichens are present and neither is abundant. The epiphytic flora is characterized by the abundance of liverworts, particularly *Bazzania* and *Herberta* spp.

Discussion

The characteristics of the cryptogamic flora of the three forest types in Hawai'i that are influenced by cloud is summarized in Table 1. Lichen epiphytes dominate in dry habitats, bryophytes in rain forest. All are poikilohydric. Lichens with chlorophyte phytobionts (green algal symbionts) are best adapted to the dry forest because photosynthesis is activated by water vapor alone (Lange and Kilian 1985; Lange, Kilian, and Ziegler 1986; Lange et al. 1993). They are soon smothered by bryophytes in wetter areas though fruticose and pendent foliose species are able to escape the overgrowth once established. Species with cyanobacterial phytobionts must be wetted in order to resume photosynthesis. They are more suited, therefore, to shaded mesic and rain forest environments, dominating the lichen flora in rain forests.

Forests that grow in dry seasonal habitats (e.g., mediterranean ecosystems) that are shrouded in frequent cloud can be found in many parts of the world (e.g., Hawai'i, Canary Islands, Mexico, Brazil). They have a high species diversity and an abundance of lichens. They also frequently have a high diversity of angiosperms. Though they may not be cloud forests in the classical sense (i.e., elfin forests with peaty soils and abundant epiphytes), they are cloud-modified ecosystems, a major portion of their water coming from cloud interception. They are rare ecosystems today in part because most are highly suitable for cattle ranching. Many have been destroyed. The remainder are in urgent need of protection.

Table 1. The abundance, dominant growth form, and species diversity of cryptogams in Hawaiian cloud forests

	Plant Type			
	Lichens		Bryophytes	
Habitat	Chlorophyt	Cyanophyt	Mosses	Liverworts
Seasonal				
Abundance	High	Low	Low	Low
Growth form	Fruticose	Foliose	Cushion	Appressed
Diversity	High	Low	Low	Low
Mesic				
Abundance	Medium	Low-medium	Medium	Medium
Growth form	Foliose	Foliose	Carpet	Carpet
Diversity	Medium	High	High	High
Rain forest				
Abundance	Low	Low-medium	High	High
Growth form	Foliose	Foliose	Carpet	Turf
Diversity	Low	Medium	Medium	Medium

References

Barkman, J. J. 1958. *Phytosociology and ecology of cryptogamic epiphytes.* VanGorcum, Assen.

Giambelluca, T. W., M. A. Nullet, and T. A. Schroeder. 1986. *Rainfall atlas of Hawai'i.* Hawai'i Department of Land and Natural Resouces Report R76. Honolulu.

Huckaby, L. S. 1993. *Lichens as bioindicators of air quality.* USDA Forest Service General Technical Report RM-224. Rocky Mountain Forest and Range Experiment Station, Fort Collins, CO.

Lange, O. L., and E. Kilian. 1985. Reaktivierung der Photosynthese trockener Flecten durch Wasserdampfaufnahme aus dem Luftraum: Artspezifisch untershiedliches verhalten. *Flora* 176:7–23.

Lange, O. L., B. Büdel, A. Meyer, and E. Kilian. 1993. Further evidence that activation of net photosynthesis by dry cyanobacterial lichens requires liquid water. *Lichenologist* 25:175–189.

Lange, O. L., E. Kilian, and H. Ziegler. 1986. Water vapour uptake and photosynthesis of lichens: Performance differences in species with green and blue-green algae as phycobionts. *Oecologia* (Berlin) 71:104–110.

Medeiros, A. C., L. L. Loope, and R. A. Holt. 1986. *Status of native flowering plant species on the south slope of Haleakalā.* Cooperative National Parks Resources Studies Unit, University of Hawai'i, Technical Report 59.

Redon, J., and O. L. Lange. 1983. Epiphytische Flechten im Bereich einer chilenischen "Nebeloase" (Fray Jorge). I. Vegetationskundliche Gliederung und Standortsbedingungen. *Flora* 174:213–243.

Rock, J. F. 1913. *The indigenous trees of the Hawaiian Islands.* Reprinted by C. E. Tuttle, Tokyo, 1974.

Rose, F. 1976. Lichenological indicators of age and environmental continuity in woodlands. In *Lichenology: Progress and problems,* edited by E. D. H. Brown, D. L. Hawksworth, and R. H. Bailey, p. 279–307. Systematics Association, Special Vol 8. Academic Press, London.

Rundel, P. W. 1978. Ecological relationships of desert fog zone lichens. *The Bryologist* 81:277–293.

Smith, D. C., and S. Molesworth. 1973. Lichen physiology. XIII. Effects of rewetting dry lichens. *New Phytologist* 72:525–533.

Tuckerman, E. 1874. Lichens. In *United States exploring expedition during the years 1838–1842, under the command of Charles Wilkes, U. S. N.* Vol. 17 *Botany.* I. *Lower Cryptogamia.* Philadelphia.

23. Cloud Forest Archipelagos: Preservation of Fragmented Montane Ecosystems in Tropical America

José Antonio Vázquez-García

The present size and distribution of tropical montane cloud forests (TMCFs) in tropical America demand special strategies for their preservation (LaBastille and Pool 1978; Vázquez-G. 1989b). TMCFs and other montane forests are known to display a discontinuous pattern (Figures 1, 2) analogous to archipelagos (Whittaker 1968; Luna-V. et al. 1989; Iltis, pers. com.) or to chains of island-like habitats (Vázquez-G. 1989b). Large preserves are usually considered effective for extensive and continuous ecosystems, and for faunal species with extensive home ranges (Shafer 1990). However, large preserves may be inappropriate for or poorly represent a regional, provincial, or even local compositional spectrum of discontinuous and unique (endemic) entities. Researchers generally agree on the discontinuous nature of cloud forests throughout the American tropics and the high endemism present in these ecosystems (Rzedowski 1978). However, current conservation efforts rarely consider the uniqueness and discontinuity of these island-like ecosystems when establishing protected areas. A clustering of CFs from northern México (Figure 1) to southern Panama and the Caribbean (Figure 2) into natural subdivisions should facilitate understanding of these mostly remnant systems and could help in defining conservation priorities with the eventual aim of integrating them into a balanced international conservation network (LaBastille and Pool 1978; Vázquez-G. 1989a).

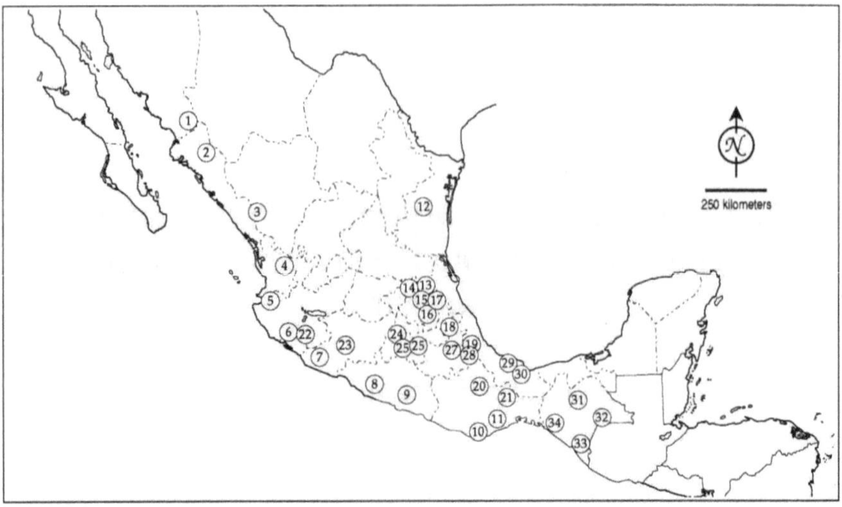

Figure 1. Montane chains of island-like cloud forest habitats in México.

NW México: 1. Saguaribo, Sonora; 2. Surotato, Sinaloa; 3. E. del Diablo, Durango; 4. El Nayar, Nayarit

S. México: 5. Cuale-Talpa, Jalisco; 6. Manántlan, Jalisco; 7. Coalcomán, Michoacán; 8. Atoyac, Guerrero; 9. Qelchultenango-Guerrero

Oaxaca: 10. Loxicha, Oaxaca; 11. Miahuatlán, Oaxaca; 20. Huautla, Oaxaca; 21. V. Nacional, Oaxaca

NE México: 12. El Cielo, Tamaulipas; 13. Xilitla, S. L. Potosí; 14. Joya del Hielo, Querètaro; 15. Chapulhuacán, Hidalgo; 16. Zacatlamaya, Hidalgo; 17. Huayacocotla, Veracruz; 18. Huachinango, Puebla; 19. Teocelo, Veracruz

Neovolcanic Chain: 22. Nevado deColima; 23. Tancítaro; 24. Toluca; 25. Popocatépetl; 26.Iztaczíhuatl;27. La Malinche; 28. Perote; 29. San Martín; 30. Santa Martha

Chiapas: 31. Jitotol, Chiapas; 32. Lagunas de Montebello; 33, 34. El Triunfo

Source: Vázquez-G. (1989b).

This study examines the distribution of neotropical cloud forest habitats and the compositional similarities among them at three different scales in order to provide a context for understanding major differences among TMCFs. This context can eventually help in improving conservation strategies for these isolated ecosystems by emphasizing the importance of establishing regional, provincial, and local archipelago preserves, defined as a set of island-like habitats that potentially depend on each other's components for maintaining biodiversity and essential ecological processes such as dispersal, gene flow, and migration.

Figure 2. Locations of cloud forests and cloud-forest parks and reserves as known or envisaged currently in the Caribbean and Middle America:

1. Quetzal Cloud-forest Reserve, Guatemala
2. Montecristo National Park (proposed), El Salvador
3. Monteverde Cloud-forest Reserve, Costa Rica
4. Volcano Poas National Park, Costa Rica
5. Volcano Chirripó National Park, Costa Rica
6. Volcano Bará National Park, Panamá
7. Blue Mountain Peak, Jamaica
8. Pico Turquino, Cuba
9. Armando J. Bermúdez National Park, Dominican Republic
10. Caribbean National Forest and Commonwealth Forests, Puerto Rico
11. Mt. Misery, St. Kitts
12. Soufrière, Guadeloupe
13. Morne Trois Pitons National Park, Dominica
14. Mt. Pelee, Martinique
15. Mt. Gimie, St. Lucia
16. Soufrière, St. Vincent
17. Mt. St. Catherine, GrenadaMethods
18. El Aripo, Trinidad

Source: LaBastille and Pool (1978). Reprinted with the kind permission of Elsevier Sequoia, Lausanne, Switzerland, publishers of *Environmental Conservation*.

Methods

The Concept Used for Cloud Forests

I used the TMCF concept in a broad sense to include *bosque mesofilo de montaña* (Miranda 1947; Rzedowski and McVaugh 1966; Rzedowski 1978), part of *bosque caducifolio* (Miranda and Hernández-X. 1963), *bosque deciduo templado* (Rzedowski 1966), evergreen cloud forests and pine-oak *Liquidambar* forest (Breedlove 1981), *bosque nuboso* and *bosque nuboso enano* (Gómez 1986), montane rain forests and elfin woodland (Beard 1955), *bosque obrofito de altura* (Lauer 1968), and, simply, cloud forest (Leopold 1950). Cloud forests in the broad sense are generally understood to occur in tropical mountains and highlands in the northern neotropics, from northern México (Gentry 1946; Miranda and Sharp 1950) to northern South America and the Caribbean Islands (LaBastille and Pool 1978), at elevations from 800–1,200 m to 2,500–2,800 m or more. They tend to be confined to mountain levels with high horizontal precipitation as intercepted cloud water (Vogelmann 1973), or in protected ravines or valleys in highlands where morning fog is high. These areas can be confined locally to a particular altitudinal zone. However, such zones vary from place to place and the forests are best considered restricted to the level of cloud formation. This cloud level should decrease with increasing humidity at the base of the mountain chain (Walter 1973) and increase with mountain chain size and the opportunity for air to warm next to ground surfaces at higher elevation (*Massenerhebung* effect, Grubb 1971). The overall mean annual precipitation (fog plus rainfall) in these forests is usually from 1,500 to 2,500 mm, and variation in species composition is said to result from temperature differences that vary with altitude (Walter 1973).

Types of Data

Floristic surveys of major neotropical TMCFs were the main source of information for examining regional and provincial variation (Table 1, Figure 3). All trees, herein defined as native woody plants taller than 5 m and with little or no branching near the base, were included. Additionally, studies of gradient analysis (Vázquez-G. 1991, 1992; Santiago-P. 1992) in the Sierra de Manantlán Biosphere Reserve of Jalisco, México, were the sources to examine local variation.

Relationships Examined Among Neotropical Cloud Forests

Variation of cloud forests was examined at three levels: (1) regional, comparing the occurrence of tree genera in major cloud forests from México, Central America, and northern South America including the Caribbean (two obviously different forests from the southeastern United States were included to examine their affinities, if any, to neotropical forests); (2) provincial, comparing the occurrence of tree species in nine major Mexican cloud forests; and (3) local, comparing quantitative data of several guilds of vascular plant species by means of gradient analysis in the cloud forests of the Sierra de Manantlán Biosphere Reserve of Jalisco, México.

Data Analysis

A correlation matrix was used for a presence and absence genera (275) table, for 15 New World CFs (mostly neotropical), to examine variation and

Table 1. Neotropical cloud forests compared in this study

Floristic Survey	Level of Comp.	Reference
Atoyac, Guerrero	Genera	Lorenzo S-A et al. 1983
Barr. de Ocuilán, Morelos	Spp.	Luna-V. et al.1989
Cerro Viejo, Jalisco	Spp.	Machuca-N. 1989
Chiriquí, Panama	Genera	D'arcy 1987
Cord. Oriental, Colombia	Genera	Cuatrecasas 1958
C. Talamanca, Costa Rica	Genera	Kappelle et al. 1991
El Cielo, Tamaulipas	Genera and spp.	Puig et al. 1983; Puig and Bracho 1986
El Triunfo, Chiapas	Genera and spp.	Long and Heath 1991
El Yunque, Puerto Rico	Genera	Howard 1968; Weaver and Murphy 1990
M. Uyuca, Honduras	Genera	Pfeifer 1961
Monte Bello, Chiapas	Spp.	Carlson 1954
Monte Verde, Costa Rica	Genera	Haber 1991
Serr. Macuira, Colombia	Genera	Sudgen 1982
S. de Manantlán E., Jalisco	Spp.	Cuevas-G. et al. 1988
S. de Manantlán W., Jalisco	Genera and spp.	Vázquez-G. et al. 1990
Sierra Maestra, Cuba	Genera	Borhidi 1991
Teocelo, Veracruz	Genera and spp.	Luna-V. et al. 1988
Valle de México, México	Spp.	Rzedowski 1970
Temperate Deciduous Forests (included as outgroups)		
Gr. Smoky M. Tennessee	Genera	Whittaker 1966
Hot Springs, Arkansas	Genera	Dale and Watts 1980

relationships among regional CFs (Figure 3). Likewise, similar data were used for a species (340) table, for nine Mexican CFs, to examine variation and relationships among provincial CFs (Figure 3). Subsequently, global nonmetric-multidimensional scaling (GN-MDS), a stress-minimization multivariate technique, was applied to both correlation matrices. The analyses for local variation in the Sierra de Manantlán Biosphere Reserve are explained in detail in Vázquez-G. (1991, 1992) for vertical separation of TMCFs, and in Santiago-P. (1992) for horizontal separation of TMCFs. Only the results are discussed here, to complement a perspective of local variation for TMCFs.

These analyses are exploratory and far from exhaustive; a more comprehensive study should improve our understanding of the relationships among neotropical and Mexican TMCFs. One limitation of the current study is that localities of the floristic surveys differ in size, elevational range, and intensity of exploration. Hence, caution must be taken when interpreting results, which are intended only to show major differences and trends among cloud forests. Some species were not considered for some areas where they occur as shrubs, but were considered in others where they occur as small trees. This error (about 5 percent or less) may slightly exaggerate compositional dissimilarities between mesic and xeric habitats.

Figure 3. Studied neotropical cloud forest floras.

Results

Regional Variation in the Neotropical Cloud Forests

The GN-MDS technique placed samples in a horseshoe-shaped arrangement (Figure 4). This curvature of the gradient is due to the high heterogeneity involved in the system, and is accentuated by the nonmonotonicity to the environment of the similarity measure used (correlation). The software available, which had better measures of similarity, was not able to handle large matrices, however, and as is well known the use of Sorensen distance should greatly reduce curvature of the gradient (Beals 1984). Being aware of the curvature of the gradient, the ordination was interpreted accordingly. Five clusters are apparent: I. temperate southeastern United States; II. northern neotropical; III. Mesoamerican; IV. Caribbean; and V. northern South American. Warm temperate deciduous forests of southeastern United States (Great Smoky Mountains, Tennessee–North Carolina; and Hot Springs National Park,

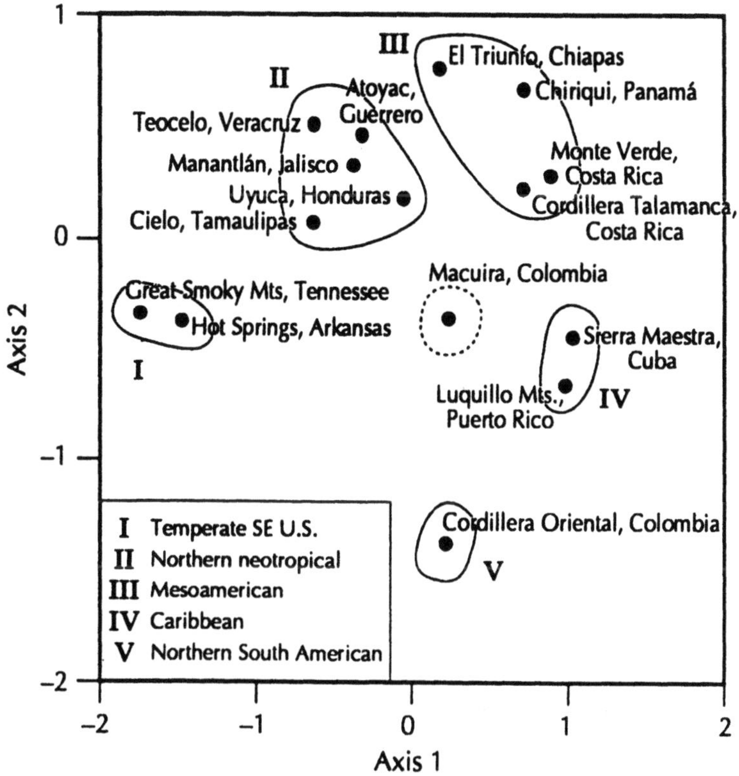

Figure 4. Global nonmetric-multidimensional scaling of neotropical cloud forests.

Arkansas), clustered at the left end of the first axis, are obviously quite distinct from neotropical cloud forests, including the geographically closest (in distance) at El Cielo Biosphere Reserve of Tamaulipas, México. Northern neotropical TMCFs (mostly Mexican ones) resulted in a compact cluster. Mesoamerican TMCFs formed a less compact but well-defined cluster, except for the one of Honduras that is perhaps more related to the northern neotropical group. The Caribbean TMCFs (Sierra Maestra, Cuba; and Luquillo Mountains, Puerto Rico) also formed a well-defined cluster. In contrast, the TMCF of Serranía Macuira, Colombia, ended up fairly isolated near the center of the ordination (Figure 4), confirming its extreme degree of isolation as reported by Sugden (1982). Finally, the northern South American group represented by Colombian-Andean TMCFs was placed at one end of the second axis, also distant from the others including the geographically closest at Serranía Macuira, Colombia. According to unpublished data collected by the author, TMCFs of Venezuelan Andes fall into this group, while TMCFs of southern Venezuela are fairly distantly related and may represent a separate region.

Provincial Variation in the Mexican Cloud Forests

The GN-MDS placed samples in a T-shaped arrangement, along the first and second axes (Figure 5). Four clusters are apparent: I. Pacific southwest;

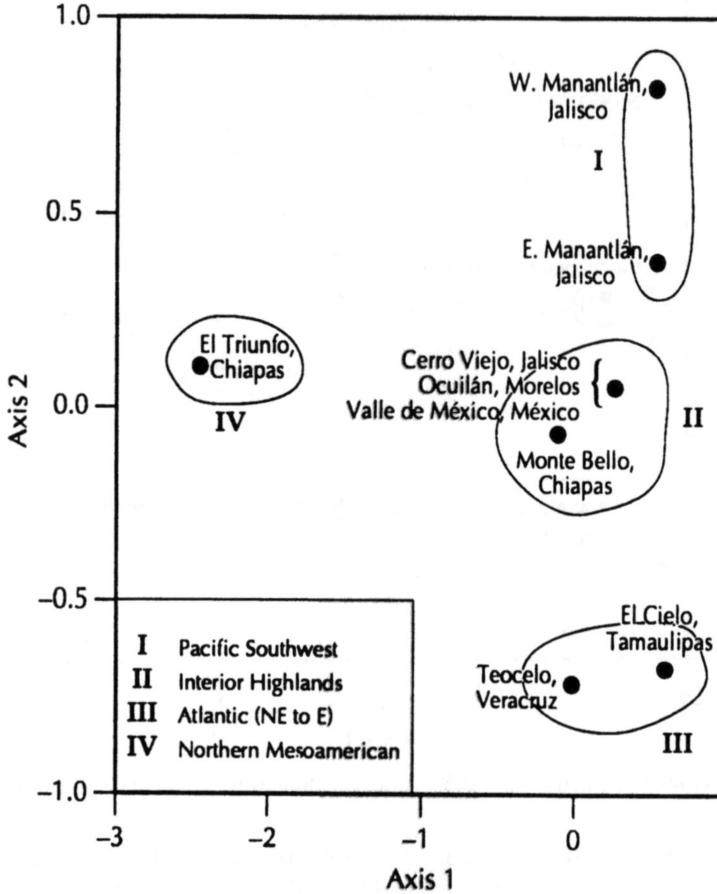

Figure 5. Global nonmetric-multidimensional scaling of Mexican cloud forests.

II. interior highlands; III. Atlantic (northern to eastern); and IV. northern Mesoamerican. Pacific southwest TMCFs (W. and E. Manantlán, Jalisco) were strongly separated from Atlantic TMCFs (El Cielo, Tamaulipas; and Teocelo, Veracruz) at opposite extremes. Interior highland TMCFs (Valle de México, México; Ocuilán, Morelos; Cerro Viejo, Jalisco; and Montebello, Chiapas) were tightly clustered between the Atlantic and Pacific southwest TMCFs, showing more similarity to each other than to geographically less distant TMCFs located in mountains closer to the oceans. Northern Mesoamerican TMCFs (El Triunfo, Chiapas) were strongly separated on the first axis and were distinct from other Mexican forests.

Local Variation of Cloud Forest in Sierra de Manantlán

Results on horizontal and vertical separation of cloud forests will not be presented since they are being prepared for separate papers by Santiago-P. and Vázquez-G., respectively. However, preliminary interpretations of those results are given under the discussion section.

Discussion

Regional Variation of Neotropical Cloud Forests

The ordination clusters or regional archipelagos at this scale appear strongly correlated with geographic location, and with the effect of seasonality along the latitudinal gradient (Figure 4). Generic diversity (Figure 6) appears not correlated with any of the ordination axes (Figure 4); in other words, generic composition is more important than generic diversity in grouping similar CFs. Thus, generic diversity was averaged for each cluster to describe them. As identified earlier, the five regional archipelagos include I. temperate southeastern United States; II. northern neotropical; III. Mesoamerican; IV. Caribbean; and V. northern South American. The southeastern U.S. forest, with relatively low tree generic diversity (μ 32) and high seasonality, clustered at one end of the first axis, confirms a clear distinction between warm temperate deciduous forests and typical neotropical TMCFs (Figure 6). Northern neotropical TMCFs are of intermediate tree generic diversity (μ 56), with high or moderate seasonality, and consist largely of Mexican forests north of the Tehuantepec Isthmus and perhaps a few from the interior upper Mesoamerican highlands. Mesoamerican TMCFs are of high tree generic diversity (μ 88), relatively low seasonality, and largely consist of Central American TMCFs,

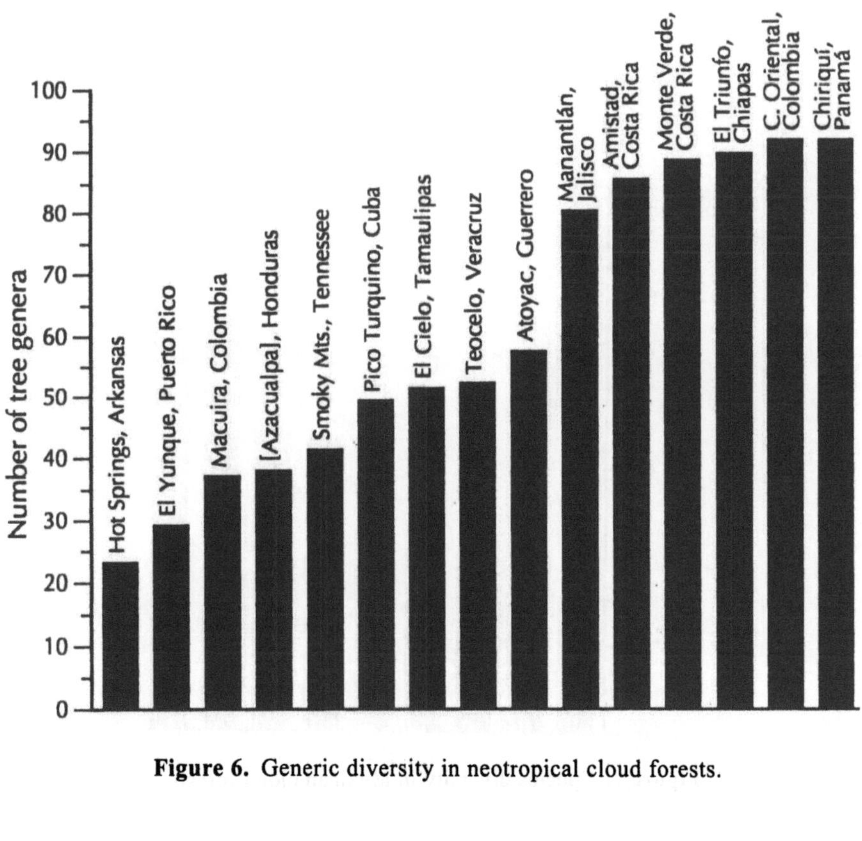

Figure 6. Generic diversity in neotropical cloud forests.

especially those located in the coastal mountains (sierras, cordilleras, or serranías), including the rich Mexican forest of El Triunfo, Chiapas. Caribbean TMCFs are of relatively low tree generic diversity (μ 39), and largely consist of Antillean montane dwarf forests, perhaps related to the elfin woodlands of Central America and northern South America. Northern South American TMCFs, including one from the Venezuelan Andes not shown in the ordination, are of high tree generic diversity (μ 89), and show little or no seasonality. Inclusion of more TMCF surveys from this area may show that these forests are even richer than Mesoamerican TMCFs.

Provincial Variation of Mexican Cloud Forests

The ordination within México can further resolve provincial archipelagos, which relate in part to geographical location and ecological similarities, perhaps corresponding to differences in rainfall and year-round precipitation (Figure 5). Species diversity (Figure 7) appears not correlated with any of the ordination axes (Figure 5); in other words, species composition is more important in determining relationships among CFs. Thus, species diversity was

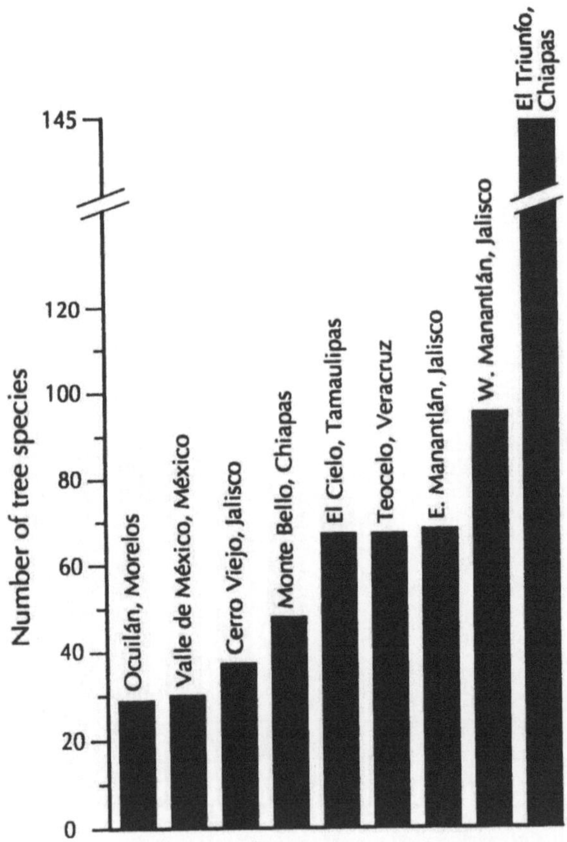

Figure 7. Species diversity in Mexican cloud forests.

averaged for each of the clusters to describe them. Apparently distinct provincial archipelagos include: I. Atlantic; II. Pacific southwest; III. interior highlands; and IV. Mesoamerican. Atlantic TMCFs (El Cielo, Tamaulipas; and Teocelo, Veracruz) have more temperate elements than any of the other TMCF provincial archipelagos, and less Mesoamerican elements; they are of intermediate tree species diversity (μ 67). The inclusion of more surveys of the Atlantic province may show that it has a greater diversity than those of the Pacific southwest. Pacific southwest TMCFs (W. and E. Sierra de Manantlán, Jalisco) are of high tree species diversity (μ 82). They keep a good proportion of both temperate elements and Mesoamerican elements, perhaps a larger proportion of the latter. Note that Atlantic and Pacific southwest TMCFs were separated as opposite marginal extremes. Interior highland TMCFs (Cerro Viejo, Jalisco; Ocuilán, Morelos; Valle de México, México; and Montebello, Chiapas) are of low tree species diversity (μ 36) and relate more to distant TMCFs of interior highlands than to closer TMCFs of seaward-facing mountains; in general these forests are quite depauperate, and it is questionable if they should be considered cloud forests. Northern Mesoamerican TMCFs such as those of El Triunfo, Chiapas, have the highest tree species diversity (μ 143) among Mexican TMCFs. Because of their affinities with Central American elements, these TMCFs can be placed into a different region (Mesoamerican), within which they are restricted to what can be considered the northern Mesoamerican province.

Local Variation of Cloud Forests in Sierra de Manantlán

Horizontal Separation

According to my interpretation of the dendrogram of TMCFs in the Sierra de Manantlán Biosphere Reserve (Santiago-P. 1992), TMCFs can be divided into five local clusters (mini-archipelagos): I. *Matudaea-Podocarpus;* II. *Ternstroemia-Quercus*; III. *Abies-Persea;* IV. *Carpinus-Cornus*; and V. *Magnolia-Acer.* In addition, a unique forest studied by Vázquez-G. et al. (1990) at Los Mazos, western Manantlán, showed the occurrence of a *Ficus-Sapium-Dipholis* association, quite analogous to that described from El Triunfo, Chiapas (Long and Heath 1991).

Vertical Separation

A clear altitudinal zonation of TMCFs was consistently found within the 600-m elevational gradient for both gentle and steep slopes of eastern Sierra de Manantlán Biosphere Reserve (Vázquez-G. 1991, 1992). Along this elevational gradient, TMCFs can be subdivided into four altitudinal bands: I. 2,400–2,500 m; II. 2,200–2,300 m; III. 2,100 m; and IV. 2,000 m (Figures 8, 9). The dominant composition of these forests shifts from trees with rather entire xeromorphic leaves such as *Quercus-Ternstroemia*, confined to upper highlands; to mesic trees such as *Quercus candicans-Abies-Tilia*, *Ostrya*, and *Oreopanax*,

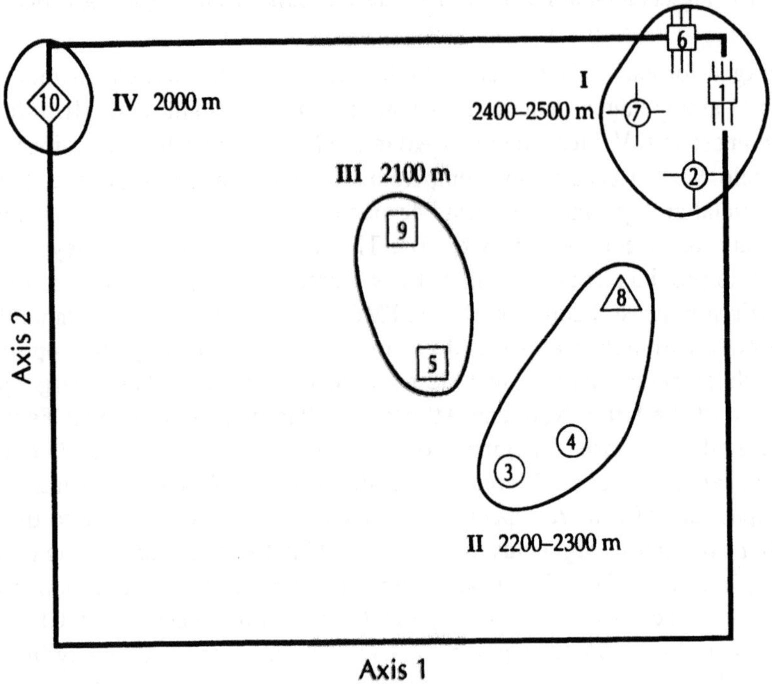

Figure 8. Bray-Curtis ordination (variance/regression) of E. Manantlán cloud forests (gentle slope).

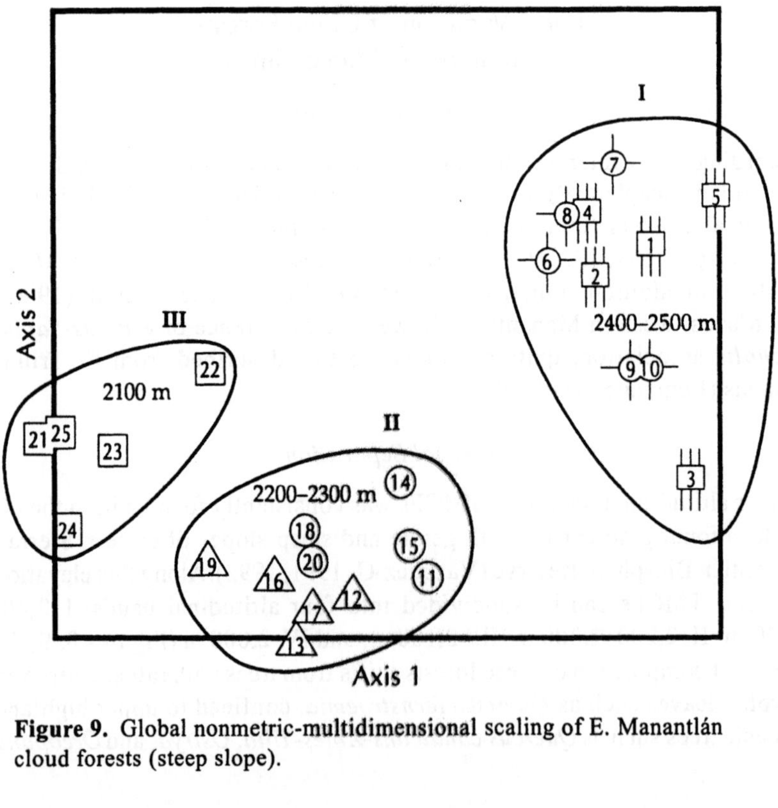

Figure 9. Global nonmetric-multidimensional scaling of E. Manantlán cloud forests (steep slope).

many of them with lobate or serrate leaves; to trees with compound or compound-like leaves such as *Trichilia*, *Zinowiewia*, *Fraxinus*, and *Picramnia* at the lower portion of this cloud forest gradient.

Species Diversity of Guilds Along Elevation

According to Vázquez-G. (1992), species diversity of all guilds (trees, shrubs, herbs, epiphytes, and vines) decreases with increasing elevation. Despite the absolute diversity of trees being greater than other guilds at any elevation, the relative diversity of vines and shrubs decreased the most throughout the gradient. Herbs show the highest variation along the gradient, confirming their stochastic nature of occurrence. Warmth-loving epiphytes (Hymenophyllaceae and other ferns) were more abundant at the core of the cloud forest, while cold mesic epiphytes (*Lycopodium* and mosses) were more abundant at the upper portion of the gradient. The epiphytes occurring at either upper or lower cloud forests are often highly restricted to tree species with barks having greater water-holding capacity, such as those of *Oreopanax*, *Clethra*, and *Quercus* at higher elevation and *Dendropanax* and *Quercus* at lower elevation.

Evergreen vs. Deciduous Species Along Elevation

Cloud forests of Sierra de Manantlán consist of varying proportions of evergreen (flowering and rarely coniferous plants) and deciduous tree species. Deciduous species like basswood (*Tilia*), sugar maple (*Acer*), ironwood (*Carpinus*), and hophornbean (*Ostrya*) tend to dominate in mesic-protected ravines, while evergreen species like *Arbutus*, *Comarostaphylis*, *Ternstroemia*, and *Symplocos* dominate either at more exposed sites or toward the upper elevational boundary of a cloud forest gradient; others like evergreen magnolia (*Magnolia iltisiana*) are found in more sandy soils. This pattern tends to occur also on a provincial scale. For example, mesic forests of the Atlantic slopes of Sierra Madre Oriental, México, tend to have more deciduous elements (such as *Liquidambar, Fagus,* and *Nyssa*) than less mesic forests of the Sierra Madre Occidental, México, where these genera do not occur. A similar pattern is shown by forests of eastern México, which have more evergreen species than temperate deciduous forests in southeastern United States (Gómez-P. 1966).

Several of these unique TMCFs studied at Sierra de Manantlán Biosphere Reserve, in addition to those yet unexplored, suggest the need for archipelago core zones (a set of areas with strict and effective protection). Efforts are being made by the Instituto Manantlán de Ecología to purchase and effectively protect the land of relict populations of *Acer saccharum* (Jardel-P., pers. com.) and other forests of special interest (Jardel-P. 1992).

Variation in Cloud Forests as Delineated by *Magnolia* Speciation

Monographic studies of *Magnolia* (Vázquez-G. 1990, in press; Howard 1948; Rockwell 1966; Lozano-C. 1983) show remarkable patterns of allopatric speciation and high endemism for a cloud forests genus (Figure 10), supporting the existence of both regional and provincial archipelagos. In addition, *Magnolia* speciation within section *Theorhodon* can further divide some provincial archipelagos into subprovincial ones. At least five regional archipelagos can be outlined by the present distribution of *Magnolia*: I. Temperate southeastern United States, containing three sections of deciduous magnolias and just one taxa of evergreen magnolias; II. northern neotropical, largely containing evergreen magnolias of section *Theorhodon* and only one section of deciduous magnolias; III. Mesoamerican, containing only evergreen magnolias of section *Theorhodon* only; IV. Caribbean, containing evergreen magnolias of section *Splendentes* only; and V. northern South American, containing no magnolias in the narrow sense, but rather having *Talauma*-like species reported from the Venezuelan Tepuis and other Colombian highlands (Vázquez-G. 1990). The distribution of *Magnolia* section *Theorhodon* in México and Central America can also outline the following provincial archipelagos (Figure 10): Pacific northwest, consisting of *M. pacifica* ssp. *tarahumara*, and *M.* sp. 1; Pacific southwest, containing *M. p.* ssp. *pacifica* and ssp. *pugana,* and *M. iltisiana;* Pacific southeast, containing *M.* sp. 2 only; Atlantic, further subdivided into northern, containing *M. tamaulipana*, and eastern, consisting of *M. schiedeana* and largely containing the deciduous *M. dealbata*; northern Mesoamerican, containing *M. sharpii*, *M. yoroconte sensu lato*, and *M. guatemalensis sensu lato*; and southern Mesoamerican, consisting of *M. poasana*, *M. sororum sensu lato*, and *M. Panamensis*. Only one region, northern South American, and one Mexican province, interior highlands, can be characterized by a nearly complete lack of magnolias. However, the overall pattern is that each region, province, or even subprovince contains its own endemic magnolias. These subdivisions, as outlined by *Magnolia* speciation and subspeciation, have little or no overlap and clearly support the existence of cloud forest archipelagos at both the regional and provincial levels.

Epilogue

In summary, the uniqueness of and relationships among neotropical TMCFs discussed here at three different scales should have direct implications in defining priorities and strategies for conservation within the context of island biogeography. I emphasize the importance of establishing regional, provincial, and local archipelago preserves as an approach that can best represent the natural distribution and heterogeneity of TMCFs in the northern neotropics. However, further analytical work is recommended to refine our understanding of endemism and relationships among neotropical TMCFs, especially at the local and provincial scale. In addition, the geographical fragmentation and inherent biological interest of TMCFs make them ideally suited to conduct relevant evolutionary and biogeographical research. In the meantime, these ecosystems deserve an immediate but effective protection.

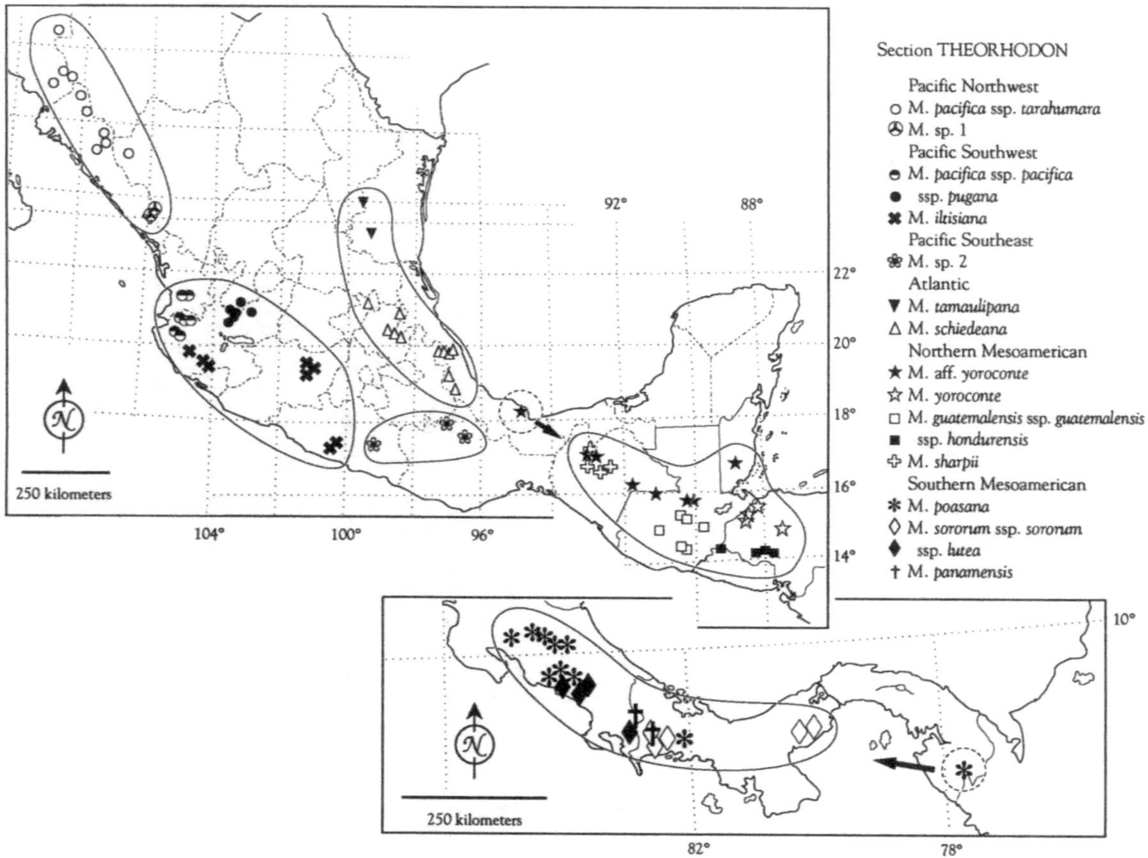

Figure 10. Distribution of the genus *Magnolia* in México and Central America. *Source: Vázquez-G. (in press).*

Acknowledgments. I thank Professor D. Waller who encouraged me to attend this symposium and to develop this manuscript from the term papers presented in the botany courses of plant geography and conservation biology at the University of Wisconsin-Madison. I thank Professors E. Beals, T. Givnish, H. Iltis, S. Solheim, G. O'Keefe, and reviewers of the East-West Center for their valuable input and comments to this manuscript. Thanks to Kandis Elliot for her excellent graphic support and to H. Ballard and L. Hernández-L. for assisting with data manipulation and analysis. Financial support was provided by CONACYT-México and the Instituto Manantlán de Ecología y Conservación de Biodiversidad (IMECBIO), Universidad de Guadalajara. Fieldwork was supported by the E. K. and O. N. Allen Herbarium Fund and the J. J. Davis Fund of the Botany Department, and the Natural History Museums Council of the University of Wisconsin-Madison. Attendance at this symposium was supported largely by the East-West Center, Program on Environment.

References

Beals, E. W. 1984. Bray-Curtis ordination: an effective strategy for analysis of multivariate ecological data. *Advances in Ecological Research* 14:1-49.

Beard, J. S. 1955. The classification of tropical American vegetation types. *Ecology* 36:89-100.

Borhidi, A. 1991. *Phytogeography and vegetation ecology of Cuba*. Budapest: Akadeiai Kiado.

Breedlove, D. 1981. *Flora of Chiapas*. San Francisco: California Academy of Sciences.

Carlson, M. 1954. Floral elements of the pine-oak-liquidambar forest of Montebello, Chiapas, México. *Bulletin of the Torrey Botanical Club* 81(5): 387-399.

Cuatrecasas, J. 1958. Aspectos de la vegetación natural de Colombia. *Revista de la Academia Columbiana de Ciencias Exactas Fisicas y Naturales* 10(40): 221-264.

Cuevas-G., R., J. A. Vázquez-G., and L. Guzmán-H. 1988. Listado preliminar de las especies vegetales en Cerro Grande. Laboratorio Natural Las Joyas, Universidad de Guadalajara. Technical document. Notas sobre la *Flora de Manantlán* 1(13): 77-83, Guadalajara, Jalisco.

Dale, E. E., and M. R. Watts. 1980. Vegetation of Hot Springs National Park, Arkansas. Department of Botany and Bacteriology, University of Arkansas. Technical Document. Arkansas.

D'arcy, W. G. 1987. *Flora of Panama. Checklist and index*. Part I: *The introduction and checklist*. Saint Louis, Missouri: Missouri Botanical Garden.

Gentry, H. S. 1946. Notes on the vegetation of Sierra Surotato in northern Sinaloa. *Bulletin of the Torrey Botanical Club* 73:451-462.

Gómez, L. D. 1986. *Vegetación de Costa Rica*. Costa Rica: Universidad Estatal a Distancia.

Gómez-P., A. 1966. *Estudios botánicos en la región de Misantla, Veracruz*. México: Edit. Inst. Nac. Rec. Reno.

Grubb, P. J. 1971. Interpretation of the "Massenerhebung effect" on tropical mountains. *Nature* 229(5279): 44-45.

Haber, W. A. 1991. Lista provisional de las plantas de Monte Verde, Costa Rica. *Brenesia* 34:63-120.

Howard, R. A. 1948. The morphology and systematics of West Indian Magnoliaceae. *Bulletin of the Torrey Botanical Club* 75(4): 335-357.

Howard, R. A. 1968. The ecology of an elfin forest in Puerto Rico. *Journal of the Arnold Arboretum* (Harvard University) 49(4): 381-418.

Jardel-P., E. J., coord. 1992. *Estrategia para la conservación de la reserva de la biosfera Sierra de Manantlán*. México: Universidad de Guadalajara.

Kappelle, M., N. Zamora, and T. Flores. 1991. Flora Leñosa de la zona alta de la Cordillera de Talamanca, Costa Rica. *Brenesia* 34:121-144.

LaBastille, A., and D. J. Pool. 1978. On the need of a system of cloud forest parks in Middle America and the Caribbean. *Environmental Conservation* 5(3): 183-190.

Lauer, W. 1968. Problemas de la división fitogeográfica de America Central. In *Geoecología de las regiones montañosas de las Americas tropicales*. Bonn: F. Duemmlers Verlag.

Leopold, A. S. 1950. Vegetation zones of México. *Ecology* 31:507-518.

Long, A., and M. Heath. 1991. Flora of the El Triunfo Biosphere Reserve, Chiapas, México: A preliminary floristic inventory of the plant communities of polygon 1. *Anales del Instituto de Biologia Universidad Nacional Autonóma de México Serie Botanica* 62(2): 133-172.

Lorenzo S-A, L., A. Ramírez-R., M. A. Soto-A., A. Breceda, M. del C. Calderon, H. Cortez, C. Puchet, M. Ramírez, R. Villalón, and E. Zapata. 1983. Notas sobre la fitogeografía del bosque mesófilo de montaña en la Sierra Madre del Sur, México. *Boletin de la Sociedad Botanica de México* 44:97-102.

Lozano-C., G. 1983. Magnoliaceae in Flora de Colombia. Universidad Nacional del Colombia 1:1-20.

Luna-V., I., L. Almeida-L., and J. Llorente-B. 1989. Florística y aspectos fitogeográficos del bosque mesófilo de montaña de las cañadas de Ocuilán, Morelos-México. *Anales del Instituto Biologia de Universidad Nacional Autonóma de México Serie Botanica* 59(1): 63-87.

Luna-V., I., L. Almeida-L., L. Villers, and L. Lorenzo. 1988. Reconocimiento florístico y consideraciones fitogeográficas del bosque mesófilo de montaña de Teocelo, Veracruz. *Boletin de la Sociedad Botanica de México* 48:35-63.

Machuca-N., J. A. 1989. *Florística y ecología de la vegetación fanerogámica de la región septentrional de Jocotepec, Jalisco, México*. Tesis de Licenciatura. México: Universidad de Guadalajara.

Miranda, F. 1947. Estudios sobre la vegetación de México. V. Rasgos de la vegetación en la cuenca del Río de las Balsas. *Revista Sociedad Mexicana de Historia Natural* 8:95-114.

Miranda, F., and E. Hernández-X. 1963. Los Tipos de vegetación en México y su clasificación. *Boletin de la Sociedad Botanica de México* 28:29-179.

Miranda, F., and A. J. Sharp. 1950. Characteristics of the vegetation in certain temperate regions of eastern México. *Ecology* 31:313-333.

Pfeifer, H. W. 1961. Vascular plants of Mount Uyuca [Azacualpa], Honduras. *Ceiba* 8(2): 102-142.

Puig, H., and R. Bracho, eds. 1986. *El bosque mesófilo de montaña de Tamaulipas*. México: Instituto de Ecología.

Puig, H., R. Bracho, and V. Sosa. 1983. Composición y estructura del bosque mesófilo en Gómez Farías, Tamaulipas, México. *Biotica* 8(4): 339-359.

Rockwell, H. C. 1966. The genus Magnolia in the United States. M. S. Thesis, University of West Virginia, Morgantown, West Virginia.

Rzedowski, J. 1966. Vegetación del estado de San Luis Potosí. *Acta Cientifica Potosina* 5:5-291.

Rzedowski, J. 1970. Nota sobre el bosque mesófilo de montaña en el Valle de México. *Anales de la Escuela Nacional de Ciencias Biologicas México* 18:91-105.

Rzedowski, J. 1978. *La vegetación de México*. México: Limusa.

Rzedowski, J., and R. McVaugh. 1966. La vegetación de Nueva Galicia. *Contributions from the University of Michigan Herbarium* 9:1-123.

Santiago-P., A. L. 1992. *Estudio fitosociológico del bosque mesófilo de montaña de la Sierra de Manantlán*. Tesis de Licenciatura. México: Universidad de Guadalajara.
Shafer, C. L. 1990. *Nature reserves: Island theory and conservation practice*. Washington: Smithsonian Institution Press.
Sugden, A. M. 1982. The vegetation of the Serranía Macuira, Guajira, Colombia: A contrast of arid lowlands and an isolated cloud forest. *Journal of the Arnold Arboretum* (Harvard University) 63(1): 1-30.
Vázquez-G., J. A. 1989a. Diversity among neotropical cloud forests. Lecture report. Department of Botany, University of Wisconsin-Madison.
Vázquez-G., J. A. 1989b. Mountain chains of island-like habitats, preservation of neotropical cloud forests. Lecture report. Department of Botany, University of Wisconsin-Madison.
Vázquez-G., J. A. 1990. *Taxonomy of the genus* Magnolia *in México and Central America*. M.S. thesis. Department of Botany, University of Wisconsin-Madison.
Vázquez-G., J. A. 1991. Patterns of vegetation along an altitudinal gradient in evaluating ecological interpretability of ordination and classification techniques. Lecture report. Department of Botany, University of Wisconsin-Madison.
Vázquez-G., J. A. 1992. Patterns of vegetation along environmental gradients. Lecture report. Department of Botany, University of Wisconsin-Madison.
Vázquez-G., J. A. In press. *Magnolia* (Magnoliaceae) in Mexico and Central America: A synopsis. *Brittonia* 45(4).
Vázquez-G., J. A., R. Cuevas-G., T. S. Cochrane, and Hugh H. Iltis. 1990. *Flora de la reserva de la biosfera Sierra de Manantlán, Jalisco-Colima, México*. México: Universidad de Guadalajara.
Vogelmann, H. W. 1973. Fog precipitation in the cloud forest of eastern México. *Bioscience* 23(2): 96-99.
Walter, H. 1973. *Vegetation of the earth, in relation to climate and the eco-physiological conditions*. Heidelberg Science Library: The English Universities Press.
Weaver, P. L., and P. G. Murphy. 1990. Forest structure and productivity in Puerto Rico's Luquillo Mountains. *Biotropica* 22(1): 69-82.
Whittaker, R. H. 1966. Vegetation of the Great Smoky Mountains. *Ecological Monographs* 26(1): 80.
Whittaker, R. H. 1968. Vegetation of the Santa Catalina Mountains, Arizona. *Journal of Ecology* 56:523-544.

24. Endangered Species in Low Elevation Cloud Forest on Gau Island, Fiji

Dick Watling and A. N. Gillison

Gau (pronounced "ngau" with a soft "ng") is Fiji's fifth largest island, with an area of 140 km^2. Its rugged inland topography comprises the northern upland, which is dominated by the twin summits of Delaco (715 m) and Delacoboni (705 m) and the main ridge running southward. Fifty-five percent of the island supports dense rain forest, while the rolling country of the eastern coast and the island's extremities are covered with grass or reed, with some recently established plantations of Caribbean pine.

In view of Gau's location (Figure 1) less than 80 km from Suva, Fiji's capital on the island of Viti Levu, it is surprising that Gau's flora has been the subject of only superficial collection and no research. No publications exist on the flora of Gau. Although the botanist A. C. Smith visited Gau island in 1953, it was clearly a very brief visit, for certain interesting species, notably *Physokentia rosea* which is locally abundant on the island, are recorded in his *Flora Vitiensis Nova* as endemic to Viti Levu (Smith 1979). No records seem to exist of other botanists visiting Gau, although some plant specimens were collected during the visit of HMS *Herald*, a British Royal Navy survey ship, in 1854 and 1855.

Until the senior author initiated a search on Gau in 1983 for the enigmatic Fiji petrel (*Pseudobulweria macgillivrayi*), the vertebrate fauna was only marginally better known, following the two visits of HMS *Herald*, and the visit of the Whitney South Sea Expedition in February 1925, which made a small collection of birds. Watling's works (1985a, b; 1986a, b) provide records and observations of the terrestrial fauna of Gau.

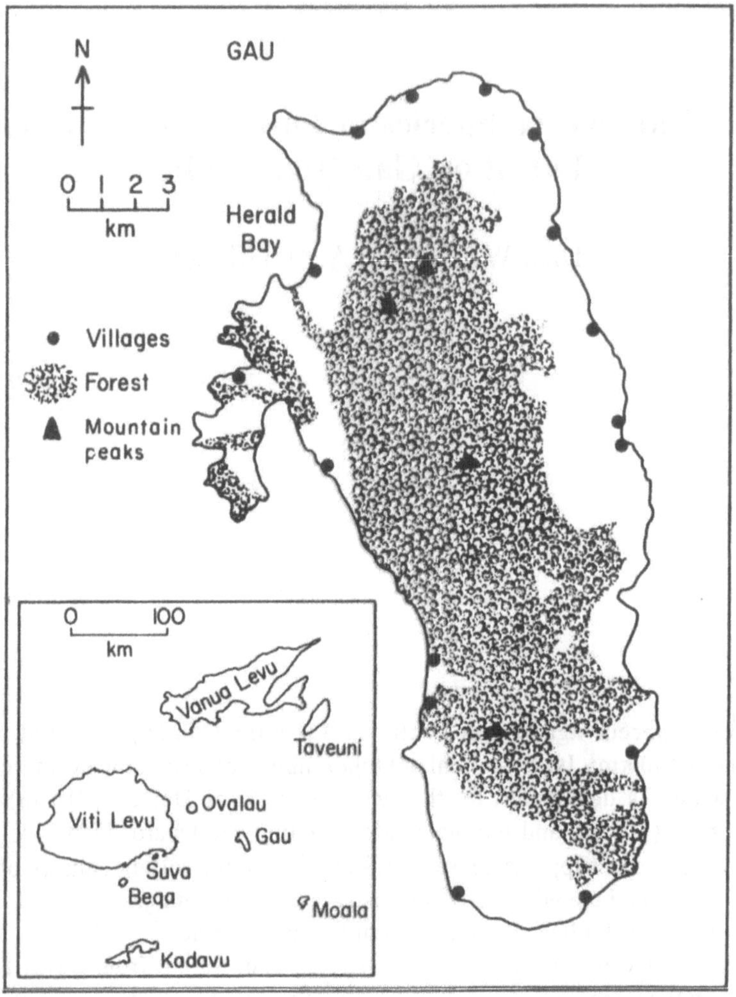

Figure 1. Gau and its location within the Fiji Islands.

Although broad regional climate patterns can be interpolated for Gau, there are no actual meteorological data (i.e., rainfall) available from the island.

Cloud Forest in Fiji

With the exception of Ash (1987) and Gillison (1992 and in press), there has been no specific research on tropical montane cloud forest (TMCF) in Fiji. Ash (1987) studied the vegetation and microclimate of a stunted ridge-top cloud forest on the island of Taveuni. In a general comment he stated: "The extent of cloud-forest in Fiji is very limited, occupying only the ridges on the highest peaks. In Taveuni TMCF occurs along about 10 km of the main range and in the other islands it is restricted to 1–2 km of ridges around isolated peaks which are separated by 30–100 km."

Berry and Howard (1973) undertook Fiji's first national forest inventory and described 41 forest types based on interpretation of aerial photographs, structure, and floristic composition. Their study distinguished "ridge thicket" as a distinct forest type "along the top of every major ridge." The trees of the ridge thicket form a dense canopy of about 7 m; the thicket is difficult to penetrate and the diameters of trees at breast height average 20 cm (Berry and Howard 1973). An examination of the series of 36 forest maps (scale 1:50,000) produced by the inventory indicates that ridge thicket is widespread and in some areas extensive on the islands of Viti Levu and Vanua Levu. Berry and Howard (1973) enumerated just under half of Fiji's estimated total forest area of 838,750 ha (Viti Levu, Vanua Levu, and Kadavu only), and of this 2,523 ha was identified as ridge thicket but with the important rider "the type is widespread but difficult to map because it occurs in long thin strips at the crest of every ridge." Although conjectural at this stage of analysis, there is, in all probability, well over 5,000 ha, perhaps 10,000 ha, of ridge thicket in Fiji's forests.

It is certain that not all ridge thicket as described by Berry and Howard (1973) is cloud forest; it may occur for other reasons, such as inadequate soil nutrients. However, it is within this forest classification that Fiji's true cloud forest will be found. Thus potential cloud forest researchers should initiate their studies with this source of information.

Kirkpatrick and Hassal (1985) and Hassal and Kirkpatrick (1985) investigated the vegetation and flora along an altitudinal transect through tropical forest on the southeast coast of Viti Levu. The transect passed over the summit of Mount Korobaba (422 m), a prominent and isolated breccia cone with stunted vegetation. Kirkpatrick and Hassal (1985) discussed the environmental relationships of the six forest types distinguished along the transect. They noted a strong development of the Massenerhebung effect (Grubb 1971) in eastern Viti Levu. At Mount Tomanjivi in the interior of the island, three of their forest types are to be found only on steep, exposed ridges above 1,000 m, whereas these types occurred below 400 m at Mount Korobaba.

Recent work by Gillison (1992 and in press) describes Fijian cloud forest according to plant functional attributes within the context of a broad range of Fijian vegetation and includes detailed descriptions of cloud forest sites in Viti Levu, Vanua Levu, and Taveuni.

Although Fijian TMCF falls more or less within a predictable biogeographic continuum, it has unique plant ecological features represented by the predominance of ferns, especially large ferns *Cyathea* spp., *Leptopteris*, and climbing Pandanaceae (*Freycinetia* spp.). In this respect it has features in common with some TMCF of Papua New Guinea and New Zealand, but the superabundance of these groups is what provides an ecological signature for Fijian cloud forest. There is no immediate explanation of this phenomenon other than perhaps that of an "ecological release" of such groups within a relatively impoverished flora (about 2,900 vascular plant species for the Fiji group).

Although data are insufficient to determine the climatic limits for Fijian cloud forest, these, in general, fall within the upper limits of megatherm to mesotherm/microtherm (Gillison, in press). The environmental envelope can be approximated within a range of 10°–20° C mean annual air temperature, where annual rainfall exceeds 3,500 mm and where the distribution of annual rainfall is not strongly seasonal (Figure 2). No data are available for mist interception, which may be a critical component for the establishment and survival of bryophyte and pteridophyte assemblages. Total precipitation (mist + rainfall) data are urgently needed to better define the physical environmental boundaries for cloud forest in Fiji.

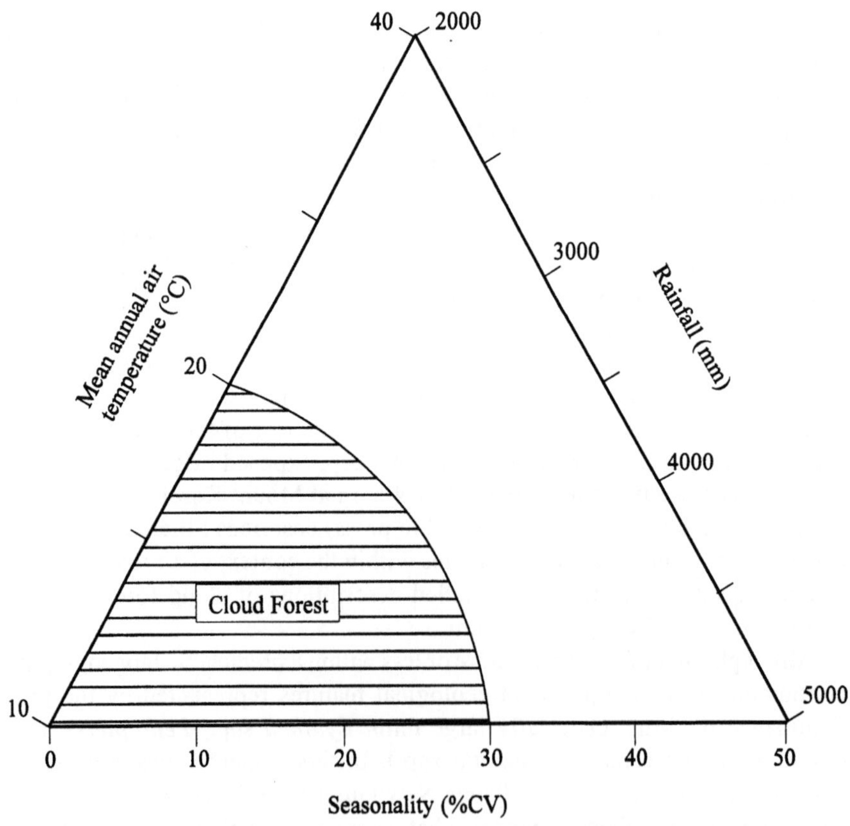

Figure 2. Approximate climate envelope for cloud forest in Fiji.

Cloud Forest on Gau Island

Gau island may prove an ideal location to investigate the relationship between Fiji's "stunted ridge thicket" and true TMCF as alluded to earlier, because of the short distances and low altitudes involved. Spur ridges on Gau's eastern and southern (windward) coasts start developing a stunted ridge thicket on any exposed ridge close to the shoreline. The stunted vegetation can form a continuum along ridges up to the central ridge and summit.

The uplands of Gau are renowned for their shroud of clouds. Although the summit is only 715 m, the ridges over 200 m intercept the southeast trade winds which blow persistently for up to 8 months of the year. The buffered coastal maritime climate, together with the frequent cloud interception, appears to be responsible for producing the cloud forest at a very low altitude on Gau (Figure 3).

The stilt-rooted palm (*Physokentia rosea*), previously regarded as endemic to "dense forest and ridge forest on Viti Levu at altitudes of 750–1,120 m" (Smith 1979), was found to be locally abundant on Gau at elevations as low as 250 m, with some individuals at even lower elevations. A similar altitudinal shift was observed for the attractive orchid *Dendrobium mohlianum*.

Cloud Forest Origins

It is generally accepted that cloud forest flora show taxonomic affinities with continental taxa at mid-to-high latitudes that are adapted to long-distance

Figure 3. Cloud forest near the summit of Gau at 600 m. *Photo by Dick Watling.*

dispersal, mainly by wind (cf. Smith and Cleef 1988). Despite this likely source there are other factors which may play a significant role in the origins and maintenance of cloud forest flora in islands. These include previous evolutionary contact with shoreline or littoral flora and the relatively well-buffered humid environments that are common to both.

As with montane plants, many pantropical littoral flora (including saline and freshwater swamps) are adapted to rapid dispersal; many plants in both environments specialize in vegetative mobility. This is advantageous under conditions of episodic "edge" effects brought about by changing environments (e.g., mobile shorelines or fire or strong wind or both; cf. Gillison 1970a, b). One can speculate that rapidly emerging islands will carry with them a suite of littoral taxa which, under the right conditions, could speciate and adapt to upland environments. In Fiji *Casuarina* and *Scaevola* are cases in point.

Littoral taxa from previously cool, postglacial climates would also likely migrate inland and upland with subsequent climatic warming, taking with them remnants of primitive as well as derived taxa (*Casuarina, Agathis, Dacrydium, Podocarpus*). In some island interiors, certain rock substrates, such as raised limestone reefs, continue to support taxa with strong affinities to present-day coral strand flora. In Fiji, an example of this can be found in the Naqalimare limestone massif of Viti Levu.

Despite the potential for spatial change (retreating or advancing forest fronts due to fire or changes in shoreline position), both littoral and cloud forest climates tend to be well-buffered (humid maritime vs. an upland saturated atmosphere). But whereas growth under warm, coastal conditions is likely to be rapid, it is relatively slow in cloud forest. Taken as a whole, cool, humid conditions are more likely to favor refugic sinks where rates of extinction and speciation are such that endemism and alpha diversity exceed those in intermediate lowland zones with more variable habitats.

Implications for Associated Fauna

From the foregoing it can be argued that, as with plant taxa, cloud forest will be among those animal habitats least subject to change over evolutionary time. It is this aspect that is of specific relevance to the Fiji petrel (*Pseudobulweria macgillivrayi*), a seabird known only from the cloud forest of Gau.

The Fiji Petrel

MacGillivray's or the Fiji petrel was known only from a single specimen collected in 1855 on Gau island during the visit of the survey ship HMS *Herald*. The petrel was not seen again either on Gau or at sea and was reported as either "lost" or extinct.

A search commenced for this "missing" petrel in May 1983. The initial phase spanned 16 months to August 1984 with visits of 1–3 weeks every 2 months. During the first two visits extensive discussions were held with Gau villagers to obtain local information and gauge traditional knowledge. The local name *Kacau* is clearly a petrel in Fijian, but the islanders appeared to have no reliable contemporary knowledge of petrels. Potential nesting habitat was assessed and trails cut across the northern ridge. The presence of petrels on the island was soon confirmed by the discovery of nesting burrows well below the cloud forest, which were subsequently identified as belonging to collared petrels (*Pterodroma brevipes*).

On subsequent visits spotlighting from prominent locations was initiated and then augmented by the use of amplified petrel calls. On 30 April 1984, a single Fiji petrel flew into the light. It was examined, measured, and weighed. The next day the petrel was photographed before being shown to villagers and then released (Figure 4).

Despite seven more spotlighting visits to Gau in the intervening years, no more Fiji petrels have been captured. However, in 1987, a second species of *Pseudobulweria*, the Tahiti petrel (*P. rostrata*), was recorded nesting on Gau (Plant et al. 1989). Because of the increased awareness among the villagers of Gau and visiting ornithologists, the number of observations of the Fiji petrel has risen to eight. The majority of these are juveniles which presumably have just left the nest and are attracted to village lights at night.

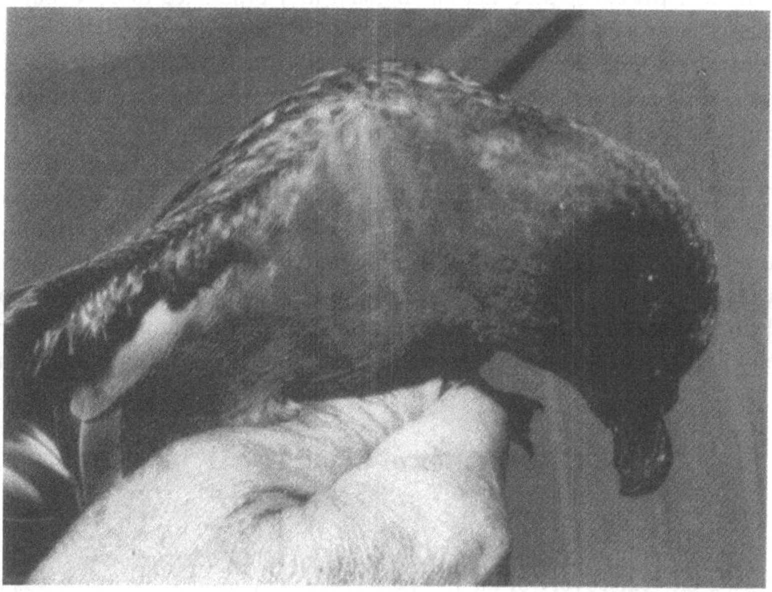

Figure 4. The Fiji petrel *Pseudobulweria macgillivrayi. Photo by Dick Watling.*

Origin of the *Pseudobulweria* Petrels

The Fiji petrel and its five relatives were for many years regarded as a distinct group of Gadfly petrels. However, Imber (1985) argued that they represent a much more ancient lineage, a lineage that diverged from the ancestors of the *Procellaria* (the central lineage of the Gadfly and related petrels) a very long time ago, perhaps during the Oligocene or Miocene epoch (Imber 1985). The *Pseudobulweria* petrels are morphologically heavier and sturdier than the Gadfly petrels, and in this manner resemble the modern *Procellaria* petrels. Unlike the Gadfly petrels, which are agile tree climbers and so can nest at any altitude, *Procellaria* petrels are relatively cumbersome and are solely ridge-top nesters with prominent take-off and landing sites. It is believed by the authors that the Fiji petrel is similar and is nesting somewhere in the cloud forest ridges of Gau. Many hundreds of hours of searching by several researchers and assistants have failed to find any nesting burrows.

With the exception of the Tahiti petrel, which is quite widespread in the Pacific, the three other extant *Pseudobulweria* petrels are very rare and one, *P. rupinarum* from St. Helena, is extinct. In addition to the Fiji petrel, which is quite obviously critically endangered, Beck's petrel (*P. becki*) is known from only two specimens taken more than 60 years ago from the waters off the northern Solomon Islands. The Reunion petrel (*P. atterima*) was "rediscovered" in the early 1970s on Reunion Island and is now known from only three specimens.

The four extant members of the group are tropical species. We speculate that, as with other petrels of procellarian ancestry, they originated in cool southern latitudes. During a glacial maximum they expanded their range into all three tropical oceans, but with the onset of warmer times survive today only in suitable refugia. It is believed that the cloud forest of Gau has provided a stable refuge for the Fiji petrel as well as for the more adventurous Tahiti petrel.

The intriguing question is: Why is the Fiji petrel apparently restricted to Gau? Two possibilities are (1) the low altitude of the cloud forest (perhaps the petrel is averse to flying far over land); and (2) the absence of feral pigs. Feral pigs are a major predator of nesting petrels and can wipe out colonies with ease. Gau is the largest and by far the most diverse island in Fiji without feral pigs. If this has always been the case then this circumstance has profound implications for the flora as well, something well worth investigating.

Conservation of Cloud Forest

Gau

Following the rediscovery of the Fiji petrel on Gau, considerable effort was diverted into conservation measures. These focused more on community awareness and education rather than on attempts to set up a protected area, because:

1. the location of the nesting site or colony is unknown;
2. the uplands are very little used by the Gau villagers (no pig hunting, just

some gathering of wild yams), and all intervillage communication is by sea or the coastal road;

3. establishment of "protected areas," a new concept in Fiji, is difficult because of the communal tenure of landholding and the lack of a dedicated government agency to undertake it; and
4. increasing the number of people and paths on the high ridges, in the absence of management, is likely to attract feral cats, a known predator of petrels on Gau, and cause damage to the cloud forest.

Community awareness has been used to engender pride in the *Kacau ni Gau* (Gau petrel) as it is referred to, and this has been very successful. Initially, discussions were held in each village; these were followed with slide presentations and lectures by the National Trust for Fiji and the distribution of a poster and pamphlet. Fiji's domestic airline, Fiji Air, which flies to Gau, has adopted a stylized *Kacau* as the company emblem. The school magazine of the Gau Secondary School is called *Kacau*.

This awareness needs to be continually reinforced, however. Researchers working on Gau are a positive way of doing this, and cloud forest research is greatly encouraged. The research on the Fiji petrel is now intermittent; as much as is considered possible has been done in the opportunistic manner to date. Funds and personnel for the concerted effort which is required are not available, either within government or elsewhere.

Other Fijian Islands

The only protected area in Fiji that is known to have cloud forest is the 1,322 ha Mount Tomaniivi Nature Reserve around the nation's highest peak on Viti Levu. However, it is limited to a ridge of less than a kilometer in length at an altitude of more than 1,200 m. The ridge thicket forest described by Berry and Howard (1973) is classified as a Protection Forest for Department of Forestry management purposes and as such cannot be logged. Enforcement of this policy is currently very poor, nonetheless there should be little reason for logging this forest type. More damage is likely to occur from forestry roads whose normal location is along ridges.

Protection of representative areas of forest types is poorly developed in Fiji, although it is now receiving attention through recent policy shifts within the Department of Forestry and the creation of a Department of Environment. Cloud forest, however, is receiving no specific attention.

References

Ash, J. 1987. Stunted cloud-forest in Taveuni, Fiji. *Pac. Sci.* 41:191–199.

Berry, M. J., and W. J. Howard. 1973. Fiji forest inventory. Land Resource Study no. 12. Vols 1–3. Land Resources Division/Overseas Development Administration/ Foreign and Commonwealth Office, London.

Gillison, A. N. 1970a. Dynamics of biotically induced forest/grassland transitions in Papua New Guinea. MSc. thesis. Forestry Dept., Australian National University, Canberra.

Gillison, A. N. 1970b. Structure and floristics of a montane grassland/forest transition, Doma Peaks region, Papua. *Blumea* 28:71–86.
Gillison, A. N. 1992. Report on terrestrial resource management. Fiji National Environmental Management Project. Asian Development Bank TA 1206, Suva.
Gillison, A.N. In press. The use of plant functional attributes in natural resource survey: A case study from Fiji. *Pac. Cons. Biol.*
Grubb, P. J. 1971. "Massenerhebung" effect on tropical mountains. *Nature London* 229:44–45.
Hassal, D. C., and J. B. Kirkpatrick. 1985. The diagnostic value and host relationships of the dependent synusia in the forests of Mt. Korobaba, Fiji. *N. Z. J. Bot.* 23:47–54.
Imber, M. J. 1985. Origins, phylogeny and taxonomy of the gadfly petrels, *Pterodroma* spp. *Ibis* 127:197–229.
Kirkpatrick, J. B., and D. C. Hassal. 1985. The vegetation and flora along an altitudinal transect through tropical forest at Mt. Korobaba, Fiji. *N. Z. J. Bot.* 23:33–46.
Plant, A. R., K. Moce, K. Vererusa, and D. Watling. 1989. A Tahiti petrel *Pseudobulweria rostrata* from Gau Island, Fiji. *Notornis* 36:149–150.
Smith, A. C. 1979. *Flora Vitiensis Nova.* Vol 1. Pacific Tropical Botanic Garden, Lawai, HI.
Smith, J. M. B., and A. M. Cleef. 1988. Composition and origins of the world's tropicalpine floras. *J. Biogeogr.* 15:631–645.
Watling, D. 1985a. The distribution of Fijian land and freshwater birds based on the collections and observations of the Whitney South Sea Expedition. *Domodomo* 3: 130–152.
Watling, D. 1985b. Notes on the birds of Gau I. Fiji. *Bull. Brit. Orn. Club* 105(3): 96–102.
Watling, D. 1986a. Notes on the Collared Petrel *Pterodroma (leucoptera) brevipes.* Bull. Brit. *Orn. Club* 106:63–70.
Watling D. 1986b. Rediscovery of a petrel and new faunal records from Gau Island. *Oryx* 20(1): 31–34.

25. Biogeography and Ecology of the Upper Montane Rain Forest of Sri Lanka (Ceylon)

Wolfgang L. Werner

Tropical evergreen rain forest forms the natural vegetation of the mountains of Sri Lanka. Lower montane rain forest covers the slopes between 900 and 1,500 m asl. The 1,500-m line is a sharp boundary between lower and upper montane rain forest, as it is the lower limit of a frequent and dense cloud cover. The highest peak, Pidurutalagala (2,524 m), is still below timberline. As conifers and Fagaceae are lacking, their ecological role is taken over by other plant families. Besides Lauraceae and Myrtaceae, Clusiaceae and Symplocaceae have an exceptionally high frequency. *Strobilanthes* species (Acanthaceae) play a special role in the ecology of these forests, as their cycles of flowering, fruiting, and dieback dominate life on the forest floor.

Since 1823, large portions of the forest in the mountains of Sri Lanka have been cleared for the British plantation industry. Nature reserves protect about 50 percent of the remaining forest. About 400 of the original 600 km^2 of upper montane rain forest remains, and there are substantial areas of tropical montane cloud forest.

The Environment

Geology

The mountains in the central part of Sri Lanka are composed of Precambrian rocks, consisting of gneisses with bands of crystalline limestone (Cooray

1967). The shape of the mountain chains in the highest parts resembles an anchor: the Peak Wilderness and Haputale ranges run from west to east in a slight curve, and the highest plains (1,800–2,100 m) are situated in the center, overtopped by Pidurutalagala. From this peak a mountain chain runs northwest (Figure 1). To the north of this mountain complex, the Knuckles range builds up into an isolated, northwest-southeast aligned group (de Rosayro 1958).

Climate

The tropical climate of Sri Lanka is influenced by the southwest monsoon (June–September) and the northeast monsoon (December–February), both of which bring rain to the exposed sides of the island. But only on southwestern or northeastern mountain slopes will the monsoons affect maximum precipitation (>5,000 mm/year); in most parts of the island the maximum is produced by convectional rains of the intermonsoonal seasons. Minimum precipitation occurs on the eastern slopes during the southwest monsoon, combined with hot and dry föhn winds. Precipitation in the highlands decreases from west to east from 5,000 to 2,000 mm. A dense fog layer commonly envelops the mountains above 1,500 m. Average annual temperature at 1,900 m (Nuwara Eliya) is 15.4° C; slight frosts may occur occasionally above 1,800 m between November and May. Very strong winds blow during the southwest monsoon, especially on west-exposed ridges. In funnel-shaped passes they may reach gale force, thus influencing vegetation stature (Mueller-Dombois 1968; Domrös 1974; Schweinfurth 1984; Werner 1988a).

Forest Type Below Upper Montane Rain Forest

The natural vegetation of the lowland is composed of evergreen rain forest in the "wet-zone" southwest of the central highlands and semideciduous (monsoon) forest types in the "dry zone" (Gaussen et al. 1965; Erdelen 1988). On the western slopes of the highlands, between 600 and 900 m, the lowland rain forest gradually changes into the lower montane rain forest. In favorable situations the trees may attain 35 m in height. Most prominent is the endemic Dipterocarp *Doona gardneri*, which is accompanied by other species of *Doona* and *Stemonoporus* (Dipterocarpaceae), *Myristica dactyloides, Fahrenheitia* (Ostodes) *zeylanica*, and various species of *Calophyllum, Syzygium,* and Lauraceae. On steep slopes, the palms *Oncosperma fasciculatum* and *Loxococcus rupicola* occur.

On the eastern slopes the lower montane rain forest exists in a dry facies without Dipterocarpaceae. Lauraceae (*Litsea, Cinnamomum, Cryptocarya, Neolitsea*), *Calophyllum, Syzygium, Myristica, Oncospermum*, and so on occur as on the western side, but with more floristic importance (cf. Werner 1989). In some parts of the eastern slopes where strong foehn winds occur,

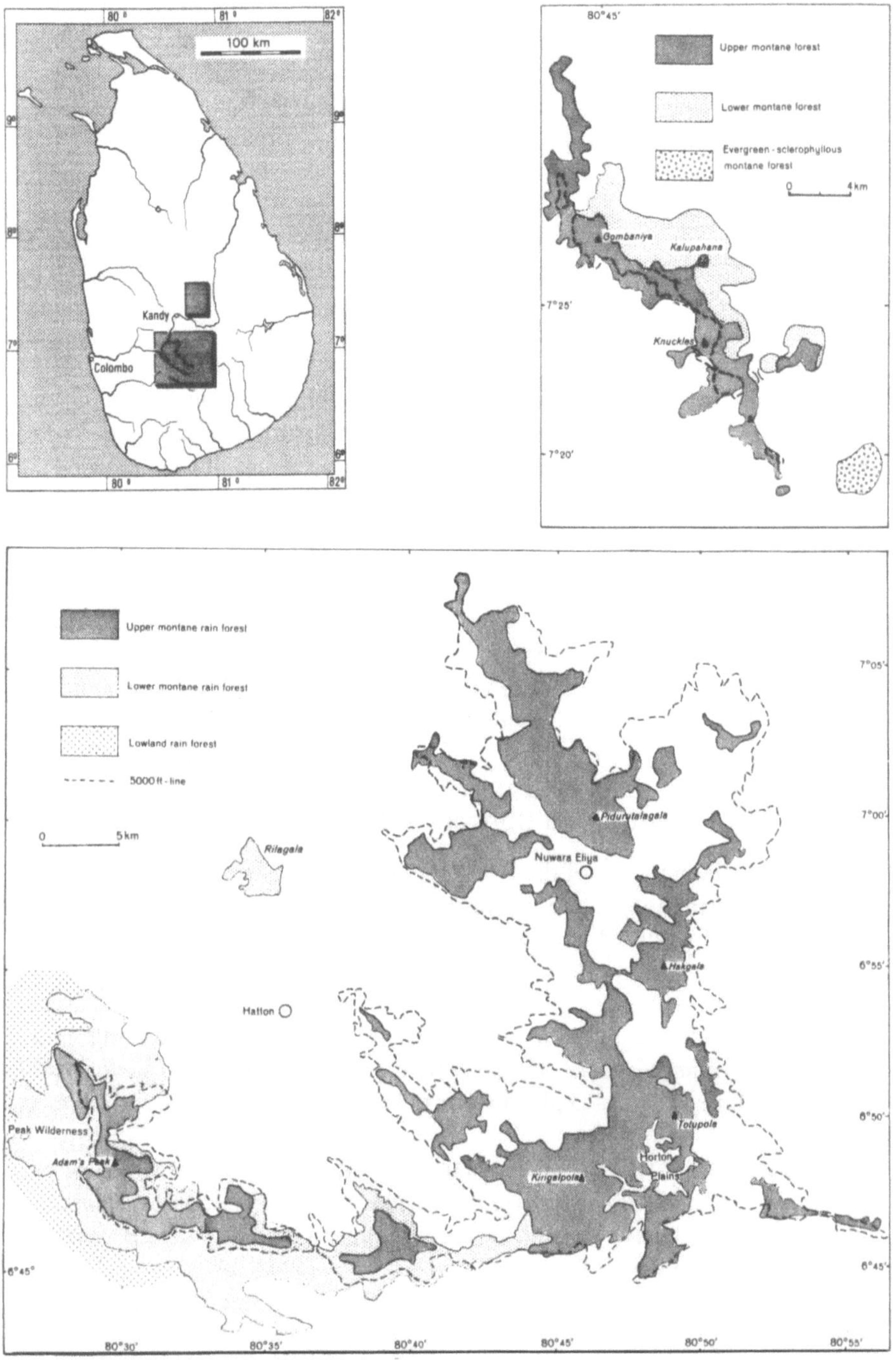

Figure 1. Forests of the mountainous area of Sri Lanka.

the natural vegetation between 600 and 1,500 m was a sclerophyllous montane forest. Today this is preserved in small remnants only. *Syzygium spathulatum* is a dominant species, accompanied by species with similar leaves: *Actinodaphne stenophylla, Calophyllum trapezifolium, Pittosporum tetraspermum, Dodonea viscosa*, and so on (Werner 1984).

Upper Montane Rain Forest

Lower Limit

At about 1,500 m the upper montane rain forest or "mist" forest begins. The change from lower to upper montane rain forest is abrupt and may be explained by the lower limit of the frequent cloud cover at that altitude, which develops during late morning and lasts until evening. During the southwest monsoon, the cloud cover often remains day and night. This is essentially cloud forest.

Upper Limit

No other vegetation type exists above the upper montane rain forest of Sri Lanka, and forest cover extends to the island's highest peaks (2,524 m at Pidurutalagala). But on the exposed peaks and ridges the trees remain stunted because of winds blowing with gale force during the southwest monsoons. A "pygmy forest" is thus formed.

Vertical and Horizontal Differentiation

No further differentiation of the forest type is justified, but some differences in vertical distribution of tree species occur.

Below 1,800 m elevation, some species of the lower montane rain forest reach into the upper montane rain forest. *Gordonia speciosa, Semecarpus nigro-viridis, Palaquium rubiginosum*, or *Ternstroemia japonica* are examples. Other tree species occur only above 1,800 m, such as *Actinodaphne ambigua* or *Syzygium rotundifolium*. This zone between 1,500 and 1,800 m forms a specific ecotone, whereas in physiognomy, the 1,500-m line is a sharp boundary.

Above 2,000–2,100 m some species disappear, such as *Symplocos cordifolia, S. cochinchinensis, Garcinia echinocarpa*, and *Gordonia zeylanica*; other species reach only to 2,400 m. This produces a gradual pauperization of the flora along the vertical gradient.

Above 2,400 m, forest is found only around the summit of Pidurutalagala, where the dwarf-bamboo *Indocalamus wightianus* replaces *Strobilanthes* as the dominant plant of the understory. *Rhododendron arboreum* is the most frequent tree in the summit area. The very top of the mountain has been cleared by humans since the last century. The more remote Mount Kirigalpota (2,389 m) gives us some idea of undisturbed vegetation on a summit, with a cover of *Rhododendron* and *Rhodomyrtus*.

Although the annual rainfall decreases from 5,000 mm to less than 2,500 mm from west to east, the composition within the upper montane rain forest of the central highlands hardly differs. The frequent mists and their additional moisture may be an explanation. Only some large-leaved *Syzygium* species of the lower montane rain forest around Adam's Peak, which range into the upper montane rain forest, are restricted to that western area. The isolated mountain range of the Knuckles, being separated from the central highlands by the Mahaweli Ganga valley, contain some endemic species of *Syzygium/ Eugenia*, whereas some species are more frequent there than in the central mountains.

Physiognomy

This forest type is rich in epiphytes. Mosses and filmy ferns cover the stems and branches. Lichens hang from the twigs. Various species of epiphytic orchids live on the branches. *Eria bicolor* is the most common species and is called "Lily of the Valley orchid" by the British. The height of the trees varies greatly and depends on wind exposure and soil.

The tallest trees, mostly *Calophyllum walkeri*, reach 20 m and overtop the dense canopy. Such tree stands can be observed in some sites of Horton Plains (Figure 2). The trees of "typical" upper montane rain forest in Sri Lanka normally reach less than 5 m high. Besides some emergents, the tree stands form a single story, and only very few young trees can be observed below the canopy. *Actinodaphne speciosa* (Elephant ear), with its large and coriaceous leaves, normally does not form part of the dense canopy. Only the uppermost branches reach that height. The crowns of most trees are umbrella-shaped; the branches are gnarled (Figure 3). The size of the leaves is small, from nanophyll to microphyll, and some even leptophyll.

Biogeography of Plants

The floristic composition of the upper montane rain forest of Sri Lanka is very peculiar, as many characteristic plants of the montane forests in tropical Asia do not occur (Whitmore 1984; Ashton and Gunatilleke 1987). Conifers such as Podocarpaceae or *Pinus*, Fagaceae (*Lithocarpus, Castanopsis*), *Schima* (Theaceae), or *Leptospermum* (Myrtaceae) are lacking. Dipterocarpaceae, playing an important role in the lower rain forest of Sri Lanka, are not represented in the cooler environment of the upper montane rain forest.

In frequency and tree size the most prominent genus is *Calophyllum* (Clusiaceae). Next to *Calophyllum* are the Lauraceae (*Cinnamomum, Litsea, Actinodaphne*), Myrtaceae (*Syzygium/Eugenia, Rhodomyrtus*), and various species of *Symplocos* or *Elaeocarpus*. *Gordonia, Adinandra,* and *Ternstroemia* (Theaceae) are frequent below 2,200 m, and *Eurya japonica* frequent above this elevation. Palaearctic elements are restricted to the montane zone above 1,500 m. These plants are typical in South India mountains or even in the Himalayas (e.g., *Michelia nilagirica, Ilex wigthtiana, Berberis aristata,* and

Figure 2. Dense canopy of forest with some emergents of *Calophyllum walkeri* in Horton Plains TMCF.

Figure 3. Umbrella-shaped crowns and gnarled stems and branches of cloud forest trees.

Rhododendron arboreum). Tree ferns (Cyatheaceae) are widespread in the upper montane rain forest and occur frequently along watercourses.

Balasubramaniam, Ratnayake, and White (in press) found that 50 percent of individual species were endemic for Sri Lanka, and more than 40 percent were endemic for the mountains of Sri Lanka and South India. This indicates that the montane vegetation of Sri Lanka is very isolated, although it is a continental island, and that the montane forests of South India and Sri Lanka form a biogeographical unit. Moreover, this proves that the indigenous element of the flora is much more vigorous than the "holarctic" elements, which probably invaded during the Pleistocene. The importance of Lauraceae and Myrtaceae is typical in a tropical montane forest, especially as Fagaceae are lacking, but the importance of Symplocaceae is surprising.

The lack of conifers and Fagaceae has consequences for the composition and importance of species, as these plant groups play an important role in the montane forests of Southeast Asia. Conifers are often emergents above the canopy. In Sri Lanka, this role is played by *Calophyllum* with its three endemic species *C. walkeri, C. trapezifolium,* and *C. cuneifolium*, making this the most conspicious tree. *C. walkeri* is the only species of the genus in the forests of Horton Plains and Pidurutalagala that grows above 2,000 m. *C. trapezifolium* grows in the western and northern parts of the central highlands but is most frequent in the Knuckles mountains north of the Mahaweli Ganga. There it grows together with *C. walkeri* and *C. cuneifolium*, the latter being endemic for the Knuckles. Some stands of the Knuckles are consociations of *C. trapezifolium* and *Syzygium fergusonii* (Werner 1984).

Note that in the Nilgiris of South India, where *Calophyllum* is not represented in the upper montane rain forest, *Syzygium calophyllifolium* plays the same role and could be easily confused with it. Lauraceae and Myrtaceae, which are important in other montane forests of tropical Asia, are still more frequent and have evolved numerous endemic species. The frequency of various species of *Symplocos* is a further peculiarity of the upper montane rain forest of Sri Lanka. All except *Symplocos cochinchinensis* are endemic. *S. elegans*, with its small leaves, has the highest density in the plots studied. For the families mentioned it is evident that species with small leaves form the canopy, whereas those with larger leaves remain in the understory or subcanopy.

Under a dense canopy, only sparse regrowth of trees can be observed. The dominant plant of the understory is *Strobilanthes* (Acanthaceae). After 6 to 12 years the plants will flower and fruit at the same time in any one area, and then die and collapse. From the seeds, a new generation will grow. While the regeneration is still short, the forest may be easily penetrated, but when it gets taller, the trees form an entangled mass. The dense undergrowth of *Strobilanthes* spp., which reaches 3 m in height, plays an important role in the dynamics of these forests, as it suppresses establishment and growth of tree saplings. Other species get an opportunity when the *Strobilanthes* population dies after flowering. Once *Strobilanthes* stands have grown, they form impenetrable forests. In former days, elephants were used to make tunnel-shaped paths through these forests. *Strobilanthes* is not only a competitor with tree saplings but also with herbs on the forest floor (such as *Impatiens* sp., *Amomum involucratum*) or with ferns. Sometimes it is replaced by dwarf-bamboo species, such as *Ochlandra stridula, Indocalamus debilis*, or *I. wightianus*, which have a similar life cycle and dieback after the flowering of a whole stand.

The nutrient content of soils in the lower and upper montane rain forest is higher than in the lowland. The content of Fe and Al is strikingly higher than in lowland soils. The high content of Al can have a poisonous and thus selective effect for various plants. Analysis of elements in leaves of forest trees from various forest types reveals an extremely high content of Al in the leaves from upper montane rain forest (Werner and Balasubramaniam 1992). This finding is also confirmed by Jayasekara (1992).

Secondary Vegetation

Where upper montane rain forest is able to regenerate, it consists of pioneer trees such as *Acronychia laurifolia, Embelia viridifolia*, and *Evodia roxburghiana,* Lauraceae-like *Litsea ovalifolia* or *L. gardneri*, *Carallia calycina* (Rhizophoraceae), *Eurya japonica, Maesa indica,* or *Agrostistachys coriacea.* Up to 1,500 m, *Macaranga peltata* is the typical pioneer tree, often accompanied by *Homalanthus populifolius* or *Mallotus albus.*

After felling or burning of natural montane forest, low secondary forest or grassland (patana) develops. *Rhododendron arboreum* is the only tree on these grasslands above 1,500 m, as it survives the naturally occurring fires and occasional frosts. The patanas consist of two dominant species of tussock grass, *Chrysopogon zeylanicus* and *Cymbopogon confertiflorus*. In wet depressions, *Arundinaria densifolia* dominates.

Regular fires on the patanas enlarge their area, as the forest fringe is often destroyed, although the wet upper montane rain forest normally does not burn. Regrowth of trees in the grassland is suppressed by various factors, such as dense grass cover, fire, frost, and browsing animals.

Human Impact and Conservation

Although prehistoric artifacts have been found on Horton Plains (2,100 m), human influence on montane vegetation was only slight until 1815. After conquest of the Kingdom of Kandy by the British, the highlands of the island were developed, and the forest was cleared for coffee estates beginning in 1823. Tea cultivation replaced coffee in 1867. In 1890 afforestation began with exotic timber trees such as *Eucalyptus*, *Cupressus, Acacia,* and later *Pinus* (Werner 1988b).

Although the lower montane rain forest was destroyed almost completely, some intact areas do exist on the slopes of Peak Wilderness. The upper montane rain forest has been better conserved since 1885, and about 400 km^2 of the original 600 km^2 remains. These forest areas occupy the upper catchments of the island's important rivers and are today threatened by illicit free-felling for firewood.

Three substantial nature reserves exist in the mountains of Sri Lanka: Hakgala Strict Natural Reserve (1,142 ha), Horton Plains National Park (3,162 ha), and Peak Wilderness Sanctuary (22,400 ha), which still contain interesting endemic faunal elements (Werner 1985a; 1986).

Fauna

The Asian elephant (*Elephas maximus*) once roamed the upper montane rain forest of Sri Lanka, where they formed tunnel-shaped paths through the undergrowth of *Strobilanthes*. These mountain herds were largely destroyed during the British colonial period. The last elephants disappeared from Horton Plains before independence in 1948, but some still remain in Peak Wilderness. The Sambhar deer (*Cervus unicolor*), the next largest ungulate, is still common in these forests, as well as the barking deer (*Muntiacus muntjak*) and wild boar (*Sus scrofa*). The leopard (*Panthera pardus*) is the largest carnivore of the upper montane rain forests of Sri Lanka and still survives even close to settlements and in smaller forest patches.

Three species of monkeys occur. These all have formed distinctive montane races with dense fur and large size. They include the bear monkey (*Trachypithecus vetulus*), which is the largest, bonnet macaque (*Macaca sinica*), and slender loris (*Loris tardigradus*). Unique mammal species include two endemic and monotypic shrews, *Feroculus feroculus* and *Solisorex pearsonii*, which inhabit swamps in and around the tropical montane cloud forest (TMCF) of Sri Lanka.

None of the endemic bird species and races are confined exclusively to the TMCF, but 14 of the 21 endemic Sri Lanka species occur there. Most common is the Ceylon white-eye (*Zosterops ceylonensis*). In addition to resident-breeding birds, various migrants from the Himalayas and northern Asia stay during winter in the montane forest zone (Werner 1985a).

Several endemic lizards of the genera *Ceratophora* and *Lyriocephalus* occupy this zone, and the widespread *Calotes* has developed into a distinct species (*C. nigrilabris*) in the upper montane rain forest.

Like plants, the animal species show a mixture of tropical elements of Oriental and of northern Holarctic latitudes. The latter category includes wild boar, nuthatch, grey tit, and wagtail. Tropical species nonetheless dominate, more than in other continental mountains of Southeast Asia.

Summary

The upper montane rain forest of Sri Lanka is comparable to other forests in tropical mountains as to its physiognomy, but it is unique in its composition of plant and animal species. Next to it is the upper montane rain forest of South India (*shola*). Both even have some locally endemic species in common. But numerous differences exist. The Sri Lankan forest is also the catchment area for the rivers of the island, and thus is very important for hydrology and soil protection.

References

Ashton, P. S., and C. V. S. Gunatilleke. 1987. New light on the plant geography of Ceylon. *Journal of Biogeography* 14:249–285.

Balasubramaniam, S., S. Ratnayake, and R. White. In press. The montane forests of the Horton Plains Nature Reserve.

Cooray, P. G. 1967. *An introduction to the geology of Ceylon.* London.

de Rosayro, R. A. 1958. The climate and vegetation of the Knuckles region of Ceylon. *Ceylon Forester* 2(3–4): 210–260.

Domrös, M. 1974. The agroclimate of Ceylon. Vol. 2, *Geoecological Research*, edited by U. Schweinfurth. Wiesbaden: Steiner Verlag.

Erdelen, W. 1988. Forest ecosystems and nature conservation in Sri Lanka. *Biol. Conserv.* 43:115–135.

Gaussen, H., et al. 1965. *Carte Internationale du Tapis Végétal 1:1.000.000—Ceylan.* Pondichérry.

Jayasekara, R. 1992. Elemental concentrations in a tropical montane rain forest in Sri Lanka. *Vegetatio* 98:73–81.

Mueller-Dombois, D. 1968. Ecogeographic analysis of a climate map of Ceylon with particular reference to vegetation. *Ceylon Forester* 8(3–4): 1–20.

Schweinfurth, U. 1984. The central highlands of Ceylon. *Trop. u. subtrop. Pflanzenwelt* 46:9–39.

Werner, W. L. 1984. Die Höhen-und Nebelwälder auf der Insel Ceylon (Sri Lanka). *Trop. u. subtrop. Pflanzenwelt* 46, edited by W. Rauh. Wiesbaden: Steiner Verlag.

Werner, W. L. 1985a. Naturreservate im Hochland der Insel Ceylon (Sri Lanka)—Peak Wilderness und Horton Plains. *Natur und Museum* 115(3): 65–76.

Werner, W. L. 1985b. The upper montane rain forests of Sri Lanka. *The Sri Lanka Forester* 15(3–4): 119–135.

Werner, W. L. 1986. A comparison between two tropical montane ecosystems in Asia: Pidurutalagala (Ceylon)—Pangrango-Gede (Java). *Mountain Research and Development* 6(4): 335–344.

Werner, W. L. 1988a. Canopy dieback in the upper montane rain forests of Sri Lanka. *Geojournal* 17(2): 245–248.

Werner, W. L. 1988b. Human impact on natural environment in the central highlands of Sri Lanka. *Universitas* 30(1): 29–37.

Werner, W. L. 1989. Die Waelder des östlichen Hochlandes von Ceylon. *Erdkundl. Wissen* 97:43–72 (Forschungen auf Ceylon 3).

Werner, W. L., and S. Balasubramaniam 1992. Structure and dynamics of the upper montane rain forests of Sri Lanka. In *Tropical forests in transition*, edited by F. G. Goldammer, 165–172. Basel; Boston: Birkhauser Verlag.

Whitmore, T. C. 1984. *Tropical rain forests of the Far East.* Oxford.

26. The Cloud Forest of Samoa

W. Arthur Whistler

Introduction

The Samoan archipelago is situated in the South Pacific Ocean at a latitude of 13°–15° S and a longitude of 168°–173° W, and runs in a west-northwest direction east of Fiji, north of Tonga, south of Tokelau, and west of Niue and the Cook Islands. Its nine inhabited islands and several uninhabited islets, plus two distant coral islands, have a total area of ca. 3,100 km^2.

The archipelago is divided politically into Western Samoa, an independent country, and American Samoa, an unincorporated territory of the United States; the two political units are separated by a strait 64 km wide. Western Samoa is by far the larger of the two with about 93 percent of the land area of the archipelago. It comprises two main islands: Savai'i (1,820 km^2 area, 1,860 m maximum elevation) and 'Upolu (1,110 km^2, 1,100 m), which are separated by a 21 km wide strait (Figure 1). American Samoa has four small high islands, the largest of which is Tutuila (124 km^2, 650 m). Approximately 100 km to the east lies the Manua group, which comprises Ta'u (39 km^2, 960 m), Ofu (5 km^2, 495 m), and Olosega (4 km^2, 640 m) (Figure 2).

The islands are volcanic in origin and were formed from basalt rising from the ocean floor of the Pacific basin, beyond the continental islands to the west. The archipelago is still volcanically active: two eruptions have been recorded in historic times on the island of Savai'i—most recently ending in 1911.

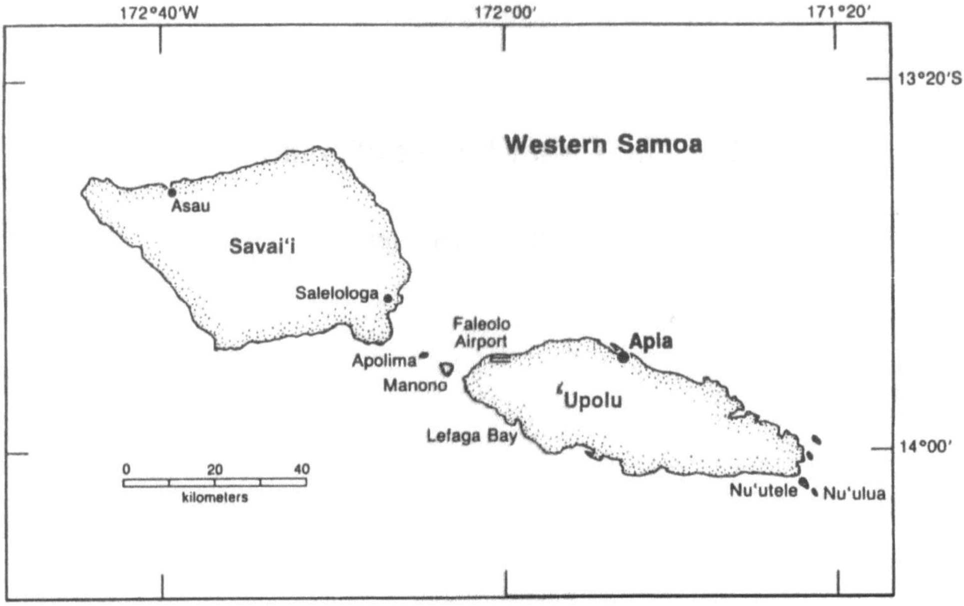

Figure 1. The islands of Western Samoa.

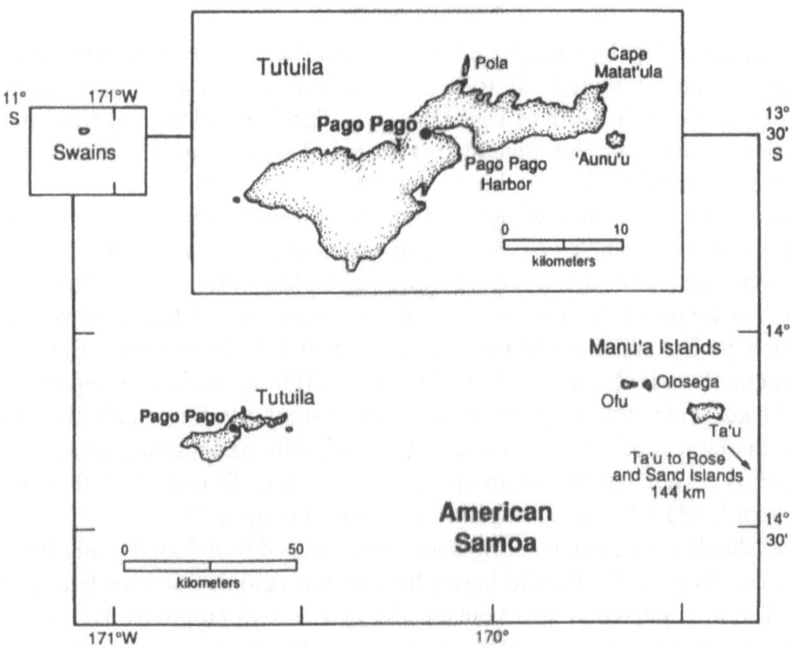

Figure 2. The islands of American Samoa.

Climate

Situated between the Tropic of Capricorn and the equator, Samoa has a tropical maritime climate. Typical temperatures (at sea level) are 23°–30° C. The difference between the mean temperatures of winter (June–September) and summer (December–March) is less than 2° C, and the average annual temperature is 26°–27° C. The relative humidity is constantly high, averaging over 80 percent throughout the year. Rainfall is heavy throughout the archipelago, with an annual minimum of 2,000 mm in all places on the main islands. Half of the rainfall occurs from December to March, but there is no real dry season, since even in the driest places (which are on the leeward or north and west sides of the islands) all months average over 50 mm of precipitation. Droughts of varying duration occasionally occur, but these do not have much permanent effect on the vegetation. Hurricanes periodically hit the islands; although there has been a rash of destructive hurricanes in the last 6 years, serious storms average less than one a decade.

Vegetation and Flora

With elevations up to 1,860 m, Samoa exhibits diverse environmental conditions and a corresponding variety of vegetation types or plant communities. The islands were originally covered almost entirely with native tropical rain forest; the only plant communities not classified as such were littoral communities (herbaceous strand, littoral shrubland, Pandanus scrub, and littoral forest), wetlands (coastal marsh, montane marsh, montane bog, freshwater swamp, and mangrove swamp), upland scrub vegetation (montane scrub and summit scrub), volcanic vegetation (lowland and upland volcanic scrub), and secondary vegetation (Whistler 1992b).

Tropical rain forest may be defined as the tall native evergreen forest occurring in the tropics in areas with sufficient precipitation throughout the year. Characteristics of rain forest include a profusion of epiphytes, a relatively open forest floor, and high species diversity. Some authors describe tropical rain forests as being at least 30 m in height, but few Samoan trees reach this height, possibly because of the periodic hurricanes. While the diversity of species in Samoa is high compared to that of temperate areas, it is much lower than in the islands to the west (Melanesia) because of the formidable oceanic barrier and the resulting "filter effect" that inhibits many species from arriving or becoming established very far from their native range.

The Samoan tropical rain forest is not homogeneous and can be subdivided into lowland rain forest, montane rain forest, and tropical montane cloud forest (TMCF). Some of these communities can be further divided into associations based on dominant species.

Environmental disturbance over the 3,000 years of human occupation has led to the loss of much of the native vegetation. Although over half of its native vegetation has been severely altered by human activity and natural catastrophes, Samoa, with its large area and rugged topography, retains more intact native forest than many other Polynesian islands. Much of the anthropogenic change has occurred in the last few decades, as an increasing population, a major forestry industry, and three destructive hurricanes have taken their toll.

The native flora of Samoa consists of approximately 536 flowering plants and 228 ferns and fern allies. The rate of endemism is only about 32 percent for flowering plants, and only a single genus, *Sarcopygme* of the Rubiaceae, is endemic. Most of the lowland species belong to wide-ranging Indo-Pacific genera, but some temperate Southern Hemisphere genera, such as *Weinmannia* and *Coriaria*, occur in the cool, high elevations of the mountains.

The Cloud Forest

As noted, the tropical rain forest of Samoa is not homogeneous, but can be divided into a number of smaller units based on forest structure and flora. One of these units, or plant communities, is cloud forest. The term "cloud forest" has been used differently by various authors, but in trying to introduce some consistency of usage, Stadtmüller (1987) defined cloud forest as "all forests of the humid tropics that are frequently covered in clouds or mist, thus receiving additional humidity, other than rainfall, through the capture and/or condensation of water droplets." TMCF typically occurs in areas where clouds and mist appear in combination with heavy orographic rainfall, and correspond to the zone of maximum cloud condensation. Stadtmüller further noted that cloud forests are characterized by the abundance of epiphytes, especially mosses and filmy ferns (family Hymenophyllaceae), and by the presence of tree ferns (typically *Cyathea* spp.).

Most of the work on TMCF has been done in tropical America, Southeast Asia, and Africa—all on continental land masses or large islands—so the situation on the small oceanic islands of the tropical Pacific may be at variance with the literature. For example, Stadtmüller described the lower limit of cloud forest as beginning between 1,500 and 2,500 m elevation. In Samoa, it is much lower than this, perhaps as low as 600 m.

The islands of Samoa are relatively small and much of their precipitation is orographic. During most of the year (April–December) the tradewinds blow from the southeast. The warm air, nearly saturated with water vapor, is forced to ascend the slopes, particularly where the longest axis of the island is perpendicular to the wind direction. The rising moist air cools, and when it reaches the dew point, water condenses to form clouds. It is these almost daily clouds that envelop the upland forest and result in the "fog drip" that helps create the cloud forest environment in Samoa.

While it is practical to define the cloud forest as Stadtmüller has done, determining its boundaries is a much more difficult and imprecise matter.

There is no distinct boundary between it and the rain forest at lower elevations, which in Samoa has usually been termed "montane forest." The elevation at which clouds build up is related to the humidity and temperature of the air. Because both of these factors vary little from day to day in Samoa, on most days there should theoretically be a relatively narrow elevational zone at which the clouds form. However, there is little climatological data away from the coast to elucidate this. At or near the summit of Ta'u, the forest is contorted and stunted, but on Savai'i changes in tree form are subtle. It is the increased epiphyte load that is most distinguishing in these situations.

Clouds form almost daily on Ta'u (960 m maximum elevation), Olosega (640 m), 'Upolu (1,110 m), and Savai'i (1,860 m), and all of these islands have TMCF at their summits. Although Tutuila has sufficient elevation (650 m), the major summits are covered with a natural fern-dominated scrubby vegetation known as "montane scrub," which appears to be limited in Samoa to that island (Figure 3). The other "high" island, Ofu, appears to lack sufficient elevation (495 m) to have an identifiable cloud forest.

Only three limited botanical studies have been conducted in the cloud forests of Samoa. The first was based on a 9-day visit to the summit of Savai'i (Whistler 1978); the second was a general study of the vegetation of American Samoa (Whistler 1980), which dealt with the cloud forests on Ta'u and Olosega; and the most recent one was a report on the vegetation and flora of Ta'u (Whistler 1992a).

The summit of Ta'u is quite different from the higher elevations of the other islands because of its unique "summit scrub" community dominated by vines, shrubs, and stunted trees. This unusual community, which covers much of Ta'u above 600 m, is similar to one recently seen by the author on Ambrym Island in Vanuatu, and may be caused by periodic hurricane damage. Another possible factor is the apparent higher-than-normal rainfall on the summit of Ta'u, caused, perhaps, by the steep south-facing side of the island that forces the tradewinds to make a rapid ascent to 960 m and drop their precipitation over a narrow zone. At similar elevations on Savai'i, the native vegetation is high forest.

The cloud forest and wet conditions extend up to the highest elevations of Samoa, but this is not the case on all tropical "high" islands. In Hawai'i, for example, an inversion layer of relatively warm air often occurs between 1,600 and 3,000 m elevation, overlying the cooler air below. This layer effectively limits the upward motion of the air, causing decreased precipitation, reduced fog drip, and relatively dry conditions at higher elevations (Wagner, Herbst, and Sohmer 1990).

There is no climatological data from Samoa to indicate that an inversion layer has much of an effect on the cloud forest. Only Savai'i is high enough to be affected, but there is cloud forest all the way up to its summit (at 1,860 m elevation). However, the rainfall there may be less than it is at lower elevations: dry conditions (wilted herbs) were noted in the summit region by the author during a June (dry season) visit in a particularly wet year (1992).

Figure 3. "Montane scrub" vegetation on Tutuila. *Photo by author.*

The dominant tree species of the Samoan cloud forest are quite different than those found in the lower elevation rain forest, and the forest canopy is not as tall, but it cannot be considered stunted. The dominant trees noted at 1,600 m on Savai'i (Whistler 1978) included, in decreasing order of importance, *Weinmannia samoense* (Cunoniaceae), *Reynoldsia pleiosperma* (Araliaceae), *Omalanthus acuminatus* (Euphorbiaceae), *Dysoxylum huntii* (Meliaceae), *Coprosma savaiiense* (Rubiaceae), *Streblus anthropophagorum* (Moraceae), *Geniostoma rupestre* (Loganiaceae), *Psychotria xanthochlora* (Rubiaceae), and *Scaevola nubigena* (Goodeniaceae). In some places, possibly those affected by recent volcanic activity, the dominant tree was *Spiraeanthemum samoense* (Cunoniaceae). In a scrubby cloud forest on top of Ta'u (Whistler 1980) at the other end of the archipelago, the dominant species recorded were *Cyathea* spp., *Syzygium samoense* (Myrtaceae),

Weinmannia affinis (Cunoniaceae), *Dysoxylum huntii, Ascarina diffusa* (Chloranthaceae), *Streblus anthropophagorum*, and *Astronidium pickeringii* (Meliaceae).

The ground cover of the Samoan cloud forest is typically very dense and is dominated by terrestrial ferns and flowering shrubs and herbs. Nearly 230 native species of ferns are found in Samoa, many of them terrestrial and restricted to the cloud forest. The most common shrubs belong to the genera *Psychotria* and *Cyrtandra,* the two largest genera in Samoa (20 species each, nearly all of them endemic). The most common nonfern herbs are orchids and several species of *Elatostema* (Urticaceae). Samoa has nearly 100 native species of orchids, many of which occur in cloud forest.

Epiphytes abound in the Samoan cloud forest (Figure 4). Samoa is well represented in mosses and filmy ferns. The filmy family Hymenophyllaceae

Figure 4. Dense epiphytic growth on tree near the summit of Ta'u. *Photo by author.*

comprises 25 species (19 *Trichomanes* and 6 *Hymenophyllum*), over half of which are found only above 400 m elevation. The tree fern genus *Cyathea* is also well represented: 9 species occur in Samoa—over half of them found only above 400 m. Characteristic nonfern epiphytes include several species of the genus *Peperomia* (Piperaceae), numerous orchids, and the endemic blueberry *Vaccinium whitmeei* (Ericaceae).

The Status of Cloud Forest in Samoa

The cloud forest of Samoa is relatively intact; it is infrequently visited and is utilized only by the occasional pig hunter. However, a poorly regulated timber industry has made inroads into its lower portion, based mainly on extraction of *Dysoxylum huntii*. The worst situation is on 'Upolu, where the majority of the archipelago's human population resides. Nearly all of the lowland and montane rain forest and cloud forest have been removed from the accessible portions of the island.

Recent hurricanes have taken their toll on the Samoan cloud forest as well. Cloud forests can recover from hurricanes, but they may not recover, or may do so only very slowly, from disturbance by man.

At present, Samoa has been spared much of the worst effects of alien animals. Although feral pigs are present, they are periodically hunted by villagers and have not caused extensive damage. Feral cattle are nonexistent or pose only a minor threat. Goats are few in number and are restricted to villages. Mongooses, deer, and sheep, which have caused such ecological damage elsewhere in the Pacific, have not been introduced. In fact, the only feral alien animals other than the pig are occasional cats.

Samoa has also been spared the worst ravages of alien plant species invasions. The most abundant weed is "mile-a-minute vine," *Mikania micrantha* (Asteraceae), which was introduced at about the turn of the century and now appears to retard forest regeneration. "Koster's curse," *Clidemia hirta* (Melastomaceae), was first reported in Samoa in 1955 and now occurs on all the main islands. It was not reported from Ta'u in 1976 (Whistler 1980), but during a recent survey was found to be invasive in the cloud forest and summit scrub (Whistler 1992a). In the 1975 survey of the summit of Savai'i (Whistler 1978) and a recent reconnaissance there in 1992, only a handful of weedy species were found, not including *Clidemia*. Perhaps the most pressing threat, at least to the lower cloud forest, is the African rubber tree, *Funtumia elastica* (Apocynaceae), which in some places on 'Upolu (mostly in the lowlands and foothills) forms monodominant forests.

Nearly all of Samoa is "customary land" owned jointly by village families, and it cannot be bought and sold. This land tenure system has both its strengths and its weaknesses. On the positive side, it has prevented wholesale alienation of the land as has happened in some other places in Polynesia (e.g., Hawai'i and New Zealand). On the negative side, consensus on land use is difficult to obtain, and useful projects, such as parks and nature preserves, are usually delayed or canceled by the customary owners.

Two areas of Samoan cloud forest have often been cited as ideal for inclusion in nature preserves or parks. Much of the island of Ta'u, including nearly all of its cloud forest, is included in a proposed U.S. national park. This rugged island is sparsely inhabited, with all the residents of the decreasing population restricted to three or four coastal villages. This would be the only U.S. national park where the land is not owned by the government: it would be leased for a long period, with the owners retaining the right to use it in traditional ways only (e.g., no commercial logging). This proposal seems to be on the way to implementation (pers. com., U.S. National Park Service, June 1993).

The other area is the upland region of Savai'i, which has been cited repeatedly as ideal for incorporation into a national park or nature preserve. Savai'i is a relatively large island (it is the largest and second highest island in Polynesia outside of Hawai'i and New Zealand) and is home to numerous endemic plants and native bird species. Landownership would be even more of a problem than on Ta'u, which has only about 1/25th the population of Savai'i.

Recommendations

The common recommendation of previous environmental studies in Samoa is further research, and this paper is certainly no exception. On the practical side, however, the other greatest needs are local education and the control of alien species. In education, much of the curriculum involving ecological issues currently taught in Samoan schools contains little information about the Samoan environment, and even less about the rain forest and cloud forest. It must be made clear that the forests are an integral part of Samoa and should be preserved for environmental reasons. There may also be economic reasons for the preservation. Properly regulated ecotourism to cloud forest areas may be a future source of income to Samoa.

The control of alien species involves preventing the arrival of potential pests (both plant and animal) and initiating biological control projects for the invasive and/or harmful species already present. Preliminary work is currently being done on the biological control of *Clidemia hirta*, and *Funtumia elastica* is a worthy candidate for scrutiny. Perhaps the best protection for Samoan cloud forests is to just leave them alone. This has been the normal course of action (or inaction) in the past. However, this traditional protection has been compromised by an expanding population and some ill-advised Western aid projects. Particularly damaging are roads cut into the interior, since these lead to rapid erosion, invasion of alien plant species, and unsustainable agriculture in ever-widening corridors along the roadside. The establishment of nature preserves, particularly in the cloud forest regions of Savai'i and Ta'u, and the effective enforcement of sound environmental policy afford the best protection for these areas.

References

Stadtmüller, T. 1987. *Cloud forests of the humid tropics*. Tokyo: United Nations University.

Wagner, W. L., D. R. Herbst, and S. H. Sohmer. 1990. *Manual of the flowering plants of Hawai'i*. 2 vols. Honolulu: Univ. of Hawai'i Press and Bishop Museum Press.

Whistler, W. A. 1978. The vegetation of the montane region of Savai'i, Western Samoa. *Pacific Science* 32(1): 79–94.

Whistler, W. A. 1980. The vegetation of eastern Samoa. *Allertonia* 2(2): 45–190.

Whistler, W. A. 1992a. *Botanical inventory of the proposed Tau unit of the National Park of American Samoa*. Honolulu: U.S. Fish and Wildlife Service, Cooperative National Park Resources Studies Unit.

Whistler, W. A. 1992b. The vegetation of Samoa and Tonga. *Pacific Science* 46(2): 159–178.

27. Distribution and Conservation of Peru's Montane Forests: Interactions Between the Biota and Human Society

Kenneth R. Young and Blanca León

The design and implementation of conservation plans for the humid montane forests (including tropical montane cloud forests—TMCFs) of Peru require information about the physical and social parameters that result in or limit the distribution, physiognomy, species composition, and diversity of those forests. Obvious physical factors include the extent and connectivity of the elevational zones where these forests are found, climatic regimes, and edaphic limitations. These in turn set the stage for ecological and evolutionary processes that influence the types of species present (e.g., whether or not there are endemics, habitat specialists, or organisms with poor dispersal abilities). Human population densities and trends (e.g., population increase, immigration/ emmigration) have strong, but not necessarily simple, relations to the degradation of natural resources (Harrison 1991). The current living standards and future aspirations of people provide a context for making predictions. Change, however, can only be induced by first recognizing the social actors (or "stakeholders") most directly involved in natural resource degradation or conservation (García-Guadilla 1992).

Habitat fragmentation is recognized as one of the most important phenomena affecting biological conservation (Simberloff 1988; Saunders, Hobbs, and Margules 1991; Terborgh 1992). In general, fragmented habitats have insufficient critical areas for species, which are left particularly exposed to predators, parasites, competition from exotic species, climatic extremes, population fluctuations, and loss of genetic diversity (Janzen 1983; Blouin and Connor

1985; Lovejoy et al. 1986; Soulé 1987; Terborgh 1989). Little recognized, however, is the fact that habitat fragmentation is also a natural consequence of the spatial distribution of populations or ecosystems in complex terrain, such as the rabbits in Mexican mountains (Velázquez 1993) or montane forest types of Peru. It is important to have an understanding of the degree and impact of this natural fragmentation, in order to have perspective when evaluating the additional fragmentation caused by human activities.

Our goal in this article is to sketch out conservation recommendations based on the extent, connectivity, and biota of the humid montane forests in Peru (with emphasis on TMCF) in relation to the social factors and actors involved. We hope our results will also allow comparisons with the situations found in tropical montane environments elsewhere in the world.

Distribution of Montane Forests in Peru

There are four general regions of humid forest above 1,500 m in the Andes Mountains of Peru (Figure 1), recognized here on the basis of topography, and predominant climatic and biogeographical influences. We define three regions in relation to their location and orientation: an eastern-slope region, a western-slope region, and a highland region. The Andes are lower in northern Peru; this requires the separation of a northern region from the previous three. Each region differs in extent and in degree of natural and human-caused forest fragmentation.

The northern montane forests are found on mountains among the relatively subdued topographic relief of northernmost Peru, mostly in the department of Cajamarca (4.5°–7° S). These forests receive climatic and biogeographic influences from the Amazon basin, the highlands of southern Ecuador, Pacific Ocean drainages, and interandean dry forests. There are many Peruvian endemic species in this region, and other north Andean species here reach their southernmost range limits (Molau 1988; Cadle 1991; Franke 1992). This is a region of great biological heterogeneity (Weberbauer 1945), given its strategic location at a major biogeographical crossroad, the Huancabamba Depression (Duellman 1979; Baumann 1988; Patterson, Pacheco, and Ashley 1992), and given the limited natural extent of many of the vegetation types. The humid montane forests are naturally isolated and have been additionally fragmented by forest clearing for agriculture and range. Because access is comparatively easy, in the last few years these forests have become increasingly threatened by logging for valuable timber species, such as podocarps (*Podocarpus* and *Prumnopitys* of the Podocarpaceae). Perhaps 75 percent of the cover of the original humid forests has been removed, with cut forests replaced by agricultural systems or scrub.

The eastern-slope montane forests occur on the Amazon side of the eastern Cordillera of the Andes above 1,500 m (Young 1991, 1992). These forests receive ample moisture from air masses uplifted from Amazonia (Johnson

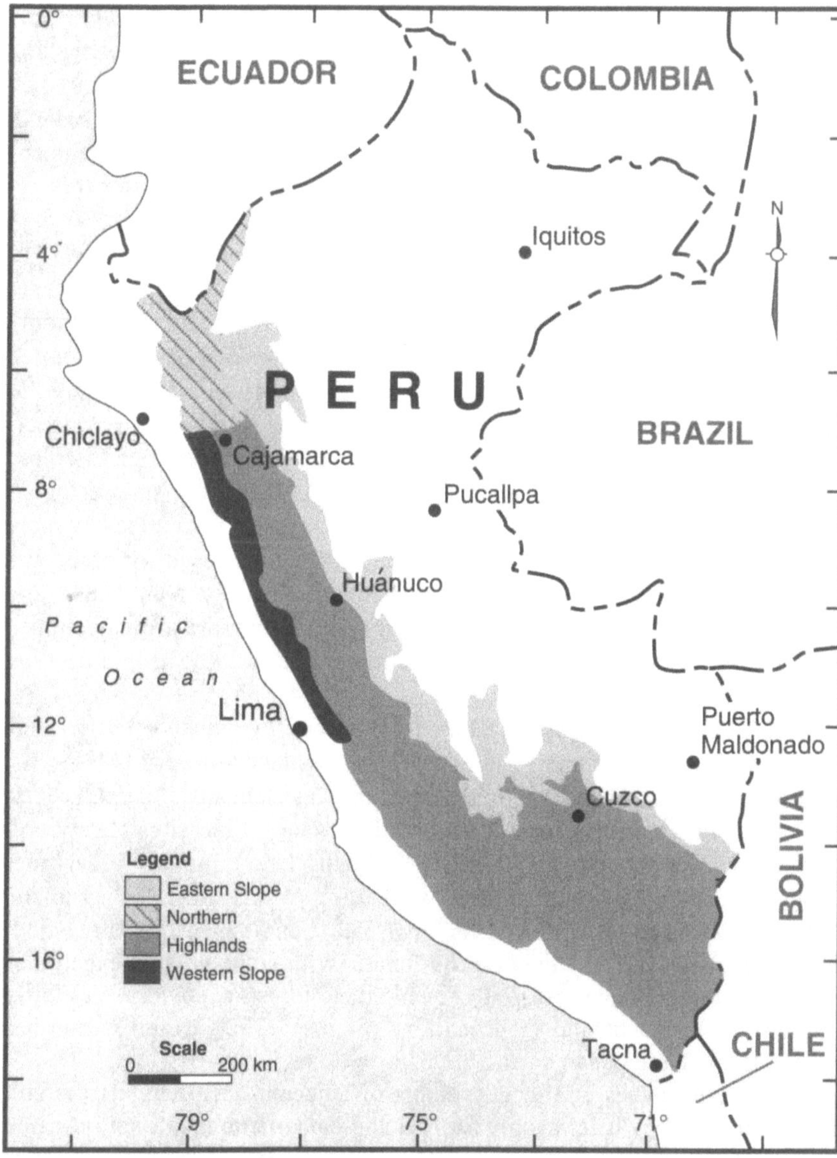

Figure 1. Distribution of humid montane forest in Peru.

1976). They naturally covered a nearly continuous zone that was 1,500 km long (5°–15° S) but often only 30 to 200 km wide because of steep elevational gradients. A substantial amount of TMCF occurred. The biota includes many altitudinal specialists, some of which are endemics due to limited mobility and the presence of topographic barriers between watersheds (Terborgh 1971, 1977; Remsen 1984; Cadle and Patton 1988; Young 1992). Because of the steep slopes and reduced agricultural potential of lands above 2,000 m, current human impact is concentrated in two areas: along roads that connect the highlands to Amazonia, and in areas below 2,000 m that have been cleared by colonists for cash and subsistence agriculture (Aramburú, Bedoya Garland, and Recharte 1982; Dourojeani 1989; Morales 1989; Young 1992). Fragmentation of continuous forest is anthropogenic in origin and has probably reduced original forest cover by about 30–40 percent.

The western-slope montane forests form a narrow belt (7°–12° S) on Pacific Ocean drainages wherever the limited atmospheric humidity condenses during several months of the year. Because of this region's drier and more seasonal climate, the original forest formations were apparently limited elevationally from between 2,500 and 3,200 m, aspect-wise to the more humid southwest-facing slopes, and latitudinally to areas north of 12° S (Valencia 1990, 1992). In addition, human populations have long been found in close association with these forest areas (Smith 1980). These low-stature forests are very fragmented, often in patches of <20 ha, and are surrounded by more arid vegetation types and by anthropogenic ecosystems used for crops and grazing (Sagástegui 1989). The forests contain less diversity (Cano and Valencia 1992) and fewer endemics (Berry 1982) than the aforementioned two regions, but they are often of more direct use to local people needing wood and other forest products.

Highland montane forests are heavily modified by human activity. They occupy humid areas of the remainder of the highlands of Peru (7°–18° S). Most woody plant species are present now only as elements of scrub vegetation, or in relict forests in ravines or other inaccessible sites (Crawford, Wishart, and Campbell 1970; Tovar 1990; Young 1993). In fact, Weberbauer (1945) describes the humid arboreal vegetation of this region as shrubland and riverine forest formations. At very high elevations (above 3,500 m), woodlands dominated by *Polylepis* can be found. With some notable exceptions (Vuilleumier and Simberloff 1980), the highland forests contain species fairly resilient to further habitat modifications (Fjeldsa 1992). Deforestation has been extensive and began during prehistoric times.

The forest patches and woody plants of adjacent open habitats are still intensively used by local people for fuel and construction materials (Ansión 1986; Reynel and Felipe-Morales 1987; Reynel 1988). Like their counterparts in the western-slope region, humid forests in the highland region only cover a small percentage of its potential area, as seen by the absence of closed forests in climatic/ecological zones (Ellenberg 1979; Tosi 1960; ONERN 1976) that could support them. This implies a net deforestation of perhaps 90–95 percent of original cover of humid montane forests, although the paleoenvironmental and experimental evidence needed to confirm this are lacking.

Biogeographical Patterns and Forest Fragmentation

The highland montane forests and those of the western slopes are located in complex and heterogeneous areas. Thus, these forests would naturally be fragmented even without human influence. They are often simpler in structure, with few strata (Valencia 1990) and a comparatively depauperate fauna. However, these forests have also been exposed to millennia of impacts by agricultural and pastoral land use systems (Lumbreras 1974). Much of the surviving biota in these forests consists of species resistant to the negative aspects of habitat fragmentation (i.e., tolerant of open environments or highly vagile). The exceptions are likely to be highly endangered species.

The northern montane forests are somewhat fragmented naturally because of the topography of the region. Anthropogenic forest fragmentation here is apparently not as ancient and pervasive as in the aforementioned regions. Rather, it is an ongoing process, with some fairly large continuous tracts, especially near the border with Ecuador. This region is known for its high endemism and regional diversity.

The eastern-slope montane forests are the most diverse (Young 1991; León, Young, and Brako 1992; O'Neill 1992) and the least fragmented by natural topographic features. They are complex forests structurally, with several strata present, even on steep slopes (Weberbauer 1945; Young 1990). Many of these are true cloud forests. Species present include those unlikely to have ever been common in the other regions, such as those with poor dispersal, those restricted to very narrow elevational belts, or those that require extensive forest-interior habitat. Most of the several dozen bird species new to science that recently have been reported from Peru come from this region (J. O'Neill, pers. com.) This is also where a large, endemic monkey species, the yellow-tailed woolly monkey, was rediscovered in the 1970s after being presumed extinct for almost a century (de Macedo and Mittermeier 1979; Leo 1984). Current deforestation in the region results in the subdivision of ranges of eastern-slope species and reduces gene flow and population sizes, in addition to preferentially eliminating forest below 2,000 m where sustained agriculture is feasible.

Important Social and Political Factors

These four regions also have different social, economic, and political realities. We will simplify by combining the four regions into two (Figure 2): a fragmented forest zone, consisting of those regions with naturally and prehistorically fragmented forests (western slope, highland, portion of the northern region); and a nonfragmented forest zone, consisting of regions that have

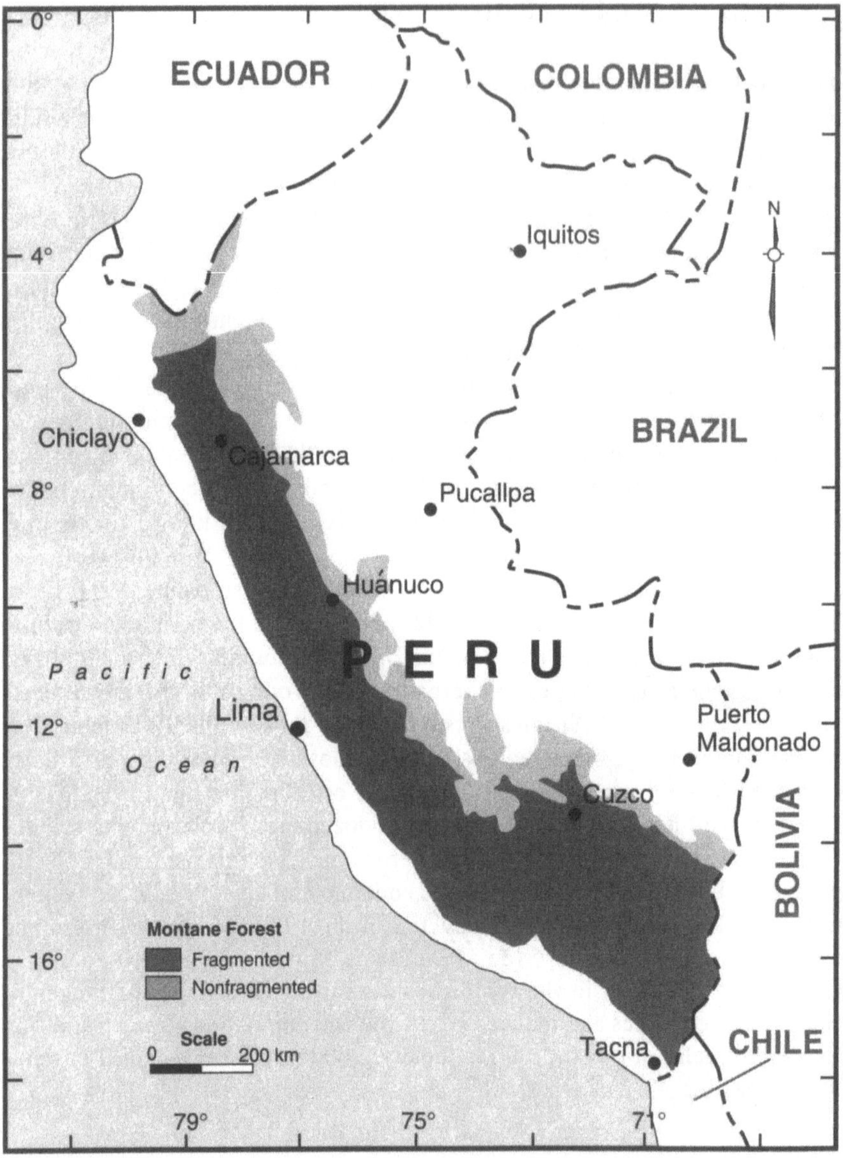

Figure 2. Forest fragmentation: naturally fragmented zone and zone of formerly nonfragmented forest now being impacted.

only recently been affected by human-caused deforestation (eastern slope, portion of the northern region).

High population densities in Peru are found on the northern and central coast, and in portions of the highlands (Figure 3; Webb and Fernández Baca 1991). The highlands are especially impacted by human populations requir-

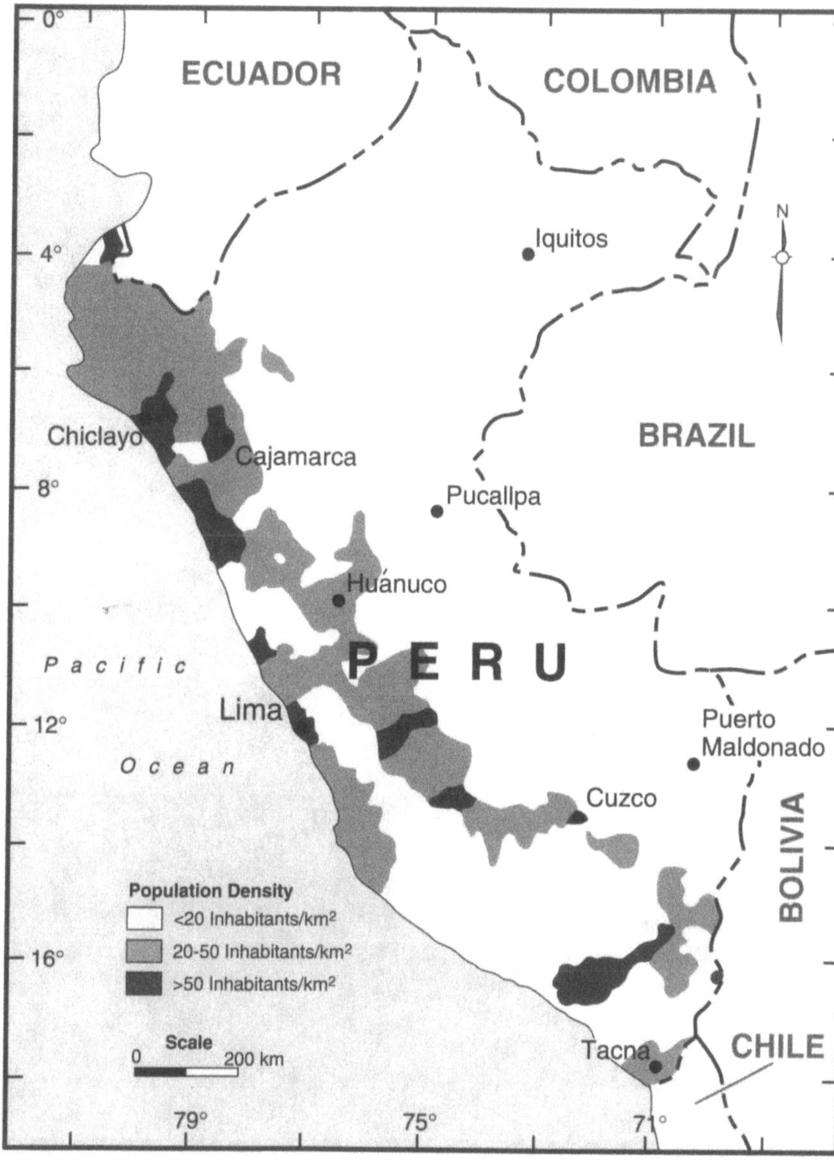

Figure 3. Population density in Peru *(after Webb and Fernández Baca, 1991).*

ing natural resources for subsistence agriculture: namely, land, water, seed (or tubers), and fuelwood. Low living standards are pervasive in the highlands, and to a lesser extent the Amazon lowlands (Figure 4; Portocarrero Maisch 1987). Hundreds of thousands of economic refugees leave the highlands each year to move to the lowlands.

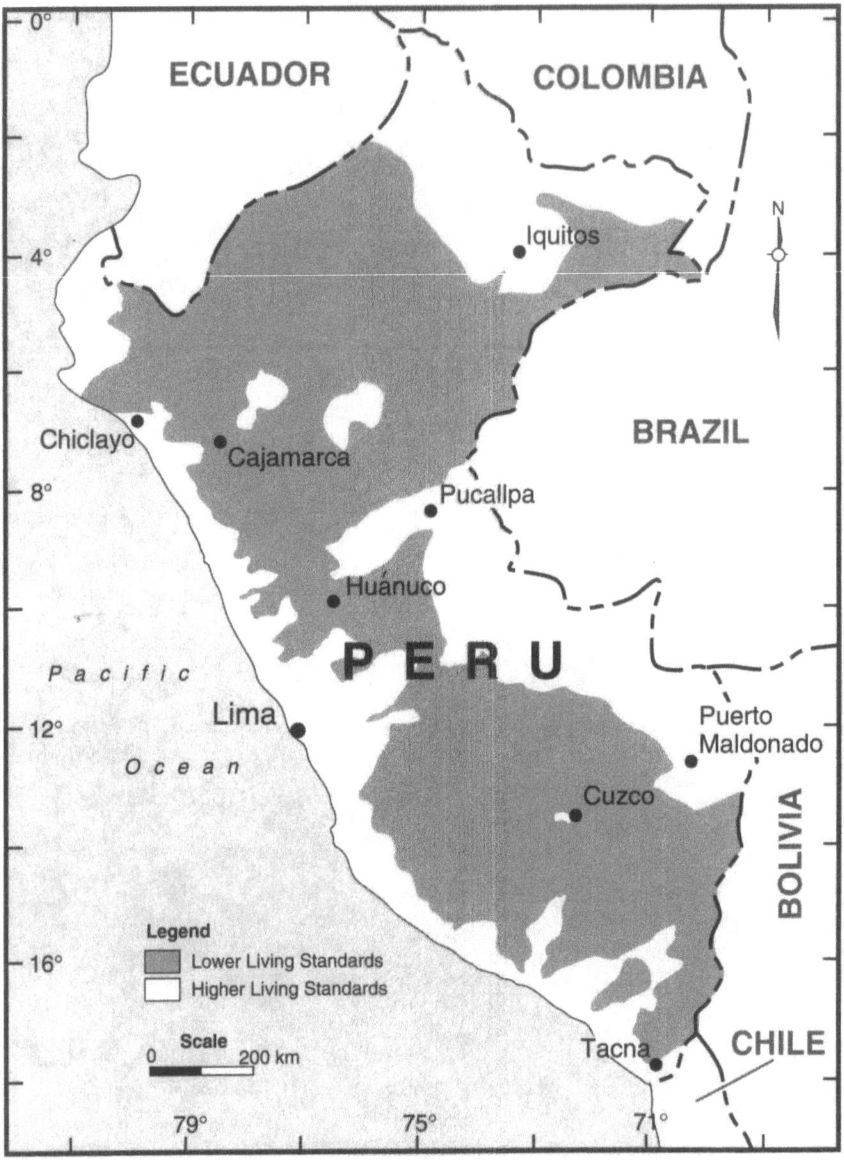

Figure 4. Living standards in Peru *(after Portocarrero Maisch, 1987).*

Fragmented and nonfragmented montane forests are influenced by these social and economic factors. However, social actors, such as governmental and international entities, non-governmental organizations (NGOs), and universities, among others, also play important direct and indirect roles in the degradation or conservation of these forests.

Fragmented TMCFs and other montane forests in Peru are influenced to differing degrees by 11 social actors (Figure 5). Eight involve local or national governmental institutions, in addition to universities. The strongest influences originate from those ministries dealing with natural resources (Agriculture, Energy and Mines) and development (Transportation, Industry and Tourism), although these influences are mainly mediated through activities (or lack of activities) at a local or regional level. In many cases the influence of these ministries has been negative as regards the conservation and wise use of these forests. The direct influence of the central goverment itself has been minimal due to the general absence of national programs directed at these forests. Universities have played an important role through research and extension programs. Unfortunately, their activities and expertise have not often been taken into account by policymakers.

The private sector is represented by three social actors: local interests, business interests, and national NGOs. The former two differ in their interests and access to power; often they have negatively influenced montane forests through the sale of land and the exploitation of natural products. National NGOs, especially those organizations working directly with environmental

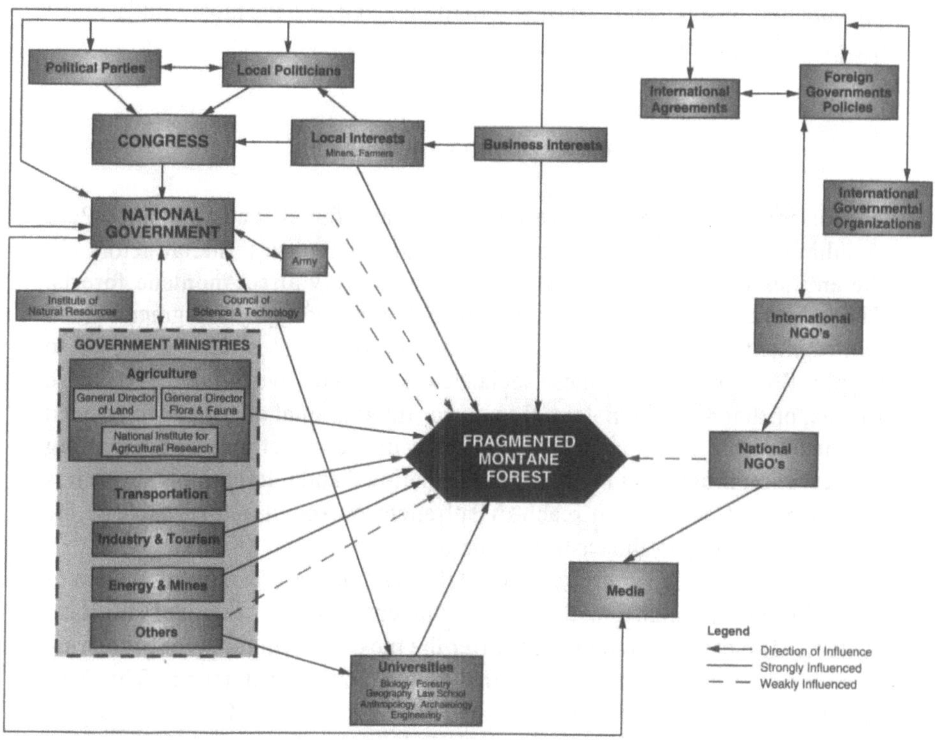

Figure 5. Actors influencing rate of fragmentation of TCMF and other montane forest.

matters, have had relatively little influence in regions with fragmented montane forests; most of their programs have instead been directed at large national parks, situated on the periphery of human settlements.

The conservation and status of nonfragmented montane forests are influenced directly by 13 social actors (Figure 6). Similar to the case of the fragmented montane forests, this includes eight different governmental entities. However, here the central government has played an important role through national development programs aimed at the Amazon lowlands or the foothills of the Andes (e.g., Alto Huallaga project); much of the funding has come from international donors. Due to poor planning and implementation, these projects have mostly resulted in continued degradation of the montane forests. The influence and impact of the universities on conservation of these forests have been similar to those in fragmented areas. The private sector's influence varies in degree. It is represented by five social actors: local interests, immigrants, business interests, and international and national NGOs. Immigrants are considered here separately from local interests because they usually have great negative environmental impact and have little say in local or regional decision-making. International demand for licit (e.g., coffee) and illicit (coca leaves, orchids) products and subsequent extraction/production and processing by business interests drives some deforestation. National environmental NGOs, financed by international NGOs, have mostly influenced the protection of several national parks.

Conservation Implications

The discussion of the conservation of montane forests ecosystems in Peru should consider social, historical, and economic realities. Social actors are the entities that are able to positively or negatively affect montane forests. The influences of the social actors vary in degree and type in fragmented and nonfragmented montane forests, respectively. In all cases there is a need to support flows that connect these social actors with positive influences on these forests, or that have the potential to do so. International donor agencies need to consider the relevant actors and their interconnections when funding conservation programs in Peru. At both a national and international level, it is important to find ways to reach disenfranchised social actors, such as the immigrants affecting the eastern-slope forests.

An expanded and improved national park and nature reserve system is needed in Peru, together with the development of similar regional and local systems. The eastern-slope forests offer the most opportunities for protecting large intact tracts of forest and for including complete watersheds. The three national parks there (Río Abiseo, Yanachaga-Chemillén, Manu) do not provide sufficient coverage of this diverse region; especially needed are new or improved protected areas in the north (department of Amazonas) and south (departments of Apurímac and Puno).

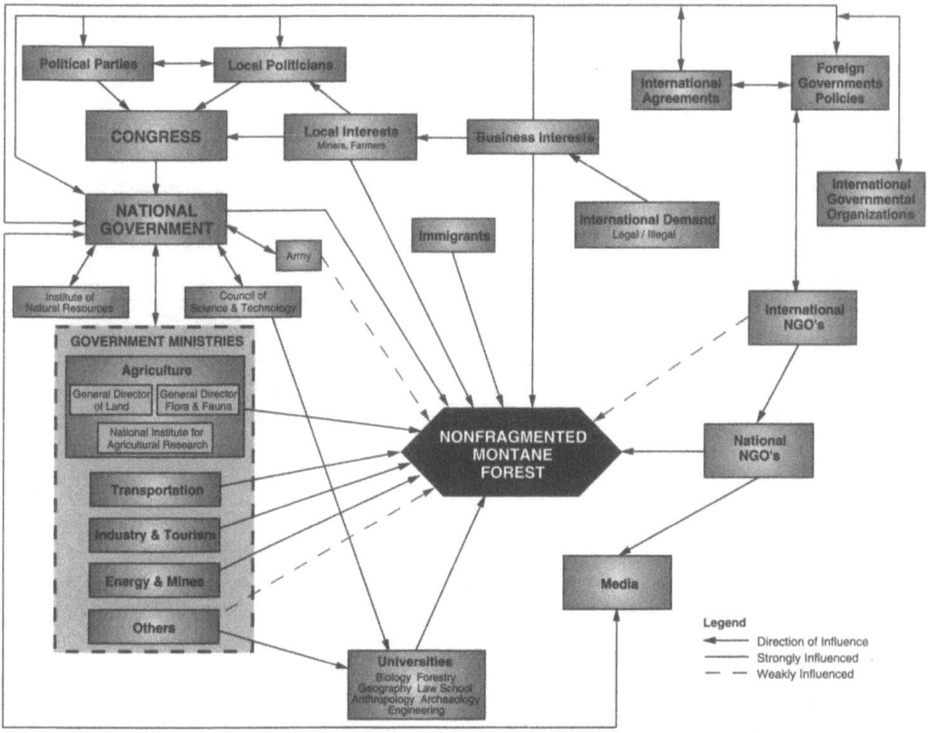

Figure 6. Actors influencing conservation status of formerly nonfragmented TCMF and other montane forest.

The northern montane forests are the most threatened in the country and the limited protection now granted (Tabaconas-Namballe Reserve, Cutervo National Park) is inadequate. Enforcement of existing laws concerning logging contracts, sale of timber products, and reforestation would be useful.

The western-slope forests are not included within any park or reserve (first noted by Valencia and Franke 1980), and the highland montane forests only receive symbolic protection by occurring in several sites within Huascarán National Park (Smith 1988). However, since these forests are often used by local people, their conservation requires programs in addition to national parks/ nature reserves, which include resource extraction (e.g., Brack 1992) and restoration projects. Needed are local programs of reforestation and watershed protection that include native tree species and the management of natural forests. These programs could be designed to increase the connectivity of isolated forest patches.

Acknowledgments. We thank the organizers of the Mountain Cloud Forest Symposium for their invitation to participate. We gratefully acknowledge the economic support of the John D. and Catherine T. MacArthur Foundation for our work in Peru. We are indebted to Joseph School, Department of Geography, University of Maryland Baltimore County, for final production of the graphics.

References

Ansión, J. 1986. *El Arbol y el Bosque en la Sociedad Andina.* Lima: Proyecto FAO/Holanda/INFOR.

Aramburú, C. E., E. Bedoya Garland, and J. Recharte B. 1982. *Colonización en la Amazonía.* Lima: Ediciones CIPA.

Baumann, F. 1988. Geographische Verbreitung und Okologie Sudamerikanisher Hochgebirgspflanzen. *Physische Geographie* (Zurich) 28:1-206.

Berry, P. E. 1982. The systematics and evolution of *Fuchsia* Sect. *Fuchsia* (Onagraceae). *Annals of the Missouri Botanical Garden* 69:1-198.

Blouin, M., and E. F. Connor. 1985. Is there a best shape for nature reserves? *Biological Conservation* 32:277-288.

Brack, A. 1992. Estrategias nuevas para la conservación del bosque montano. In *Biogeografía, Ecología y Conservación del Bosque Montano en el Perú*, ed. K. R. Young and N. Valencia, 223-227. Lima: Memorias del Museo de Historia Natural, UNMSM, Vol. 21.

Cadle, J. E. 1991. Systematics of lizards of the genus *Stenocercus* (Iguania: Tropiduridae) from northern Peru: new species and comments on relationships and distribution patterns. *Proceedings of the Academy of Natural Sciences of Philadelphia* 143:1-96.

Cadle, J. E., and J. L. Patton. 1988. Distribution patterns of some amphibians, reptiles, and mammals of the eastern Andean slope of southern Peru. In *Proceedings of a Workshop on Neotropical Distribution Patterns*, ed. W. R. Heyer and P. E. Vanzolini, 225-244. Rio de Janeiro: Academia Brasileira de Ciencias.

Cano, A., and N. Valencia. 1992. Composición florística de los bosques nublados secos de la vertiente Occidental de los Andes peruanos. In *Biogeografía, Ecología y Conservación del Bosque Montano en el Perú*, ed. K. R. Young and N. Valencia, 171-180. Lima: Memorias del Museo de Historia Natural, UNMSM, Vol. 21.

Crawford, R. M., D. Wishart, and R. M. Campbell. 1970. A numerical analysis of high altitude scrub vegetation in relation to soil erosion in the eastern Cordillera of Peru. *Journal of Ecology* 58:173-191.

de Macedo, H., and R. Mittermeir. 1979. Redescubrimiento del primate peruano *Lagothrix flavicauda* (Humboldt 1812) y primeras observaciones sobre su biología. *Revista de Ciencias, UNMSM* 71:78-92.

Dourojeani, M. J. 1989. Impactos ambientales del cultivo de la coca y la producción de cocaina en la Amazonía peruana. In *Pasta Basica de Cocaina: un estudio multidisciplinario,* ed. F. R. León and R. Castro de la Mata, 281-299. Lima: CEDRO.

Duellman, W. E. 1979. The herpetofauna of the Andes: patterns of distribution, origin, differentiation, and present communities. *Monographs of the Museum of Natural History* (Lawrence, Kansas) 7:371-459.

Ellenberg, H. 1979. Man's influence on tropical mountain ecosystems in South America. *Journal of Ecology* 67:401-416.

Fjeldsa, J. 1992. Un análisis biogeográfico de la avifauna de los bosques de queñoa (*Polylepis*) de los Andes y su relevancia para establecer prioridades de conservación. In *Biogeografía, Ecología y Conservación del Bosque Montano en el Perú*, ed. K. R. Young and N. Valencia, 207-221. Lima: Memorias del Museo de Historia Natural, UNMSM, Vol. 21.

Franke, I. 1992. Biogeografía y ecología de las aves de los bosques montanos del

Perú occidental. In *Biogeografía, Ecología y Conservación del Bosque Montano en el Perú*, ed. K. R. Young and N. Valencia, 181-188. Lima: Memorias del Museo de Historia Natural, UNMSM, Vol. 21.

García-Guadilla, M. P. 1992. Identity, strategy and symbolic effectiveness: the Venezuelan environmentalist movement. *International Journal of Sociology and Social Policy* 12:78-109.

Harrison, S. 1991. Population growth, land use and deforestation in Costa Rica, 1950-1984. *Interciencia* 16:83-93.

INE. 1989. *Peru: Población y Superficie a Nivel Distrital.* Lima: Instituto Nacional de Estadística.

Janzen, D. H. 1983. No park is an island: increase in interference from outside as park size decreases. *Oikos* 41:402-410.

Johnson, A. M. 1976. The climate of Peru, Bolivia and Ecuador. In *Climates of Central and South America*, ed. W. Schwerdtfeger, 147-218. Amsterdam: Elsevier.

Leo, M. 1984. The effect of hunting, selective logging and clear-cutting on the conservation of the yellow-tailed woolly monkey (*Lagothrix flavicauda*). M.A. thesis, University of Florida, Gainesville.

León, B., K. R. Young, and L. Brako. 1992. Análisis de la composición florística del bosque montano oriental del Perú. In *Biogeografía, Ecología y Conservación del Bosque Montano en el Perú*, ed. K. R. Young and N. Valencia, 141-154. Lima: Memorias del Museo de Historia Natural, UNMSM, Vol. 21.

Lovejoy, T. E., R. O. Bierregaard, Jr., A. B. Rylands, J. R. Malcolm, E. C. Quintela, L. H. Harper, K. S. Brown, Jr., A. H. Powell, G. V. N. Powell, H. O. R. Shubart, and M. B. Hays. 1986. Edge and other effects of isolation on Amazon forest fragments. In *Conservation Biology: the Science of Scarcity and Diversity*, ed. M. E. Soulé, 257-285. Sunderland: Sinauer Associates.

Lumbreras, L. G. 1974. *The Peoples and Cultures of Ancient Peru.* Washington, D.C.: Smithsonian Institution Press.

Molau, U. 1988. Scrophulariaceae—Part I. Calceolarieae. *Flora Neotropica Monograph* 47.

Morales, E. 1989. *Cocaine: White Gold Rush in Peru.* Tucson: University of Arizona Press.

O'Neill, J. P. 1992. A general overview of the montane avifauna of Peru. In *Biogeografía, Ecología y Conservación del Bosque Montano en el Perú*, ed. K. R. Young and N. Valencia, 47-55. Lima: Memorias del Museo de Historia Natural, UNMSM, Vol. 21.

ONERN. 1976. *Mapa Ecológico del Perú y Guía Explicativa.* Lima: Oficina Nacional de Evaluación de Recursos Naturales.

Patterson, B. D., V. Pacheco, and M. V. Ashley. 1992. On the origins of the western slope region of endemism: systematics of fig-eating bats, genus *Artibeus*. In *Biogeografía, Ecología y Conservación del Bosque Montano en el Perú*, ed. K. R. Young and N. Valencia, 189-205. Lima: Memorias del Museo de Historia Natural, UNMSM, Vol. 21.

Peñaherrera, C. 1989. *Atlas del Perú.* Lima: Instituto Geográfico Nacional.

Portocarrero Maisch, J. (ed.) 1987. *Los Hogares Rurales en el.Perú: Importancia y articulación con el desarrollo agrario.* Lima: Grupo de Análisis de Política Agrícola, Proyecto PADI-Ministerio de Agricultura, Fundación Friedrich Ebert.

Remsen, J.V., Jr. 1984. High incidence of "leapfrog" pattern of geographic variation in Andean birds: implications for the speciation process. *Science* 224:171-173.

Reynel, C. 1988. *Plantas para Leña en el Sur-occidente de Puno.* Puno, Peru: Proyecto Arbolandino.

Reynel, C., and C. Felipe-Morales. 1987. *Agroforestería Tradicional en los Andes del Perú.* Lima: Proyecto FAO/Holanda/INFOR.

Sagástegui, A., A. 1989. *Vegetación y Flora de la Provincia de Contumazá*. Trujillo, Peru: CONCYTEC.

Saunders, D. A., R. J. Hobbs, and C. R. Margules. 1991. Biological consequences of ecosystem fragmentation: a review. *Conservation Biology* 5:18-32.

Simberloff, D. 1988. The contribution of population and community biology to conservation science. *Annual Review of Ecology and Systematics* 19:473-511.

Smith, C. E. 1980. Plant remains from Guitarrero cave. In *Guitarrero Cave: Early Man in the Andes*, ed. T. Lynch, 87-119. New York: Academic Press.

Smith, D. N. 1988. Flora and vegetation of the Huascarán National Park, Ancash, Peru, with preliminary taxonomic studies for a manual of the flora. Ph.D. dissertation, University of Iowa, Ames.

Soulé, M. (ed.) 1987. *Viable Populations for Conservation*. Cambridge: Cambridge University Press.

Terborgh, J. 1971. Distribution on environmental gradients: theory and a preliminary interpretation of distributional patterns in the avifauna of the Cordillera Vilcabamba, Peru. *Ecology* 52:23-40.

Terborgh, J. 1977. Bird species diversity on an Andean elevational gradient. *Ecology* 58:1007-1019.

Terborgh, J. 1989. *Where Have All the Birds Gone?* Princeton: Princeton University Press.

Terborgh, J. 1992. Maintenance of diversity in tropical forests. *Biotropica* 24:283-292.

Tosi, J., Jr. 1960. *Zonas de Vida Natural en el Perú*. Lima: Organización de Estados Americanos.

Tovar S., O. 1990. *Tipos de Vegetación, Diversidad Florística y Estado de Conservación de la Cuenca del Mantaro*. Lima: Centro de Datos para la Conservación-Universidad Nacional Agraria.

Valencia, N. 1990. Ecology of the forests on the western slopes of the Peruvian Andes. Ph.D. dissertation, University of Aberdeen, Scotland, UK.

Valencia, N. 1992. Los bosques de la vertiente Occidental del Perú. In *Biogeografía, Ecología y Conservación del Bosque Montano en el Perú*, ed. K. R. Young and N. Valencia, 155-170. Lima: Memorias del Museo de Historia Natural, UNMSM, Vol. 21.

Valencia, N., and I. Franke. 1980. El Bosque de Zárate y su conservación. *Boletín de Lima* 7:76-86, 8:26-35.

Velázquez, A. 1993. Man-made and ecological habitat fragmentation: study case of the volcano rabbit (*Romerolagus diazi*). *Zeitschrift für Saugetierkunde* 58:54-61.

Vuilleumier, F., and D. Simberloff. 1980. Ecology versus history as determinants of patchy and insular distributions in high Andean birds. *Evolutionary Biology* 12:235-379.

Webb, R., and G. Fernández Baca. 1991. *Almaneque Estadístico: Perú en Números 1991*. Lima: Cuánto.

Weberbauer, A. 1945. *El Mundo Vegetal de los Andes Peruanos*. Lima: Ministerio de Agricultura.

Young, K. R. 1990. Biogeography and ecology of a timberline forest in north-central Peru. Ph.D. dissertation, University of Colorado, Boulder.

Young, K. R. 1991. Floristic diversity on the eastern slopes of the Peruvian Andes. *Candollea* 46:125-143.

Young, K. R. 1992. Biogeography of the montane forest zone of the eastern slopes of Peru. In *Biogeografía, Ecología y Conservación del Bosque Montano en el Perú*, ed. K. R. Young and N. Valencia, 119-154. Lima: Memorias del Museo de Historia Natural, UNMSM, Vol. 21.

Young, K. R. 1993. National park protection in relation to the ecological zonation of a neighboring human community: an example from northern Peru. *Mountain Research and Development* 13 (in press).

Young, K. R., and N. Valencia. 1992. Introducción. In *Biogeografía, Ecología y Conservación del Bosque Montano en el Perú*, ed. K. R. Young and N. Valencia, 5-9. Lima: Memorias del Museo de Historia Natural, UNMSM, Vol. 21.

Comprehensive Bibliography

The following list of references was compiled from various sources and was not cross-referenced for accuracy or content. Selections were made based on their perceived relevance to the study of tropical montane cloud forests.

Aitken, P., F.M.J. Ribeny, and J.A.H. Brown. 1972. The estimation of mean annual runoff over the territory of Papua New Guinea. *Civil Engineering Transactions* (Australia) 14:49–56.

Almeda, F. 1989. *Tessmannianthus*, an arborescent genus of Melastomataceae new to Panama. *Annals of the Missouri Botanical Garden* 76(1): 1–6.

Arriaga, L. 1988. Gap dynamics of a tropical cloud forest in northeastern Mexico. *Biotropica* 20(3): 178–184.

Asbury, C.E., W.H. McDowell, R. Trinid-Pizarro, and S. Berreos. 1994. Solute deposition from cloud water to the canopy of a Puerto Rican montane forest. *Atmospheric Environment* 28(10):1773–1780.

Ash, J. 1987. Stunted cloud-forest in Taveuni, Fiji. *Pacific Science* 41(1–4): 191–199.

Askew, C.P. 1964. The mountain soils of the east ridges of Mount Kinabalu. *Proceedings of the Royal Society of Botany* 161:65–74.

Aylett, G.P. 1985. Irradiance interception, leaf conductance and photosynthesis in Jamaican upper montane rain forest trees. *Photosynthetica* 19:323–327.

Barradas, V.L. 1983. Capacidad de captación de agua a partir de niebla en *Pinus montezumae* Lambert, de la región de las Grandes Montañas del Estado de Veracruz. *Biotica* 8(4): 427–430.

Baynton, H.W. 1968. The ecology of an elfin forest in Puerto Rico, 2. The microclimate of Pico del Oeste. *Journal of the Arnold Arboretum* 49:419–430.

Baynton, H.W. 1969. The ecology of an elfin forest in Puerto Rico, 3. Hilltop and forest influences on the microclimate of Pico del Oeste. *Journal of the Arnold Arboretum* 50(1): 80–92.

Baynton, H.W., H.L. Hamilton Jr., P.E. Sherr, and J.J.B. Worth. 1965. Temperature structure in and above a tropical forest. *Quarterly Journal of the Royal Meteorological Society* 91:225–232.

Beebe, William, and Jocelyn Crane. 1947. Ecology of Rancho Grande: a subtropical cloud forest in northern Venezuela. *Zoológica* 32:43–60.

Benedict, Faye F. 1976. Herbivory rates and leaf properties in four forests in Puerto Rico and Florida. M.S. thesis, University of Florida, Gainesville. 78 pp.

Bentley, B.L. 1987. Nitrogen fixation by epiphylls in a tropical rainforest. *Annals of the Missouri Botanical Garden* 74:234–241.

Bentley, B.L., and F.J. Carpenter. 1984. Direct transfer of newly-fixed nitrogen from free-living epiphyllous microorganisms to their host plant. *Oecologia* 63:52–56.

Benzing, D.H. 1980. *The biology of the bromeliads*. Mad River Press, Inc., Eureka. 305 pp.

Benzing, D.H. 1981. Mineral nutrition of epiphytes: an appraisal of adaptive features. *Selbyana* 5:219–223.

Benzing, D.H. 1990. *Vascular epiphytes*. Cambridge University Press, Cambridge. 354 pp.

Benzing, D.H., and J. Seeman. 1978. Nutritional piracy and host tree decline. *Selbyana* 2:133–148.

Bisang, I. 1985. Zur Verbreitung und Okologie der Frullania-Arten der Schweiz. *Botanica Helvetica* 95:247–278.

Blackie, J.R. 1979. The water balance of the Kericho catchments. *East African Agricultural and Forestry Journal* 43:55–84.

Brack, A. 1992. Estrategias nuevas para la conservación del bosque montano. Pages 223–227 in K.R. Young and N. Valencia, editors. *Biogeografia, ecología y conservación del bosque montano en el Perú*. Memorias del Museo de Historia Naturel, UNMSM 21, Lima.

Braak, C. 1922. The climate of the Dutch East Indies. *Verhandelingen van het Koninklijk Magnetisch en Meteorologisch Observatorium te Batavia* 8(1): 1–497.

Brewer-Carias, Charles. 1973. Plantas carnívoras del cerro de la neblina, estudio y fotografía sobre tres mecanismos de trampas. *Defensa de la Naturaleza* (Venezuela) 2(6): 17–21, 23–26.

Briscoe, C.B. 1966. Weather in the Luquillo Mountains of Puerto Rico. U.S. Dept. of Agriculture Forest Service Research Paper ITF–3, Río Piedras, PR. 250 pp.

Brown, A.D. 1995. Ecology and conservation of the Argentine montane forest. Pages 107–115 in L.S. Hamilton, J.O. Juvik, and F.N. Scatena, editors. *Tropical Montane Cloud Forests*. Springer-Verlag, New York.

Brown, S., A.E. Lugo, S. Silander, and Liegel. 1983. Research history and opportunities in the Luquillo Experimental Forest. U.S. Forest Service General Technical Report SO–44, Southern Forest Experiment Station, New Orleans. 128 pp.

Brown, William H. 1919. Vegetation of Philippine mountains. Department of Agriculture and Natural Resources, Bureau of Science, Manila.

Bruijnzeel, L.A. 1989. (De)forestation and dry season flow in the humid tropics: a closer look. *Journal of Tropical Forest Science* 1:229–243.

Bruijnzeel, L.A. 1989. Nutrient content of bulk precipitation in south central Java, Indonesia. *Journal of Tropical Ecology* 5:187–202.

Bruijnzeel, L.A. 1989. Nutrient cycling in moist tropical forests: the hydrological framework. Pages 383–415 in J. Proctor, editor. *Mineral nutrients in tropical forest and savanna ecosystems*. Blackwell Scientific Press, Oxford.

Bruijnzeel, L.A. 1990. Hydrology of moist tropical forests and effects on conversion: a state of knowledge review. IHP-UNESCO Humid Tropical Programme, Paris.

Bruijnzeel, L.A., and J. Proctor. 1995. Hydrology and biogeochemistry of tropical montane cloud forests: what do we really know? Pages 38–78 in L.S. Hamilton, J.O. Juvik, and F.N. Scatena, editors. *Tropical Montane Cloud Forests*. Springer-Verlag, New York.

Bruijnzeel, L.A., M.J. Waterloo, J. Proctor, A.T. Kuiters, and B. Kotterink. 1993. Hydrological observations in montane rain forests on Gunung Silam, Sabah, Malaysia, with special reference to the "Massenerhebung" Effect. *Journal of Ecology* 81:145–167.

Bruijnzeel, L.A., and K.F. Wiersum. 1987. Rainfall interception by a young *Acacia auriculiformis* plantation forest in West Java: application of Gash's analytical model. *Hydrological Processes* 1:309–319.

Brun, R. 1976. Methodik und Ergebnisse zur Biomassenbestimmung eines Nebelwaldokosystems in den venezolanischen Anden. XVI IUFRO World Congress Division I, Oslo. Pp. 490–499. (Cited in Medina, 1986.)

Buckley, R.C., R.T. Corlett, and P.J. Grubb. 1980. Are the xeromorphic trees of tropi-

cal upper montane rain forests drought-resistant? *Biotropica* 12(2): 124–136.
Burgess, P.F. 1969. Ecological factors in hill and mountain forests of the states of Malaya. *Malay Nature Journal* 22:119–128.
Burnham, C.P. 1974. The role of the soil-forming factors in controlling altitudinal zonation on granite in Malaysia. Pages 59–74 in J.R. Flenley, editor. *Altitudinal zonation in Malesia.* Transactions of the Third Aberdeen-Hull Symposium on Malesian Ecology, University of Hull, Department of Geography, 1973, Hull.
Burrowes, Patricia A. 1987. An ecological study of a cloud forest herpetofauna in southern Colombia. M.A. thesis, University of Kansas, Lawrence. 93 pp.
Byer, Michael D., and Peter L. Weaver. 1977. Early secondary succession in an elfin woodland in the Luquillo Mountains of Puerto Rico. *Biotropica* 9(1): 35–47.
Caceres, G. 1981. Inportancia hidrológica de la intercepción horizontal en un bosque muy humedo premontano en Balalaica, Turrialba, Costa Rica. MSc thesis, University of Costa Rica, Turrialba.
Calder, I.R., I.R. Wright, and D. Murdiyarso. 1986. A study of evaporation from a tropical rainforest—West Java. *Journal of Hydrology* 89:13–31.
Calvo, J. 1986. An evaluation of Thornthwaite's water balance technique in predicting stream runoff in Costa Rica. *Hydrological Sciences Journal* 31:51–60.
Campanella, R. 1995. The role of GIS in evaluating contour-based limits of cloud forest reserves in Honduras. Pages 116–124 in L.S. Hamilton, J.O. Juvik, and F.N. Scatena, editors. *Tropical Montane Cloud Forests.* Springer-Verlag, New York.
Cano, A., and N. Valencia. 1992. Composición florística de los bosques nublados secos de la vertiente Occidental de los Andes peruanos. Pages 171–180 in K.R. Young and N. Valencia, editors. *Biogeografia, ecología y conservación del bosque montano en el Perú.* Memorias del Museo de Historia Naturel, UNMSM 21, Lima.
Catling, Paul M., and Leonard P. Lefkovitch. 1989. Associations of vascular epiphytes in a Guatemalan cloud forest. *Biotropica* 21(1): 35–40.
Cavelier, J. 1988. The ecology of elfin cloud forests in northern South America. Dissertation submitted for the annual research fellowship competition. Trinity College, University of Cambridge, Cambridge.
Cavelier, J. 1989. Root biomass, production and the effect of fertilization in two tropical rain forests. PhD thesis, University of Cambridge, Cambridge.
Cavelier, J. 1990. Tissue water relations in elfin cloud forest tree species of Serrania de Macuira, Guajira, Colombia. *Tree* 4:155–163.
Cavelier, J. 1995. Reforestation with the native tree *Alnus acuminata*: effects on phytodiversity and species richness in an upper montane rain forest area of Colombia. Pages 125–137 in L.S. Hamilton, J.O. Juvik, and F.N. Scatena, editors. *Tropical Montane Cloud Forests.* Springer-Verlag, New York.
Cavelier, J. (in press). *El ciclo del agua en bosques montanos y consecuencias de la deforestacion.* C. En Uribe, editor. Bosques Montanos de Colombia. Banco de Occidente, Bogota.
Cavelier, Jaime, and Guillermo Goldstein. 1989. Mist and fog interception in elfin cloud forests in Colombia and Venezuela. *Journal of Tropical Ecology* 5:309–322.
Cavelier, Jaime, and C.A. Mejia. 1990. Climatic factors and tree stature in the elfin cloud forest of Serrania de Macuira, Colombia. *Agricultural and Forest Meteorology* 53(1–2): 105–124.
Cavelier, Jaime, and María C. Peñuela. 1990. Soil respiration in the cloud forest and dry deciduous forest of Serrania de Macuira, Colombia. *Biotropica* 22(4): 346–352.
Cereceda, P., and R.S. Schemenauer. 1989. The role of topography and wind flow in high elevation fog water collection. Page 198 in Third International Conference on Southern Hemisphere Meteorology and Oceanography, 13–17 November, Buenos Aires.

Cereceda, P., and R.S. Schemenauer. 1991. The occurrence of fog in Chile. *Journal of Applied Meteorology* 30:1097–1105.

Cereceda, P., R.S. Schemenauer, and M. Suit (in press). Producción de agua de niebla en Peru. *Atmosphera.*

Cereceda, P., R.S. Schemenauer, and R. Valencia (in press). Posibilidades de abastecimiento de agua de niebla en localidades costeras de Antofagasta. *Revista de Geografia Norte Grande.*

Cereceda-Troncoso, P. 1989. La destribución de la niebla en Chile. *Revista de Geografia Norte Grande* 16:43–49.

Cereceda-Troncoso, P. 1989. Las nieblas costeras como recurso hidrológico. *Revista Geografica de Chile Terra Australis* 31:29–33.

Cereceda-Troncoso, P., J. Barros, and R.S. Schemenauer. 1990. Las nieblas costeras de Chile y Oman similitudes y deferencias. *Revista Geografica de Chile Terra Australis* 33:49–60.

Cleef, Antoine M., Orlando Rangel-Ch, and Sonia Salamanca-V. 1983. Reconocimiento de la vegetación de la parte alta del transecto Parque los Nevados. Pages 150–173 in T. van der Hammen, A. Pérez Preciado, and Polidoro Pinto, editors. La Cordillera Central Colombiana transecto Parque los Nevados: Introducción y datos inicales. *Studies on tropical Andean ecosystems*, Volume 1. J. Cramer, Vaduz, Liechtenstein.

Cleef, Antoine M., Orlando Rangel-Ch, and Sonia Salamanca-V. (in press). Los bosques andinos del transecto Parque los Nevados (Cordillera Central). In T. van der Hammen and A.G. dos Santos, editors. La Cordillera Central Colombiana transecto Parque los Nevados (cuarta parte). *Studies on tropical Andean ecosystems.* Volume 4. J. Cramer, Berlin-Stuttgart.

Cockburn, P.F. 1969. Gunung Mandi Angin, Trenggaun: a botanical visit. *Malayan Nature Journal* 22:164–170.

Coxson, D.S. 1989. Effects of desiccation on net photosynthetic activity in the Basidiomycete lichen *Cora pavonia* E. Fries from the cloud/mist zone of the tropical volcano La Sonfriere (Guadeloupe). *The Bryologist* 90(3): 241–245.

Coxson, D.S. 1991. Nutrient release from epiphytic bryophytes in tropical montane rain forest, Guadeloupe. *Canadian Journal of Botany* 69(10): 2122–2129.

Coxson, D.S., D.D. McIntyre, and H.J. Vogel. 1992. Pulse release of sugars and polyols from canopy bryophytes in tropical montane rain forest, Guadeloupe, French West Indies. *Biotropica* 24(2a): 121–133.

Crawford, R.M., D. Wishart, and R.M. Campbell. 1970. A numerical analysis of high altitude scrub vegetation in relation to soil erosion in the eastern Cordillera of Peru. *Journal of Ecology* 58:173–191.

Curtis, J.T. 1946. Nutrient supply of epiphytic orchids in the mountains of Haiti. *Ecology* 27(3): 264–266.

Daugherty, Howard E. 1973. The Montecristo cloud-forest of El Salvador—a chance for protection. *Biological Conservation* 5(1): 227–230.

De Barcellos Falkenberg, D., and J.C. Voltolini. 1995. The montane cloud forest in southern Brazil. Pages 138–149 in L.S. Hamilton, J.O. Juvik, and F.N. Scatena, editors. *Tropical Montane Cloud Forests*. Springer-Verlag, New York.

De los Santos, A.D. 1981. Water budget and nutrient fluxes in mossy forest on Mountain Data, Mountain Province, Philippines. MSc thesis, University of the Philippines at Los Banos, Laguna.

Del Moral, Roger, and Cornelius H. Muller. 1969. Fog drip: a mechanism of toxin transport from *Eucalyptus globulus*. *Bulletin of the Torrey Botanical Club* 96(4): 467–475.

Dilks, T.J.K., and M.C.F. Proctor. 1975. Comparative experiments on temperature responses of bryophytes: assimilation, respiration and freezing damage. *Journal of Bryology* 8:317–336.

Dohrenwend, R.E. 1979. Hydrologic behavior at the top of a tropical mountain. Research Note 29. Michigan Technological University, Houghton.

Dollard, G.J., and M.H. Unsworth. 1983. Field measurements of turbulent fluxes of wind-driven fogs to a grass surface. *Atmospheric Environment* 17:775–780.

Doumenge, C., D. Gilmour, M.R. Perez, and J. Blockhus. 1995. Tropical montane cloud forests: conservation status and management issues. Pages 24–37 in L.S. Hamilton, J.O. Juvik, and F.N. Scatena, editors. *Tropical Montane Cloud Forests*. Springer-Verlag, New York.

Duellman, W.E. 1988. Patterns of species diversity in anuran amphibians in the American tropics. *Annals of the Missouri Botanical Garden* 75(1): 79–104.

Dye, P.J., B.W. Olbrich, and I.R. Calder. 1992. A comparison of the heat pulse method and deuterium tracing methods for measuring transpiration from *Eucalyptus grandis* trees. *Journal of Experimental Botany* 43:337–343.

Edwards, I.D., R.W. Payton, J. Proctor, and S. Riswan. 1990. Altitudinal zonation of the rain forests in the Manusela National Park, Seram, Maluka, Indonesia. Pages 161–175 in P. Baas, K. Kalkman, and R. Geesink, editors. *The plant diversity of Malesia*. Kluwer, Dordrecht.

Edwards, P.J. 1973. Nutrient cycling in a New Guinea montane forest. PhD thesis, University of Cambridge, Cambridge. 198 pp.

Edwards, P.J. 1977. Studies of mineral cycling in a New Guinea montane forest. II. The production and disappearance of litter. *Journal of Ecology* 65:971–992.

Edwards, P.J. 1982. Studies on mineral cycling in a montane rain forest in New Guinea. V. Rates of cycling in throughfall and litter fall. *Journal of Ecology* 70:807–827.

Edwards, P.J., and P.J. Grubb. 1977. Studies of mineral cycling in a montane rain forest in New Guinea. I. The distribution of organic matter in the soil and vegetation. *Journal of Ecology* 65:943–969.

Edwards, P.J., and P.J. Grubb. 1982. Studies of mineral cycling in a montane rain forest in New Guinea. IV. Soil characteristics and the division of mineral elements between the vegetation and soil. *Journal of Ecology* 70:649–666.

Ekern, P.C. 1964. Direct interception of cloud water on Lana'ihale, Hawaii. *Soil Science Society of America Proceedings* 28:419–421.

Ellenberg, H. 1979. Man's influence on tropical mountain ecosystems in South America. *Journal of Ecology* 67:401–416.

Ellenberg, H. 1985. Site conditions under which leaves develop so called drip tips. *Flora (Jena)* 176(3–4): 169–188.

Etherington, John R. 1984. Comparative studies of plant growth and distribution in relation to waterlogging. X. Differential formation of adventitious roots and their experimental excision in *Epilobium hirsutum* and *Chamerion angustifolium*. *Journal of Ecology* 72:389–404.

Ezcurra, Exequiel, and Judith Becerra. 1987. Experimental decomposition of litter from the Tamaulipan cloud forest: a comparison of four simple models. *Biotropica* 19(4): 290–296.

Fassbender, H.W., and U. Grimm. 1981. Ciclos bioquímicos en un ecosistema forestal de los Andes Occidentales de Venezuela II. Producción y descomposición de los residuous vegetales. *Turrialba* 31:39–48.

Fauth, J.E., B.I. Crother, and J.B. Slowinski. 1989. Elevational patterns of species richness, evenness, and abundance of the Costa Rican leaf-litter herpetofauna. *Biotropica* 21:178–185.

Fishlock, C.W.L., and Hancock. 1933. Notes on the flora and fauna of Ruwenzori with special reference to Bujuku Valley. *Journal of the East African & Ugandan Natural History Society* 44:205–229.

Fleming, T.H. 1986. Secular changes in Costa Rican rainfall: correlation with elevation. *Journal of Tropical Ecology* 2:87–91.

Flenley, J.R. 1991. Ultraviolet light and the altitudinal zonation of tropical rain forests. Pages 27–33 in T.R.R. Johnston and J.R. Flenley, editors. *Aspects of Environmental Change*. Misc. Series 91/1, Massey University, Department of Geography, Palmerston North, New Zealand.

Flenley, J.R. 1992. Ultraviolet-B insolation and the altitudinal forest limit. Pages 273–282 in P. Furley, J. Proctor, and J.A. Ratter, editors. *The nature and dynamics of forest-savanna boundaries*. Chapman and Hall, London.

Flenley, J.R. 1995. Cloud forest, the Massenerhebung effect, and ultraviolet insolation. Pages 150–164 in L.S. Hamilton, J.O. Juvik, and F.N. Scatena, editors. *Tropical Montane Cloud Forests*. Springer-Verlag, New York.

Flores, Saul. 1992. Growth and seasonality of seedlings and juveniles of primary species of a cloud forest in northern Venezuela. *Journal of Tropical Ecology* 8:299–305.

Foerster, John W. 1971. The ecology of an elfin forest in Puerto Rico, 14. The algae of Pico del Oeste. *Journal of the Arnold Arboretum* 52(1): 86–109.

Fogden, M., and P. Fogden. 1984. All that glitters may be toads. *Natural History* 5:47–50.

Folster, H., and H.W. Fassbender. 1978. Untersuchungen uber Bodenstandorte der humiden Bergwalder in der nordlichen Andenkordillere. Pages 101–110 in Wissenschaftliche Tagung "Landnutzung in den humiden Gebieten der Tropen," Institute of Soil Science and Forest Nutrition, Gottingen.

Forman, Richard T.T. 1975. Canopy lichens with blue-green algae: a nitrogen source in a Colombian rain forest. *Ecology* 56(5): 1176–1184.

Frahm, J.P. 1987. Okologische Studien uber die epiphytische Moosvegetation in Regenwalder NO-Perus. *Beihefte zur Nova Hedwigia* 88:143–158.

Frahm, J.P. 1987. Struktur und Zusammensetzung der epiphytischen Moosvegetation in Regenwalder NO-Perus. *Beihefte zur Nova Hedwigia* 88:115–141.

Frahm, J.P. 1987. Which factors control the growth of epiphytic bryophytes in tropical rainforests? *Symposia Biologica Hungarica* 35:639–648.

Frahm, J.P. 1990. The altitudinal zonation of bryophytes on Mt. Kinabalu. *Nova Hedwigia* 51:133–149.

Frahm, J.P. 1990. The ecology of epiphytic bryophytes on Mt. Kinabalu, Sabah (Malaysia). *Nova Hedwigia* 51:121–132.

Frahm, Jan Peter, Wolfgang Frey, Harald Kuschner, and Mario Menzel. 1990. Mosses and liverworts of Mt. Kinabalu. Sabah Parks Publication No. 12. Kota Kinabalu, Malaysia.

Frahm, Jan Peter, and S.R. Gradstein. 1991. An altitudinal zonation of tropical rain forests using bryophytes. *Journal of Biogeography* 18:669–678.

Frangi, J.L., and A.E. Lugo. 1985. Ecosystem dynamics of a subtropical floodplain forest. *Ecological Monographs* 53:351–369.

Frangi, Jorge L., and Ariel E. Lugo. 1992. Biomass and nutrient accumulation in ten year old bryophyte communities inside a flood plain in the Luquillo Experimental Forest, Puerto Rico. *Biotropica* 24(2a): 106–112.

Frei, Erwin. 1978. Andepts in some high mountains of east Africa. *Geoderma* 21:119–121.

Fritz-Sheridan, R.P., and D.S. Coxson. 1988. Nitrogen fixation on the tropical volcano, La Soufriere (Guadeloupe): nitrogen fixation, photosynthesis and respiration during the prevailing cloud/shroud climate by *Stereocaulon virgatum*. *Lichenologist* 20(1): 41–61.

Fritz-Sheridan, Richard P., and Jacques Portecop. 1987. Nitrogen fixation on the tropical volcano, La Soufriere (Guadeloupe): I. A survey of nitrogen fixation by blue-green algal microepiphytes and lichen endophytes. *Biotropica* 19(3): 194–199.

Fuenzalida, H.A., R.S. Schemenauer, and P. Cereceda. 1989. Subtropical Stratocumuli as a water resource. Pages 199–200 in the Third International Conference on Southern Hemisphere Meteorology and Oceanography, 13–17 November, Buenos Aires.

Fulford, Margaret, Barbara Crandall, and Ray Stotler. 1970. The ecology of an elfin forest in Puerto Rico, 11. The leafy Hepaticae of Pico del Oeste. *Journal of the Arnold Arboretum* 51(1): 56–69.

Fulford, Margaret, Barbara Crandall, and Raymond Stotler. 1971. The ecology of an elfin forest in Puerto Rico, 15. A study of the leafy hepatic flora of the Luquillo Mountains. *Journal of the Arnold Arboretum* 52(3): 435–458.

Gallagher, M.W., T.W. Choularton, A.P. Morse, and D. Fowler. 1988. Measurements of the size dependence of cloud droplet deposition at a hill site. *Quarterly Journal of the Royal Meteorological Society* 114:1291–1303.

Galloway, D.J. 1991. *Tropical lichens: their systematics, conservation and ecology*. Clarendon Press, Oxford.

Gates, David M. 1969. The ecology of an elfin forest in Puerto Rico, 4. Transpiration rates and temperatures of leaves in cool humid environment. *Journal of the Arnold Arboretum* 50(1): 93–98.

Gautam-Basak, M., and J. Proctor. 1983. Micronutrients, aluminium, silicon and ash in leaf litterfall from forests in Gunung Mulu National Park, Sarawak. *Malaysian Forester* 46:224–232.

Gentry, A.G. 1982. Patterns of neotropical plant species diversity. *Evolutionary Biology* 15:1–84.

Gentry, A.G., and C.H. Dodson. 1987. Diversity and biogeography of neotropical vascular epiphytes. *Annals of the Missouri Botanical Garden* 74:205–233.

Gentry, A.H. 1992. Tropical forest biodiversity distributional patterns and their conservational significance. *Oikos* 63(1): 19–28.

Gentry, Alwyn H., and Calaway Dodson. 1987. Contribution of nontrees to species richness of a tropical rain forest. *Biotropica* 19(2): 149–156.

Gerard, Robert D., and J. Lamar Worzel. 1967. Condensation of atmospheric moisture from tropical maritime air masses as a freshwater resource. *Science* 157 (3794): 1300–1302.

Gill, A.M. 1969. The ecology of an elfin forest in Puerto Rico, 5. Aerial roots. *Journal of the Arnold Arboretum* 50(2): 197–209.

Gioda, Alain, A. Acosta Baladón, Pierre Fontanel, Zósimo Hernández Martín, and Arnoldo Santos. 1993. El árbol fuente. *Mundo Científico* 132(13): 126–134.

Gioda, A., Z. Hernandez Martin, and E. Gonzales. 1993. Observatoires, brouillards et arbres fontaines aux Canaries. *Veille Climatique Satellitaire* 46:38–49.

Gioda, A., J. Maley, R.E. Guasp, and A.A. Baladón. 1995. Some low elevation fog forests of dry environments: applications to African paleoenvironments. Pages 156-164 in L.S. Hamilton, J.O. Juvik, and F.N. Scatena, editors. *Tropical Montane Cloud Forests*. Springer-Verlag, New York.

Gonggrijp, L. 1941. Evaporation from montane forest in West Java at an altitude of 1,750–2,000 m above sea level. *Tectona* 34:437–447.

Gradstein, S.R. 1992. The vanishing tropical rain forest as an environment for bryophytes and lichens. Pages 232–256 in J.W. Bates and A.R. Farmer, editors. *Bryophytes and Lichens in a Changing Environment*. Oxford University Press, London.

Gradstein, S.R., and J.P. Frahm. 1987. Die floristisch Hohengliederung der Moose entlang des BRYOTROP-Transektes in NO Peru. *Beihefte zur Nova Hedwigia* 88:105–113.

Gradstein, S.R., and W.H.A. Hekking. 1979. Studies on Colombian cryptograms IV. A catalogue of the Hepaticae of Colombia. *Journal of the Hattori Botanical Laboratory* 45:93–144.

Gradstein, S.R., G.B.A. van Reenen, and D. Griffin III. 1989. Species richness and origin of the bryophyte flora of the Colombian Andes. *Acta Botanica Neerlandica* 38:439–448.

Gradstein, S.R., and W.A. Weber. 1982. Bryogeography of the Galapagos Islands. *Journal of the Hattori Botanical Laboratory* 52:127–152.

Green, T.G.A., and G.T. Jane. 1983. Diurnal patterns of water potential in the evergreen cloud forests of the Kaimai Ranges, North Island, New Zealand. *New Zealand Journal of Botany* 21:379–389.

Grieve, I.C., J. Proctor, and S.A. Cousins. 1990. Soil variation with altitude on Volcan Barva, Costa Rica. *Catena* 17:525–534.

Grimm, U., and H.W. Fassbender. 1981. Ciclos biquímicos en un ecosistema forestal de los Andes Occidentales de Venezuela. I. Inventario de las reservas orgánicas y minerales (N, P, K, Ca, Mg, Mn, Fe, Al, Na). *Turrialba* 31:27–37.

Grubb, P.J. 1971. Interpretation of the "Massenerhebung" effect on tropical mountains. *Nature* 229:44–45.

Grubb, P.J. 1974. Factors controlling the distribution of forest types on tropical mountains: new facts and a new perspective. Pages 13–46 in J.R. Flenley, editor. *Altitudinal zonation in Malesia.* Transactions of the Third Aberdeen-Hull Symposium on Malesian Ecology, University of Hull, Department of Geography, 1973, Hull.

Grubb, P.J. 1977. Control of forest growth and distribution on wet tropical mountains: with special reference to mineral nutrition. *Annual Review of Ecology and Systematics* 8:83–107.

Grubb, P.J. 1989. The role of mineral nutrients in the tropics: a plant ecologist's view. Pages 417–439 in J. Proctor, editor. *Mineral nutrients in tropical forest and savanna ecosystems.* Blackwell Scientific Press, Oxford.

Grubb, P.J., J.R. Lloyd, T.D. Pennington, and T.C. Whitmore. 1963. A comparison of montane and lowland rain forest in Ecuador. I. The forest structure, physiognomy, and floristics. *Journal of Ecology* 51:567–701.

Grubb, P.J., and E.V.J. Tanner. 1976. The montane forests and soils of Jamaica: a reassessment. *Journal of the Arnold Arboretum* 57(3): 313–368.

Grubb, P.J., and T.C. Whitmore. 1966. A comparison of montane and lowland rain forest in Ecuador. II. The climate and its effects on the distribution and physiognomy of the forests. *Journal of Ecology* 54:303–333.

Grubb, P.J., and T.C. Whitmore. 1967. A comparison of montane and lowland forest in Ecuador. III. The light reaching the ground vegetation. *Journal of Ecology* 55:33–57.

Haggar, J.P. 1988. The structure, composition, and status of the cloud forests of Pico Island in the Azores, Portugal. *Biological Conservation* 46(1): 7–22.

Hamilton, A.C., and R.A. Perrott. 1981. A study of altitudinal zonation in the montane forest belt of Mt. Elgon, Kenya. *Vegetation* 45:107–125.

Hamilton, L.S., J.O. Juvik, and F.N. Scatena. 1995. The Puerto Rico Tropical Cloud Forest Symposium: introduction and workshop synthesis. Pages 1–23 in L.S. Hamilton, J.O. Juvik, and F.N. Scatena, editors. *Tropical Montane Cloud Forests.* Springer-Verlag, New York.

Hamilton, L.S., with P.N. King. 1983. *Tropical forested watersheds: hydrologic and soils response to major uses or conversions.* Westview Press, Boulder. 168 pp.

Harr, R.D. 1982. Fog drip in the Bull Run watershed, Oregon. *Water Resources Bulletin* 18:785–789.

Heaney, A., and J. Proctor. 1989. Chemical elements in litter in forests on Volcan Barva, Costa Rica. Pages 255–271 in J. Proctor, editor. *Mineral nutrients in tropical forest and savanna ecosystems.* Blackwell Scientific Publications, Oxford.

Heaney, Angela, and John Proctor. 1990. Preliminary studies on forest structure and floristics on Volcan Barva, Costa Rica. *Journal of Tropical Ecology* 6:307–320.

Hedberg, O. 1961. The phytogeographical position of the afroalpine flora. Pages 914–919 in Vol. 1, *Recent advances in botany.* Lectures and symposia presented to the IX International Botanical Congress, Montreal, 1959. University of Toronto Press, Toronto.

Hedberg, O. 1964. Features of afroalpine plant ecology. *Acta Phytogeographica Suecica* 49:1–144.

Hedberg, O. 1986. Origins of the afroalpine flora. In P. Vuilleumier and M. Monasterio, editors. *High altitude tropical biogeography*. Oxford University Press, London.

Herrmann, R. 1971. The temporal variation of soil moisture tension under different altitudinal vegetation formations on a high tropical mountain (Sierra Nevada de Santa Marta, Colombia). *Erdkunde* 25(2): 90–102.

Hetsch, W., and H. Hoheisel. 1976. Standorts-und Vegetations-gliederung in einem tropischen Nebelwald. *Allgemeine Forst-und Jagd Zeitung* 147:200–209.

Hofstede, R.G.M., J. Wolf, and D.H. Benzing (in press). Epiphytic biomass and nutrient status of a Colombian upper montane rain forest. *Selbyana*.

Holttum, R.E. 1924. The vegetation of Gunung Belumut in Johore. *Gardens' Bulletin Singapore* 3:245–257.

Hommel, P.F.M. 1987. Landscape ecology of Ujung Kulon (West Java, Indonesia). PhD thesis, Wageningen Agricultural University, Wageningen.

Horn, Sally P. 1988. Effect of burning on a montane mire in the Cordillera de Talamanca, Costa Rica. *Brenesia* 30:81–92.

Horner, J.D., J.R. Gosz, and R.G. Cates. 1988. The role of carbon-based plant secondary metabolites in decomposition in terrestrial ecosystems. *American Naturalist* 132:869–883.

Hosokawa, T. 1952. A plant sociological study in the mossy forests of Micronesian islands. *Memoirs of the Faculty of Science, Kyushu University, Series E (Biology)* 1:65–82.

Hosokawa, T., N. Odani, and T. Tagawa. 1964. Causality of the distribution of corticulous species in forests with special reference to the eco-physiological approach. *Bryologist* 67:396–411.

Howard, Richard A. 1966. Notes on some plants of Puerto Rico. *Journal of the Arnold Arboretum* 47:137–144.

Howard, Richard A. 1968. The ecology of an elfin forest in Puerto Rico, 1. Introduction and composition studies. *Journal of the Arnold Arboretum* 49(4): 381–418.

Howard, Richard A. 1969. The ecology of an elfin forest in Puerto Rico, 8. Studies of stem growth and form and of leaf structure. *Journal of the Arnold Arboretum* 50(2): 225–262.

Howard, Richard A. 1970. The "alpine" plants of the Antilles. *Biotropica* 2(1): 24–28.

Howard, Richard A. 1970. The ecology of an elfin forest in Puerto Rico, 10. Notes on two species of Marcgravia. *Journal of the Arnold Arboretum* 51(1): 41–55.

Howard, Richard A. 1970. The summit forest of Pico del Oeste, Puerto Rico. Chapter B-20 in Howard T. Odum and Robert F. Pigeon, editors. *A tropical rain forest, a study of irradiation and ecology at El Verde, Puerto Rico*. U.S. Atomic Energy Commission, Division of Technical Information, Oak Ridge. (Available from the National Technical Information Service, Springfield, VA 22151.)

Huber, O. 1976. *Pflanzenökologische Untersuchungen im Gebirgsnebelwald von Rancho Grande (Venezolanische Küstenkordillere)*. Dissertation [PhD-Thesis], Leopold-Franzens-Universität, Innsbruck (Austria). 127pp.

Huber, O. 1978. Light compensation point of vascular plants of a tropical cloud forest and an ecological interpretation. *Photosynthetica* 12(4): 382–390.

Huber, O. 1986. El clima. In: *La selva nublada de Rancho Grande, Parque Nacional 'Henri Pittier'* (O. Huber, ed.), pp. 17-29. Fondo Editoiral 'Acta Científica Venezolana', Caracas.

Huber, Otto, editor. 1986. *La selva nublada de Rancho Grande parque nacional "Henri Pittier."* Fondo Editorial Acta Científica Venezolana y Seguros Anauco C.A. Caracas. 289 pp.

Huber, O. 1986. Las selvas nubladas de Rancho Grande: observaciones sobre su fisionomía, estructura y fenología. In: *La selva nublada de Rancho Grande, Parque Nacional 'Henri Pittier'* (O. Huber, ed.), pp. 131-170. Fondo Editoiral 'Acta Científica Venezolana', Caracas.

Huber, O. 1986. Punto de compensación de luz en plantas vasculares del bosque

nublado de Rancho Grande y una interpretación ecológica. In: *La selva nublada de Rancho Grande, Parque Nacional 'Henri Pittier'* (O. Huber, ed.), pp. 197-202. Fondo Editoiral 'Acta Científica Venezolana', Caracas.
Huber, O. 1986. Ubicacíon geográfica y fisiografía. In: *La selva nublada de Rancho Grande, Parque Nacional 'Henri Pittier'* (O. Huber, ed.), pp. 13-15. Fondo Editoiral 'Acta Científica Venezolana', Caracas.
Huey, B.R. 1978. Latitudinal pattern of altitude faunal similarity: mountains might be higher in the tropics. *American Naturalist* 112:225–229.
Ingraham, N.L., and R.A. Matthews. 1988. Fog drip as a source of groundwater recharge in northern Kenya. *Water Resources Research* 24:1406–1410.
Ingraham, N.L., and R.A. Matthews. 1988. The role of fog drip in the hydrologic cycle. Pages 670–677 in Volume 2, *Proceedings of the IWRA Sixth World Congress on Water Resources, Water for World Development,* Ottawa.
Ingraham, N.L., and R.A. Matthews. 1990. A stable isotopic study of fog: the Point Reyes Peninsula, California. *Chemical Geology* (Isotope Geoscience Section) 80:281–290.
Jacobson, S.K., and J.J. Vandenberg. 1991. Reproductive ecology of the endangered golden toad (*Bufo periglenes*). *Journal of Herpetology* 25:321–327.
Jaffe, M.J. 1980. Morphogenetic responses of plants to mechanical stimuli or stress. *BioScience* 30:239–243.
Jane, G.T., and T.G.A. Green. 1985. Patterns of stomatal conductance in six evergreen tree species from a New Zealand cloud forest. *Botanical Gazette* 46:413–420.
Janzen, D.H. 1967. Why mountain passes are higher in the tropics. *American Naturalist* 101:233–249.
Johansson, D. 1974. Ecology of vascular epiphytes in West African rain forest. *Acta Phytogeographica Suecica* 59:1–136.
Johns, R.J. 1982. Plant zonation. Pages 309–330 in J.L. Gressitt, editor. *Biogeography and ecology of New Guinea,* Volume 1. Dr. W. Junk Publishers, The Hague.
Jordan, C.F. 1978. Stemflow and nutrient transfer in a tropical rain forest. *Oikos* 31:257–263.
Jordan, C.F., and J. Heuveldop. 1991. The water budget of an Amazonian rainforest. *Acta Amazonica* 11:87–92.
Juvik, J.O., and P.C. Ekern. 1978. A climatology of mountain fog on Mauna Loa, Hawai'i island. Technical Report No. 118, Water Resources Research Center, University of Hawai'i, Honolulu. 63 pp.
Juvik, J.O., and D. Nullet. 1995. Relationships between rainfall, cloud-water interception, and canopy throughfall in a Hawaiian montane forest. Pages 165–182 in L.S. Hamilton, J.O. Juvik, and F.N. Scatena, editors. *Tropical Montane Cloud Forests*. Springer-Verlag, New York.
Juvik, James O., and Douglas J. Perreira. 1974. Fog interception on Mauna Loa, Hawai'i. *Proceedings of the Association of American Geographers* 6:22–25.
Kapos, V., and E.V.J. Tanner. 1985. Water relations of Jamaican upper montane rain forest trees. *Ecology* 66:241–250.
Kashiyama, T. 1956. Decrease of sea-fog density by a model shelterbelt. In *Proceedings of the 12th Congress of the International Union of Forestry Research Organizations*, Oxford.
Kerfoot, O. 1968. Mist precipitation on vegetation. *Forestry Abstracts* 29(1): 8–20.
Kinsman, S. 1990. Regeneration by fragmentation in tropical montane forest shrubs. *American Journal of Botany* 77(12): 1626–1633.
Kitayama, K. 1987. Vegetation of Mount Kinabalu. *Review of Forestry Culture* 8:103–113.
Kitayama, K. 1992. An altitudinal transect study of the vegetation on Mount Kinabalu, Borneo. *Vegetatio* 102:149–171.

Kitayama, K. 1995. Biophysical conditions of the montane cloud forests of Mount Kinabalu, Sabah, Malaysia. Pages 183–197 in L.S. Hamilton, J.O. Juvik, and F.N. Scatena, editors. *Tropical Montane Cloud Forests*. Springer-Verlag, New York.

Kitayama, Kanehiro, and D. Mueller-Dumbois. 1992. Vegetation of the wet windward slope of Haleakala, Maui, Hawai'i. *Pacific Science* 46:197–220.

Klein, R.M. 1980. Ecologia da flora e vegetacao do Vale do Itaja (continuacao). *Sellowia* 32(32): 165–389.

Kochummen, K.M. 1982. Effects of elevation on vegetation on Gunung Jerai, Kedah. Research Pamphlet No. 87, Forest Research Institute, Peninsular Malaysia, Kepong, Malaya. 124 pp.

Koptur, Suzanne, William A. Haber, Gordon W. Frankie, and Herbert G. Baker. 1988. Phenological studies of shrub and treelet species in tropical cloud forests of Costa Rica. *Journal of Tropical Ecology* 4:323–346.

Korner, C., A. Allison, and H. Hilscher. 1983. Altitudinal variation of leaf conductance and leaf anatomy in heliophytes of montane New Guinea and their interrelation with microclimate. *Flora* 174:91–135.

LaBastille, A. 1974. Use of artificial nest boxes by quetzals in Guatemala. *Biological Conservation* 5:183–190.

LaBastille, A. 1983. *Pharomachrus mocinno* (quetzal). Pages 599–601 in D.H. Janzen, editor. *Costa Rican natural history*. University of Chicago Press, Chicago.

LaBastille, Anne. 1973. An ecological survey of the proposed Volcano Baru National Park, Republic of Panama. IUCN Occasional Paper No. 6, Morges, Switzerland.

LaBastille, Anne. 1973. Establishment of a Quetzal cloud-forest reserve in Guatemala. *Biological Conservation* 5:60–62.

LaBastille, Anne, and Douglas J. Pool. 1978. On the need for a system of cloud-forest parks in middle America and the Caribbean. *Environmental Conservation* 5(3): 183–190.

Lamprecht, H. 1958. Der Gebirg-Nebelwald der Venezolanischen Anden. *Schweizerische Aeitschrift fur Forstwesen* 2:1–16.

Lane-Poole, C.F. 1925. The forests of Papua and New Guinea. *Empire Forestry Journal* 4:206–234. (Cited in Stadtmüller, 1987.)

Langley, A.M. 1985. Altitudinal changes in climate and litter on Hunung Silam, a small ultrabasic mountain on Sabah, east Malaysia. BSc thesis, University of Stirling, Stirling.

Lawton, R. 1980. Wind and the ontogeny of elfin stature in a Costa Rican lower montane rain forest. PhD thesis, University of Chicago, Chicago.

Lawton, R.O. 1990. Canopy gaps and light penetration into a wind-exposed tropical lower montane rain forest. *Canadian Journal of Forest Research* 20:659–667.

Lawton, R.O., and F.E. Putz. 1988. Natural disturbance and gap-phase regeneration in a wind-exposed tropical cloud forest. *Ecology* 69(3): 764–777.

Lawton, Robert, and Valerie Dryer. 1980. The vegetation of the Monteverde Cloud Forest Reserve. *Brenesia* 18:101–116.

Lawton, Robert O. 1982. Wind stress and elfin stature in a montane rain forest tree: an adaptive explanation. *American Journal of Botany* 69(8): 1224–1230.

Lawton, Robert O. 1984. Ecological constraints on wood density in a tropical montane rain forest. *American Journal of Botany* 71(2): 261–267.

Leigh, E.G. 1975. Structure and climate in tropical rain forest. *Annual Review of Ecology and Systematics* 6:67–86.

Leigh, Egbert Giles. 1990. Tree shape and leaf arrangement: a quantitative comparison of montane forests, with emphasis on Malaysia and South India. Pages 119–174 in J.C. Daniel and J.S. Serrao, editors. *Conservation in developing countries: problems and prospects*. Proceedings of the Centenary Seminar of the Bombay Natural History Society. Oxford University Press, Bombay.

Leighton, M., and N. Wirawan. 1986. Catastrophic drought and fire in Borneo tropical rain forest associated with the 1982–83 El Niño southern oscillation. Pages 75–102 in G.T. Prance, editor. *Tropical rain forest and the world atmosphere*. Westview Press, Boulder.

Leo, M. 1995. The importance of tropical montane cloud forest for preserving vertebrate endemism in Peru: the Río Abiseo National Park as a case study. Pages 198–211 in L.S. Hamilton, J.O. Juvik, and F.N. Scatena, editors. *Tropical Montane Cloud Forests*. Springer-Verlag, New York.

Lewis, Walter H. 1971. High floristic endemism in low cloud forests of Panama. *Biotropica* 3(1): 78–80.

Lloyd, C.R., J.H.C. Gash, W.J. Shuttleworth, and A. de O.F. Marques. 1988. The measurement and modeling of rainfall interception by Amazonian rain forest. *Forest and Agricultural Meteorology* 43:277–294.

Lloyd, C.R., and A. de O.F. Marques. 1988. Spatial variability of throughfall and stemflow measurements in Amazonian rain forest. *Forest and Agricultural Meteorology* 42:63–73.

Loiselle, B.A., and J.G. Blake. 1991. Temporal variation in birds and fruits along an elevational gradient in Costa Rica. *Ecology* 72:180–193.

Long, A.J. 1995. The importance of tropical montane cloud forests for endemic and threatened birds. Pages 79–106 in L.S. Hamilton, J.O. Juvik, and F.N. Scatena, editors. *Tropical Montane Cloud Forests*. Springer-Verlag, New York.

Lowry, J.B., D.W. Lee, and B.C. Stone. 1973. Effects of drought on Mount Kinabalu. *Malayan Nature Journal* 26:178–179.

Lugo, A.E. 1986. Water and the ecosystems of the Luquillo experimental forest. General Technical Report SO 63, Institute of Tropical Forestry, Rio Piedras, Puerto Rico.

Lugo, Ariel E., and F.N. Scatena. 1992. Epiphytes and climate change research in the Caribbean: a proposal. *Selbyana* 13:123–130.

Lumer, C. 1980. Rodent pollination of *Blakea* (Melastomataceae) in a Costa Rican cloud forest. *Brittonia* 32:512–517.

Lumer, C., and R.D. Schoer. 1986. Pollination of *Blakea austin-smithii* and *B. penduliflora* (Melastomataceae) by small rodents in Costa Rica. *Biotropica* 18:363–364.

Lundgren, L., and B. Lundgren. 1979. Rainfall interception and evaporation in the Mazumbai forest reserve, West Usambara Mountains, Tanzania. *Geografiska Annaler* 61A:155–177.

Lush, C. 1995. Cloud forests of the Rwenzori Mountains, Uganda: research and management possibilities. Pages 212–222 in L.S. Hamilton, J.O. Juvik, and F.N. Scatena, editors. *Tropical Montane Cloud Forests*. Springer-Verlag, New York.

Lyford, Walter H. 1969. The ecology of an elfin forest in Puerto Rico, 7. Soil, root, and earthworm relationships. *Journal of the Arnold Arboretum* 50(2): 210–224.

McDowell, W.H., C.G. Sánchez, C.E. Asbury, and C.R. Ramos Perez. 1990. Influence of sea salt aerosols and long range transport on precipitation chemistry at El Verde, Puerto Rico. *Atmospheric Environment* 24A:2813–2821.

McVean, D.N. 1974. Mountain climates of the Southwest Pacific. Pages 47–58 in J.R. Flenley, editor. *Altitudinal zonation in Malesia.* Transaction of the Third Aberdeen-Hull Symposium on Malesian Ecology, University of Hull, Department of Geography, 1973, Hull.

Magdefrau, K. 1983. The bryophyte vegetation of the forests and paramos of Venezuela and Colombia. *Nova Hedwigia* 38:1–63.

Mamanteo, B.P., and V.P. Veracion. 1985. Measurements of fogdrip, throughfall and stemflow in the mossy and Benguet pine (*Pinus kesiya* Royle ex Gordon) forests in the upper Agno river basin. *Sylvatrop* (Philippines Forest Research Journal) 10:271–282.

Marrs, R.H., J. Proctor, A. Henney, and M.D. Mountford. 1988. Changes in soil nitrogen-mineralization and nitrification along an altitudinal transect in tropical rainforest in Costa Rica. *Journal of Ecology* 76:466–482.

Martin, Gary J., and A. de Avila. 1990. Exploring the cloud forests of Oaxaca, Mexico. World Wildlife Fund Report, October–December, Gland, Switzerland.

Martin, Paul S. 1955. Zonal distribution of vertebrates in a Mexican cloud forest. *American Naturalist* 89(849): 347–361.

Martin, Thomas E. 1984. Impact of livestock grazing on birds of a Colombian cloud forest. *Tropical Ecology* 25:158–171.

Maxwell, B. 1982. Floristic description of native upland forests on Kosrae, eastern Caroline Islands. *Micronesia* 18(2): 109.

Medeiros, A.C., L.L. Loope, and R.W. Hobdy. 1995. Conservation of cloud forests in Maui County (Maui, Moloka'i, and Lana'i), Hawaiian Islands. Pages 223–233 in L.S. Hamilton, J.O. Juvik, and F.N. Scatena, editors. *Tropical Montane Cloud Forests*. Springer-Verlag, New York.

Medina, E. 1986. Aspectos ecofisiológicos de plantas de bosques nublados tropicales: el bosque nublado de rancho grande. Pages 189–196 in O. Huber, editor. *La selva nublada de Rancho Grande parque nacional "Henri Pittier."* Fondo Editorial Acta Científica Venezolana y Seguros Anauco C.A. Caracas.

Medina, E., E. Cuevas, and P.L. Weaver. 1981. Composición foliar y transpiracion de especies leñosas de Pico del Este, Sierra de Luquillo, Puerto Rico. *Acta Científica Venezolana* 32:159–165.

Medina, E., M. Sobrado, and R. Herrera. 1978. Significance of leaf orientation for leaf temperature in an Amazonian sclerophyll vegetation. *Radiation and Environmental Biophysics* 15:131–140.

Menzel, M. 1988. The bryophytes of Sabah (North Borneo) with special reference to the BRYOTROP transect of Mount Kinabalu. 1. The study area and its bryological exploration. *Willdenowia* 18:281–300.

Merlin, M.D. 1985. Woody vegetation in the upland region of Rarotonga, Cook Islands, South Pacific. *Pacific Science* 39(1): 81–99.

Merlin, M.D., and J.O. Juvik. 1995. Montane cloud forest in the tropical Pacific: some aspects of their floristics, biogeography, ecology, and conservation. Pages 234–253 in L.S. Hamilton, J.O. Juvik, and F.N. Scatena, editors. *Tropical Montane Cloud Forests*. Springer-Verlag, New York.

Merriam, Robert A. 1973. Fog drip from artificial leaves in a fog wind tunnel. *Water Resources Research* 9(6): 1591–1596.

Moll, E.J. 1972. The current status of Mistbelt Mixed *Podocarpus* forest in Natal. *Bothalia* 10(4): 595–598.

Monteith, J.L. 1965. Evaporation and environment. Pages 205–234 in *The state and movement of water in living organisms*. Symposia of the Society for Experimental Biology No. 19. Academic Press, New York.

Monteith, J.L., editor. 1976. *Vegetation and the atmosphere*. Volume 2. Case studies. Academic Press, New York.

Mueller-Dombois, D. 1967. Ecological relations in the alpine and subalpine vegetation on Mauna Loa, Hawai'i. *Journal of the Indian Botanical Society* 46:403–411.

Myers, Charles W. 1969. The ecological geography of cloud forest in Panama. American Museum Novitates No. 2396, American Museum of Natural History, New York. 52 pp.

Nadkarni, N.M. 1986. The nutritional effects of epiphytes on host trees with special reference to alteration of precipitation chemistry. *Selbyana* 9:44–51.

Nadkarni, N.M., and T.J. Matelson. 1992. Biomass and nutrient dynamics of epiphytic litterfall in a neotropical montane forest, Costa Rica. *Biotropica* 24(1): 24–30.

Nadkarni, N.M., and T.J. Matelson. 1992. Biomass and nutrient dynamics of fine litter of terrestrially rooted material in a neotropical montane forest, Costa Rica. *Biotropica* 24(2a): 113–120.

Nadkarni, Nalini. 1988. Use of a portable platform for observations of tropical forest canopy animals. *Biotropica* 20(4): 350–351.

Nadkarni, Nalini M. 1981. Canopy roots: convergent evolution in rainforest nutrient cycles. *Science* 213(4524): 1023–1024.

Nadkarni, Nalini M. 1984. Epiphyte biomass and nutrient capital of a neotropical elfin forest. *Biotropica* 16(4): 249–256.

Nadkarni, Nalini M., and John T. Longino. 1990. Invertebrates in canopy and ground organic matter in a neotropical montane forest, Costa Rica. *Biotropica* 22(3): 286–289.

Nadkarni, Nalini M., and T.J. Matelson. 1991. Fine litter dynamics within the tree canopy of a tropical cloud forest. *Ecology* 72:2071–2082.

Nadkarni, Nalini M., and Teri J. Matelson. 1989. Bird use of epiphyte resources in neotropical trees. *The Condor* 91:891–907.

Nadkarni, Nalini M., and R.B. Primack. 1989. A comparison of mineral uptake and translocation by above-ground and below-ground root systems of *Salix syringiana*. *Plant and Soil* 113:39–45.

Nagel, J.F. 1956. Fog precipitation on Table Mountain. *Quarterly Journal of the Royal Meteorological Society* 82:452–460.

Nevling, Lorin I., Jr. 1969. The ecology of an elfin forest in Puerto Rico, 5. Chromosome numbers of some flowering plants. *Journal of the Arnold Arboretum* 50(1): 99–103.

Nevling, Lorin I., Jr. 1970. The ecology of an elfin forest in Puerto Rico, 12. A new species of Gonocalyx (Ericaceae). *Journal of the Arnold Arboretum* 51(2): 221–227.

Nevling, Lorin I., Jr. 1971. The ecology of an elfin forest in Puerto Rico, 16. The flowering cycle and an interpretation of its seasonality. *Journal of the Arnold Arboretum* 52(4): 586–613.

Newsome, D. 1986. Aspects of soil surface characteristics on Rakata and Anak Krakatau. Pages 134–156 in M. Bush, P. Jones, and K. Richards, editors. *Krakatoa centenary expedition 1983 final report.* University of Hull, Department of Geography, Miscellaneous Series 33, Hull.

Newton, Adrian C., and John R. Healey. 1989. Establishment of *Clethra occidentalis* on stems of the tree-fern *Cyathea pubescens* in a Jamaican montane rain forest. *Journal of Tropical Ecology* 5:441–445.

Odum, H.T. 1970. Rain forest structure and mineral cycling homeostasis. Pages H3–52 in Howard T. Odum and Robert F. Pigeon, editors. *A tropical rain forest, a study of irradiation and ecology at El Verde, Puerto Rico.* U.S. Atomic Energy Commission, Division of Technical Information, Oak Ridge, TN. (Available from the National Technical Information Service, Springfield, VA 22151.)

Odum, H.T., George Ann Briscoe, and C.B. Briscoe. 1970. Fallout radioactivity and epiphytes. Chapter H–13 in Howard T. Odum and Robert F. Pigeon, editors. *A tropical rain forest, a study of irradiation and ecology at El Verde, Puerto Rico.* U.S. Atomic Energy Commission, Division of Technical Information, Oak Ridge, TN. (Available from the National Technical Information Service, Springfield, VA 22151.)

Odum, Howard T., and Robert F. Pigeon, editors. 1970. *A tropical rain forest, a study of irradiation and ecology at El Verde, Puerto Rico.* U.S. Atomic Energy Commission, Division of Technical Information, Oak Ridge, TN. (Available from the National Technical Information Service, Springfield, VA 22151.)

Ohsawa, M. 1982. Botanical notes on vegetation zonation in Mt. Kerintji, Sumatra. *Biological Science* (Tokyo) 34:48–56.

Ohsawa, M. 1990. An interpretation of latitudinal patterns of forest limits in south and east Asian mountains. *Journal of Ecology* 78:326–339.

Ohsawa, M. 1991. Structural comparison of tropical montane rain forests along latitudinal and altitudinal gradients in south and east Asia. *Vegetatio* 97:1–10.

Ohsawa, M. 1993. Latitudinal pattern of mountain vegetation zonation in southern and eastern Asia. *Journal of Vegetation Science* 4:13–18.

Ohsawa, M. 1995. The montane cloud forest and its gradational changes in Southeast Asia. Pages 254–265 in L.S. Hamilton, J.O. Juvik, and F.N. Scatena, editors. *Tropical Montane Cloud Forests*. Springer-Verlag, New York.

Ohsawa, M., P.H. Nainggolan, N. Tanaka, and C. Anwar. 1985. Altitudinal zonation of forest vegetation on Mount Kerinci, Sumatra: with comparisons to zonation in the temperate region of East Asia. *Journal of Tropical Ecology* 1:193–216.

Okanoue, M., and Y. Ohtani. 1981. On estimation for altitude of cloud belt. *Shinrin Ritchi* 23:31–34.

O'Neill, J.P. 1992. A general overview of the montane avifauna of Peru. Pages 47–55 in K.R. Young and N. Valencia, editors. *Biogeografía, ecología y conservación del bosque montano en el Peru*. Memorias del Museo de Historia Naturel, UNMSM 21, Lima.

Paoletti, Maurizio G., R.A.J. Taylor, Benjamin R. Stinner, Deborah H. Stinner, and David H. Benzing. 1991. Diversity of soil fauna in the canopy and forest floor of a Venezuelan cloud forest. *Journal of Tropical Ecology* 7:373–383.

Payton, R.O. 1993. Soils of the Manusela National Park. Pages 19–61 in I.D. Edwards, A. MacDonald, and J. Proctor, editors. *Natural history of Seram*. Intercept, Andover, U.K.

Penafiel, S.R. 1980. The mossy forests of the Central Cordillera Ranges and the Philippines. *Canopy* 6:8.

Penafiel, S.R. 1995. Biological and hydrological values of the mossy forests in the Central Cordillera Mountains, Philippines. Pages 266–273 in L.S. Hamilton, J.O. Juvik, and F.N. Scatena, editors. *Tropical Montane Cloud Forests*. Springer-Verlag, New York.

Penman, H.L. 1956. Evaporation: an introductory survey. *Netherlands Journal of Agricultural Science* 4:9–29.

Pérez, Francisco L. 1987. Soil moisture and the upper altitudinal limit of giant paramo rosettes. *Journal of Biogeography* 14:173–186.

Perry, D.R. 1978. Factors influencing arboreal epiphytic sociology in Central America. *Biotropica* 10:235–237.

Persinos, G.J., S.K. Christie, J.M. Bidinger, and M.J. Lapiana. 1970. The ecology of an elfin forest in Puerto Rico, 13. Phytochemical screening and literature survey. *Journal of the Arnold Arboretum* 51(4): 540–546.

Persson, R. 1974. World forest resources. Royal College of Forestry, Stockholm.

Pfadenheuer, J., and S.C. Boechat. 1981. Vegetation und Okologie sines Sphagnum-Moores in Sudbrasilien. *Vegetatio* 44:177–187.

Pocs, T. 1980. The epiphytic biomass and its effect on the water balance of two rain forest types in the Uluguru Mountains. *Acta Botanica Academiae Scientarium Hungaricae* 26:143–167.

Pocs, T. 1982. Tropical forest bryophytes. Pages 59–104 in A.J.E. Smith, editor. *Bryophyte ecology*. Chapman and Hall, London.

Podolsky, Robert D. 1992. Strange floral attractors: pollinator attraction and the evolution of plant sexual systems. *Science* 258:791–793.

Proctor, J. 1984. Tropical forest litterfall II: the data set. Pages 83–113 in L.L. Sutton and A.C. Chadwick, editors. *Tropical rain-forest: The Leeds Symposium*. Leeds Philosophical and Literary Society, Leeds.

Proctor, J., J.M. Anderson, and H.W. Vallack. 1983. Comparative studies on forests, soils and litterfall at four altitudes on Gunung Mulu, Sarawak. *Malaysian Forester* 46(1): 60–76.

Proctor, J., Y.F. Lee, A.M. Langley, W.R.C. Munro, and T. Nelson. 1988. Ecological studies on Gunung Silam, a small ultrabasic mountain in Sabah, Malaysia. I. Environment, forest structure and floristics. *Journal of Ecology* 76:320–340.

Proctor, J., C. Phillipps, G.K. Duff, A. Heaney, and F.M. Robertson. 1989. Ecological studies on Gunung Silam, a small ultrabasic mountain in Sabah, Malaysia. II. Some forest processes. *Journal of Ecology* 77:317–331.

Proctor, M.E.D. 1968. Studies in Malaysian rain forest. I. The forest on Triassic sediments in Jenka Forest Reserve. *Journal of Ecology* 56:143–196.

Puig, H., R. Bracho, and V. Sosa. 1983. Composición florística y estructura del bosque mesofilo en Gomez Farias, Tamaulipas, Mexico. *Biotica* 8(4): 339–360.

Rangel-Ch, J.O., A.M. Cleef, and S. Salamanca-V. (in press). Los bosques subandinos y ecuatoriales del transecto Parque Los Nevados (Cordillera Central). In T. van der Hammen and A.G. dos Santos, editors. La Cordillera Central Colombiana, Transecto Parque los Nevados (cuarta parte). *Studies on tropical Andean ecosystems*. Volume 4. J. Cramer, Berlin-Stuttgart.

Raynor, B. 1995. Montane cloud forests in Micronesia: status and future management. Pages 274–283 in L.S. Hamilton, J.O. Juvik, and F.N. Scatena, editors. *Tropical Montane Cloud Forests*. Springer-Verlag, New York.

Rebertus, Alan J. 1988. Crown shyness in a tropical cloud forest. *Biotropica* 20(4): 338–339.

Richards, P.W. 1952. *The tropical rain forest*. Cambridge University Press, Cambridge.

Richards, P.W. 1984. The ecology of tropical forest bryophytes. Pages 1233–1270 in R.M. Schuster, editor. *New manual of bryology,* Volume 2. The Hattori Botanical Laboratory, Nichinan, Japan.

Richards, P.W. 1991. A bryologist in British Guiana and the West Indies. *Journal of Bryology* 16:437–441.

Richardson, J.H. 1982. Some implications of tropical forest replacement in Jamaica. *Zeitschrift fur Geomorphologie* 44:107–118.

Rijsdijk, A., and L.A. Bruijnzeel. 1990. Erosion, sediment yield and land-use patterns in the upper Konto watershed, East Java, Indonesia. Konto River Project Communication 18(1–3). Konto River Project, Malang, Indonesia.

Roth, I. 1980. Leaf structure of plants from humid tropical forests. *Botanische Jahrbuecher fuer Systematik Pflanzengeschlichte und Pflanzengeographie* 101(4): 489–525.

Roth, I. 1990. *Leaf structure of a Venezuelan cloud forest in relation to the microclimate*. Gebruder Borntraeger, Berlin. 244 pp.

Roth, Ingrid, and Tatiana Merida de Bifano. 1971. Morphological and anatomical studies of leaves of the plants of a Venezuelan cloud forest. I. Shape and size of the leaves. *Acta Biologica Venezolana* 7(2): 127–155.

Russell, Keith W., and H.A. Miller. 1977. The ecology of an elfin forest in Puerto Rico, 17. Epiphytic mossy vegetation of Pico del Oeste. *Journal of the Arnold Arboretum* 58(1): 1–24.

Rutter, A.J. 1975. The hydrological cycle in vegetation. Pages 111–154 in J.L. Monteith, editor. *Vegetation and the atmosphere*. Academic Press, New York.

Sarmiento, F.O. 1995. Human impacts on the cloud forests of the Upper Guayllabamba River Basin, Ecuador, and suggested management responses. Pages 284–295 in L.S. Hamilton, J.O. Juvik, and F.N. Scatena, editors. *Tropical Montane Cloud Forests*. Springer-Verlag, New York. 399 pp.

Savage, Jay M. 1966. An extraordinary new toad (Bufo) from Costa Rica. *Revista de Biologia Tropical* 14:153–167.

Scatena, F.N. 1995. The management of Luquillo elfin cloud forest ecosystems: irreversible decisions in a nonsubstitutable ecosystem. Pages 296–308 in L.S. Hamilton, J.O. Juvik, and F.N. Scatena, editors. *Tropical Montane Cloud Forests*. Springer-Verlag, New York.

Schemenauer, R.S., and P. Cereceda. 1991. Pages 836-842 in *The chemical composition of fog water supplies in rural areas of Chile, Peru and Oman.* International Seminar on Efficient Water Use, 23–25 October, Mexico City.

Schemenauer, R.S., and P. Cereceda. 1991. Fog as a permanent water resource in arid lands. VIIth IWRA World Congress on Water Resources, 13–18 May, Rabat, Morocco. A2:23–25.

Schemenauer, R.S., and P. Cereceda. 1992. Fog as an alternative to rainwater collection. Conference on Rainwater Catchment Systems, 4–7 October, Kyoto, Japan. Volume 2, pages 593–601.

Schemenauer, R.S., and P. Cereceda (in press). Meteorological conditions at a coastal fog collection site in Peru. *Atmosfera.*

Schemenauer, R.S., P. Cereceda, and N. Carvajal. 1987. Measurements of fog water deposition and their relationships to terrain features. *Journal of Climate and Applied Meteorology* 26:1285–1291.

Schemenauer, Robert S., and P. Cereceda. 1991. Fog-water collection in arid coastal locations. *Ambio* 20(7): 303–308.

Schemenauer, Robert S., and P. Cereceda. 1992. Monsoon cloudwater chemistry on the Arabian Peninsula. *Atmospheric Environment* 26A:1583–1587.

Schemenauer, Robert S., and P. Cereceda. 1992. The quality of fog water collected for domestic and agricultural use in Chile. *Journal of Applied Meteorology* 31:276–290.

Schemenauer, Robert S., and P. Cereceda. 1992. Water from fog-covered mountains. *Waterlines* 10:10–13.

Schemenauer, Robert S., and Pilar Cereceda. 1992. Fog as an alternative to rainwater collection. International Rainwater Catchment Systems Association Regional Conference, 4–10 October, Kyoto. Volume 2, pages 593–601.

Schemenauer, Robert S., and Pilar Cereceda. 1992. The use of fog for groundwater recharge in arid regions. International Seminar on Groundwater and the Environment in Arid and Semiarid Areas, 16–20 August, Beijing. Pages 84–91.

Schemenauer, Robert S., H. Fuenzalida, and P. Cereceda. 1988. A neglected water resource: the Camanchaca of South America. *Bulletin of the American Meteorological Society* 69:138–147.

Schemenauer, Robert S., and P.I. Joe. 1989. The collection efficiency of a massive fog collector. *Atmospheric Research* 24:53–69.

Schimper, A.F.W. 1888. Die Epiphytische Vegetation Amerikas. *Botanische Mitteilungen aus den Tropen.* Volume 2. G. Fischer, Jena, Germany.

Schmidt, F.H. 1950. On the distribution of sunshine in Java. *Verhandelingen van het Meteorologisch en Geofysisch Observatorium te Djakarta* 42:1–77.

Seifritz, W. 1924. The altitudinal distribution of lichens and mosses on Mt. Gedeh, Java. *Journal of Ecology* 12:307–313.

Sevink, J. 1984. An altitudinal sequence of soils in the Sierra Nevada de Santa Marta. Pages 131–138 in T. van der Hammen and P.M. Ruiz, editors. *Studies on tropical Andean ecosystems.* Volume 2. La Sierra Nevada de Santa Marta (Colombia). J. Cramer, Berlin-Stuttgart.

Sheridan, Richard P. 1991. Nitrogenase activity by *Hapalosiphon flexuosus* associated with *Sphaqnum erythrocalvx* mats in the cloud forest on the Volcano La Soufriere, French West Indies. *Biotropica* 23(2): 134–140.

Shreve, Forrest. 1914. A montane rain-forest: a contribution to the physiological plant geography of Jamaica. Carnegie Institution, Washington. 110 pp.

Shuttleworth, W.J. 1988. Evaporation from Amazonian rain forest. *Philosophical Transactions of the Royal Society Series B* 233:321–346.

Smith, C.W. 1995. Lichens as indicators of cloud forest in Hawai'i. Pages 309–314 in L.S. Hamilton, J.O. Juvik, and F.N. Scatena, editors. *Tropical Montane Cloud Forests.* Springer-Verlag, New York.

Sobrevila, C., and M.T.K. Arroyo. 1982. Breeding systems in a montane tropical cloud forest in Venezuela. *Plant Systematics and Evolution* 140(1): 19–38.

Stadtmüller, Thomas. 1987. *Los bosques nublados en el tropico humedo: una revisión bibliográfica.* Universidad de las Naciones Unidas y Centro Agronómico Tropical de Investigación y Ensenanza, Turrialba, Costa Rica. 85 pp. Also published in English, as *Cloud forests in the humid tropics: a bibliographic review.* The United Nations University, Tokyo, and Centro Agronómico Tropical de Investigación y Enseñanza, Turrialba, Costa Rica. 81 pp.

Stadtmüller, Thomas, and N. Agudelo. 1990. Amount and variability of cloud moisture input in a tropical cloud forest. International Association of Hydrological Sciences Publication 193:25–32, Washington.

Steinhardt, U. 1979. Untersuchungen uber den Wasser-und Nahrstoffhaushalt eines andinen Wolkenwaldes in Venezuela. *Gottinger Bodenkundliche Berichte* 56:1–185.

Steyermark, Julian A. 1974. The summit vegetation of Cerro Autana. *Biotropica* 6(1): 7–13.

Steyermark, Julian A., and Otto Huber. 1978. Flora del Avila. Vollmer Foundation and Ministerio del Ambiente y de los Recursos Naturales Renovables, Caracas. (Available from Sociedad Venezolana de Ciencias Naturales, Apartado de Correos 76771, Caracas 107, Venezuela.)

Stone, B.C. 1983. Studies in Malesian Rutaceae. III. *Melicope suberosa,* a new species and new generic record for the Malayan (Peninsular) flora. *Gardens' Bulletin* 36(1): 93–100.

Stratton, Donald A. 1989. Longevity of individual flowers in a Costa Rican cloud forest: ecological correlates and phylogenetic constraints. *Biotropica* 21(4): 308–318.

Sugden, A.M. 1980. Neotropical cloud forest, with particular reference to Colombia. PhD dissertation, Oxford University, Oxford.

Sugden, A.M. 1981. Aspects of the ecology of vascular epiphytes in Colombian cloud forests. II. Habitat preference of Bromeliaceae in the Serrania de Macuira. *Selbyana* 5:264–273.

Sugden, A.M. 1982. The ecological, geographic and taxonomic relationships of an isolated Colombia cloud forest, with some implications for island biogeography. *Journal of the Arnold Arboretum* 63(1): 31–61.

Sugden, A.M. 1982. The vegetation of the Serrania de Macuira, Colombia: a contrast of arid lowlands and an isolated cloud forest. *Journal of the Arnold Arboretum* 63(1): 1–30.

Sugden, A.M. 1985. Leaf anatomy in a Venezuelan montane forest. *Botanical Journal of the Linnean Society* 90(4): 231–241.

Sugden, A.M., E.V.J. Tanner, and V. Kapos. 1985. Regeneration following clearing in a Jamaican montane forest: results of a ten-year study. *Journal of Tropical Ecology* 1:329–351.

Sugden, Andrew M. 1982. Long-distance dispersal, isolation, and the cloud forest flora of the Serrania de Macuira, Guajira, Colombia. *Biotropica* 14(3): 208–219.

Sugden, Andrew M. 1986. The montane vegetation and flora of Margarita Island, Venezuela. *Journal of the Arnold Arboretum* 67(2): 187–232.

Sugden, Andrew M., and Richard J. Robins. 1979. Aspects of the ecology of vascular epiphytes in Colombian cloud forests, I. The distribution of the epiphytic flora. *Biotropica* 11(3): 173–188.

Sullivan, J.H., A.H. Teramura, and L.H. Ziska. Variation in UV-B sensitivity in plants from a 3,000 m elevational gradient in Hawai'i. *American Journal of Botany* 79:737–743.

Talbot, R.W., R.C. Harris, E.V. Browell, G.L. Gregory, D.I. Sabacher, and S.M. Beck. 1986. Distribution and geochemistry of aerosols in the tropical North Atlantic troposphere: relationship to Saharan dust. *Journal of Geophysical Research* 91:5173–5182.

Tanner, E.V.J. 1977. Four montane rain forests of Jamaica: a quantitative characterization of the floristics, the soils and the foliar mineral levels, and a discussion of the interrelations. *Journal of Ecology* 65:883–918.

Tanner, E.V.J. 1977. Mineral cycling in montane rain forests in Jamaica. PhD thesis, University of Cambridge, Cambridge.

Tanner, E.V.J. 1980. Litterfall in montane rain forests of Jamaica and its relation to climate. *Journal of Ecology* 68:833–848.

Tanner, E.V.J. 1980. Studies on the biomass and productivity in a series of montane rain forests in Jamaica. *Journal of Ecology* 68:573–588.

Tanner, E.V.J. 1981. The decomposition of leaf litter in Jamaican montane rain forests. *Journal of Ecology* 69:263–275.

Tanner, E.V.J. 1985. Jamaican montane forests: nutrient capital and cost of growth. *Journal of Ecology* 73:553–568.

Tanner, E.V.J., and V. Kapos. 1982. Leaf structure of Jamaican upper montane rainforest trees. *Biotropica* 14:16–24.

Tanner, E.V.J., V. Kapos, and W. Franco. 1992. Nitrogen and phosphorus fertilization effects of Venezuelan montane forest trunk growth and litterfall. *Ecology* 73:78–86.

Tanner, E.V.J., V. Kapos, S. Freskos, J.R. Healey, and A.M. Theobald. 1990. Nitrogen and phosphorus fertilization of Jamaican montane forest trees. *Journal of Tropical Ecology* 6:231–238.

Teramura, A.H. 1983. Effects of ultraviolet-B radiation on the growth and yield of crop plants. *Physiologica Plantarum* 58:415–427.

Terborgh, J. 1977. Bird species diversity on an Andean elevational gradient. *Ecology* 58:1007–1019.

Thompson, Donald A. 1983. Effects of Hurricane Allen on some Jamaican forests. *Commonwealth Forestry Review* 62(2): 107–115.

Thouret, J.C., and T. van der Hammen. 1983. La sequencia Holocenica y Tardiglacial en el Parque los Nevados. Pages 262–276 in T. van der Hammen, A. Perez, and P. Pinto, editors. La Cordillera Central Colombiana, Transecto Parque los Nevados (Introducción y datos iniciales). *Studies on tropical Andean ecosystems*, Volume 1. J. Cramer, Vaduz, Liechtenstein.

Tie, Y.L., I.C. Baillie, C.M.S. Phang, and C.P. Lim. 1979. Soils of Gunung Mulu National Park. Department of Agriculture, Sarawak.

Timm, R.M., D.E. Wilson, B.L. Clauson, R.K. Laval, and C.S. Vaughan. 1989. Mammals of the La Selva-Braulio Carrillo complex, Costa Rica. *North American Fauna* 75:1–162.

Troll, C. 1956. Der Klima-und Vegetationsaufbau der Erde im Lichte neuer Forschungen. *Journal of the Academy of Science, Mainz* 1956:216–229.

Troll, C. 1958. Tropical mountain vegetation. *Proceedings of the 9th Pacific Science Congress* 20:37–45.

Twomey, S. 1957. Precipitation by direct interception of cloud-water. *Weather* 12(4): 120–122.

Ufano, G. 1990. Determination of rainfall and streamflow patterns in a gauged mossy forest watershed. MS thesis, Benguet State University, La Trinidad, Benguet, Philippines.

van der Hammen, T., S. Diaz-Piedrahita, and V.J. Alvarez, editors. 1989. La Cordillera Central Colombiana, Transecto Parque los Nevados (segunda parte). *Studies on tropical Andean ecosystems*, Volume 3. J. Cramer, Berlin-Stuttgart.

van der Hammen, T., and A.G. dos Santos, editors (in press). La Cordillera Central Colombiana, Transecto Parque los Nevados (cuarta parte). *Studies on tropical Andean ecosystems*, Volume 4. J. Cramer, Berlin-Stuttgart.

van der Hammen, T., A. Perez, and P. Pinto, editors. 1983. La Cordillera Central

Colombiana, Transecto Parque Los Nevados (Introducción y datos iniciales). *Studies on tropical Andean ecosystems*, Volume 1. J. Cramer, Vaduz, Liechtenstein.
van der Hammen, T., and P.M. Ruiz, editors. 1984. La Sierra Nevada de Santa Marta (Colombia): Transecto Buritaca-La Cumbre. *Studies on tropical Andean ecosystems*, Volume 2. J. Cramer, Berlin-Stuttgart.
van der Hammen, T., and H. Witte (in press). Ecoclimatological aspects of the Parque los Nevados transect area. In T. van der Hammen and A.G. dos Santos, editors. La Cordillera Central Colombiana, Transecto Parque los Nevados (cuarta parte). *Studies on tropical Andean ecosystems*, Volume 4. J. Cramer, Berlin-Stuttgart.
Van der Plas, M.C. 1992. Hydrological characteristics of the forest floor in undisturbed and logged-over rain forest, Danum Valley, Sabah, Malaysia. MSc thesis, Free University, Amsterdam.
Van Leerdam, A., R.J. Zagt, and E.J. Veneklaas. 1990. The distribution of epiphyte growth-forms in the canopy of a Colombian cloudforest. *Vegetatio* 87:59–71.
Van Oye, P. 1924. Sur l'ecologie des epiphytes a la surface des troncs d'arbres a Java. *Revue Generale de Botanique* 36:1–30, 68–83.
Van Reenen, G.B.A. 1983. Distribución y ecología de musgos y hepaticas (datos iniciales.) Pages 206–209 in T. van der Hammen, A. Pérez, and P. Pinto, editors. La Cordillera Central Colombiana, Transecto Parque Los Nevados (Introducción y datos iniciales). *Studies on tropical Andean ecosystems*, Volume 1. J. Cramer, Vaduz, Liechtenstein.
Van Reenen, G.B.A., and S.R. Gradstein. 1983. A transect analysis of the bryophyte vegetation along an altitudinal gradient on the Sierra Nevada de Santa Marta, Colombia. *Acta Botanica Neerlandica* 32:163–175.
Van Reuler, G. 1987. Soil studies in the Bukit Raya nature reserve. Pages 7–23 in H.P. Nooteboom, editor. *Report on the 1982–1983 Bukit Raya expedition*. Rijksherbarium, Leiden, the Netherlands.
Van Steenis, C.G.G.J. 1961. An attempt towards an explanation of the effect of mountain mass elevation. *Proceedings of the Royal Academy of Science of the Netherlands* 64:435–442.
Van Steenis, C.G.G.J. 1972. *The mountain flora of Java*. E.J. Brill, Leiden.
Vance, E.D., and N.M. Nadkarni. 1990. Microbial biomass and activity in canopy organic matter and the forest floor of a tropical cloud forest. *Soil Biology and Biochemistry* 22(5): 677–684.
Vaughan, R.E., and P.O. Wiehe. 1947. Studies on the vegetation of Mauritius. IV. Some notes on the internal climate of the upland climax forest. *Journal of Ecology* 34:126–136.
Vázquez-García, J.A. 1995. Cloud forest archipelagos: preservation of fragmented montane ecosystems in tropical America. Pages 315–332 in L.S. Hamilton, J.O. Juvik, and F.N. Scatena, editors. *Tropical Montane Cloud Forests*. Springer-Verlag, New York.
Veneklaas, E.J. 1990. Nutrient fluxes in bulk precipitation and throughfall in two montane tropical rain forests, Colombia. *Journal of Ecology* 78:974–992.
Veneklaas, E.J. 1990. Rainfall interception and above-ground nutrient fluxes in Colombian montane tropical rain forest. PhD thesis, University of Utrecht, Utrecht.
Veneklaas, E.J. 1991. Litterfall and nutrient fluxes in two montane tropical rain forests, Colombia. *Journal of Tropical Ecology* 7(3): 319–336.
Veneklaas, E.J., and R. Van Ek. 1990. Rainfall interception in two tropical montane rain forests, Colombia. *Hydrological Processes* 4:311–326.
Veneklaas, E.J., R.J. Zagt, A. Van Leerdam, R. Van Ek, A.J. Broekhoven, and M. Van Genderen. 1990. Hydrological properties of the epiphyte mass of a montane tropical rain forest. *Vegetatio* 89:183–192.
Venkataramanan, C., and S. Chinnamani. 1978. A preliminary note on the return of

nutrient by the leaf litter of wet (montane) evergreen Shola forests of Nilgiris. *Indian Forester* 104:450–456.

Vis, M. 1986. Interception, drop size distributions and rainfall kinetic energy in four Colombian forest ecosystems. *Earth Surface Processes and Landforms* 11:591–603.

Vitousek, P.M., P.A. Matson, and R.A. Turner. 1988. Elevation and age gradients in Hawaiian montane rainforest: foliar and soil nutrients. *Oecologia* 77:565–570.

Vogelmann, H.W. 1973. Fog precipitation in the cloud forests of eastern Mexico. *BioScience* 23(2): 96–100.

Vuilleumier, P., and M. Monasterio, editors. 1986. *High altitude tropical biogeography*. Oxford University Press, London.

Wadsworth, F.H., and J.A. Bonnet. 1951. Soil as a factor in the occurrence of two types of montane forest in Puerto Rico. *Caribbean Forester* 12:67–70.

Wadsworth, Frank H. 1951. Forest management in the Luquillo Mountains, I. The setting. *Caribbean Forester* 12(3): 93–114.

Wagner, Richard J., Anstiss B. Wagner, and Richard A. Howard. 1969. The ecology of an elfin forest in Puerto Rico, 9. Chemical studies of colored leaves. *Journal of the Arnold Arboretum* 50(4): 556–565.

Wake, D.B., and J.F. Lynch. 1976. The distribution, ecology, and evolutionary history of Plethodontid salamanders in tropical America. *Science Bulletin* (Los Angeles County Museum of Natural History) 25:1–65.

Walter, H. 1979. *Vegetation und Klimazonen*. Ulmer Verlag, Stuttgart.

Waterloo, M.J. 1989. A hydrological study of the mass elevation effect on Gunung Silam, a small coastal ultrabasic mountain in Sabah, East Malaysia. MSc thesis, Free University, Amsterdam.

Watling, D., and A.N. Gillison. 1995. Endangered species in low elevation cloud forest on Gau Island, Fiji. Pages 333–342 in L.S. Hamilton, J.O. Juvik, and F.N. Scatena, editors. *Tropical Montane Cloud Forests*. Springer-Verlag, New York.

Weathers, Kathleen C., Gene E. Likens, F. Herbert Bormann, Susan H. Bicknell, Bernard T. Bormann, Bruce C. Daube Jr., John S. Eaton, James N. Galloway, William C. Keene, Kenneth D. Kimball, William H. McDowell, Thomas G. Siccama, Daniel Smiley, and Robert A. Tarrant. 1988. Cloudwater chemistry from ten sites in North America. *Environmental Science and Technology* 22:1018–1026.

Weaver, P.L. 1972. Cloud moisture interception in the Luquillo Mountains of Puerto Rico. *Caribbean Journal of Science* 12(34): 129–144.

Weaver, P.L. 1972. The dwarf forest of Pico del Este and the Luquillo Mountains. *Revista InterAmericana Review* 2(2): 174–186.

Weaver, P.L. 1981. Ecological aspects and value of dwarf forests in the Caribbean. *Caribbean Conservation News* 2(8): 13–16.

Weaver, P.L. 1983. Tree growth and stand changes in the subtropical life zones of the Luquillo Mountains of Puerto Rico. U.S. Department of Agriculture, Forest Service Research Paper S0–190, New Orleans. 24 pp.

Weaver, P.L. 1990. Succession in the elfin woodland of the Luquillo Mountains of Puerto Rico. *Biotropica* 22(1): 83–89.

Weaver, P.L., M.D. Byer, and D.L. Bruck. 1973. Transpiration rates in the Luquillo Mountains of Puerto Rico. *Biotropica* 5(2): 123–133.

Weaver, P.L., E. Medina, D. Pool, K. Dugger, J. Gonzáles-Liboy, and E. Cuevas. 1986. Ecological observations in the dwarf cloud forest of the Luquillo Mountains. *Biotropica* 18(1): 79–85.

Weaver, Peter L. 1975. Transpiration rates in the elfin forest of the Luquillo Mountains of Puerto Rico. *Caribbean Journal of Science* 15(1–2): 21–30.

Werner, Wolfgang L. 1988. Canopy dieback in the upper montane rain forests of Sri Lanka. *GeoJournal* 17:245–248.

Werner, W.L. 1995. Biogeography and ecology of the upper montane rain forest of Sri

Lanka (Ceylon). Pages 343–352 in L.S. Hamilton, J.O. Juvik, and F.N. Scatena, editors. *Tropical Montane Cloud Forests*. Springer-Verlag, New York.

Werner, Wolfgang L., and S. Balasubramaniam. 1992. Structure and dynamics of the upper montane rain forests of Sri Lanka. Pages 165–172 in J.G. Goldammer, editor. *Tropical forests in transition*. Birkhhauser Verlag, Basel.

Wheelwright, N.T. 1983. Fruits and the ecology of resplendent quetzals. *The Auk* 100:286–301.

Wheelwright, N.T. 1985. Competition for dispersers, and the timing of flowering and fruiting in a guild of tropical trees. *Oikos* 44:465–477.

Wheelwright, N.T., W.A. Haber, K.G. Murray, and C. Guindon. 1984. Tropical fruit-eating birds and their food plants: a survey of a Costa Rican lower montane forest. *Biotropica* 16:173–192.

Whistler, W. Arthur. 1978. Vegetation of the montane region of Savai'i, Western Samoa. *Pacific Science* 32(1): 79–94.

Whistler, W. Arthur. 1992. Vegetation of Samoa and Tonga. *Pacific Science* 46:159–178.

Whistler, W.A. 1995. The cloud forest of Samoa. Pages 353–362 in L.S. Hamilton, J.O. Juvik, and F.N. Scatena, editors. *Tropical Montane Cloud Forests*. Springer-Verlag, New York.

Whitmore, T.C. 1969. The vegetation of the Solomon Islands. *Philosophical Transactions of the Royal Society* (London) 255:259–270.

Whitmore, T.C. 1975. *Tropical rain forests of the Far East*. Oxford University Press, London.

Whitmore, T.C. 1984. A new vegetation map of Malesia at scale 1:5 million. *Journal of Biogeography* 11:461–471.

Whitmore, T.C. 1984. *Tropical rain forests of the Far East*. 2nd edition. Clarendon Press, Oxford. (Translated into Malay, 1991.)

Whitmore, T.C. 1989. Tropical forest nutrients, where do we stand? Pages 1–13 in J. Proctor, editor. *Mineral nutrients in tropical forest and savanna ecosystems*. Blackwell Scientific Publications, Oxford.

Whitmore, T.C. 1990. *An introduction to tropical rain forests*. Clarendon Press, Oxford.

Whitmore, T.C., and C.P. Burnham. 1969. The altitudinal sequence of forest and soils on granite near Kuala Lumpur. *Malayan Nature Journal* 22:99–118.

Whittaker, R.J., B. Bush, and K. Richards. 1989. Plant recolonization and vegetation succession on the Krakatau islands, Indonesia. *Ecological Monographs* 59:59–123.

Witte, H.J.L. (in press). Seasonal and altitudinal distribution of precipitation, temperature and humidity in the Parque los Nevados transect (Cordillera Central, Colombia). In T. van der Hammen and A.G. dos Santos, editors. La Cordillera Central Colombiana, Transecto Parque los Nevados (cuarta parte). *Studies on tropical Andean ecosystems*, Volume 4. J. Cramer, Berlin-Stuttgart.

Wolf, J. 1993. Ecology of epiphytes and epiphyte communities in montane rain forests, Colombia. PhD thesis, University of Amsterdam, Amsterdam.

Wolf, J.H.D. 1989. Comunidades epifitas en un transecto altitudinal en la Cordillera Central, Colombia: datos iniciales sobre la cantidad de especies de briofitos y liquenes. Pages 461–483 in T. van der Hammen, S. Diaz-Piedrahita, and V.J. Alvarez, editors. La Cordillera Central Colombiana, Transecto Parque los Nevados (segunda parte). *Studies on tropical Andean ecosystems*, Volume 3. J. Cramer, Berlin-Stuttgart.

Wolf, J.H.D. (in press). Diversity patterns and biomass of epiphytic bryophytes and lichens along an altitudinal gradient in the northern Andes. *Annals of the Missouri Botanical Garden*.

Wolf, J.H.D. (in press). Epiphyte communities of tropical montane rain forests in the northern Andes. II. Upper montane communities. *Phytocoenologia*.

Yamada, I. 1975. Forest ecological studies of the montane forest of Mt. Pangrango, West Java. I. Stratification and floristic composition of the montane rain forest near Cibodas. *South East Asian Studies* 13:402–426.

Yamada, I. 1976. Forest ecological studies of the montane forest of Mt. Pangrango, West Java. II. Stratification and floristic composition of the forest vegetation of the higher part of Mt. Pangrango. *South East Asian Studies* 13:513–534.

Yamada, I. 1976. Forest ecological studies of the montane forest of Mt. Pangrango, West Java. III. Litter fall of the tropical montane forest near Cibodas. *South East Asian Studies* 14:194–229.

Young, K.R., and B. León. 1988. Vegetación de la zona alta del Parque Nacional Río Abiseo, San Martin. *Revista Forestal del Perú* 15:3–20.

Young, Kenneth R. 1991. Floristic diversity on the eastern slopes of the Peruvian Andes. *Candollea* 46:125–143.

Young, Kenneth R. 1991. Natural history of an understory bamboo (*Chusguea* sp.) in a tropical timberline forest. *Biotropica* 23(4b): 542–554.

Young, Kenneth R., and B. León. 1990. Curvature of woody plants on slopes of a timberline montane forest in Peru. *Physical Geography* 11:66–74.

Young, Kenneth R., and B. León. 1991. Diversity, ecology and distribution of high-elevation pteridophytes within Rio Abiseo National Park, north-central Peru. *Fern Gazette* 14:25–42.

Young, K.R., and B. León. 1995. Distribution and conservation of Peru's montane forests: interactions between the Biota and human society. Pages 363–376 in L.S. Hamilton, J.O. Juvik, and F.N. Scatena, editors. *Tropical Montane Cloud Forests.* Springer-Verlag, New York.

Young, Kenneth R., and N. Valencia, editors. 1992. *Biogeografía, ecología y conservación del bosque montano en el Perú.* Memorias del Museo de Historia Naturel, UNMSM 21, Lima.

Zadroga, F. 1981. The hydrological importance of a montane cloud forest area of Costa Rica. Pages 59–73 in R. Lal and E.W. Russell, editors. *Tropical agricultural hydrology*. John Wiley and Sons, Chichester.

Zadroga, Frank. 1981. Interceptación horizontal en los bosques de altura de Costa Rica y su importancia hidrológica. *Ciencias Ambientales* (Costa Rica) 2:21–28.

Zamierowski, E.E. 1975. Leaching losses of minerals from leaves of trees in montane forest in Kenya. *Journal of Ecology* 63:679–687.

Zinck, A. 1986. Los Suelos. Caracteristicas y fragilidad de los suelos en ambiente de Selva Nublada: el ejemplo de Rancho Grande. Pages 31–66 in O. Huber, editor. *La selva nublada de Rancho Grande parque nacional "Henri Pittier."* Fondo Editorial Acta Científica Venezolana, Caracas.

Zinck, A. and O. Huber. 1978. La fragilidad de los suelos en la selva nublada de Rancho Grande - Cordillera de la Costa. *5° Congresso Venezolano de la Ciencia del Suelo*, Sociedad Venezolana de la Ciencia del Suelo, Cagua (Venezuela). 15 pp.

Ziska, L.H., A.H. Teramura, and J.H. Sullivan. 1992. Physiological sensitivity of plants along an elevational gradient to UV-B radiation. *American Journal of Botany* 79:863–871.

Index

Ecological Studies

Volume 68
Stable Isotopes in Ecological Research (1989)
P.W. Rundel, J.R. Ehleringer, and K.A. Nagy (Eds.)

Volume 69
Vertebrates in Complex Tropical Systems (1989)
M.L. Harmelin-Vivien and F. Bourlière (Eds.)

Volume 70
The Northern Forest Border in Canada and Alaska (1989)
J.A. Larsen

Volume 71
Tidal Flat Estuaries: Simulation and Analysis of the Ems Estuary (1988)
J. Baretta and P. Ruardij (Eds.)

Volume 72
Acidic Deposition and Forest Soils (1989)
D. Binkley, C.T. Driscoll, H.L. Allen, P. Schoeneberger, and D. McAvoy

Volume 73
Toxic Organic Chemicals in Porous Media (1989)
Z. Gerstl, Y. Chen, U. Mingelgrin, and B. Yaron (Eds.)

Volume 74
Inorganic Contaminants in the Vadose Zone (1989)
B. Bar-Yosef, N.J. Barrow, and J. Goldshmid (Eds.)

Volume 75
The Grazing Land Ecosystems of the African Sahel (1989)
H.N. Le Houérou

Volume 76
Vascular Plants as Epiphytes: Evolution and Ecophysiology (1989)
U. Lüttge (Ed.)

Volume 77
Air Pollution and Forest Decline: A Study of Spruce (*Picea abies*) on Acid Soils (1989)
E.-D. Schulze, O.L. Lange, and R. Oren (Eds.)

Volume 78
Agroecology: Researching the Ecological Basis for Sustainable Agriculture (1990)
S.R. Gliessman (Ed.)

Volume 79
Remote Sensing of Biosphere Functioning (1990)
R.J. Hobbs and H.A. Mooney (Eds.)

Volume 80
Plant Biology of the Basin and Range (1990)
C.B. Osmond, G.M. Hidy, and L.F. Pitelka (Eds.)

Volume 81
Nitrogen in Terrestrial Ecosystem: Questions of Productivity, Vegetational Changes, and Ecosystem Stability (1990)
C.O. Tamm

Volume 82
Quantitative Methods in Landscape Ecology: The Analysis and Interpretation of Landscape Heterogeneity (1990)
M.G. Turner and R.H. Gardner (Eds.)

Volume 83
The Rivers of Florida (1990)
R.J. Livingston (Ed.)

Volume 84
Fire in the Tropical Biota: Ecosystem Processes and Global Challenges (1990)
J.G. Goldammer (Ed.)

Volume 85
The Mosaic-Cycle Concept of Ecosystems (1991)
H. Remmert (Ed.)

Volume 86
Ecological Heterogeneity (1991)
J. Kolasa and S.T.A. Pickett (Eds.)

Volume 87
Horses and Grasses: The Nutritional Ecology of Equids and Their Impact on the Camargue (1992)
P. Duncan

Volume 88
Pinnipeds and El Niño: Responses to Environmental Stress (1992)
F. Trillmich and K.A. Ono (Eds.)

Ecological Studies

Volume 89
Plantago: A Multidisciplinary Study (1992)
P.J.C. Kuiper and M. Bos (Eds.)

Volume 90
Biogeochemistry of a Subalpine Ecosystem: Loch Vale Watershed (1992)
J. Baron (Ed.)

Volume 91
Atmospheric Deposition and Forest Nutrient Cycling (1992)
D.W. Johnson and S.E. Lindberg (Eds.)

Volume 92
Landscape Boundaries: Consequences for Biotic Diversity and Ecological Flows (1992)
A.J. Hansen and F. di Castri (Eds.)

Volume 93
Fire in South African Mountain Fynbos: Ecosystem, Community, and Species Response at Swartboskloof (1992)
B.W. van Wilgen et al. (Eds.)

Volume 94
The Ecology of Aquatic Hyphomycetes (1992)
F. Bärlocher (Ed.)

Volume 95
Palms in Forest-Ecosystems of Amazonia (1992)
F. Kahn and J.-J. DeGranville

Volume 96
Ecology and Decline of Red Spruce in the Eastern United States (1992)
C. Eagar and M.B. Adams (Eds.)

Volume 97
The Response of Western Forests to Air Pollution (1992)
R.K. Olson, D. Binkley, and M. Böhm (Eds.)

Volume 98
Plankton Regulation Dynamics (1993)
N. Walz (Ed.)

Volume 99
Biodiversity and Ecosystem Function (1993)
E.-D. Schulze and H.A. Mooney (Eds.)

Volume 100
Ecophysiology of Photosynthesis (1994)
E.-D. Schulze and M.M. Caldwell (Eds.)

Volume 101
Effects of Land-Use Change on Atmospheric CO_2 Concentrations: South and South East Asia as a Case Study (1993)
V.H. Dale (Ed.)

Volume 102
Coral Reef Ecology (1993)
Y.I. Sorokin (Ed.)

Volume 103
Rocky Shores: Exploitation in Chile and South Africa (1993)
W.R. Siegfried (Ed.)

Volume 104
Long-Term Experiments With Acid Rain in Norwegian Forest Ecosystems (1993)
G. Abrahamsen et al. (Eds.)

Volume 105
Microbial Ecology of Lake Plußsee (1993)
J. Overbeck and R.J. Chrost (Eds.)

Volume 106
Minimum Animal Populations (1994)
H. Remmert (Ed.)

Volume 107
The Role of Fire in Mediterranean-Type Ecosystems (1994)
J.M. Moreno and W.C. Oechel (Eds.)

Volume 108
Ecology and Biogeography of Mediterranean Ecosystems in Chile, California, and Australia (1994)
M.T.K. Arroyo, P.H. Zedler, and M.D. Fox (Eds.)

Ecological Studies

Volume 109
Mediterranean Type Ecosystems: The Function of Biodiversity (1994)
G.W. Davis and D.M. Richardson (Eds.)

Volume 110
Tropical Montane Cloud Forests (1994)
L.S. Hamilton, J.O. Juvik, and F.N. Scatena (Eds.)

MIX
Papier aus verantwortungsvollen Quellen
Paper from responsible sources
FSC® C105338

If you have any concerns about our products,
you can contact us on
ProductSafety@springernature.com

In case Publisher is established outside the EU,
the EU authorized representative is:
Springer Nature Customer Service Center GmbH
Europaplatz 3, 69115 Heidelberg, Germany

Printed by Libri Plureos GmbH
in Hamburg, Germany